AF704243

Compressor Handbook for the Hydrocarbon Processing Industries

Gulf Publishing Company • Book Division • Houston, London, Paris, Tokyo

Compressor Handbook for the Hydrocarbon Processing Industries

Prepared by
Gulf Publishing Company
Book Division
Selected from the pages of
Hydrocarbon Processing

Compressor Handbook for the Hydrocarbon Processing Industries

Copyright © 1979 by Gulf Publishing Company, Houston, Texas. All rights reserved. Printed in the United States of America. This book, or parts thereof, may not be reproduced in any form without permission of the publisher.

Library of Congress Cataloging in Publication Data

Main entry under title:

Compressor handbook for the hydrocarbon processing industries.

1. Compressors—Handbooks, manuals, etc.
I. Gulf Publishing Company. Book Division.
II. Hydrocarbon processing catalog.
TP159.C57C65   660.2'83   79-50252
ISBN 0-87201-131-3

Contents

What Process Engineers Need to Know About Compressors

Process design information is collected here without mechanical details to make process studies easier when compressor sizing is required.

William Dimoplon, Crawford & Russell, Inc., Stamford, Conn.

FAST, RELIABLE estimates of compressor horsepower and operating conditions are frequently needed for process studies, economic evaluations and plant engineering. Such estimates need not be of the highest accuracy because exact designs will ultimately be made by the manufacturer based on actual mechanical designs and operating conditions. Methods for making such quick estimates are collected here for use also in initial stages of process design before manufacturer's quotations have been received. Guides for the selection of compressor type are presented based on gas thruputs and discharge pressures. Basic compression equations are summarized and convenient charts are given for simplifying calculations.

METHODS

A given compressor can be analyzed using the adiabadic, polytropic or isothermal formulas. The choice usually depends on which efficiency is most readily available.

In recent years, the manufacturers of centrifugal and axial-flow compressors have generally switched to the polytropic formulas (Eq. 19-29). These machines operate at essentially constant polytropic efficiency and polytropic head, regardless of the gas compressed, provided there is no change in inlet flow acfm, and rotating speed. Thus, the polytropic formulas are more convenient when a given machine's operating data for one gas must be used to predict its performance for gases with different C_p/C_r ratios and densities.

Another important advantage of the polytropic formulas is the ease in estimating the outlet temperature. Using either the adiabadic or isothermal formulas (with efficiency corrections), the brake horsepower must first be calculated, and the outlet temperature determined by energy balance (Eq. 11-18, 30-32).

Most industrial-scale compressors approach adiabadic operation. For this reason, the adiabadic efficiency is more theoretically meaningful than either the polytropic or isothermal efficiencies.[8]

Although especially useful for centrifugal and axial-flow compressors, the polytropic calculation model is less basic because it considers the compressor as a heat transfer device.[8]

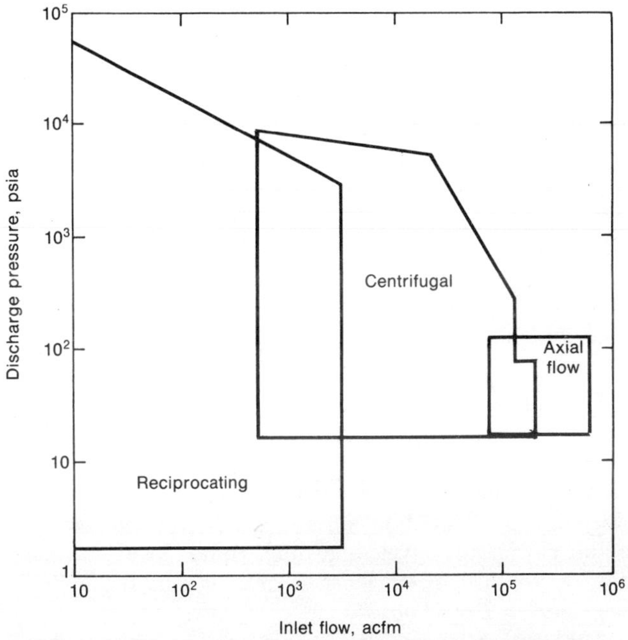

Fig. 1—Approximate ranges of application for reciprocating, centrifugal and axial-flow compressors.

The isothermal calculation model has the advantage of supplying the minimum theoretical power requirement, assuming the heat of compression is fully removed by cooling. To the design engineer, this theoretical limit is of more than academic interest. However, few, if any, industrial-scale compressors approach isothermal operation; thus, this model is less appropriate than the adiabadic model.

The adiabadic formulas, corrected for efficiency, are widely used for reciprocating and rotary compressors (Eq. 11-18).

COMPRESSOR TYPES

Principal compressor types used in the industry are: reciprocating, centrifugal, and axial flow designs. In addition, rotary compressors are used for special services. The generally applied limits for operation of the prime types are shown in Fig. 1 based on today's practices.

Reciprocating compressors are considered for applications where the inlet gas rate is about 3,000 acfm or less. They are favored for low-flow, high-pressure services.

The maximum compression ratio per stage is usually about 3:1 to 4:1. Higher compression ratios can result in reduced volumetric and mechanical efficiencies and greater mechanical stress. The outlet temperature also limits the compression ratio. The mechanical design limit is usually 350° to 400° F, but lower temperatures may be required depending on the gas properties. In addition, the pressure rise per stage is often limited by compressor valve design considerations to 1,000 psi or less.

Fig. 1 shows approximate application ranges for reciprocating compressors. Maximum discharge pressures range from more than 60,000 psia for small machines, to about 3,000 psia for larger units.

Correlations for over-all adiabadic efficiency, E_{ao}, are given in the *NGPSA Engineering Data Book.*[2] Typical values range from 0.75 to 0.85. For "quicky" estimates, an E_{ao} of 0.8 may be used. Horsepower and discharge temperature are calculated using adiabadic equations.

The design inlet flow (in acfm) is related to the compressor volumetric displacement by the volumetric efficiency, E_v.

$$E_v = \frac{\text{Design inlet flow}}{\text{Volumetric piston displacement}} \quad (1)$$

The volumetric efficiency is principally a function of the clearance volume within the compressor. Gas within the clearance pockets expands and contracts with each stroke, thereby reducing the net compressor thruput. Volumetric efficiency is affected to a lesser extent by slippage past piston rings and valves, and other minor losses. Calculation of volumetric efficiency is discussed in the *NGPSA Engineering Data Book.*[2]

A reciprocating machine operating at constant speed will compress essentially the same inlet acfm, regardless of the gas compostion, suction conditions or discharge pressure. This, of course, applies within the design limits of the compressor and drive. Thus, at fixed speed, the inlet acfm remain essentially unchanged, even when compressing against a variable discharge pressure. The inlet acfm is essentially proportional to rotating speed.

Due to fairly high maintenance requirements for reciprocating compressors, it is often desirable to either provide two half-capacity machines in parallel, or a full-capacity spare compressor. This is good practice for continuous process applications where a compressor shutdown would in turn shut down other major equipment.

Centrifugal compressors are used for process applications from about 500 to 200,000 acfm (inlet), and axial-flow compressors from about 75,000 to 600,000 acfm (inlet). Smaller axials are available (down to 20,000 acfm), but centrifugals are preferred at these capacities.

Axial-flow compressors are more efficient than centrifugals, and can be provided in higher capacities. However, centrifugals are far more common in the process industry. Centrifugal compressors have wider operating ranges. Axials are generally used only for air, or clean, non-corrosive gases, since these machines are more susceptible to deposits, corrosion and erosion than centrifugals.

The maximum compression ratio per stage is limited by the outlet temperature. The usual limit is 400° to 450° F for centrifugals and about 600° F for axials. The temperature limit is lower for centrifugals because they have diaphragms between impellers, which tend to expand radially due to the temperature rise of the compressed gas.

For some applications, the compression ratio is set by the maximum number of impellers the manufacturer can provide within a single casing (up to about 6 to 8 for centrifugals and about 12 to 24 for axials).

For centrifugal machines, a single impeller will usually develop up to 10,000 to 12,000 ft.-lbs./lb. of polytropic head. The corresponding value for an axial is about half (4,500 to 6,000 ft.-lbs./lb. per impeller). To compensate for this, an axial unit will usually have about twice as many impeller-stages (sets of rotating and stationary blades) as a corresponding centrifugal.

Fig. 1 shows approximate ranges of application for centrifugal and axial flow machines. Maximum discharge pressures range from about 10,000 psia for small centrifugals down to considerably lower pressures for large machines. Axials above 75,000 acfm are usually limited to about 130 psia at the outlet. As noted earlier, Fig. 1 should be regarded only as a rough guide to current process industry practice.

Polytropic efficiency is widely used to describe the performance of centrifugal and axial-flow compressors. Approximate polytropic efficiencies (E_p), sufficiently accurate for estimating purposes, are plotted in Fig. 2 as a function of gas thruput. The actual polytropic efficiency will vary with speed, impeller design, compression ratio and other factors. Fig. 2 should be used with caution at high or low compression ratios. The mechanical efficiency, E_m, may be taken as 0.99 for preliminary estimates. Horsepower, polytropic head and discharge temperature can be computed using polytropic equations.

A centrifugal or axial-flow compressor is normally engineered to operate at or near its highest efficiency at the design flow rate. The operating range for a centrifugal running at fixed speed is typically from 100% to about 50% of the design flow. The low end of this range is set by the surge point, where operation becomes unstable due to flow oscillation and vibration. The operating range for an axial running at fixed speed is much narrower, since the surge point is usually 85 to 95% of design capacity. A centrifugal's range of operation can be increased using speed variation or adjustable inlet vanes. An axial's operating range is extended by providing speed variation and a means of adjusting the stator blade angle.

In general, the head versus capacity curve for a centrifugal compressor operating at fixed speed is quite flat. At minimum thruput (the surge point), the total head is typically only 105 to 115% of the head at design thruput. The corresponding curve for an axial is much steeper. Although an axial machine operates over a significantly narrower capacity range, its head at the surge point is

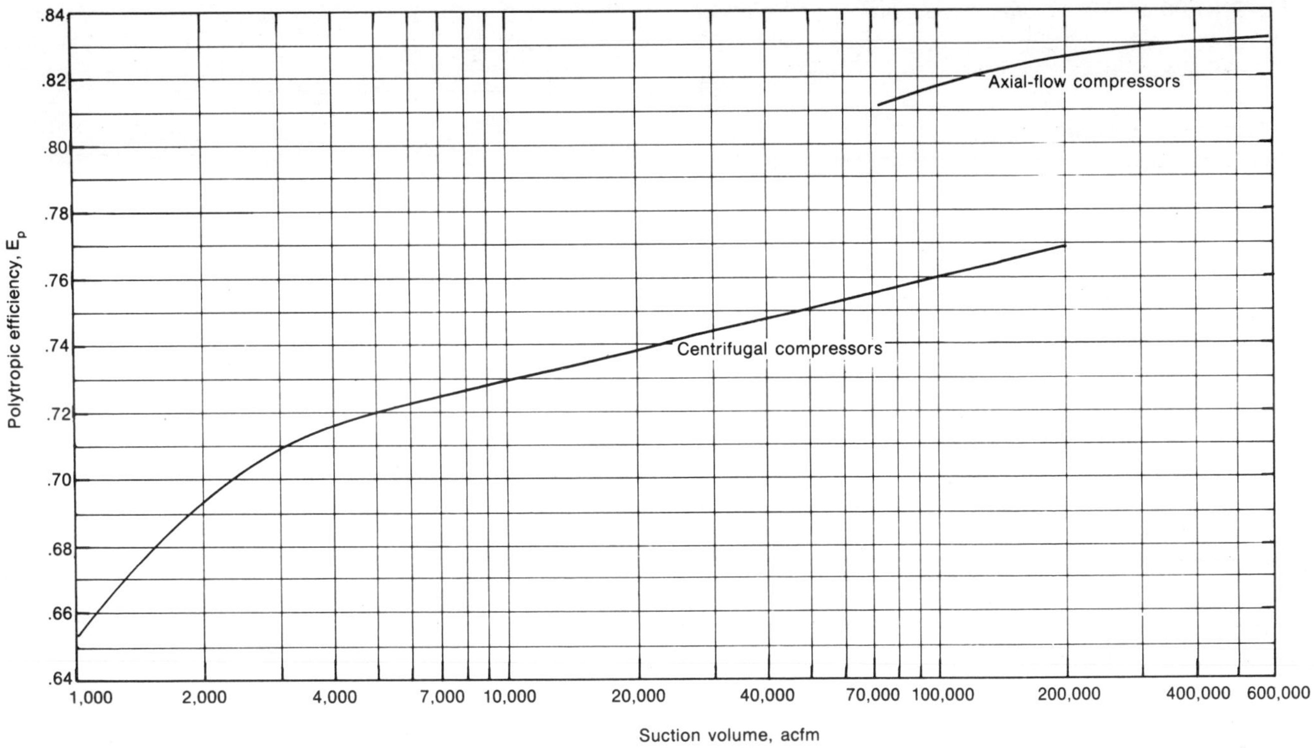

Fig. 2—Approximate polytropic efficiencies for centrifugal and axial-flow compressors.

typically 120 to 130% of design. With its steeper head-capacity curve, the capacity of an axial is much less affected by discharge pressure variations than a centrifugal.

The performance of centrifugal compressors can usually be approximated by the "fan laws." The volumetric flow rate varies approximately with rotating speed, the head with the square of speed and the power input with the cube of speed.

Axial compressors are sometimes placed in series with centrifugals, with the axial stages at the inlet end. In addition, axial and centrifugal impellers are also provided in the same casing. This permits large gas volumes to be compressed by the inlet, axial portion of the unit, until the gas acfm is within the centrifugal operating range.

Both centrifugal and axial-flow compressors generally require less maintenance than reciprocating machines. Except for special situations, a single centrifugal or axial-flow unit is normally used, rather than providing a spare, or two half-capacity machines in parallel. When a single centrifugal or axial-flow compressor is used for a critical, continuous service, it is good practice to stock a spare rotor assembly.

Rotary compressors. Table 1 shows ranges of application and typical over-all efficiencies for common types of rotary compressors. Over-all adiabadic efficiency curves may be found in a review article by Scheel,[5] correlated as a function of compression ratio.

Adiabadic equations can be applied to compute the horsepower and discharge temperature.

TABLE 1—Rotary compressor data summary[5]

Type	Maximum Displacement acfm (inlet)	Normal Compression Ratios	Typical Over-all Efficiency, %
Helical screw	20,000	2:1 to 4:1	75
Spiral axial	13,000	3:1	70
Straight lobe	30,000	1.7:1	68
Sliding vane	6,000	2:1 to 4:1	72
Liquid ring	13,000	5:1	50

Multistage compression. The equations for isentropic and polytropic compressions apply to a single stage of compression. For a given compression, the power requirement is less when two or more stages are provided in series, with intercoolers between stages to remove the heat of compression.

For these purposes, a stage of compression may be defined as one or more units in series, with no intercoolers between units. A centrifugal or axial-flow compressor casing with multiple impellers and no intermediate coolers is thus a single stage of compression. Confusion sometimes arises because each impeller is often referred to as a "stage." These are not stages in the same sense we use here, unless an intercooler is provided for each intermediate impeller. In theory, a stage of compression could consist of more than one casing in series, in the unusual event that no intercoolers are provided between casings.

For s stages of compression, with the same compression ratio for each stage:

$$\frac{P_d}{P_s} = \left(\frac{P^o{}_d}{P^o{}_s}\right)^{\frac{1}{s}} \tag{2}$$

where: P_d/P_s is the compression ratio for each stage.

$P^o{}_d/P^o{}_s$ is the over-all compression ratio for the multistage unit.

Note that Equation 2 neglects pressure drop across the intercooler(s), separator(s), interstage piping, etc., which will increase P_d/P_s for each stage.

The compression ratio per stage from Equation 2 is a useful first approximation. Whenever possible, however, the actual interstage pressures should be defined by the compressor manufacturer, who will take into account differences in equipment design and gas properties for each stage, along with intercooler and aftercooler pressure drops. If the interstage pressures are not dictated by process considerations, the manufacturer should be given acceptable interstage pressure ranges whenever possible, rather than setting specific interstage pressures for him to meet.

Limiting Discharge Temperature. The maximum discharge temperature limit often sets the number of intercooled stages for a given application. For most compressors built of standard materials, the mechanical limit is usually 350 to 450° F, and up to 600° F for axials. Lower temperatures may be required, however, depending on the properties of the compressed gas (corrosiveness; tendency to form polymer or gum, or undergo other chemical reactions; reactivity with the compressor lubricating oil; oxidation or explosion possibilities; etc.).

COMPRESSOR COSTS

Preliminary cost curves for reciprocating and centrifugal compressors are reported in the literature.[10, 11] For rough estimates, an axial machine can be priced using centrifugal compressor cost curves.

ISENTROPIC COMPRESSION

The reversible adiabadic (isentropic) compression of an ideal gas follows the equation $Pv^k =$ constant, where P is the absolute pressure, v is the gas molal volume and $k = C_p/C_v$, the specific heat ratio.

The theoretical horsepower requirement is given by:

$$\text{IHP} = \frac{.0642\,Q}{m}\left[\left(\frac{P_d}{P_s}\right)^m - 1\right]\left(\frac{T_s}{520}\right)(Z_m) \qquad (3)$$

where: $Q =$ Standard cubic feet per minute of gas (60° F, 14.7 psia)

$m = \dfrac{k-1}{k}$

$k = \dfrac{C_p}{C_v} = \dfrac{C_p}{C_p - 1.987}$ (molal specific heat ratio, using Btu/lb. mol °F as units of C_p and C_v)

$P_s, P_d =$ Absolute pressure at suction and discharge, respectively, psia

$T_s =$ Absolute temperature at suction, °R

$Z_m = \dfrac{Z_s + Z_d}{2} =$ mean compressibility factor

$Z_s, Z_d =$ Gas law compressibility factor at suction and discharge, respectively.

For 100 scfm of gas thruput, Equation 3 becomes:

$$\text{IHP per 100 scfm} = \frac{6.42}{m}\left[\left(\frac{P_d}{P_s}\right)^m - 1\right]\left(\frac{T_s}{520}\right)(Z_m) \qquad (4)$$

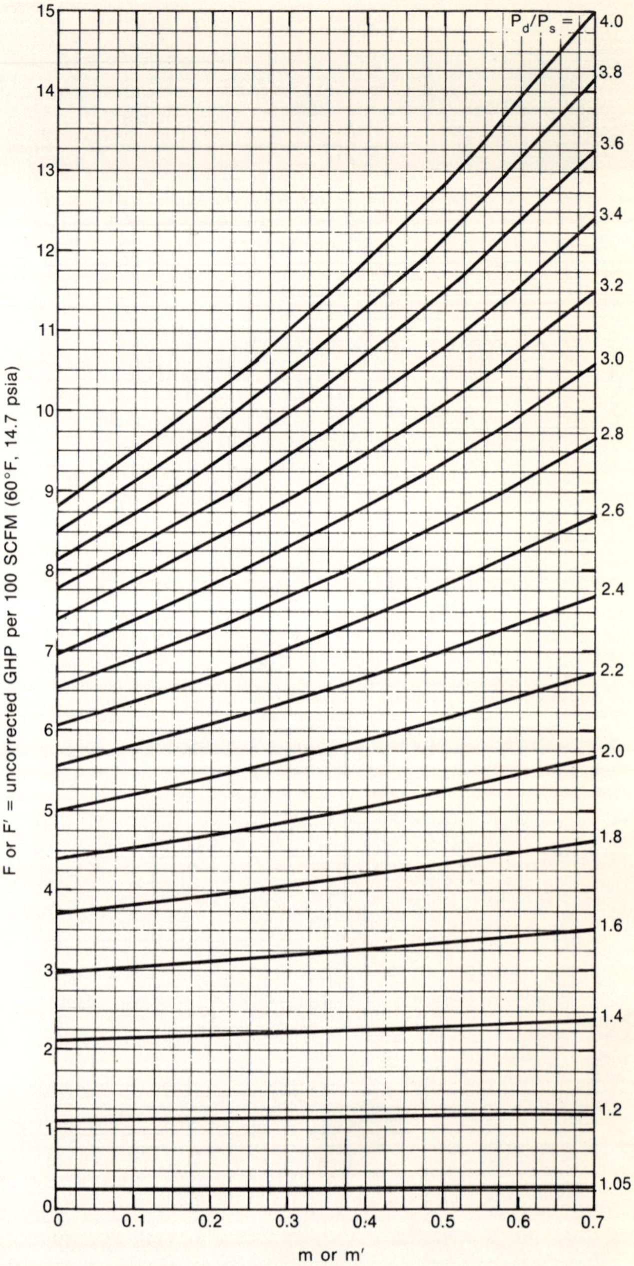

Fig. 3—Relationship of horsepower requirement per 100 scfm and isentropic exponent, F vs. m (or F′ vs. m′)—low range.

Equation 4 may be put in the form:

$$\text{IPH per 100 scfm} = (F)\left(\frac{T_s}{520}\right)(Z_m) \qquad (5)$$

$$\text{where: } F = \frac{6.42}{m}\left[\left(\frac{P_d}{P_s}\right)^m - 1\right] \qquad (6)$$

F represents the isentropic horsepower requirement per 100 scfm, when the inlet temperature is 60° F, and the mean compressibility factor is 1, as for an ideal gas. F is conveniently estimated from Fig. 3 or 4, and the ideal horsepower requirement obtained using Equation 5, to correct for T_s and Z_m.

Although these equations are based on an ideal gas with a constant specific heat, they give fairly accurate

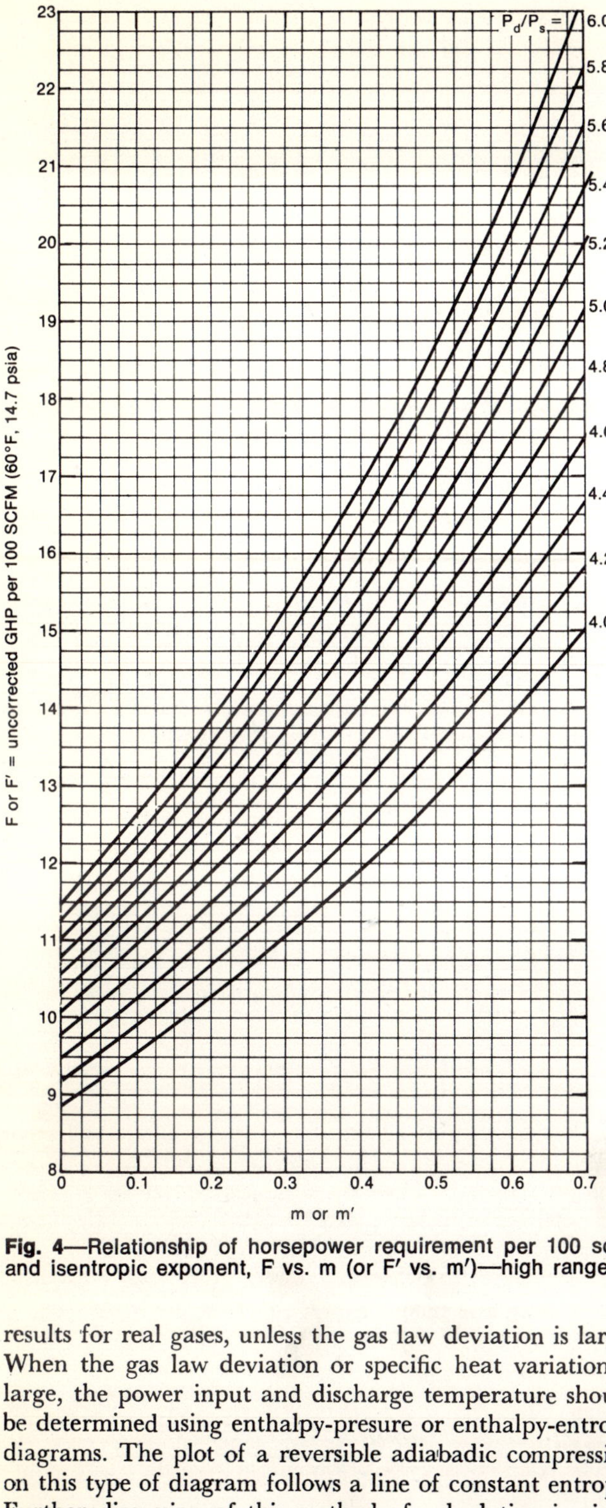

Fig. 4—Relationship of horsepower requirement per 100 scfm and isentropic exponent, F vs. m (or F' vs. m')—high range.

results for real gases, unless the gas law deviation is large. When the gas law deviation or specific heat variation is large, the power input and discharge temperature should be determined using enthalpy-presure or enthalpy-entropy diagrams. The plot of a reversible adiabadic compression on this type of diagram follows a line of constant entropy. Further discussion of this method of calculation is given elsewhere.[1, 4]

The discharge temperature, T_d, is obtained by

$$\frac{T_d}{T_s} = \left(\frac{P_d}{P_s}\right)^m \quad (7)$$

This relation is plotted in Fig. 5, for convenience in estimating the discharge temperature.

Applying the isentropic formulas. Although the isentropic compression is an ideal case, the resulting equations can be applied to real compressors, using appropriate efficiency corrections. The following definitions of efficiency are used in computing the power consumption:

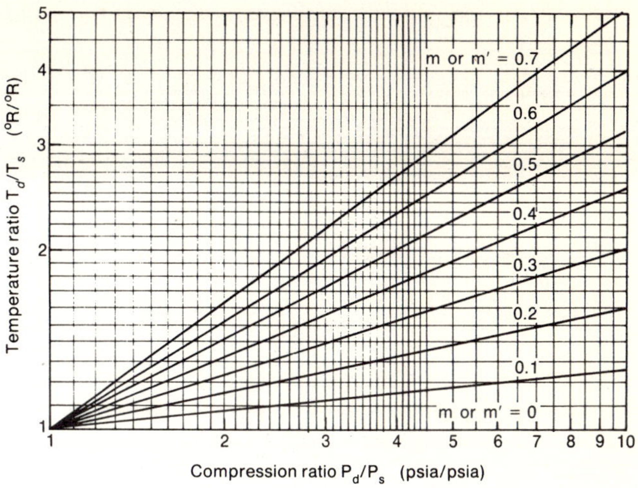

Fig. 5—Temperature ratio (T_d/T_s) versus compression ratio (P_d/P_s).

- Adiabadic efficiency, E_a: The isentropic horsepower, IHP, from Equation 3, divided by the actual horsepower delivered to the gas, GHP.

$$E_a = \frac{\text{IHP}}{\text{GHP}} \quad (8)$$

- Mechanical efficiency, E_m: The fraction of the driver brake horsepower, BHP, actually transmitted to the gas. (This allows for mechanical losses.)

$$E_m = \frac{\text{GHP}}{\text{BHP}} \quad (9)$$

- Over-all adiabadic efficiency, E_{ao}: The isentropic horsepower, IHP, for a stage of compression divided by the drive brake horsepower, BHP.

$$E_{ao} = \frac{\text{IHP}}{\text{BHP}} = E_a E_m \quad (10)$$

Using the over-all efficiency, E_{ao}, the basic equations describing the horsepower input for one stage of compression are:

$$\text{BHP} = \frac{\text{IHP}}{E_{ao}} = \frac{.0642\,Q}{m}\left[\left(\frac{P_d}{P_s}\right)^m - 1\right]\left(\frac{T_s}{520}\right)(Z_m)\left(\frac{1}{E_{ao}}\right) \quad (11)$$

$$\text{BHP per 100 scfm} = F\left(\frac{T_s}{520}\right)(Z_m)\left(\frac{1}{E_{ao}}\right) \quad (12)$$

Note that F is the same quantity used for an isentropic compression, and may be evaluated using Figs. 3 or 4.

The adiabadic head, H_a, represents the isentropic energy supplied to the compressed gas, in ft.-lbs./lb. of gas. H_a is calculated by the following relation, which applies to one stage of compression.

$$H_a = \left(\frac{RT_s}{M}\right)\left(\frac{1}{m}\right)\left[\left(\frac{P_d}{P_s}\right)^m - 1\right](Z_m) \quad (13)$$

where M is the gas molecular weight, and R is the gas constant, 1,545 ft.-lbs./lb. mol ° F.

This simplifies to:

$$H_a = \left(\frac{1545\,T_s}{M}\right)\left(\frac{F}{6.42}\right)(Z_m) \tag{14}$$

Figs. 3 or 4 can be used to estimate F.

The compressor horsepower (GHP or BHP) is related to the adiabadic head, H_a, weight flow, W (lbs./min.), and adiabadic efficiency by the following relations:

$$\text{GHP} = \frac{W\,H_a}{33{,}000\,E_a} \tag{15}$$

$$\text{BHP} = \frac{W\,H_a}{33{,}000\,E_{ao}} \tag{16}$$

Adiabadic efficiency can be estimated when polytropic efficiency is known. Polytropic efficiency, E_p, and the polytropic exponent, m', are defined in Equations 23 and 24.

$$E_a = \left[\left(\frac{P_d}{P_s}\right)^m - 1\right] \Big/ \left[\left(\frac{P_d}{P_s}\right)^{m'} - 1\right] \tag{17}$$

The outlet temperature should be computed by energy balance. The following relation is obtained by applying the first law of thermodynamics to a single stage of compression, neglecting changes in kinetic energy and potential energy.

$$T_d - T_s = \frac{6.33\,(2547\,\text{BHP} - L)}{Q\,C_p} \tag{18}$$

L represents the total heat lost from one stage of compression to cooling water (if any), and to the surroundings, in Btu/hr. L includes the duty of heat transfer jackets, lubricating oil coolers, etc., but does not include compressor intercoolers or aftercoolers.

Note that Equations 11 through 18 must be applied separately to each stage in a multistage system.

POLYTROPIC COMPRESSION

A polytropic compression follows the equation $Pv^n =$ constant. The exponent n is a constant that depends on gas properties, the amount of cooling (if any) supplied to the compressor body and the energy losses due to friction and irreversibility.

The basic equations are analagous to those for a reversible adiabadic compression. The power equations for a one-stage polytropic compression are:

$$\text{GHP} = \frac{0.0642\,Q}{m'}\left[\left(\frac{P_d}{P_s}\right)^{m'} - 1\right]\left(\frac{T_s}{520}\right)(Z_m)\left(\frac{1}{E_p}\right) \tag{19}$$

$$\text{GHP per 100 scfm} = F'\left(\frac{T_s}{520}\right)(Z_m)\left(\frac{1}{E_p}\right) \tag{20}$$

$$\text{where } F' = \frac{6.42}{m'}\left[\left(\frac{P_d}{P_s}\right)^{m'} - 1\right] \tag{21}$$

$$\text{and } m' = \frac{n-1}{n} \tag{22}$$

The polytropic efficiency E_p is defined by Equation 24.

The polytropic exponent m' is computed from the specific heat ratio k, and the polytropic efficiency E_p.

$$m' = \frac{1}{E_p}\left(\frac{k-1}{k}\right) = \frac{m}{E_p} \tag{23}$$

$$E_p = \frac{m}{m'} = \left(\frac{k-1}{k}\right) \Big/ \left(\frac{n-1}{n}\right) \tag{24}$$

F' can be estimated from Figs. 3 or 4.

Note the similiarity of polytropic Equations 19 through 21 to isentropic Equations 3, 5 and 6.

The equations for polytropic head, H, are analagous to those for adiabatic head (Equations 13 and 14). Polytropic head is the energy supplied to the compressed gas in ft.-lbs./lb. For one stage of compression:

$$H = \left(\frac{RT_s}{M}\right)\left(\frac{1}{m'}\right)\left[\left(\frac{P_d}{P_s}\right)^{m'} - 1\right]Z_m \tag{25}$$

This simplifies to:

$$H = \left(\frac{1545\,T_s}{M}\right)\left(\frac{F'}{6.42}\right)Z_m \tag{26}$$

If polytropic head has been computed, or is otherwise known, the compressor horsepower (GHP or BHP), is readily calculated from the following relations:

$$\text{GHP} = \frac{WH}{33{,}000\,E_p} \tag{27}$$

$$\text{BHP} = \frac{WH}{33{,}000\,E_{po}} \tag{28}$$

where the over-all polytropic efficiency $E_{po} = E_pE_m$. Note the similarity to Equations 15 and 16.

The outlet temperature is computed by the following relation, which is analogous to isentropic Equation 7.

$$\frac{T_d}{T_s} = \left(\frac{P_d}{P_s}\right)^{m'} \tag{29}$$

Polytropic formulas 19 through 29 must be applied separately to each stage in a multistage system.

If operating data are available for a given compressor, the polytropic efficiency can be calculated directly, using Equations 23 and 29 (providing T_s, T_d, P_s, P_d and k are known). For the usual design case, however, E_p is estimated from the manufacturer's empirical correlations, or from a similar compressor application. For preliminary estimates Fig. 2 can be used.

ISOTHERMAL COMPRESSION

The theoretical horsepower for a reversible isothermal compression, THP, is:

$$\text{THP} = \frac{Q\,T\,Z_m}{8110}\,ln\,\frac{P_d}{P_s} \tag{30}$$

THP represents the power requirement for the idealized case where the heat of compression is fully removed by cooling. The basic assumption is that the temperature of the compressed gas never increases above the inlet temperature. In general, this is not achieved, or even approached, in an industrial-scale compressor. The heat of compression causes the gas temperature to rise, increasing the power requirement over the isothermal value computed by Equation 30.

The performance of a real compressor can be related to a reversible isothermal compression by the relations:

$$\text{GHP} = \frac{\text{THP}}{E_t} \tag{31}$$

$$\text{BHP} = \frac{\text{THP}}{E_t E_m} = \frac{\text{THP}}{E_{to}} \tag{32}$$

where E_t is the isothermal efficiency; the over-all isothermal efficiency $E_{to} = E_t E_m$.

Once the brake horsepower for one state of compression, BHP, is calculated, the discharge temperature can be computed by energy balance (Equation 18).

GAS PROPERTIES

The specific heat ratio, k for the gas or gas mixture can be determined from the relation:

$$k = \frac{C_p}{C_v} = \frac{C_p}{C_p - 1.987} \tag{33}$$

where C_p and C_v are specific heats at constant pressure and volume, respectively (Btu/lb. mol °F).

It is useful to know that C_p is usually about 7 Btu/lb. mol °F for most diatomic gases, and 4.97 Btu/lb. mol °F for ideal monatomic gases.

For a gas mixture, C_p is evaluated by taking a weighted mole fraction average.

$$C_p = y_1 C_{p1} + y_2 C_{p2} + y_3 C_{p3} + \ldots \tag{34}$$

C_p should logically be evaluated at the average of the suction and discharge temperatures. However, in practice C_p at the inlet condition is often used. This normally introduces little error, and gives slightly conservative results, since the specific heats of most gases increase with both temperature and pressure. (An exception is monatomic gases at low pressure, where specific heat is independent of temperature.)

The gas law compressibility factors, Z_s and Z_d, may be determined from generalized correlations. For many applications at low pressures, Z_m may be taken as unity.

EXAMPLE

Problem:

75,000 acfm of dry air at 100° F and 14.4 psia are to be compressed to 45 psia. The specific heat may be taken as 7 Btu/lb. mole °F, and gas compressibility Z as 1. Compare centrifugal and axial-flow compressors with an isentropic compression. Estimate:

a. Polytropic and adiabadic efficiency

b. Outlet temperature

c. Gas horsepower

Solution:

	Isentropic Compression	Centrifugal Compressor	Axial-Flow Compressor
Specific heat ratio, k (Equation 33)	$\frac{7.00}{7.00-1.987} = 1.3964$	1.3964	1.3964
Polytropic efficiency, E_p (Fig. 2)	1.000	0.756	0.812
m or m' (Equation 3 or 22)	$\frac{0.3964}{1.3964} = 0.2839$	$\frac{0.2839}{0.756} = 0.3755$	$\frac{0.2839}{0.812} = 0.3496$
Inlet temperature, °R	$100 + 460 = 560$	560	560
Compression ratio, P_d/P_s	$\frac{45}{14.4} = 3.125$	3.125	3.125
Temperature ratio, T_d/T_s (Fig. 5 or Equations 7 and 29)	$(3.125)^{0.2839} = 1.382$	$(3.125)^{0.3755} = 1.534$	$(3.125)^{0.3496} = 1.489$
Discharge temperature, T_d, °R	$(560)(1.382) = 774$	$(560)(1.534) = 859$	$(560)(1.489) = 834$
Discharge temperature, °F	314	399	374
Scfm (60°F, 1 atm	$(75{,}000)\left(\frac{460+60°F}{460+100°F}\right)\left(\frac{14.4}{14.7}\right) = 68{,}222$	68,222	68,222
$F = \frac{6.42}{m}\left[\left(\frac{P_d}{P_s}\right)^m - 1\right]$ (or Fig. 3)	8.637	....	....
$F' = \frac{6.42}{m'}\left[\left(\frac{P_d}{P_s}\right)^{m'} - 1\right]$ (or Fig. 3)	....	9.129	8.986
Gas horsepower, IHP or GHP (Equations 5 or 20)	$\left(\frac{68{,}222}{100}\right)(8.637)\left(\frac{560}{520}\right)(1.00) = 6{,}345$	$\left(\frac{68{,}222}{100}\right)(9.129)\left(\frac{560}{520}\right)(1.00)\left(\frac{1}{.756}\right) = 8{,}872$	$\left(\frac{68{,}222}{100}\right)(8.986)\left(\frac{560}{520}\right)(1.00)\left(\frac{1}{.812}\right) = 8{,}131$
Adiabadic efficiency, E_a	1.000	$\frac{6{,}345}{8{,}872} = 0.716$	$\frac{6{,}345}{8{,}131} = 0.780$

NOMENCLATURE

acfm Gas inlet volumetric flow actual ft.^{3}/min.
BHP Brake horsepower delivered by the driver
C_p, C_v Molar specific heats at constant pressure or volume, respectively, Btu/lb. mole °F
E_a Adiabadic efficiency (see Equation 8)
E_{ao} Over-all adiabadic efficiency = $E_a E_m$
E_m Mechanical efficiency
E_p Polytropic efficiency (see Equation 24)
E_{po} Over-all polytropic efficiency = $E_p E_m$
E_t Isothermal efficiency (see Equation 31)
E_{to} Over-all isothermal efficiency = $E_t E_m$
E_v Volumetric efficiency of a reciprocating compressor (see Equation 1)
F or F' Uncorrected horsepower per 100 scfm (see Equations 6 and 21)
GHP Horsepower delivered to the gas
H Polytropic head, ft.-lbs./lb. of gas
H_a Adiabadic head, ft.-lbs./lb. of gas
IHP Theoretical isentropic horsepower delivered to the gas
k Specific heat ratio, C_p/C_v
L Heat loss from the compressor body to cooling water (if any) and to the surroundings, Btu/hr.
m Isentropic exponent = $(k-1)/k$
m' Polytropic exponent = $(k-1)/(kE_p)$
M Gas molecular weight, lbs./mol
n Exponent in the basic polytropic relation, Pv^n = constant
P Gas absolute pressure, psia
P_s, P_d Gas absolute pressure at suction and discharge, respectively, psia
P_s^o, P_d^o Gas absolute pressures at inlet to first stage and discharge from last stage of a multistage compressor, repectively, psia
Q Gas volumetric flow in standard cubic feet/minute at 60° F and 14.7 psia
R Gas constant, 1545 ft.-lbs./lb. mole °R
s Number of stages of compression (with intercoolers between stages)
scfm Standard cubic feet/minute at 60° F and 14.7 psia
T Absolute temperature, °R
T_s, T_d Gas absolute temperatures at suction and discharge, respectively, ° R
THP Theoretical isothermal horsepower delivered to the gas
v Volume of one mole of gas
W Gas flow in lbs./min.
y Mole fraction of a given component in the gas
Z_s, Z_d Gas law compressibility factor at suction and discharge, respectively
Z_m Mean compressibility factor = $\frac{1}{2}(Z_s + Z_d)$

LITERATURE CITED

[1] Neerken, R. F., "Compressor Selection for the Process Industries," *Chemical Engineering*, Jan. 20, 1975, pp. 78-94.
[2] *Engineering Data Book*, 8th Edition, Natural Gas Processors Suppliers Association, Tulsa, Okla., 1966, pp. 40-41.
[3] Rollins, J. P., (editor), *Compressed Air and Gas Handbook*, 4th Edition, Compressed Air and Gas Institute, New York, 1973, Fig. 3-72, pp. 3-71.
[4] *Elliott Multistage Centrifugal Compressors*, Bulletin P-11B, Elliott Co., Jeannette, Pa., 1970, pp. 33-35.
[5] Scheel, L. F., "New Ideas on Rotary Compressors," *Hydrocarbon Processing*, November 1969, pp. 271-278 (note especially Table 3 and Fig. 8).
[6] Clarke, L. and Davidson, R. L., *Manual for Process Engineering Calculations*, 2nd Edition, McGraw-Hill, New York, 1962, pp. 360-378.
[7] Church, D. M., "Numerical Integration Simplifies Nonideal Gas Calculations," *Chemical Engineering*, March 19, 1962, Fig. 3, pp. 169.
[8] Scheel, L. F., "New Ideas on Centrifugal Compressors," *Hydrocarbon Processing*, September 1968, pp. 253-260.
[9] Turnham, C. G., "Axial Compressors Can Help to Improve Gas Processing Economics," *Oil and Gas Journal*, March 8, 1976, pp. 51-55.
[10] Bresler, S. A. and Smith, J. H., "Guide to Trouble-Free Compressors," *Chemical Engineering*, June 1, 1970, pp. 161-170.
[11] Guthrie, K. M., "Capital Cost Estimating," *Chemical Engineering*, March 24, 1969, pp. 128.
[12] Kennedy, J. L., "What You Should Know about Reciprocating Compressors—Part 1," *Oil and Gas Journal*, Nov. 13, 1967, pp. 105-107.

Compressor Problems: Causes and Cures

Most compressor problems can be prevented with a good purchaser specification based on field experience. Here are some experiences and solutions.

John J. Dwyer,
Air Products & Chemicals, Inc., Allentown, Pa.

SUCCESSFUL COMPRESSOR operation not only depends on an experienced and capable maintenance team but to a greater extent on a good set of purchaser specifications. Too often the buyer relies on specifications, written by the contractor or an industry group such as the API. Extraneous words and phrases like *rugged, heavy duty, refinery type* and *long life* may sound nice in a specification but do nothing for the operating superintendent. A good set of specifications not only defines the duty and equipment to be supplied, but also is a history of the operator's past problems and their cures. Every compressor problem and its solution should be fed back to the specifying engineer so future specifications reflect and help prevent past problems.

Some compressor problems are unique to a particular location or set of conditions. An example is the time an operator ran headfirst, with his safety helmet, into a lube oil pressure switch causing the compressor to trip and shut down the entire plant. But most problems are caused by design features that can be corrected in future installations.

By far the most important type compressor installed in the process industry today is the centrifugal machine. It is the product of more than 40 years of industrial experience and development. Most problems in earlier years resulted from poor balancing and high vibration. It was not unusual for a repair shop to statically balance the rotor in a lathe. The compressor was assembled and started up. If a coin could be balanced on the bearing housing, the job was acceptable. Later, small hand-held mechanical vibrometers were used to measure bearing housing vibrations. Setting a limit on bearing housing vibration appeared to be a good idea until proximity-type probes were developed. On one high-speed machine that had several bearing failures, we measured housing vibrations of about 1 mil. We installed one of our first proximity type probes and found the shaft was moving 5 mils. In another application where we had installed proximity probes, although the total amplitude of vibration was low, most of the vibration was at approximately 40 percent of operating speed indicating oil whip or whirl in a fixed-pad type bearing. An analysis of compressor vibration with proximity probes and an oscilloscope picture will give a good understanding of equipment conditions and need for maintenance.

Dynamic balancing of rotors is now a routine procedure for almost all manufacturers but there are still occasional problems. A few years ago we had a small rotor dynamically balanced, shipped to the field, assembled and started up. We immediately lost a set of bearings and damaged the rotor. The rotor was repaired and rebalanced but the same failure happened. We then discovered that the balancing machine was not properly calibrated.

As a result of these and similar problems we have improved our compressor specifications. We require all rotors to be dynamically balanced with not more than two impellers added to the rotor between each balancing step, and require the balancing machine to be calibrated at the conclusion of the operation.

Vibration specs. We specify the amount of residual unbalance that we will accept. We also require proximity probes at each radial bearing and require the surface under the probe to be finished to the same standard as the bearing journal. Maximum shaft vibration limits at operating speed are specified and filtered vibration amplitude at frequencies other than operating speed are limited to a small percentage of over-all amplitude. In critical services, two probes at 90 degrees are required at each bearing so an orbit analysis may be made on an oscilloscope. On critical high-speed units we insist on tilting-pad bearings to eliminate the possibility of oil whip.

Bearing failures. Although many problems associated with bearings can be analyzed and prevented, it is almost impossible to eliminate human error. Several years ago at one of our plants a large centrifugal compressor was turning backwards at low speed, without lubrication, due to a check valve leak. The bearings were removed, inspected and were satisfactory. However, on reassembly one pivoted shoe thrust bearing was installed backward so the steel retainer rode on the steel thrust collar. The machine was started and ran satisfactorily for about a year but subsequently the failure did occur, at night, on a weekend.

Insurance company statistics have shown that more than 50 percent of centrifugal compressor outages in the past few years have either been caused by lube oil failures or directly affected by the lubricating system. Compressor manufacturers tend to treat lube oil systems as one of the necessary evils that go along with the sale of a compressor. For the past several years we have been purchasing our lube consoles separately. We have developed a set of specifications to assure the quality of individual system components and purchase the system from one of a number of firms that specialize in fabricating this class of product. By this method we have not only eliminated many field problems but have also obtained interchangeability of filter cartridges, control valves and switches which reduces our spare parts inventory.

Lateral and torsional problems. Most centrifugal compressor built today are flexible shaft machines; that is, they operate above the first natural critical speed. They are also subject to torsional critical problems. To avoid these problems, most manufacturers conduct lateral and torsional studies concurrent with the design of each new machine, and the first lateral critical is normally checked on the manufacturer's test stand. Torsional problems are prevalent with synchronous motor drivers since the instantaneous torque output varies considerably during motor acceleration with an exciting frequency starting at 120 hertz and diminishing to zero at synchronous speed. We have tried to eliminate lateral and torsional problems by specifying the acceptable range of lateral critical speeds as a percent of operating speed, and requesting that torsional nodes of resonance be a given percentage margin away from operating speed. On new designs or critical installations, at times, we have employed consultants to analyze the lateral and torsional characteristics of the complete train including the instantaneous torque characteristics of the motor.

High-speed couplings have been the subject of much discussion over the past several years. Problems associated with flexible coupling lock-up, where they transmit axial forces, have resulted in many purchasers requiring the compressor and driver vendor to design thrust bearings to accommodate this possibility. In addition, minute dust particles in the lubricant have blocked oil passages by separating-out through centrifugal force within the coupling. One widely used method of minimizing dust problems, especially on large compressors, is to install filters (one micron or less) in the coupling oil feed line. One major vendor completely eliminates the flexible coupling problem by using solid couplings. However, extra precautions must be exercised when using this approach to assure near perfect hot alignment and sufficient axial clearances on all labyrinth seals, especially in long compressor trains. Another promising solution to the coupling lock-up and dirt problem is the flexible diaphragm coupling. For the present, we are specifying that compressor thrust bearings be designed to accept some axially transmitted force and require sludge drain holes in gear-type couplings.

Labyrinth seals. The design and fitting of interstage and casing seals is an important part of over-all design and an area that causes its share of problems. Interlocking labyrinths where both the rotating and stationary members have labyrinth teeth are superior to a single labyrinth in reducing leakage. However, axial clearances and float are critical. In one design it was possible to install the stationary part backwards which resulted in almost line-to-line contact of the labyrinth teeth. Since this type of assembly error was possible, it naturally occurred, damaging the balance drum on startup. We have since modified our specifications to require the stationary parts of interlocking labyrinths to be designed with an off-center key.

It is easy to improve compressor efficiency by minimizing interstage labyrinth clearances. The new compressor will operate with minimum leakage, but when the rotor becomes slightly unbalanced or vibrations occur from another source, the seals will wear rapidly, opening the clearance and allowing greater leakage with a commensurate decrease in efficiency. To prevent this situation, we specify that the minimum clearance shall be equal to the bearing clearance plus rotor run-out at the seal due to its weight plus a given amount of mass unbalance.

Materials of construction. The selection of materials of construction is left almost solely to the purchaser when corrosive gases are to be compressed because he frequently has more experience with the gas than the machinery manufacturer. For most gases, the casing and diaphragm castings offer few material problems; but the impeller, shaft sleeve and labyrinth materials are more of a challenge. Straight carbon steel impellers will frequently pit in acid gas service and bronze labyrinths are to be avoided if the gas might contain moist ammonia. In some air compressors, we have found it necessary to use 400 Series stainless impellers and aluminum labyrinths because of corrosive fumes in the atmosphere.

When discussing materials we should not overlook the inter and aftercooler. Several years ago we ran into a series of failures on cooler tubes in air service. After a short investigation we found that condensate pH was below 5. Further investigation showed a high concentration of H_2S in the atmosphere. Unknowingly we were running a small sulfuric acid plant in the intercoolers by reacting hot moisture and oxygen from the atmosphere with the H_2S and then condensing the product. The problem was partially eliminated by changing to copper-nickel cooler tubes.

Moisture separators. No discussion of centrifugal compressor problems would be complete without mentioning the need of adequate moisture separators following intercoolers or at the inlet of process machines where liquid may be present in the gas stream. High vibration, rotor unbalance, pitting and corrosion of impellers, clogged diffuser passages and a host of other problems can many times be traced to inadequate moisture separators. Subcooled mositure vapor will normally condense on a particle of dirt or other foreign matter in the gas stream as a nuclei. The moisture particle entering the compressor will impact on the impeller. The moisture vaporizes, depositing the dirt particle. It is not unusual in a four-stage intercooled air compressor to have deposits build up to

1/16-inch thick or more on the third or fourth-stage wheel. As the deposits build up, performance deteriorates until a large section of deposit breaks off the wheel, causes unbalance, a sudden rise in vibration and an unscheduled shutdown. We have been insisting on high-efficiency moisture separators; and on moist gas process machines we install a large low-velocity suction drum with a moisture-separating pad immediately before the compressor inlet.

Reciprocating compressors. Most reciprocating compressor failures can be related to high vibration, pulsation, lubrication and wear. Analyzing each type of failure and placing it into a general category assists in developing a cure.

Vibration switches. Several years ago we experienced a number of different types of problems all resulting in a major failure. A seized crosshead caused by insufficient clearances resulted in a broken piston rod. Piston ring breakage scored a cylinder that took several days to hone. A broken valve fell into a cylinder and bent a piston rod. Although the cause and ultimate cure for these and several other problems were all different, they had one factor in common; if the compressor was shut down immediately, damage would be minor. We equipped all of our large balanced-opposed compressors with acceleration type vibration switches. At first we had some difficulty with spurious shutdowns, but the problem was eliminated after we learned proper switch positioning and setting. We had two problems shortly after installing vibration switches on a 2,500-hp five-stage machine. Once we lost lubrication to a cyclinder. This caused the piston to chatter and the machine was tripped by the vibration switch. Damage was limited to light scoring of the cylinder which was repaired in a day. Some time later the machine again shut down on vibration. After about a 4-hour check of the cylinders adjacent to the switch, operators were about to consider the switch faulty when they found a cracked valve casting. The valve was still functioning but the vibration of broken casting under cyclic loading was enough to trip the vibration switch. Our specifications now require vibration switches on all reciprocating machines and the switches are set in the field after initial startup.

Gas pulsation. The major cause of reciprocating compressor problems can probably be traced to excessive gas pulsation. Valve breakage, piping cracks and cooler tube leaks are some of the more common problems caused by high pressure fluctuations. Installing specially designed pulsation snubbers or volume bottles at the design stage is fairly simple; but what can be done when the machine exhibits high pulsation levels in the field? In one installation where we had excessive pulsation in a 3,000-psi line, we were able to damp the pulsation by installing an orifice in the line. This is a fairly simple solution, but power cost is usually high. A little more sophisticated solution that does not add power is to install a Helmholtz resonator in the line. This is simply a dead-end leg installed with a tee in the pulsating line. The major trick is to know the pulsation frequency and calculate the resonator frequency to damp the primary pulsating frequency.

To minimize pulsation problems and facilitate problem checks in the field we specify a maximum inlet, interstage, and discharge pulsation level. In addition we require installation of pipe taps in the inlet and outlet nozzles of all pulsation snubbers so we may easily check the actual pulsation levels in the field with a pressure pick-up and oscilloscope.

Compressor valves are also subject to a type of pulsation failure. Many designers associate high valve velocities with efficiency loss and are extra generous with low valve velocities. This generosity can spell failure with low molecular weight gases. With low lifting pressure and low velocities, the valve disc will flutter and eventually break. In some installations we have corrected this problem in the field by installing thicker valve plates when available. Otherwise, we machine down the valve keeper to reduce the lift. We have established a set of guide lines for valve velocities with different molecular weight gases and require that the manufacturers proposed valve velocities be submitted with the equipment proposal.

Lubricants. Air compressor cylinder lubrication and compressor fires were once practically synonymous. Lubricants would carbonize on valves and a hot spot on a valve could cause a fire. The fire danger was practically eliminated on air compressors by converting to a specially compounded synthetic lubricant with a high ignition point. It did not take long to realize that the solution of one problem can be the cause of another. We found that the synthetic lubricant is an excellent paint remover. Lubricant that traveled along the rod entered the crank case and the crank case paint began to soften and peel. As a temporary measure we installed an extra set of rod scraper rings to reduce the lubricant carryover into the crank case, and monitored the percentage of synthetic lubricant in the crank case oil. When the percentage of synthetic got too high we would simply drain a barrel of the mixture and add a barrel of fresh lubricant during operation. We now specify all reciprocating compressors where we might want to use synthetic cylinder lubricants, to have special lubricator seals and a coating of epoxy base paint inside the crank case.

Lubricators. Cylinder lubrication is another potential problem area. Some cylinders had only one lubrication point in the bore. A cylinder failure would be caused by a failure of the pump feeding that point or a line leak. We partially corrected this problem on some units by installing a feed point in the cylinder inlet pipe. Although this does not result in an even lubricant distribution it can prevent a major failure. The lubricator itself may also give rise to a problem. If the operator neglects to fill the lubricator, failure will result when the reservoir runs dry. This type of failure could be protected by a reservoir level switch, but the level switch is of no help if there is a failure in the lubricator drive mechanism.

Another method of protection is to install pressure switches in each lubricator feed line, but on large units with 20 or more lubricator points this means 20 pressure switches, alarm lights and high installation costs. We use a short suction pump and pressure switch mounted in the end position of the lubricator reservoir opposite the driver.

If the lubricant level runs low or the drive mechanism fails, this one pumping unit will activate an alarm.

Packing. Breaking in a new set of high pressure packing used to be an experience capable of causing the operator and mechanics no end of grief. One method was to increase the pressure slowly over many hours. If the packing started to smoke, the pressure was reduced. What no one wanted to admit was that if the packing started to smoke it was probably too late. On one particularly troublesome installation, after installing the packing we removed one of the packing case tie rods and installed a thermocouple adjacent to the first seal ring. The compressor was gradually brought up to pressure by monitoring the packing case temperature. In a later installation, we converted to a lubricated-filled Teflon breaker and seal-ring design and built up pressure at a much greater rate. Our specifications for high pressure packing now require cooled packing cases and Teflon rings.

Nonlube compressors. Some operators have reported fantastic life from what they term nonlube service, but investigation shows that they do lubricate the packing or have lubricant carryover from the crankcase. In other nonlube compressors the compressed gas was wet. Even atmospheric air has some lubricant value from the moisture in the air. We have a number of reciprocating compressors in nonlube service handling dry nitrogen. The compressors are fitted with extended-distance pieces with flingers mounted on the rod to prevent oil carryover. At first, we found a great deal of variance in service life between different units. We have been able to correlate the service life with what we term a PV factor; that is, the mean cylinder pressure multiplied by the average piston velocity. Although this correlation is not precise, we are able to predict with some accuracy the possibility of satisfactory operation. To improve nonlube operation we specify maximum rider-band loading, cylinder-bore finish, and equip all horizontal cylinders with rod-drop detectors to monitor ring wear in operation. In some designs, we have found excessive rider-band loading caused when gas pressure gets under the band and causes it to act like a seal ring. We have corrected this in the field by either drilling holes in the rider band to relieve the pressure or by segmenting the ring.

Maintenance. Not all problems associated with compressors can be attributed strictly to design factors. Preventive and major maintenance also play an important role in achieving high onstream factors. We have established a centralized control system for all compressor maintenance and a computerized preventive maintenance system. Each month a deck of computer cards indicating preventive maintenance jobs to be performed are mailed to each plant. The majority can be accomplished with the compressor in operation, such as sampling lube oil for analysis, inspecting filters and testing alarm devices.

Major maintenance on centrifugal compressors is only scheduled after a thorough analysis of operating parameters. We have developed computer programs to check actual compressor performance data against design values. We also analyze shaft vibration data and look for trends of increasing vibration or indications of bearing instability. If either performance or vibration analysis indicates deterioration, major maintenance is planned. Since most of our facilities are of a size that do not warrant a complete maintenance staff, major maintenance is handled by specialists from our Central Maintenance staff. By this method we can limit our local maintenance personnel to a minimum, yet provide the expertise for major overhauls.

The final and perhaps most important aspect of major compressor maintenance is that we do not open a centrifugal compressor on some hypothetical periodic schedule. It is opened only when analysis of performance or vibration indicates a need. Following this approach, we have consistently achieved an onstream factor above 98 percent.

Part I.

Centrifugal Compressors—Sizing, Specifying, Selecting

Quick Method for Centrifugal Compressor Estimates

Centrifugal compressor size, price, and driver requirements are often needed in a hurry during project planning. Chart method streamlines estimating procedure.

Don Hallock, Elliott Co., Jeannette, Pa.

In the early planning stage of process unit projects, quick estimates of compressor size, price, and driver requirements are desirable. Typical questions asked by process engineers are: We want to increase the discharge pressure of our existing process. Would it cost more to supercharge the existing compressor with a high flow, low pressure compressor, or to "top" the existing compressor with a low flow, high pressure compressor? We plan to buy new compressor capacity and would like the flexibility of multiple part-flow units. What premium do we pay for this flexibility? Which approach lends itself best to available turbines?

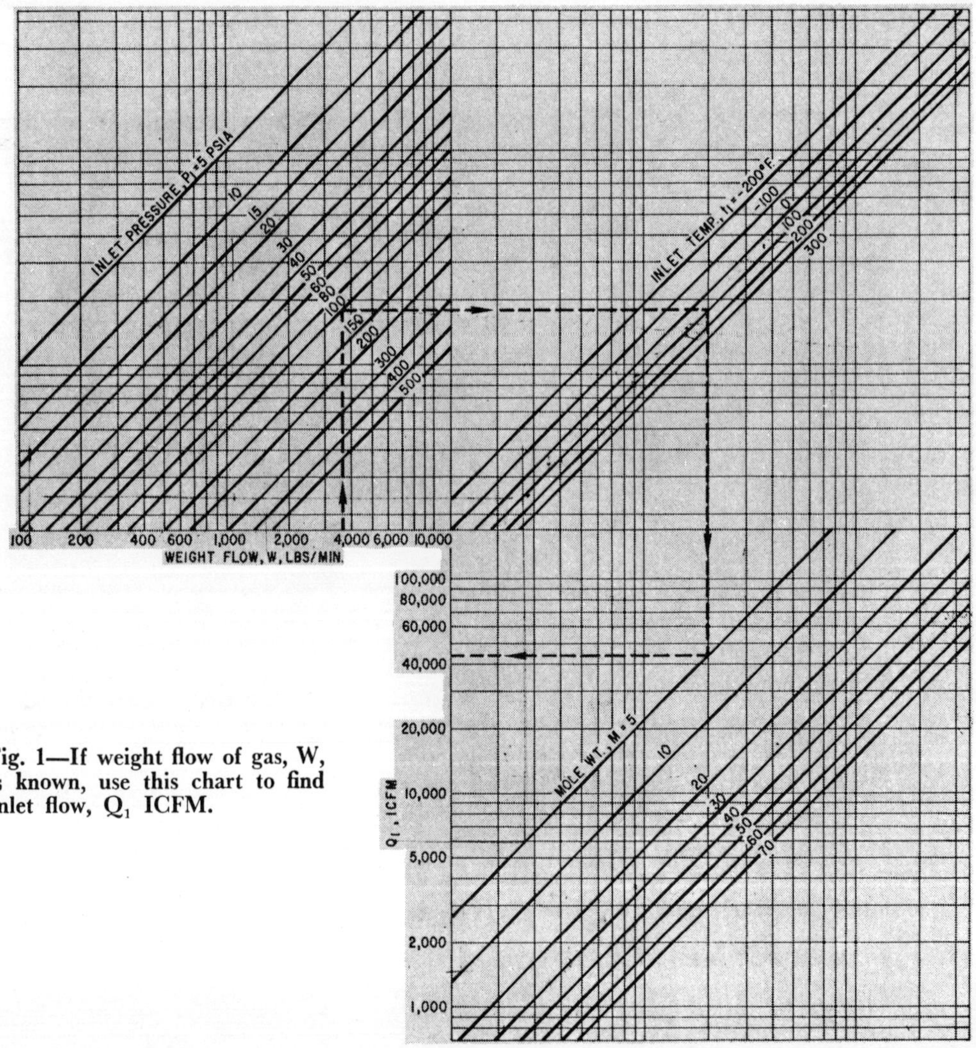

Fig. 1—If weight flow of gas, W, is known, use this chart to find inlet flow, Q_1 ICFM.

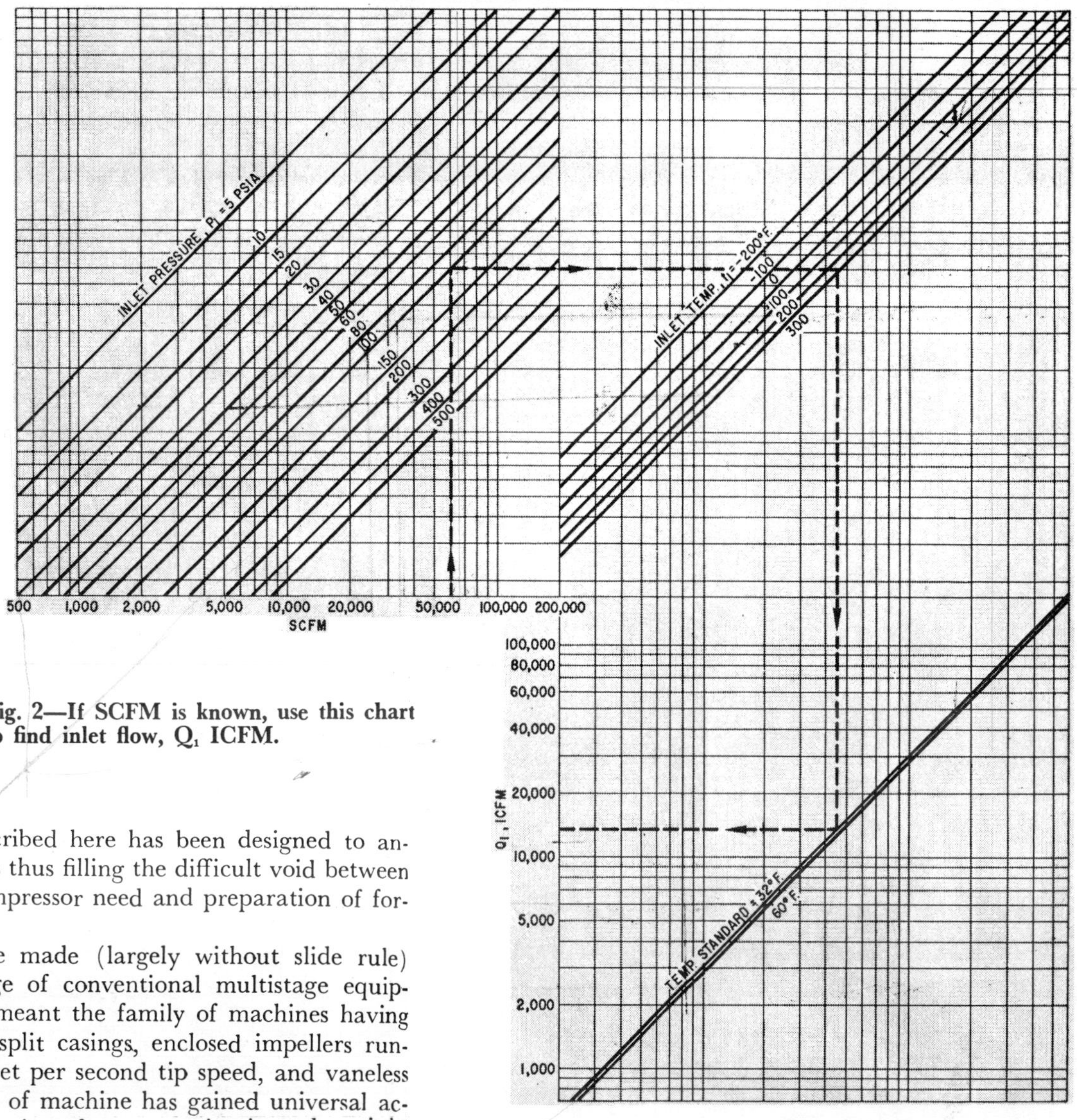

Fig. 2—If SCFM is known, use this chart to find inlet flow, Q_1 ICFM.

The method described here has been designed to answer these questions thus filling the difficult void between conception of a compressor need and preparation of formal inquiries.

Selections can be made (largely without slide rule) for the entire range of conventional multistage equipment, by which is meant the family of machines having heavy horizontally split casings, enclosed impellers running under 1,000 feet per second tip speed, and vaneless diffusion. This type of machine has gained universal acceptance for compression of gases ranging in mole weight from hydrogen to chlorine, and ranging in flow from 700 cfm recycle service to 170,000 cfm air service.

Barrel Casing. The figures can also be used for the vertically split machine (barrel casing) normally used for hydrogen service above 400 or 500 psig, and for any service above 750 psig. Barrels are presently available under 10,000 cfm for 2,500 psig and higher.

Horizontally Split Casing. Representative pressure ratings of horizontally split steel casings are 750 psig to 8,000 cfm, 600 psig to 15,000 cfm, and 300 psig to 60,000 cfm. Cast iron pressure ratings are 40 to 50 percent of the foregoing steel ratings. Above 60,000 cfm, casings are generally available in cast iron only, with ratings between 50 and 100 psig.

Flow Limits. Although flow limits for centrifugals are dictated somewhat by the specific conditions involved and by the available hardware of different manufacturers, generally positive displacement equipment is indicated below 500 cfm, and above 200,000 cfm axial equipment is indicated. The flow regions from 500 to 2000 cfm can be considered a "gray area" between positive displacement and centrifugal, and from 50,000 to 200,000 cfm a "gray area" between centrifugal and axial.

Gas flows other than air above 50,000 or 60,000 cfm have not appeared until recently, but increases in plant size are now pushing flows above this figure, particularly on propane or propylene refrigeration service. Pressure limitations of large casings and physical size of oil seals are moving compressor vendors to double-flow (two gas streams in parallel), multibody selections for applications above these flows.

Selection Accuracy. The selection charts are based generally on Elliott Co. equipment, but the user will find that they apply reasonably well for any conventional machine. The charts are constructed mostly in log-log form, so that consistent accuracy is maintained over the entire flow range. Accuracy of 10 to 15 percent is attainable, sufficient for most estimating purposes.

It must be emphasized that the charts are *not* applicable for any type of positive displacement machine, for axial machines, or for the recent generation of high tip speed, vaned diffusion centrifugal equipment sometimes used for air service. Also, the charts are of necessity based on uncooled compression at constant weight flow.

To handle cooled compression (air over 50 psig, chlorine, ethylene plant feed gas), or side load compression (any refrigeration process having economizers or extraction), it is necessary to divide the total compression into a series of uncooled, constant weight flow compressions. The charts are then applied separately to each of these compression requirements. This procedure will be explored more fully after the use of the charts is described.

Use of Charts

The following quantities must be known:

1. W—weight flow in lbs. per min., or scfm—standard cu. ft. per min.
2. P_1—inlet pressure in psia
3. R_p—pressure ratio (discharge psia ÷ inlet psia)
4. t_1—inlet temp., °F
5. M—mole. weight
6. K—ratio of specific heats

Determine Inlet CFM, Q_1. If W is known, use Figure 1, proceeding through P_1, t_1 and M to find Q_1.

If SCFM is known, use Figure 2, proceeding through P_1, t_1 and "temperature standard" to find Q_1.

Determine Head H. On Figure 3, enter R_p and proceed through K, t_1 and M as shown. If head H exceeds 80,000 to 90,000, more than one compressor body will be required.

Determine Number of Stages Required. On Figure 4, enter head H and proceed through M to read the number of stages required. Round this off to the next higher even number.

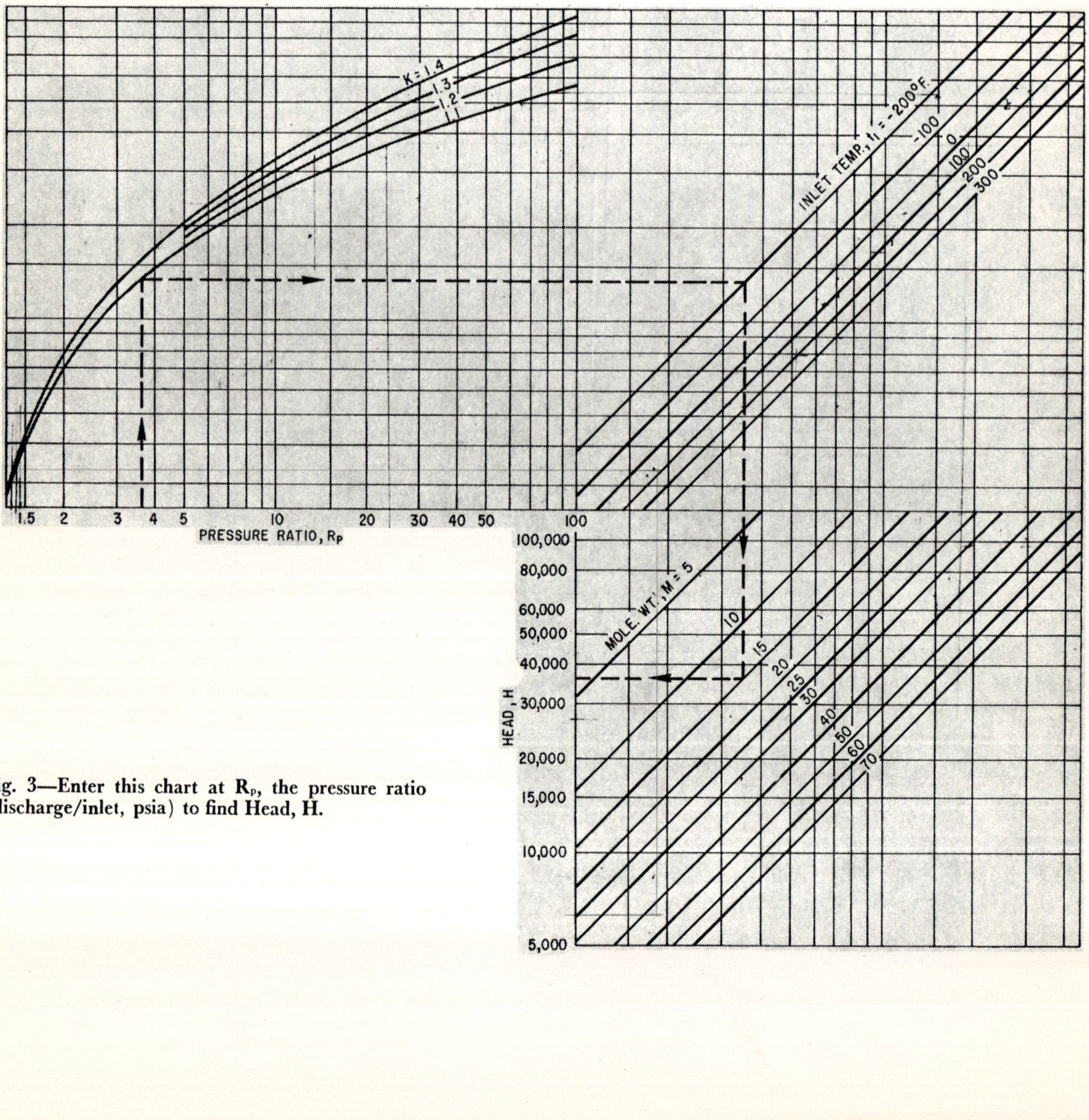

Fig. 3—Enter this chart at R_p, the pressure ratio (discharge/inlet, psia) to find Head, H.

Determine Speed and Size of Machine. On Figure 5, enter Q_1 and read maximum width in inches. Proceed to the stepped lines and read rpm and flange sizes. Proceed through number of stages and read length of machine in inches. In the example shown, the ICFM is 45,000 and the gas is between propane and chlorine in mole weight. The speed is shown to be 4,000 rpm and the flanges are 36 and 24 inches. A slightly higher flow requires 3,500 rpm and 42 and 30 inch flanges.

Determine Approximate Price of Machine. On Figure 6, enter Q_1, proceed through H and read dollars. Correct this figure by the multipliers shown. The resulting price does not include lube system, driver, baseplate or special features. Because of changing market conditions and the wide difference in customer specifications, the accuracy of this particular curve must be considered no better than 15 to 20 percent, and must be considered subject to change.

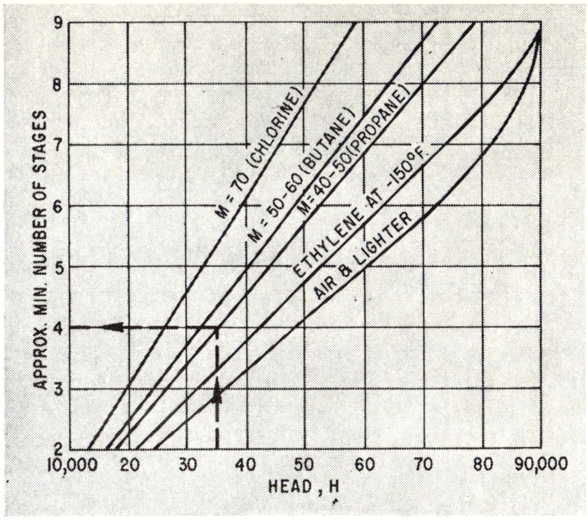

Fig. 4—Enter this chart with H found on Figure 3 to find number of stages required.

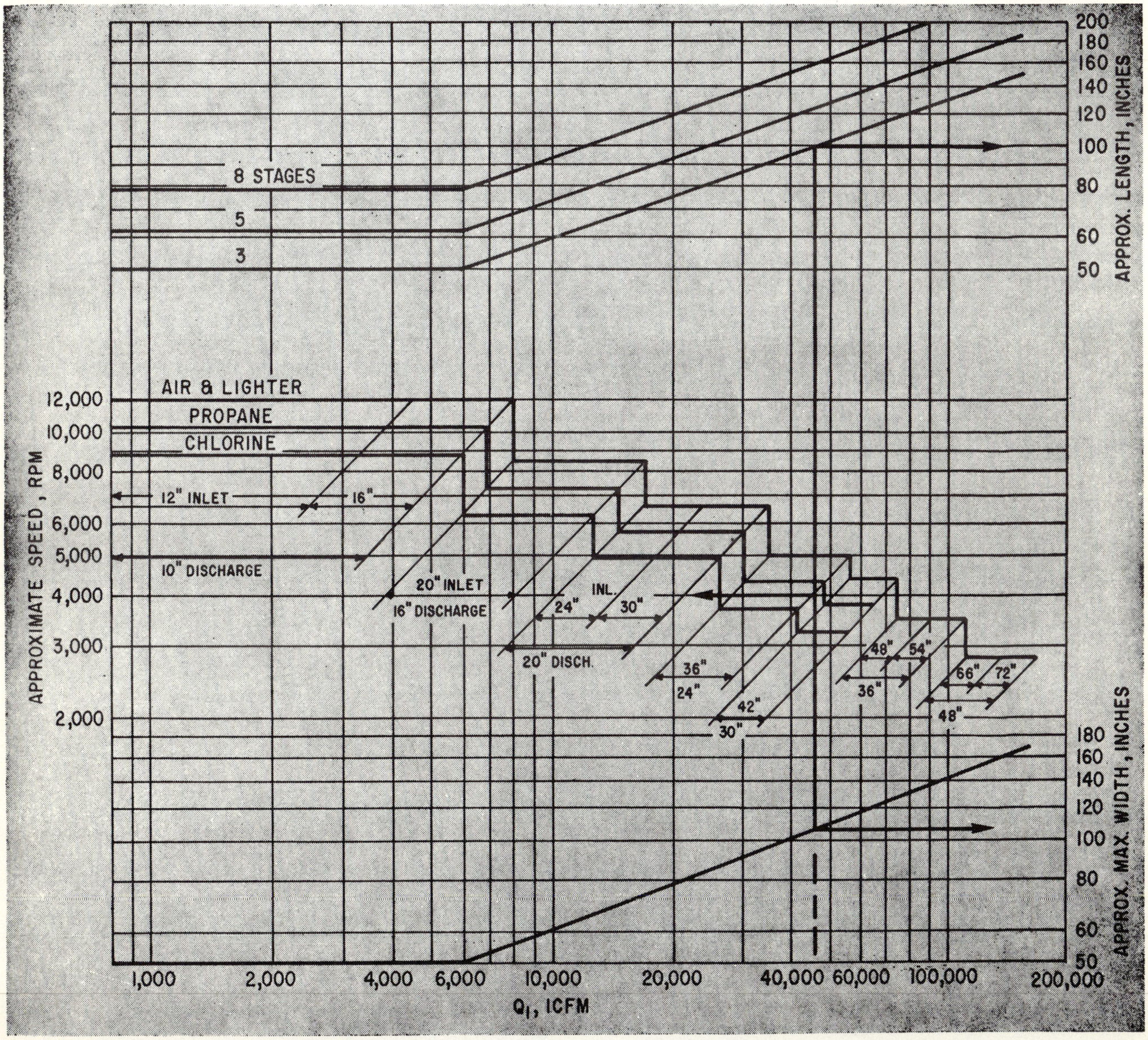

Fig. 5—Enter this chart at Q_1 found from Figure 1 or 2 and find speed, width, length and flange sizes.

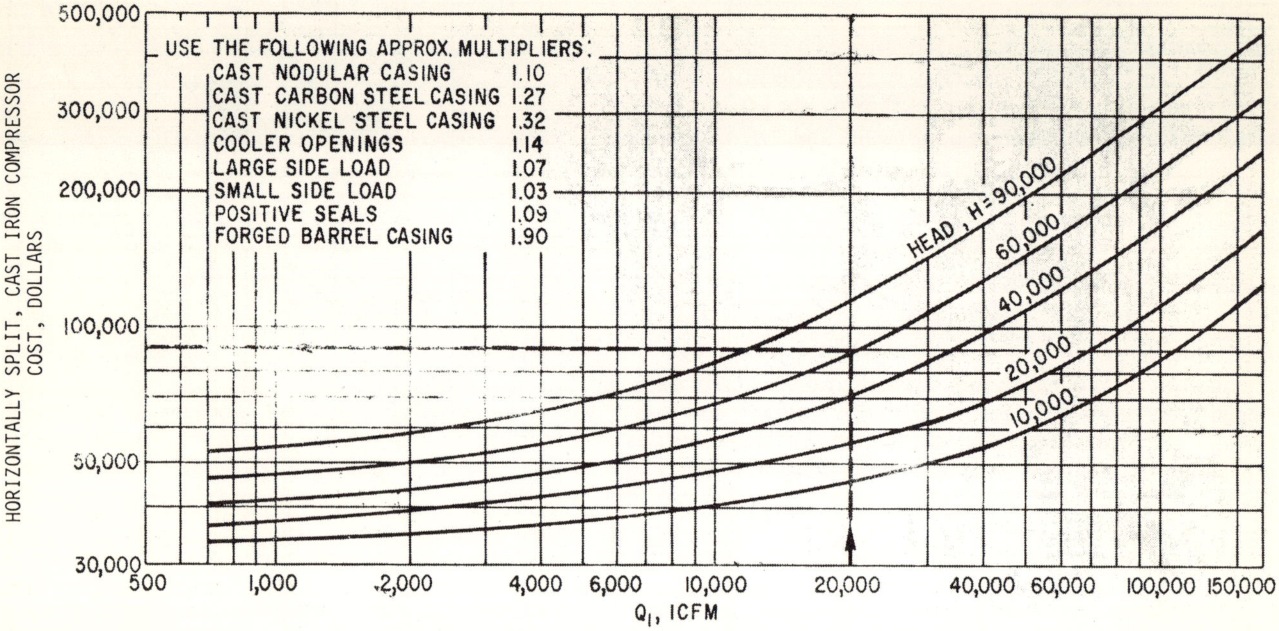

Fig. 6—Enter this chart at Q_1 found from Figure 1 to head H from Figure 3 and find compressor cost in dollars.

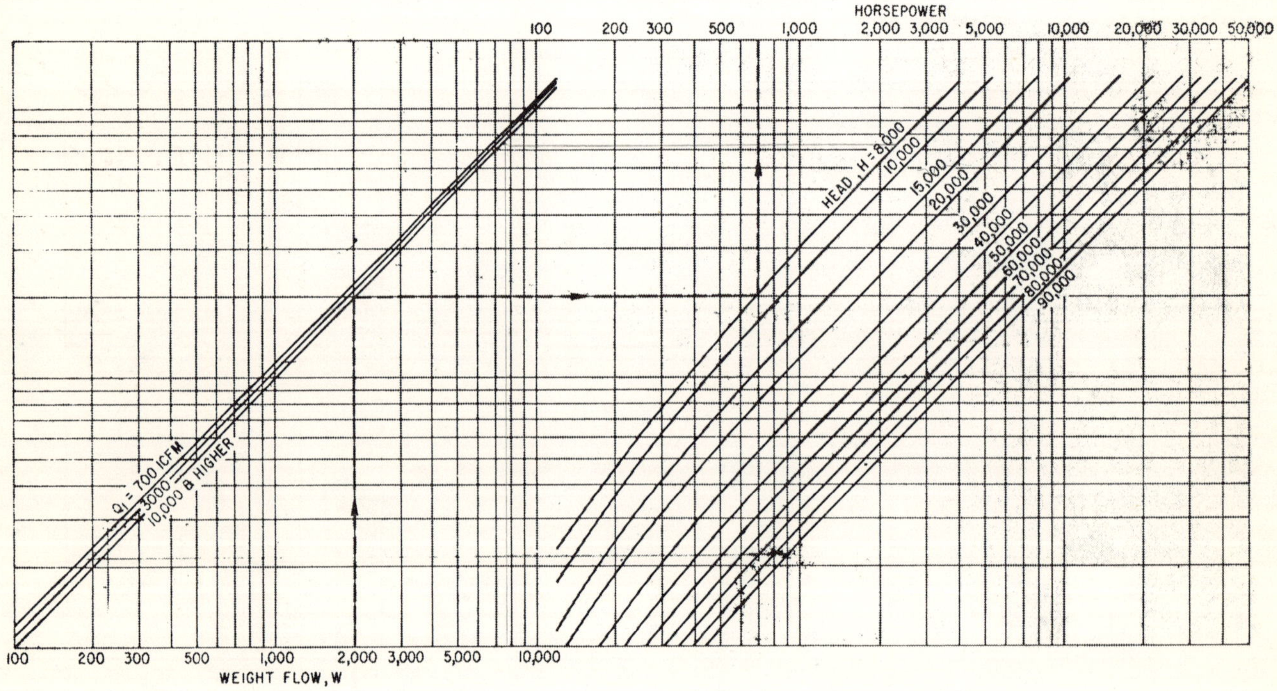

Fig. 7—Enter this chart at weight flow of gas, W, and proceed to find compressor horsepower required.

Determine Horsepower Requirement. On Figure 7, enter W, proceed through Q_1 and H and read HP. If W is not known, work backward from Q_1 on Figure 1 to find W before using Figure 7.

For uncooled, constant weight flow compression such as alkylation, wet gas, recycle, or air under 50 psig, the foregoing is sufficient to determine price, size, and driver requirement. For cooled or variable weight flow compression, proceed as follows:

Cooled Compression. Assume one cool and two compression sections, each section handling a pressure ratio equal to the square root of the overall pressure ratio.

- Determine discharge temperature t_2 from Figure 8, proceeding through R_p, Q_1, K, and t_1.

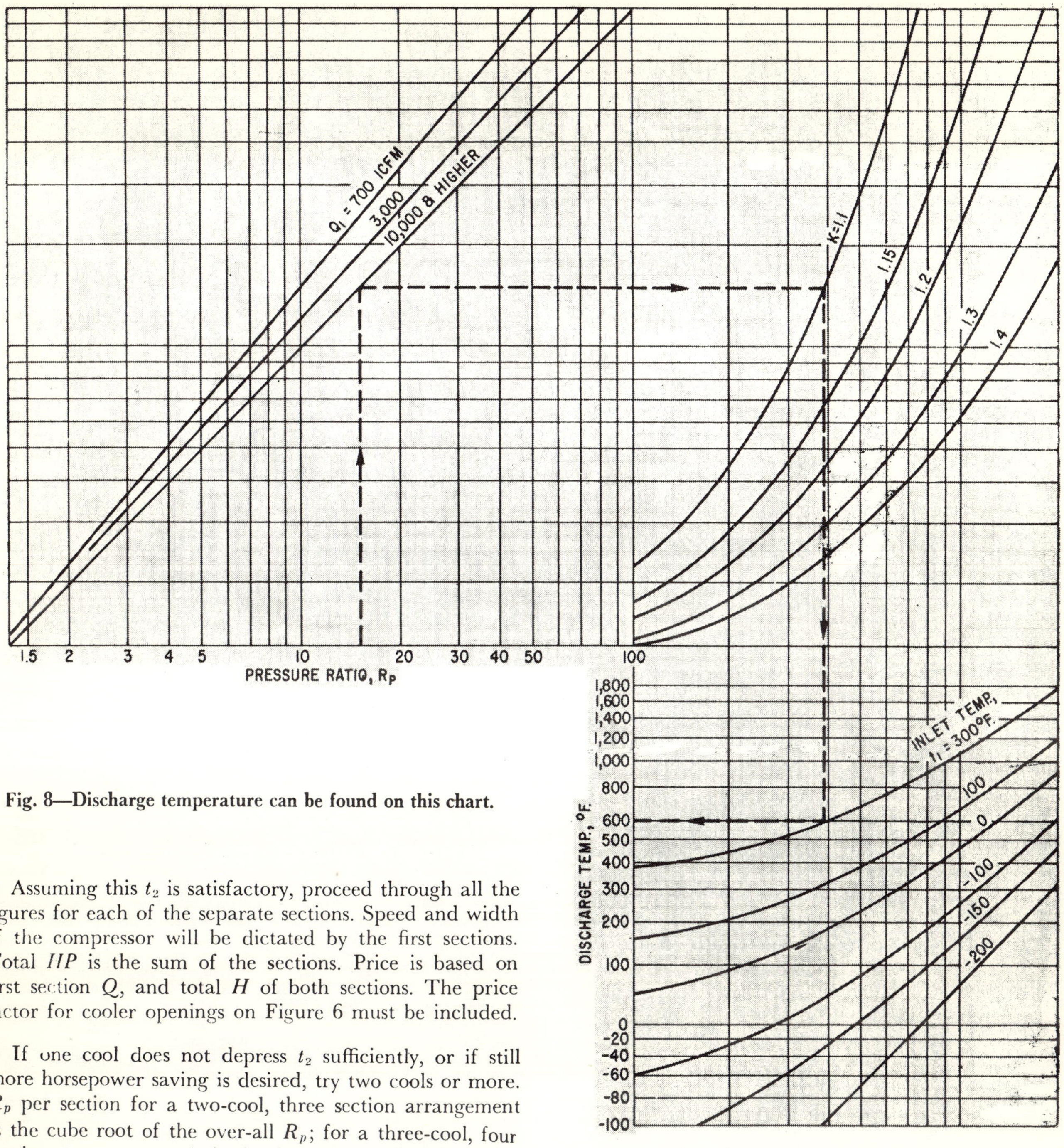

Fig. 8—Discharge temperature can be found on this chart.

- Assuming this t_2 is satisfactory, proceed through all the figures for each of the separate sections. Speed and width of the compressor will be dictated by the first sections. Total HP is the sum of the sections. Price is based on first section Q, and total H of both sections. The price factor for cooler openings on Figure 6 must be included.

- If one cool does not depress t_2 sufficiently, or if still more horsepower saving is desired, try two cools or more. R_p per section for a two-cool, three section arrangement is the cube root of the over-all R_p; for a three-cool, four section arrangement, it is the fourth root. Bear in mind that more than one set of cooler openings is seldom available on a single compressor body. When more than one cooler is chosen, therefore, more than one compressor is dictated.

Considerable judgment is required in choosing the number of coolers to use. Once temperature limits are satisfied, the use of additional coolers becomes a matter of economics between compressor and cooler cost and horsepower evaluation.

Variable Weight Flow. For applications having side flows either in or out, it is necessary to consider each constant flow compression section separately. Mixture temperature to the second section after the first "in" side flow must be calculated by finding the discharge temperature of the first section from Figure 8, multiplying by the first section weight flow, adding in the product of the side stream temperature and weight flow, and dividing by the sum of the weight flows. With mixture t_1, P_1, W, M and K known, the figures can now be used for the second section, and so on through the machine.

M and K of the side stream will generally be the same or quite close to those of the inlet, so mixture calculations for these quantities will normally be unnecessary. For extraction side flows, the second section inlet conditions are the same as the first section discharge conditions, except for W.

Normally the first section will "see" the largest Q_1, in which case the first section Q_1 will dictate the size and speed of the machine. An occasional refrigeration process, however, will show the second section Q_1 to be the greatest. In this case that Q_1 will dictate machine size and speed.

To determine the number of stages required, add the stages for each compression section and add in a blank stage for each large side load. It is impossible to give criteria for exactly what constitutes a "large" side load, but experience has shown that a typical propylene unit will require a blank stage for the first side load only, whereas a typical ethylene machine may require two blank stages. If the total number of stages, including blanks, exceeds nine, a second machine will probably be required.

Sizing Centrifugal Compressors

There are three basic reasons for using centrifugal compressors instead of reciprocating compressors in hydrocarbon processing plants.

1. Environmental. They occupy less space; they operate with minimum attention and are quieter.
2. Lower Operating Cost. They will run 12 to 30 months without mechanical repair. The maintenance cost is about one-third the cost of piston compressor operation, excluding the drivers.
3. Lower Initial Investment. The lowest cost turbo-driver is either a gear-motor or a steam turbine. The cost of such a driver and centrifugal compressor is approximately equal to the least costly type of synchronous motor driven reciprocating compressor in the 2,500 hp category. The centrifugal machinery cost is about two-thirds that of the piston machinery in the 5,000 hp bracket. The cost advantage of the turbomachinery is enhanced as the power demand is increased.

The reciprocating machine and motor drivers are priced on a cost per horsepower basis. Centrifugal machines and steam turbines are sold on a lump sum, plus a small unit cost per hp for the driver. The cost of a 12,000-hp centrifugal compressor may not be much more than the same casing rated for 1,000-hp service. Confronted with such an economy, there is an understandable demand on the part of the hydrocarbon processors for turbomachinery. The purpose of this chapter is to review the prevalent *state of the art* as to the compressors capability and a method of power rating and sizing the machine.

Compression Head

The compression head is the intangible measurement of the energy density imparted to a gas stream by a compressor. It may also be defined as the enthalpy added to the gas by the compressor. It may be observed by the increase in the gage pressure as the gas passes through the machine. Some engineers choose to evaluate the compression head as an isothermal function. A great many European compressors are evaluated on this basis. It presumes that the gas leaves the compressor at the entering temperature. This is an obvious false premise.

Isothermal Head

The isothermal head is determined from the equation:

$$L_{iso} = 144(\mathrm{Log}_e R_c) P_o (v_{so}), \text{ ft-lb/lb.} \qquad (1)$$

The product of this gas head and the weight flow, pps, divided by 550 gives hp $_{iso}$.

$$\eta_{iso} = L_{iso}(W)/\text{Gas hp}(550). \qquad (2)$$

The resultant efficiency is related to the reference gas head. The term *gas horsepower* refers to the input shaft power, free of mechanical losses which have no thermal effect on the gas. This value is attained by subtracting the 0.4 root of the *Brake horsepower* from that Bhp value. For example, the Bhp is 5,000; the 0.4 root is 30 hp of friction; the Gas hp is 5,000 − 30 = 4,970. The isothermal head and efficiency serve no particular significance or reference value. It is 6 to 9 percent less than the adiabatic efficiency. It is about 10 percent less than the polytropic efficiency.

Adiabatic Head

The adiabatic head is determined from the equation:

$$L_{ad} = (R_c^{\sigma} - 1)1{,}545(T_o)\ Z_a/\overline{m}\ (\sigma) \qquad (3)$$

The power developed with the adiabatic head represents the minimum realistic power requirement. Instances where a piston compressor has a generous valve area and is operated at reduced speed, the power tends to approach and almost equal the adiabatic power. The temperature of compression follows the same adiabatic pattern. As the piston speed is increased and the valve area reduced, the power requirement tends to exceed the adiabatic value, reflecting the dynamic losses. The temperature rise generally follows the adiabatic function. The reason is that the gas velocity in a piston machine rarely exceeds 0.2 Mach through the minimum restricted valve lift area. The velocities through a centrifugal machine are seldom under 0.2 Mach and are usually greater than 0.3 Mach. The friction of these high velocities against the obstructions and the resistances that exist in the impeller and the diffuser creates heat. These frictional losses contribute to the abnormal heating above the adiabatic. The loss is directly proportional to the degree of superheating the adiabatic temperature.

Adiabatic versus Isentropic

The adiabatic head and efficiency are often referred to as the isentropic values. The terms adiabatic and isentropic are used interchangeably. The thermodynamic definition of an adiabatic process requires that no heat be added or withdrawn from a facility where a change of state occurs. It may or may not be reversible to qualify as an adiabatic process. If the process is reversible, it is truly an isentropic operation. An isentropic change of state occurs at constant entropy. This identifies isentropic analysis with Mollier charts and tables of gas properties. For lack of a better mark of distinction, we refer to such enthalpy calculations as isentropic and the exponential R_c values as adiabatic processes.

Polytropic Head

The polytropic head is determined from the equation:

$$L_p = (R_c^{\sigma'} - 1)1{,}545(T_o)Z_a/\bar{m}\,(\sigma'). \qquad (4)$$

The polytropic efficiency can be determined from the Equation 6 relationship. The unique and only justification for the polytropic efficiency is its identity with the discharge temperature.

$$T_e = T_o R_c^{\sigma'} \text{ and} \qquad (5)$$
$$\sigma' = (k-1)/k\eta_p = \sigma/\eta_p. \qquad (6)$$

The polytropic efficiency can be determined for any operating point, where the suction and discharge temperature and the k value of the gas are known. This ratio on the Rankine scale and the knowledge of the gas k value applied to Equation 6 give the polytropic efficiency. The usual connotation of the word *efficiency* refers to the degree of perfection.

The adiabatic process represents the ideal, the minimal energy requirement to effect a change of state for any gas. All reference material, Mollier Charts and Gas Tables are so dedicated. The isothermal and polytropic efficiencies are *pseudo* values. The former presumes that the discharge temperature is equal to the suction temperature. There is no experience in nature or by man to sustain this premise. There are numerous industrial and laboratory tests to sustain the adiabatic process.

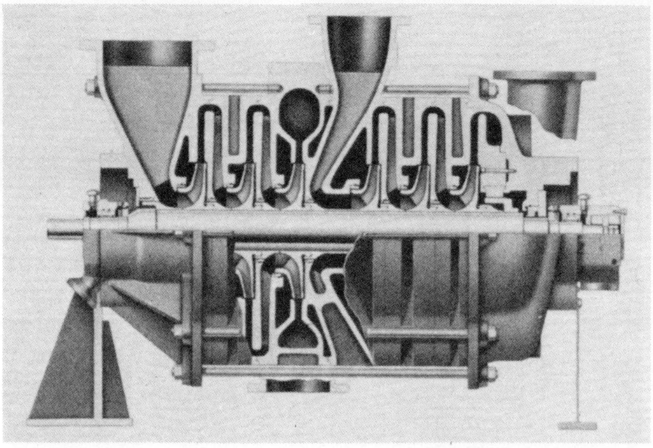

Fig. 1—A half-sectional view of a six-stage module assembled centrifugal compressor. The radial diaphragms are stacked together over the shaft and impellers. The outer circumference of the diaphragm is the casing shell. Draw bolts lock the diaphragms and the manifold into a unit. The above model shows a mid-point bleeder and return manifold. (Courtesy of Allis-Chalmers Corporation.)

Fig. 2—Versatile model multistage centrifugal compressor in service.

Fig. 3—Three-unit train of ammonia synthesis gas compressor requiring 20,000 hp. This 2,800 psig process has been a most popular application during the past several years. (Courtesy of Cooper Energy Services.)

The polytropic phenomenon is the thermal evidence of dynamic gas losses that are the resultant of the compressors' mechanical deficiencies. It is customary for some manufacturers to present the head-capacity performance curve using the polytropic head. A true presentation requires an established efficiency curve and an individual calculation for each point and each efficiency. This procedure is not always followed. Another disadvantage is individual performance curves are not comparable unless the subject curves have identical efficiencies at each respective flow quantity.

Figure 1 illustrates a versatile model multistage centrifugal compressor. The lower discharge flange illustrates the method by which the gas may be withdrawn for intercooling or other process purposes. The gas is reentered through the center vertical flanged manifold. Figure 2 shows such a compressor in service. Figure 3 illustrates a three-unit train of an ammonia synthesis gas compressor.

Efficiency Conversion Chart

Figure 4 presents a simplified correction factor for converting polytropic efficiencies to adiabatic efficiency. It was prepared from ratios of compression of 1.2 to 3.7 for air, natural gas, LPG and hydrogen mixtures. The correction values are not absolute, but are close approximations. The integrated radical ($R_c^g - 1$) is used as the abscissa, and the ratio of the polytropic to the adiabatic efficiency is the ordinate.

The *isoeffic* (constant efficiency) lines cover the normal range of adiabatic efficiencies. For an illustration, an air compressor is operated at $R_c = 2$, the integrated radical ($R_c - 1$) is equal to 0.22 when η =

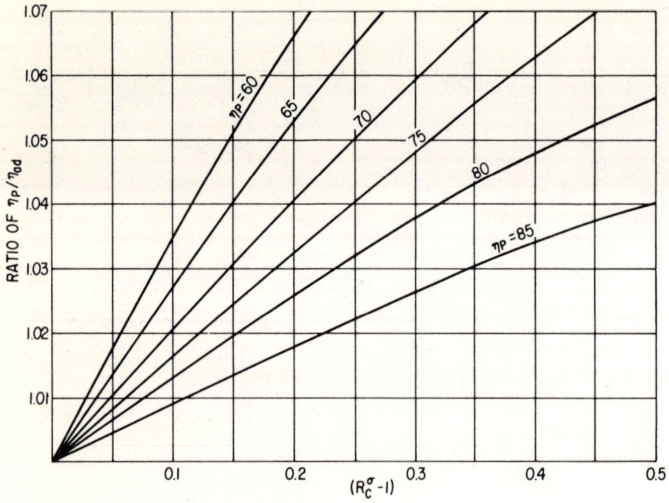

Fig. 4—Ratio of polytropic to adiabatic efficiency, with reference to the adiabatic exponential function and the polytropic efficiency.

(1.40 — 1.0)/1.40 = 0.286. Then if the polytropic efficiency was 65 percent, the correction ratio is 1.057 and the adiabatic efficiency would be (65/1.057) = 61.5 percent. If the polytropic efficiency was 85 percent, the adiabatic efficiency would be (85/1.022) = 83 percent. These correction ratios are obtained from the equation:

$$R\eta = \eta_p\,(R_c^{g/\eta p} - 1)\;/\;(R_c^g - 1). \tag{7}$$

The equivalent isothermal efficiency for the above 83 percentage case can be determined by the inverse proportion of the respective heads and efficiencies. The isothermal head is

$$L_{iso} = 144(\text{Log}_e R_c) P_o v_{so},\ \text{ft-lb/lb} \qquad \text{(repeating 1)}$$
$$L_{iso} = 144(0.693)14.7(13.1) = 19{,}200\ \text{ft-lb/lb.}$$

The adiabatic head is

$$L_{ad} = (R_c^g - 1)1{,}545(T_o)Z_a/\overline{m}\sigma \qquad \text{(repeating 3)}$$
$$L_{ad} = (0.22)1{,}545(520)1.00/29(0.286) = 21{,}300\ \text{ft-lb/lb}$$

The equivalent isothermal efficiency is 83(19,200/21,300) = 75 percent.

Example Head Calculation

The following example illustrates the derivation and application of the three reference heads and efficiencies. A shop performance test demonstrated that a centrifugal compressor was handling 36,200 cfm of ambient air at 14.7 psia and 60°F to 29.4 psia. The driving motor has an efficiency of 93.5 percent and demands 2,000 kw of power for the test. The discharge temperature is 215°F and the mechanical efficiency of the compressor had been previously determined to be 98 percent. The Gas hp is 0.935(0.98)(2,000/0.746) = 2,450 hp. This value and the discharge temperature represent the vital data pertinent to the plant process. The respective efficiency used to project this information is only incidental to the main objective of fixing the power requirement. The weight flow is

$$v_{so} = 10.73 T_o Z_o/\overline{m}P_o \tag{8}$$
$$v_{so} = 10.73(520)1.00/29(14.7) = 13.1\ \text{cf/lb.}$$
$$w = 36{,}200/13.1(60) = 46\ \text{lb/sec (pps).}$$

The isothermal head for R_c = (29.4/14.7) was determined to be 19,200 ft-lb/lb in the previous paragraph. Transposition of the fundamental power equation produces the isothermal efficiency in the following manner:

$$\text{Gas hp} = wL_{iso}/550\eta_{iso} \tag{9}$$
$$\eta_{iso} = wL_{iso}/\text{Gas hp}(550)$$
$$\eta_{iso} = 46(19{,}200)/2{,}450(550) = 65.5\ \text{percent.}$$

The adiabatic head for R_c was determined to be 21,200 ft-lb/lb. The adiabatic efficiency equation has the same form as given above.

$$\eta_{ad} = wL_{ad}/\text{Gas hp}(550) \tag{10}$$
$$\eta_{ad} = 46(21{,}200)/550(2{,}450) = 72.3\ \text{percent.}$$

The isentropic efficiency (η_i) should equal the adiabatic efficiency. The enthalpy difference between suction conditions of 60°F and 14.7 to 29.4 psia at constant entropy should be 21,200/778 or 27.2 Btu/lb.

$$\text{Gas hp} = w(H_e - H_o)/2544\eta_i \tag{11}$$
$$\eta_i = 46(3{,}600)27.2/2{,}544(2{,}450) = 72.2\ \text{percent.}$$

The fundamental design of the impeller is represented by the vector diagram of Figure 5. The Euler equation, Equation 12, resolves these vectors into an energy density, gas head:

$$L_{eu} = (V_{u2}U_2 - V_{u1}V_1)/g. \tag{12}$$

The Euler efficiency will approach the adiabatic equation if the flows typify the actual flow and the Euler efficiency will be

$$\eta_{eu} = wL_{eu}/550\ \text{Gas hp.} \tag{13}$$

In addition to the vector analysis, there are the flow losses through the suction eye, the vanes, the diaphragms, the scroll or diffuser, the guide vanes, the seal leakage and disk friction which contribute to the inefficiencies of the compressor and raising the

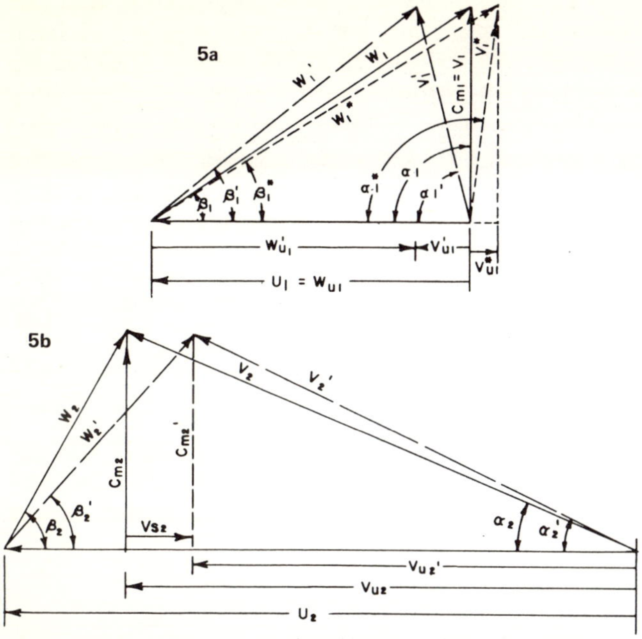

Fig. 5a—Vector diagram depicting typical suction velocities. C_m is the meridional velocity, U is the peripheral velocity of the impeller, V is the absolute velocity of gas flow, W is relative velocity of the gas flow and α_1 is the eye entrance angle.
Fig. 5b—Vector diagram depicting typical discharge velocities. β_2 is the backlay angle of the vane at the impeller tip.

discharge temperature. The reason for making a shop performance test is to confirm the accuracy of the design premise.

Temperature

The polytropic efficiency is an overall thermal evaluation of the design. The discharge temperature is calculated thus:

$$T_e = T_o R_c^{\sigma'} \tag{14}$$

$$\sigma' = (k-1)\, k\eta_p = \sigma/\eta_p \tag{15}$$

$$R_c^{\sigma'} = 2.00^{\sigma'} = 675/520 = 1.298$$

$$\sigma' = \log 1.298/\log 2.000 = 0.376$$

$$\eta_p = \sigma/\sigma' = 0.286/0.376 = 76 \text{ percent.}$$

Without the discharge temperature, the power requirement should have reflected the machines efficiency and the polytropic efficiency can be developed from Figure 4. The polytropic efficiency is usually 2 to 4 percent greater than the adiabatic efficiency. Taking the abscissa as 0.22 (from previous data) and estimating the polytropic η_p as 75 percent, the $R\eta$ is 1.035 and the resolved adiabatic efficiency is 72.5 percent. The polytropic efficiency has merit in establishing a realistic discharge temperature. It remains constant at optimized specific speeds and specific diameters for any gas providing there is no abnormal leakage and the Mach number and the Reynolds affect are equal or adequately corrected. The polytropic head is as useless and misleading as the isothermal head. But most manufacturers use the term in describing performance of their machinery.

It is determined from the equation:

$$L_p = (R_c^{\sigma'} - 1)1{,}545 T_o Z_a/\bar{m}\,\sigma' \tag{16}$$

$$L_p = (0.298)1{,}545\,(520)/29(0.376) = 22{,}000 \text{ ft}$$

$$\eta_p = 46(22{,}000)/550(2{,}450) = 75 \text{ percent.}$$

Caution

There are two important points to remember concerning the various reference efficiencies.

1. The realistic discharge temperature can only be determined from the polytropic efficiency as applied to Equation 14 or to the ΔH in a Mollier chart solution.
2. The realistic power can be determined from any reference efficiency, if these efficiencies are referenced to the same real power data. The adiabatic references are the most professional and useful for this purpose.

Design Criteria

The maximum obtainable efficiencies of turbomachines, together with optimum design geometry, based on the state-of-the-art knowledge can be presented as a function of the similarity parameters, specific speed, N_s, and specific diameters, D_s, for constant Reynolds and Mach numbers. The specific speed and specific diameters are

$$N_s = N(Q^{0.5})/L^{0.75} \tag{17}$$

$$D_s = D(L^{0.25})/Q^{0.5} \tag{18}$$

$$U = d(N)/229 \tag{19}$$

$$d_r = 1{,}980 L^{0.5}/N \tag{20}$$

$$d_b = 1{,}760 L^{0.5}/N \tag{21}$$

$$q_{ad} = g(L_{ad})/U^2 \tag{22}$$

where d_r is the outside diameter of radial vaned impeller. The pressure coefficient is 0.65 for a single stage machine. This coefficient is depreciated about 2 percent per stage for the resistance of each return diaphragm. The D_sN_s product is 165 for the radial vane. The nominal 45 degree backlay vane has a D_sN_s product of 147 and a pressure coefficient of 0.55 which is subject to the same frictional depreciation. The quick solution for the backlay vaned impeller diameter is given in Equation 21. U is the peripheral tip speed (fps); Q is the volume flow in cubic feet per second (cfs); L is the gas head; N is the rpm and g is the gravitational acceleration, 32.2 ft/sec^2. The head coefficient may have numerous reference bases, such as the adiabatic given in Equation 21, where L_{ad} is derived from Equation 3. The Euler coefficient is sometimes referred to as the geometric or theoretical pressure coefficient, q_{th}. There are design data available for correcting the geometric (vector analysis) to the realisitc q_{ad} values. These factors correct for the number of vanes, the slip, the various inefficiencies of the guide vanes, the rotor, the scroll or diffuser and the discharge velocity conversion.

Mach Numbers

All significant velocities in aerodynamic design are referenced as decimal Mach numbers. Velocities of equivalent Mach numbers have approximately equal resistances in terms of percent of the system pressures. Table 1 illustrates the behavior.

The velad is the gas head (ft) required to support the 0.2 M velocity, $V^2/2g$, or 480(480)/64.4 = 3,580 ft. The *head per psi* is determined from the specific volume (144 v_{so}), when divided into the velad, gives the pressure drop, Δp *per velad*. The flow-coefficient (ϕ_2) at the point of *surge* is shown on Figure 6 to be about 0.2. If the tip speed U_2 is taken as 0.9 Mach, the meridional suction eye velocity (C_{m_1}) for an air compressor would be C_{m_1} = 0.2(0.9)1,120 = 202 fps. The same equivalent *surge* would be realized if the impeller had a tip speed of 2,160 fps and the *eye* velocity was 431 fps. If the tip speed remained the same 1,020 fps as for the air case, the flow coefficient for the greater flow of 431 fps, ϕ_2 would be 0.42, and the impeller would experience the opposite flow characteristic where *choke* exists. The *eye* velocities for the other gases are given on the bottom line of Table 1 for 0.2 and U_2 = 0.9 Mach.

Flow Characteristics

The design conditions consist of adiabatic (or polytropic) head and the flow, in actual cfm for a given operation. Figure 7 illustrates a typical master impeller characteristic curve which has been developed by thorough testing. The abscissa is plotted as a unit of flow per revolution, usually cfm/rpm, Q/N or other similar flow references. The solid line represents a single-stage performance. The dash line shows the loss from the return diaphragm in multistage operation. The pressure coefficient establishes the head from Equation 21. The dotted line is the product of q_{ad} and η_{ad}. The crest of this curve represents the Best Operating Point (BOP). Impellers are selected so that the design point is as close to this crest as practical. The performance

Table 1
An array of gases flowing at 0.2 mach

Gas	Hydrogen Mixture	Natural Gas	Air	Propane	Butane
$\overline{m}$	6.2	18.2	29	44	58
k	1.38	1.26	1.40	1.13	1.09
Sonic, fps	2,400	1,350	1,120	760	700
0.2 *M*, Velocity	480	270	224	152	140
Velad, ft., ($V^2/2g$)	3,580	1,130	780	360	305
Ft./psi	8,750	3,010	1,890	1,240	945
Δ psi/Velad	0.41	0.38	0.41	0.29	0.32
Equivalent C_m, fps	431	242	202	136	126

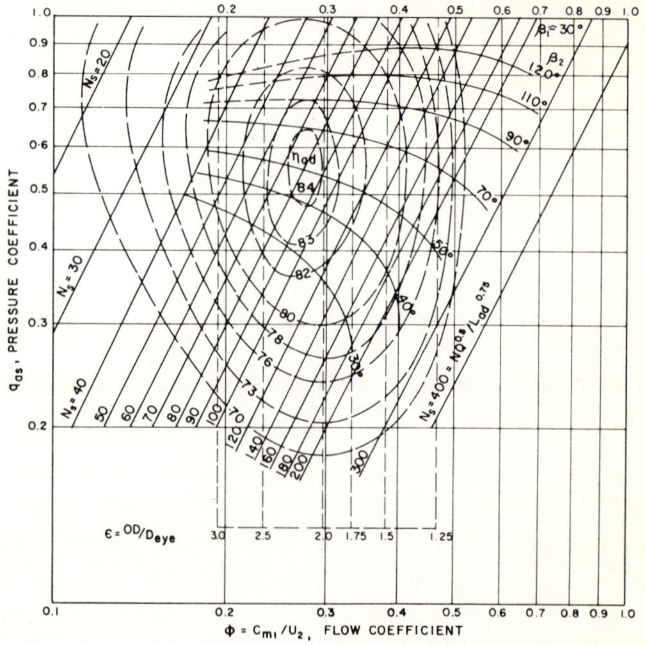

Fig. 6—Eckert chart giving pertinent adiabatic performance data for impellers having various vane angles.

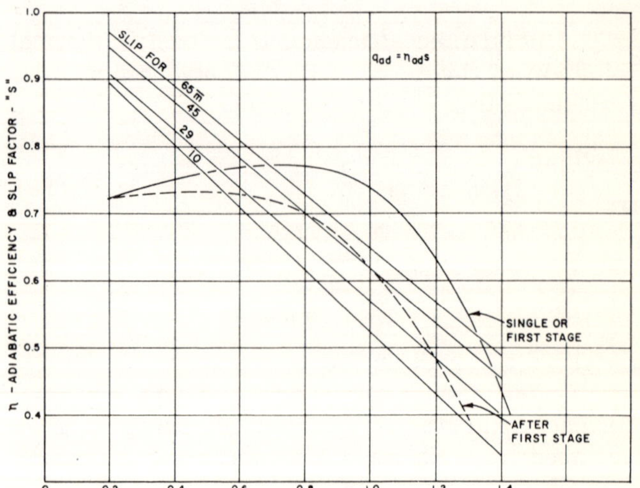

Fig. 7—Typical basic performance curve of a commercial centrifugal impeller, depicting the pressure coefficient, efficiency and slip plotted against the flow coefficient Q/N. The solid lines represent single stage performance. The broken lines represent multistage factors. The pressure coefficient, qad = SLIP (η_{ad}).

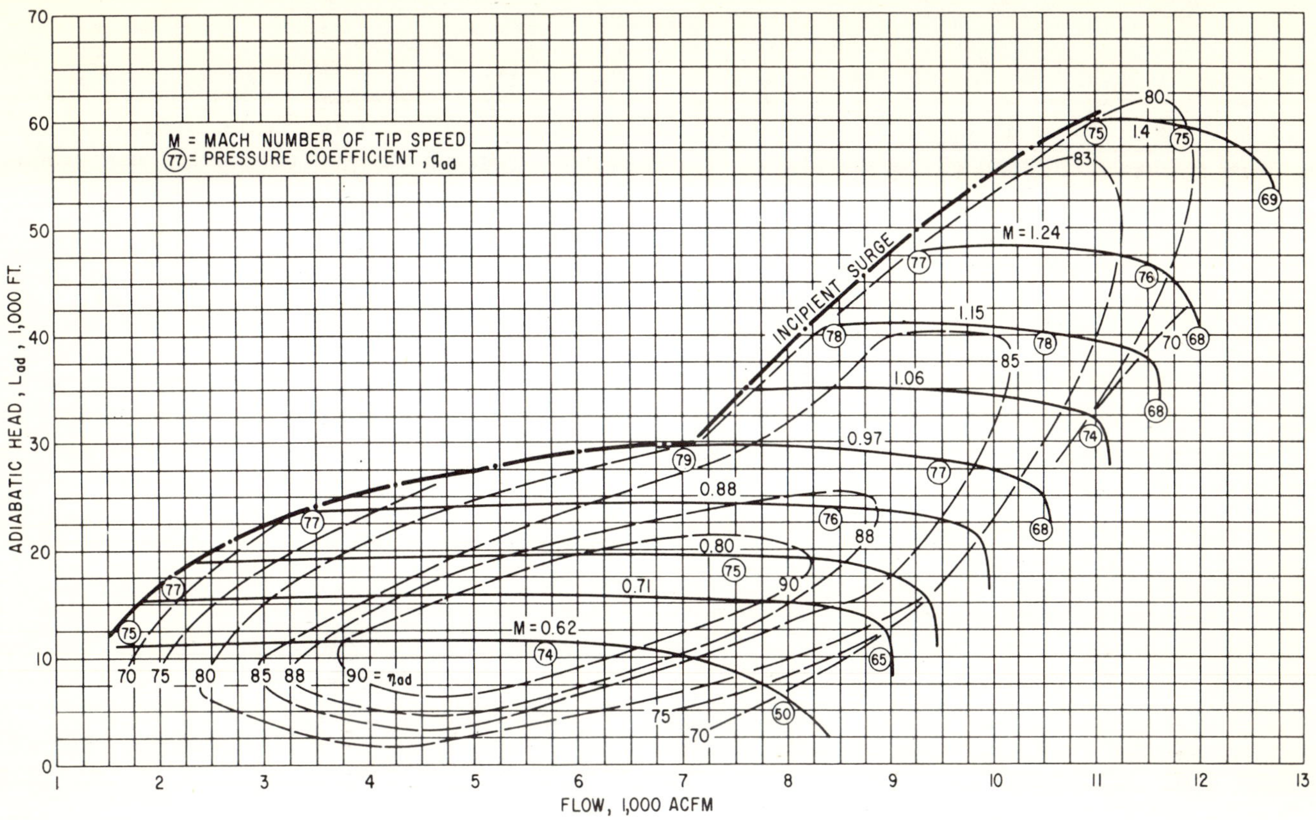

Fig. 8—Performance curve of highly perfected NACA impeller.

represented by the $D_S N_S$ intersection on Figure 7 are BOP. The BOP performance for Figure 6, 8 and 9 are obviously the zones of the highest efficiency.

The pump head (and q_{ad} values) will increase some 5 to 8 percent from BOP as the flow is retarded to the point of *stall* or *surge*. The *surge* zone follows the N_S lines of 40 to 60 on Figure 6. It is well identified on Figure 6.8. It follows the flow coefficient of 0.10 to 0.15 on Figure 9. The other extreme from *stall* is the maximum flow condition of *choke* or *stonewall*. This flow represents the maximum flow that can be drawn through the impeller eye. It is the flow experienced to the right of ϕ_2 = 0.35 to 0.50 in Figure 6. The abrupt *knee* breaks on the H-Q curves illustrate the *choke* affect most emphatically. Particular notice should be made of the high q_{ad} values shown in the small circles, ranging from 0.79 at *surge* to 0.77 at BOP and 0.68 at the *knee*. This represents the degree of perfection attained by NACA.

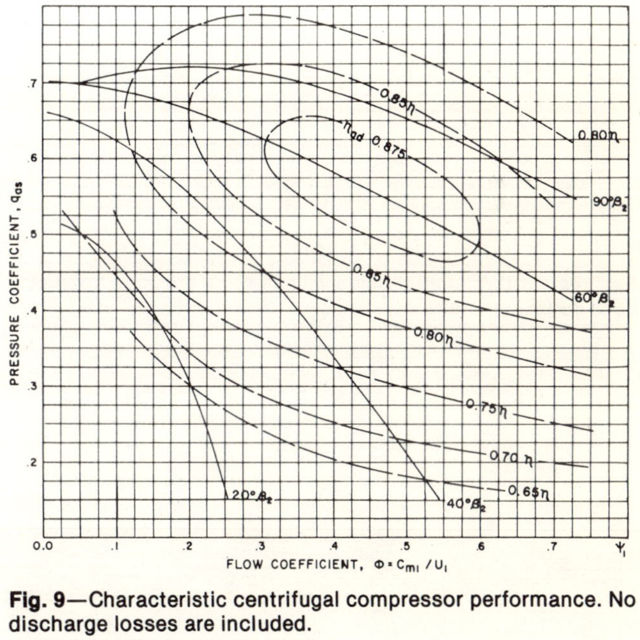

Fig. 9—Characteristic centrifugal compressor performance. No discharge losses are included.

Slip

Gas that is approaching the eye of an impeller will take on a prerotation in the direction of the impeller turning if the casing does not contain inlet guide vanes. The prerotation tends to reduce the compressor capacity and increase the slope of the head curve. Moveable Inlet Guide Vanes (IGV) are sometimes installed to unload the compressor by exaggerating this characteristic. The reverse action of the IGV can create a counter-rotation which has the effect of supercharging each individual impeller so equipped. This tends to reduce the H-Q and the power curve. These effects are illustrated in Figure 10. The impact of the gas molecules against the back side of a rotating vane distorts the radial flow pattern. The effect of this distortion on the ideal relative velocity vector, W_2 in Figure 5*b* causes the β_2 to be reduced to β_2' and W_2 to increase to W_2'. When this vector is resolved with the tip speed U_2, the resultant absolute velocity is V_2'. The tangential component of the ideal absolute vector is C_{u2} and for the distorted flow is C_{u2}'. The difference between these two vectors is the *Slip* V_{s2}. It is usually about 15 percent of the absolute velocity. The meridional velocity leaving the tip annulus is C_{m2}, which is presumed to equal C_{m2} and C_{m1}.

Figure 5*a* shows the same flow behavior for the gas entering the impeller. U_1 is the peripheral speed of the impeller eye. C_{m1} is the meridional vector and represents the average impeller eye velocity. The flow coefficient ϕ is C_{m1}/U_2 and is applied as the abscicca on Figure 6. There are occasions when a similar *gulp* factor (ϕ') is taken as C_{m1}/U_1 and is so applied in Figure 9. The blade angle at the eye β_1 includes the relative vector W_1 and the eye velocity U_1. When the guide vanes direct the meridional flow normal to the eye velocity U_1, the horizontal component W_{u1} is equal to U_1. When the inlet guide vanes are adjustable and the gas is given a prewhirl (rotation in the direction of rotation), the relative velocity follows the angle β_1' and vector W_1'. The absolute vector is V_1' and the horizontal component and slip is V_{u1}'. When the incoming gas is given a counter rotation with regard to the impeller, the flow follows angle $\beta_1{}^*$ and the relative vector $W_2{}^*$. The absolute velocity is $V_1{}^*$ and the supercharging head beneficient is $V_{u1}{}^*$. The magnitude of the head correction can be evaluated from the Euler equation by applying the following guide angle A degrees of deviation from normal.

$$L_{eu} = [V_{u2}U_2 - 1 + A(U_1 C_{m1} \operatorname{Sin} A)]/g. \qquad (23)$$

The counter-rotation should be treated as a negative quantity, when corrected by the equation minus sign, it becomes additive.

The *slip* is largely affected by the number of vanes (n). The effect on the Euler equation for an impeller having back-lay vanes is

$$L_{eu} = [U_2 V_{u2} (1 - 2/n) - U_1 V_{u1}]/g. \qquad (24)$$

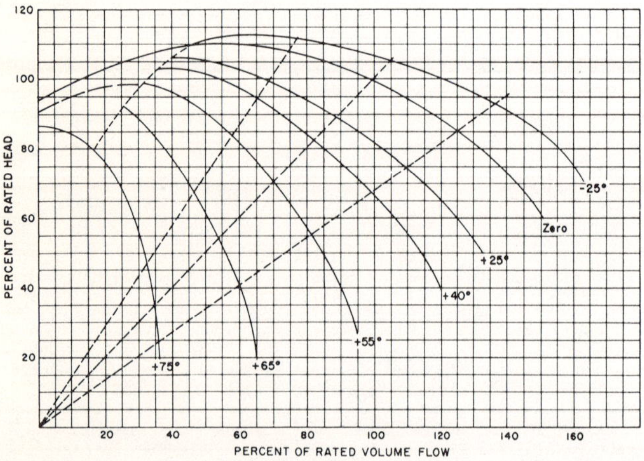

Fig. 10a—Effect of inlet guide vanes on the pressure head of a centrifugal compressor.

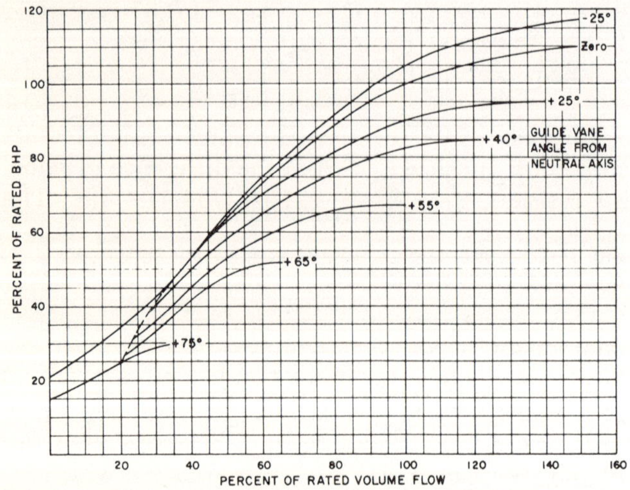

Fig. 10b—Effect of inlet guide vanes on the power requirement for centrifugal compressors.

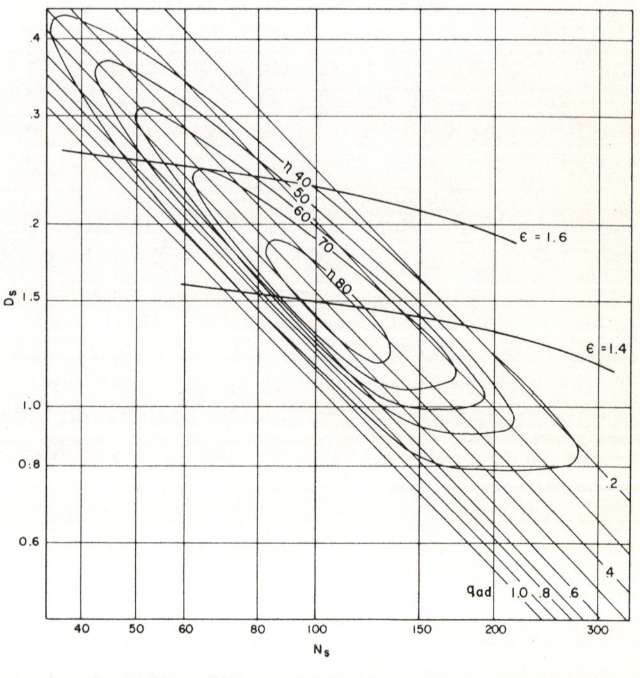

Fig. 11—Baljé generalized performance curves, orienting the adiabatic-dynamic efficiency and the pressure coefficient as a function of the specific speed and specific diameter.

The equation for a radial vane impeller is

$$L_{eu} = (1 - 2/n)\, U_2^{\,2}/g. \qquad (25)$$

Flow Coefficient

The two heavy lines striking horizontally across Figure 11 marked E is 1.4 represent the approximate ratio of the OD to the eye diameter. Where N_S is taken as 100 and D_S as 1.47, E is 1.4 (presuming the impeller is designed for end suction with no shaft extension through the eye and is 24 in. in diameter). The suction eye is 24/1.4 = 17 inches and has 1.585 sq ft area. Presume further that the tip speed U_2 is 1,000 fps and U_1 is (1,000/1.4) = 715 fps. An optimized flow coefficient ϕ_2, value from Figure 6 for a 50-degree β_2 back-lay vane impeller is 0.275. The meridional velocity is 0.275 (1,000) = 275 fps. The impeller design capacity is 275 (1.585) = 435 cfs or 26,000 acfm. The flow coefficient ϕ_1 is (275/715) = 0.385. This number is a reasonable value corresponding to a 55-degree back-lay impeller in Figure 9.

Application Example Problem

The design flow rate is 18,000 mols per hour of 22 mol weight gas, having a k value of 1.26 and both Z values are unity. The gas is compressed from 85° F and 40 psia to 51.6 psia. Develop the approximate size impeller, connections, speed, efficiency and power.

Solution. The specific volume is

$$vs = 10.73\,(545)\,(1.00)/22\,(40.0)$$
$$= 6.64 \text{ cf/lb} \qquad (8)$$
$$w = 18{,}000\,(22)/3{,}600 = 110 \text{ pps.}$$
$$Q = 110\,(6.64) = 730 \text{ cfs or } 43{,}800 \text{ acfm}$$
$$Q^{0.5} = 27.0$$
$$\sigma = (1.26 - 1.0)/1.26 = 0.206$$
$$L_{ad} = [(51.6/40)^{0.206} - 1]\; 1{,}545\,(545)\,(1.0)/22\,(0.206).$$
$$L_{ad} = 10.000 \text{ ft-lb/lb.} \qquad (3)$$

An inspection of the Baljé and Eckert charts (Figures 6 and 9) shows that 100 N_S is an optimum design point, and 1.47 D_S is a matching ordinate for a reasonable q_{ad} value of 0.55.

$$L^{0.75} = 1{,}000,\; L^{0.5} = 100 \text{ and } L^{0.25} = 10.$$
$$N = N_S\, L^{0.75}/Q^{0.5} = 100\,(1{,}000)/27.0$$
$$= 3{,}700 \text{ rpm.} \qquad (17)$$
$$D = D_S\, Q^{0.5}/L^{0.25} = 12\,(1.47)\,27.0/10$$
$$= 47.5 \text{ in.} \qquad (18)$$
$$dia = 1{,}760\,(100/3{,}700) = 47.5 \text{ in.} \qquad (20)$$

The last equation is a simple check of the equation above. The mean tip speed U_2 is 47.5 (3,700)/229 = 769 fps. The flow coefficient is read as 0.32 on Figure 5 at the intersection of N_S = 100 and q_{ad} = 0.55. The meridional velocity C_{m1} entering the eye is 0.32 (769) = 246 fps. The eye area is 730 (144)/246 = 427 sq in or 20.6 in in diameter. The A diameter ratio is 47.5/20.6 = 2.31. The effect of the tip speed Mach numbers on the adiabatic efficiencies is shown in Figure 8.

The efficiencies (η) in Figure 11 relate the L_{ad} to the total inlet pressure and the static outlet pressure. The conditions further presume that the meridional flow entering the impeller eye (C_{m1}) is equal to the velocity leaving the machine (C_{m3}) and that the Reynolds number is 10^6 or greater. The usual practice in multistaged compressors is to relate the compression head to the total discharge pressure (ηt). These values are greater than η, especially for N_S values > 60 and for D_S values < 2.0. The difference between η and ηt is that the former discharge pressure does not include the discharge nozzle velocity head which may exceed the suction nozzle velocity that is presumed to be equal to C_{m1}. The process gas line velocities are usually about 0.08 Mach or 70 to 100 fps for common (20 to 30 $\overline{m}$) gases. The nozzles on multistage compressors are operated at velocities of 85 to 150 fps at design loads. The discharge nozzel is two thirds the diameter of the suction nozzle on multistage casings. The suction and discharge nozzles are usually the same size on single or two-stage cantilevered type compressors.

The discharge temperature for this example at 80% adiabatic efficiency is 545 $(51.6/40)^{0.206/0.8}$ = 582 °R or 122° F. The discharge volume is 730 cfs (1.0677/1.29) = 604 cfs, v_{se} = 5.48 cf/lb.

The process suction line would be sized at 110 fps: 730 (144)/110 = 960 sq. in. or 35 inches ID and 36 inches OD. The discharge line is 0.67 (35) = 23 inches ID and 24 inches OD.

The compressor casing for this volume has 36 by 24-inch nozzles. The discharge velocity is 60.4 (144)/450 = 193 fps. The discharge static pressure correction is calculated from the difference in velocity (velads): $[(193)^2 - (110)^2]/64.4 = 392$. The energy density on the discharge line is 392 ft-lb/lb short of the objective 10,000 ft-lb/lb. Nearly 4 percent of the machine's output is in the discharge nozzle. The conditions for using the efficiency values from Figure 10 required C_{m1} to be equal to C_{m3}. The corrected efficiency from conditions of static discharge to total discharge conditions is (10,000 − 392) 80.0/10.000 = 76.6 percent.

This condition can be corrected by providing a 27-inch discharge nozzle on the compressor casing, or by installing a 24 x 32 divergence tube 100 inches long to convert the excessive discharge velocity to static pressure. The velocity in the 32-inch tube is 110 fps and equal to the suction velocity. This restores the dyanmic efficiency to 80 percent. The power required to operate the single-stage compressor is

$$\text{Gas } hp = w\, L_{ad}/550n \qquad (26)$$
$$= 110\ (10{,}000)/550\ (0.766) = 2{,}610.$$

The mechanical losses from the bearings, etc., which do not affect the temperature of the gas, are not included in Figure 11. It is proposed that the mechanical losses are well within the frictional hp of the equation: fph = (dynamic hp)$^{0.4}$.

The frictional hp for this case would be (2,610)$^{0.4}$ = 25. The driver should include this value, plus a 10 percent power tolerance, which totals 2,895 bhp at 3,700 rpm.

Corrections

The data plotted on Figure 11 included as much test and operating performance as available. The curves were developed from loss analysis, simplifying assumptions and averaging. The correction for the high discharge velocity is an example of how other corrections may be applied. The Baljé efficiencies include the normal interstage leakage and disk friction.

Basic Specification Considerations

A compressor will perform properly only if accurate and realistic design data are given to the manufacturer. Here is a guide to specifying.

John E. Stryker
Compressed Air and Gas Institute
Cleveland

THE MOST IMPORTANT considerations in specifying a centrifugal compressor for refinery applications are a thorough knowledge of the gas itself, the compression ratio, and the quantity to be handled. The hydraulic performance, mechanical design, materials of construction, type of seal and control are all directly dependent on this information.

While the foregoing is an important and essential phase of the application, it is only about half of the over-all problem and the selection of the type, size and characteristics of the driver completes the unit. This article, however, deals with the factors affecting the compressor only.

The centrifugal compressor manufacturer is somewhat limited in the assistance he can render by the quality and quantity of the performance data, conditions and requirements submitted and, consequently, the buyer shares heavily in the responsibility for a suitable selection. It should always be kept in mind that since the hydraulic design is predicated on the following few conditions and gas characteristics, namely, intake pressure, intake temperature and intake capacity, molecular weight, compressibility factor, ratio of specific heats and compression ratio each value should be established only after a thorough and complete study. A seemingly small change in any of the gas properties may well mean that an entirely different unit is required in order to produce the same capacity and pressure.

The required performance is generally established by a nominal set of conditions termed "Design," "Normal," "Rated," or "Guarantee," which is the best estimate of what is anticipated and is usually on the conservative side. As a safety margin, an overload performance is frequently specified as "Maximum" and has the order of magnitude of 105 percent rated speed (corresponding to approximately 110 percent of rated head at the rated capacity) and 110 or 115 percent of rated power (corresponding to approximately 110 or 115 percent excess weight flow at the rated head). The judical choice of these values will be reflected in the ability to perform the required service, first cost of the equipment, economy of operation, degree of flexibility of operation, and adaptability to future increased or decreased base load. Proper emphasis must be placed on a realistic determination of the size of compressor; pyramiding safety factors may result in a costly unit operating at light loads while skimping may result in a unit which operates inefficiently at overload and seriously limits the process output.

Fluid to Be Compressed—Whenever it is possible to do so, the gas stream should be free of liquids, corrosive components, abrasive material, particles likely to deposit in the impellers or diaphragms, and compounds which polymerize at conditions encountered anywhere within the unit. Actually, the compressor itself is inherently a centrifuge and will effectively remove liquids and much foreign materials from the gas stream, but this is usually not a desirable manner of operation due to the problem of disposal of this material which tends to clog the passages. The effective removal of all or most such material ahead of the compressor will contribute to a minimum of maintenance and long trouble-free operation.

Centrifugal compressors are, of course, basically gas handling units but within limits may lend themselves to rather severe service by means of special designs, special materials, and special operating procedures, all of which may require adjustment to the particular local conditions. Very often, for cases of non-ideal conditions, the determining factor of whether a particular unit will give satisfactory or unsatisfactory service is the degree to which an operating technique and schedule is developed. This requires a certain amount of ingenuity, patience, experimenting, and then strict and unceasing adherence to the successful routine.

For cases where it is not possible or practical to eliminate these undesirable components, certain design features may be employed to minimize their ill-effects. Departures from conventional materials and construction entail compromises in performance and increase in the cost. The following will serve to illustrate what can be done along these lines:

Abrasive Dust—Impeller may be designed with heavier metal thickness, resistant metals may be selected, and even chrome plating or other surface coatings have merit in some cases. If the unit handles air from an atmosphere containing appreciable dust an intake filter can be justified.

Deposits are likely to build up in impellers with resultant unbalance and restriction to flow. Impellers with radial type vanes are fairly self-cleaning by virtue of the centrifiugal force being in the same direction as the blade and the flow. Backward bent vanes are less satisfactory under these conditions as the centrifugal effect tends to hold the particle against the underside of the vane. Impellers of the unshrouded radial or "paddle wheel" type have also a scraping action on the casing wall which is quite effective in preventing build-ups and in addition is accessible for manual cleaning. Figure 1 shows this type of construction. Deposits within the impeller and casing passages have in some cases been removed by washing, i.e., atomizing water or a suitable solvent, but such a procedure should involve a minimum of injection and careful, frequent, and regular internal inspection of the unit.

FIGURE 1—"Paddle wheel" type of impeller construction.

Corrosive Constituents—Depending on the nature of the contaminant, amount of moisture present, and temperature encountered, standard materials of construction may not be adequate, and impellers, shaft sleeves, and other rotating parts may require bronze, stainless steels, or monel materials. Generally, the stationary parts may be carbon steel or cast iron, as these parts can tolerate without distress a much greater degree of corrosion than the highly stressed, cynamically balanced rotor. For severe conditions the casing may be designed initially with a "corrosion allowance" such that the casing may be serviceable over a long period of time before reaching a predetermined retirement thickness. In the case of conditions below the dew point at the intake and in the presence of some corrodents it is feasible to use special resistant materials in the first stage and sometimes even in the second stage. The heat of compression will, of course, raise the gas temperature above the dew point in the succeeding stages and permit the use of standard materials. In some cases of corrosive constituents, it may be necessary to avoid copper bearing metals, aluminum or lead.

Liquids—Entrained liquid in a gas stream already saturated will increase appreciably the power required to drive the compressor, may cause serious damage to the rotating element and thrust bearing if the moisture enters in slugs, or cause pitting and erosion of the impellers due to the continuous impingement of droplets of liquid against the rapidly rotating impellers. Also, as the mixture is compressed and the temperature increases as the moisture evaporates any residue may deposit within the unit. If moisture separators cannot be used, then every effort should be made to drain the intake pipe ahead of the unit, the intake head (if the intake connection is up) and the individual stages.

Seals—The type of fluid being compressed and the pressure level will determine to a great extent the shaft seal requirements. Such a seal is provided where the shaft passes through the casing to atmosphere or through the bearing housing to atmosphere on the drive end of compressors having the bearing integral with the casing and at the intake pressure level.

Labyrinth Seal—For air compressors and others handling gas of a non-toxic, non-inflammable and possibly a non-obnoxious nature at moderate pressures, slight leakage of the fluid from the casing is of no consequence and the conventional labyrinth seal is used. This seal is shown in Figure 2. The packing box is a soft non-galling aluminum or bronze alloy and sufficient clearance is provided so that the shaft will not be in contact with the labyrinths.

Ejector Seal—Where leakage of gas to atmosphere cannot be tolerated, a seal similar to the conventional labyrinth described above can be used if provided with an injection (or ejection) point at an intermediate location. A source of "inert" sealing gas or air can be injected at this point at a pressure greater than that inside the casing causing flow into the unit and outward to atmosphere. In this manner, the sealing gas only will leak to the atmosphere and, of course, will slightly dilute the product stream or by connecting this intermediate connection to a steam or gas ejector, a negative pressure can be produced within the seal and the

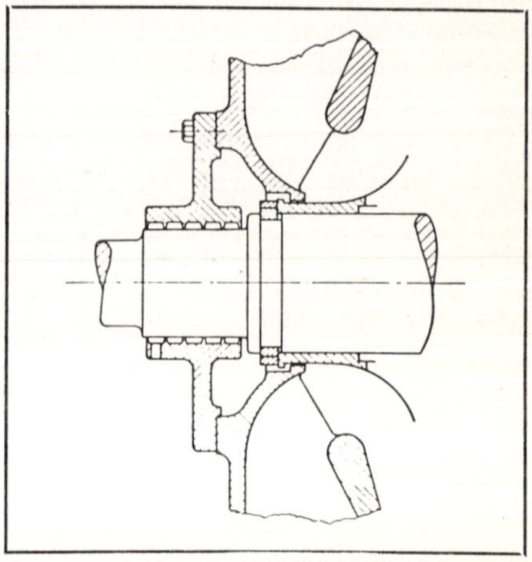

FIGURE 2—Labyrinth shaft seal.

net result will be a slight loss of some product gas and a mixture of this product gas, atmospheric air, and the ejector motive fluid to be disposed of possibly back into the compressor intake or to a flare line.

Similar seals using carbon rings instead of metalic labyrinths will accomplish exactly the same function with less leakage but the carbon seal is most suitable for units handling clean gas or where clean sealing gas may be injected between two carbon rings. This seal is not adaptable to compressors handling dirty gas where the ejector system is used because of the flow of dirty gas across the inner carbon ring.

To insure good sealing under variable conditions and to minimize the quantity of injection gas or motive gas for the ejector a differential control may be used to maintain a fixed differential above or below the pressure being sealed against.

Oil Seal—For cases where the above described "dry" type seals are not adaptable, the more elaborate oil seal is available. It will produce a positive seal preventing the leakage of gas from or atmospheric air into the casing. In general, due to its appreciable expense and maintenance requirements, usage of the oil seal is limited to applications where the pressure level is high, where no leakage is tolerable or where for reasons of unavailability of sealing gas, dilution of the product gas, etc., the other types of seals are not adequate. The seal oil system consists of a reservoir, main and standby oil pumps, coolers, filters, and control valves. Depending on the gas, contamination of the oil may be a problem to a greater or lesser degree not only from the standpoint of foreign matter from the gas and sludge or solids produced by combination of gas with the oil but also from dilution with condensibles from the gas stream. Consequently, it is sometimes necessary to employ oil conditioning equipment and for some designs the oil which has come in contact with the gas is thrown away rather than being returned to the system.

Compression Ratio and Quantity to Be Handled—Centrifugal compressors must be designed to produce a pressure at least equal to the maximum system pressure encountered. Unlike the positive displacement unit, which will discharge against any pressure (providing the compressor is mechanically suitable and the driver is capable of delivering the required power), the centrifugal will have no output unless designed to produce a pressure exceeding the pressure level inside the system to which it is connected. If the system pressure is higher than that produced by the compressor and no check valve is used gas will flow back through the unit and out its intake.

The following general comments may be helpful in an evaluation of maximum possible deviations from the specified performance requirements:

- One performance point only is guaranteed.
- Other performance points are "expected" but not guaranteed.
- The shape of the performance curve and pumping point are also "expected" but not guaranteed.
- The guarantee requires meeting the specified capacity and discharge pressure without tolerance and meeting the power within a 4 percent tolerance.
- European practice has been less stringent on the guarantee and a tolerance in capacity and pressure is permitted in addition to that in power.

It should, however, be appreciated that independently of the legal obligations of the performance guarantee it is entirely realistic to assume that the unit will be capable of meeting all of the quoted conditions.

Pressure and Quantity—The process system resistance curve, i.e., pressure vs. quantity frequently is most difficult to determine with accuracy. Yet for a centrifugal which has essentially a constant pressure variable capacity characteristic the compression ratio is quite critical and among the few difficulties encountered in applications many are inadequate pressure. A careful analysis is required of the pressure drops in intake filters, intake piping and orifice plates as well as that in the discharge piping, all valves, fittings, coolers, heaters, separators and the like. For cases where the output is controlled at constant pressure or a hydrostatic head or the equivalent is involved or where there is a combination of this and friction drop, the problem is simplified due to the greater certainty of the controlled factor. The compressor must, however, be selected to have sufficient margin in compression ratio to offset the possible accumulation of errors in evaluating the system characteristics.

With the above in mind, it becomes apparent that once the system is defined, it still remains to specify for the machine the compression ratio at the various conditions and gas properties encountered. It should be understood that compressors are rated on the intake flange to discharge flange compression ratio based on total pressures (static plus velocity head) and no allowance exists for pressure losses in the intake or discharge piping. The unit must be designed to develop the required discharge pressure and capacity for the most severe combination of circumstances, i.e., minimum intake pressure, maximum intake temperature, minimum molecular weight, maximum ratio of specific heats. If all of these conditions can not occur at the same time then, of course, only the worst set which can exist simultaneously need be considered.

It is obvious now that there is some degree of uncertainty from two standpoints (a) the actual system characteristic and (b) the actual pressure produced by the compressor when handling the process gas at the existing conditions. Note system and compressor characteristics plotted in Figure 3. The only point at which

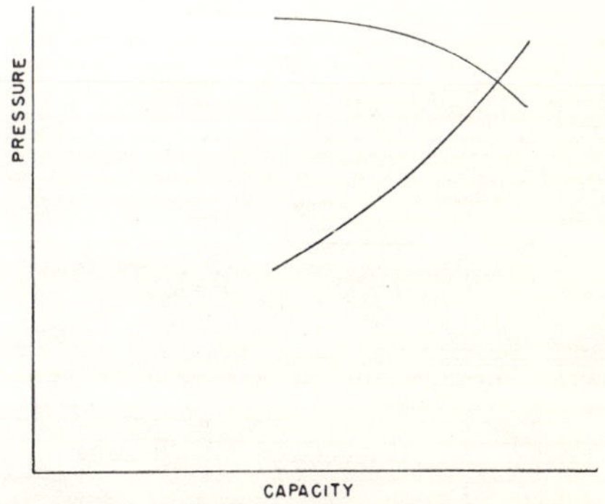

FIGURE 3—Typical pressure-volume relationship for centrifugal compressor and system frictional resistance curve.

this unit will operate when discharging into this system is the intersection of the two curves.

Size and Quantity—At this point it perhaps would be helpful to discuss briefly the compressor size in terms of increased or decreased capacity and/or increased or decreased pressure. In Figure 4 curve *A* represents a particular size of unit. To increase the capacity at the same compression ratio curve *B* would apply and to increase the compression ratio at the same capacity curve *C* would apply. All three of these units are of basically different size. Curve *B* might represent a unit of the same number of stages as *A* but with impellers of greater width and diameter running at a lower speed in a larger casing. Curve *C* might represent a unit of twice the number of stages as *A* and with impellers of the same diameter and about the same width and running at the same speed. For the unit having performance curve *A*, an increase in speed (if the unit and driver are suitable for the higher speed) would result in curve *A′* and a decrease in speed curve *A″*. The compressor changes in pressure and capacity simultaneously but only within rather narrow limits. To obtain the most efficient and economical unit for any particular set of conditions, the centrifugal is designed with the minimum number of stages required to produce the rated head. In other words centrifugals are rated at or near their maximum allowable speed and if the compression ratio is shy it is seldom possible to compensate for this by increased speed. Finally the effect of the power wheel or of variable intake guide vanes on the size of unit as compared with variable speed is shown in Figure 5. Curve *A* represents 100 percent speed as well as the radial position of the guide vanes. As the vane angle is changed, curves *D, E,* and *F* are produced. The curves become steeper, the pumping limit improved and the unit becomes substantially smaller. Figure 6 illustrates a multistage compressor with power wheel and guide vanes including some of the constructional details. In the case of variable intake guide vanes the direction of flow entering the impeller is changed so as to cause prerotation and thereby decrease the pressure rise through the impeller. The power wheel is in effect a turbine wheel located so as to be an extension of the first stage impeller inlet and is arranged with variable guide vanes at its periphery. When the vane direction is changed from the radial a pressure drop is created and the flow is directed into the power wheel in such a manner as to drive it and impart torque to the shaft reducing the net power required to produce a lower discharge pressure. With the guide vanes in the radial position the performance is substantially the same as for an equivalent unit without the power wheel.

Excess Capability—From the foregoing it can be seen that virtually no possibility exists for increasing the pressure of a centrifugal at the same capacity for any constant set of intake conditions. Similarly, while the capacity may be increased at a sacrifice of pressure, the system characteristics usually require more, not less, pressure at an increased quantity of flow.

As a result of the uncertainties described above as well as the relatively fixed nature of compressor and process characteristic, the compressor must be built with some excess capability. The exact amount of this margin is governed by the degree of accuracy to which the particular process conditions are known. Another consideration is possible increased production. The capacity of any plant is limited of necessity to the capacity of one or more elements of the equipment. If indications are that the plant may have some extra capacity the centrifugal should be amply sized since the additional cost involved is moderate compared to the major expense involved in rebuilding or supplementing an existing unit for more capacity or pressure. In all cases the driver must be selected for the horsepower required for the maximum conditions because any limit in power will be as serious as a small compressor.

Several possibilities exist for matching the compressor to the process but in each case the result is to get the curve of the unit to intersect the process curve at the desired quantity of flow. If the intersection occurs at too low a capacity the desired capacity cannot be realized, if (by coincidence) the intersection occurs at exactly the desired capacity that is an ideal situation in one sense although it leaves no reserve for varying conditions and finally if this intersection occurs at a greater capacity than desired, then either the compressor or process characteristic must be

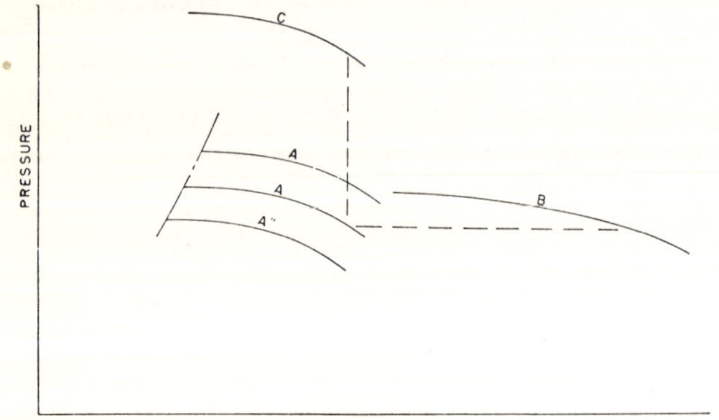

FIGURE 4—Performance curves for three compressors of different basic size.

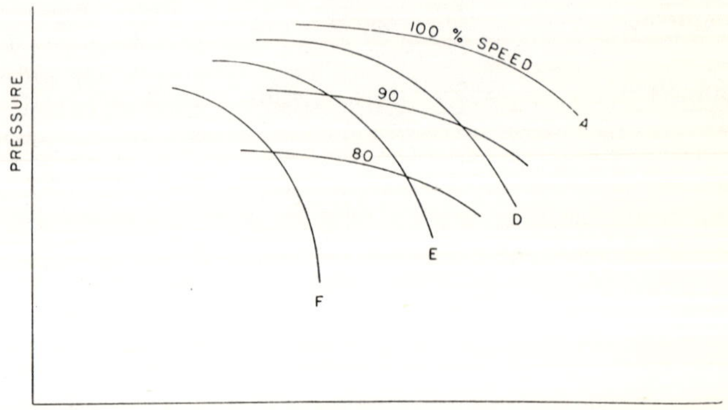

FIGURE 5—Variable intake guide vane performance curves superimposed on variable speed curves.

FIGURE 6—(A) View of compressor with casing upper half removed showing power wheel and guide vanes.

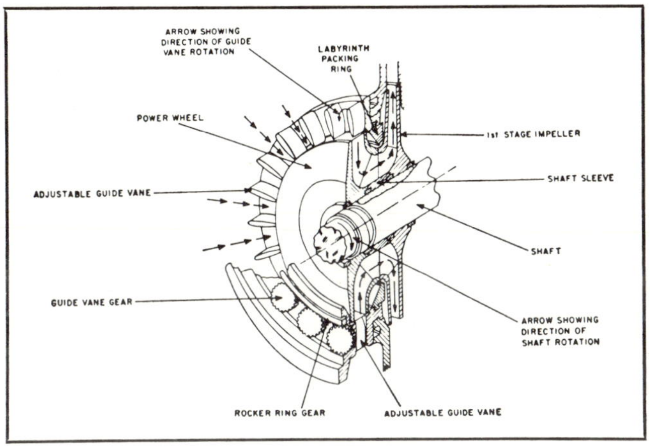

FIGURE 6 — (B) Sketch of power wheel and vane mechanism with direction of flow indicated.

FIGURE 6 — (C) Guide vane assembly with vanes in a partially throttled position.

altered. The centrifugal may, if the driver is of the variable speed type, be slowed down as shown in Figure 7 (a) to a speed at which the characteristic curve just provides the necessary pressure at the desired capacity. Similarly this curve can be lowered by means of a power wheel or variable intake guide vanes, Figure 7 (b). In each of these cases the margin in compressor size is eliminated in an economical manner, i.e., the power consumed, Figure 7 (a) and (b), is that required to provide only the pressure level utilized. The process curve, on the other hand, may be raised by throttling across a valve at the discharge of the compressor as per Figure 7 (c). Thus the intersection is effected at the desired capacity while the compressor curve remains unchanged, i.e., the power consumed corresponds to that required to produce a pressure exceeding the normal process pressure level and consequently the excess pressure represents a continuous waste of some power. Note that the relative merits of variable speed, variable intake vanes or power wheel depend on many factors but the only significant comparison here is between any one of these means of adjusting the compressor characteristic curve and throttling at the discharge. It is true that intake throttling is somewhat more economical than discharge throttling depending in part on the particular design of the unit; however, in both cases of throttling practically speaking the loss is complete. For any proposed installation an analysis of the economics will indicate whether throttling or other provisions are justified.

Supplementing the previous section on "fluid to be compressed" it must be realized that while variable vanes might be desirable from the standpoint of economy due consideration must be given to the effect of the gas on these movable parts. For air or clean gas no problems are to be expected but if the gas is dirty or tends to create build-ups inside the casing the vanes may become frozen in some position in a short time. Under adverse conditions much can be done from the operating and maintenance standpoints to keep the vanes free but in general unless the process gas is fairly clean it is well to consider variable speed or throttling.

In exploring more thoroughly the quantity to be compressed independent of the process pressure-capacity characteristic it is apparent that the centrifugal is flexible in its capacity handling ability. This capacity will vary from the pumping limit (averages approximately 50 to 70 percent of rated flow at intake conditions for a single casing) to the rated flow, and beyond, at substantially the same pressure level. The pumping limit or surge point occurs at or near the peak of the pressure curve and is the lowest capacity for which the unit will deliver a continuous flow of gas at a steady pressure. In the region below the pumping limit the flow will tem-

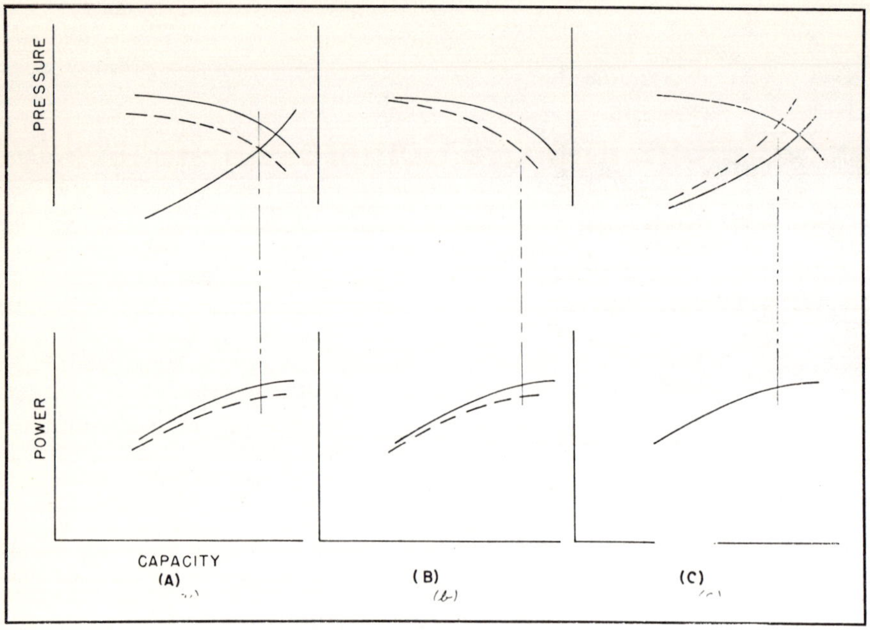

FIGURE 7—(A), (B) and (C) Power consumption for varying compressor characteristics compared with throttling at the discharge.

porarily reverse creating surges in pressure the frequency and severity of which are a function of the design of compressor, the number of stages, the gas being compressed, the pressure, and the particular process system. Operation in this range is not recommended and provisions must be made to increase the flow to something above the pumping limit. "Minimum Volume" controls are available and can be arranged to recirculate gas back to the intake or for air units waste air to the atmosphere. If recirculation is contemplated provision must be made to prevent overheating the compressor.

Efficiency—Figure 8 shows that for only one capacity does the best efficiency occur. With a fairly flat efficiency characteristic, however, a reasonably good efficiency can be obtained for quite a wide range of capacity on either side of the peak efficiency. It might be added that for various hydraulic designs the efficiency curve will tend to peak more or less sharply but unless the capacity is known precisely and the load constant, a flatter efficiency characteristic is most desirable. In practice the rated capacity is located at or just beyond (higher flow) the peak efficiency point so as to provide as wide a stable operating range as possible. If the unit ultimately operates at an average capacity slightly less than the rated point it will tend toward the peak efficiency.

Flow Conditions—For the centrifugal compressor the quantity of flow is dependent on the impeller capacity per revolution. It is, therefore, wise to express the flow in terms of volume at intake conditions in order to avoid any possible misunderstandings as to the exact conditions which are to be the basis of design. While the conversion calculations from weight flow or "standard" conditions are reasonably well known and understood it occasionally occurs that the data provided is not clearly defined or complete and may lead to flow quantities which differ significantly from the true value. The same general philosophy applies equally well to the determination of the gas physical properties and no one is in a better position than the user to state the intake capacity and gas properties accurately. In the last analysis the compressor manufacturer has control only over the design and the unit will perform properly only if subjected to the conditions for which it was built.

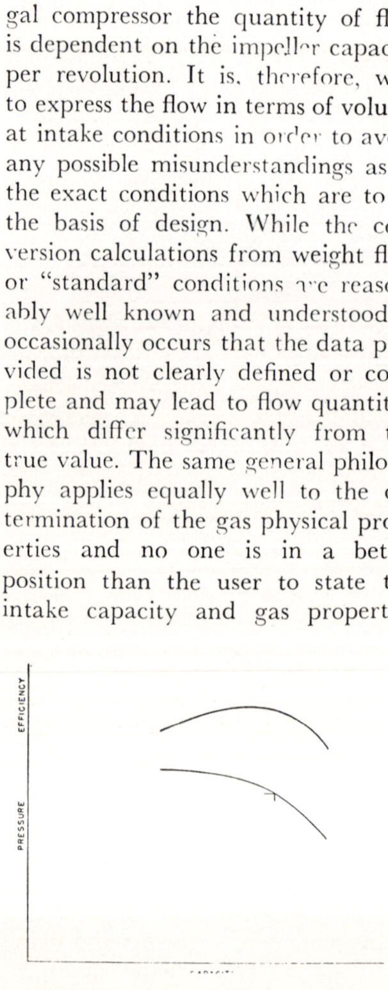

FIGURE 8—Compressor efficiency curve.

As in the case for the compression ratio the unit is rated on the basis of the quantity of flow measured at the discharge flange (usually expressed in capacity at the intake conditions) and if any of this net output of gas is diverted from the process suitable adjustment must be made in the compressor rating. Leakage from shaft seals and internal recirculation are, of course, taken into account in the design in as much as they occur between the intake and discharge flanges and are chargeable to the compressor. Gas used as motive power for shaft seal ejectors, for example, must be accounted for in the determination of the compressor capacity rating.

Sidestreams can be accommodated in many designs but exactly the same considerations apply and in fact the problem of determining precisely the quantities of flow and compression ratios is even more complex. Allowances must be made to insure pressure levels compatible with the entrance or exit of the required quantities at intermediate points. Where sidestreams of relatively high flow rates leave the unit any percentage deviation in this quantity could result in tremendous percentage of overload or underload of the impellers handling the balance of the flow.

For compressors operating in series with components removed during cooling between units or with other gas entering between units every effort must be made to accurately identify conditions and gas properties at the entrance of each unit. For single state units the behavior of the gas is less significant than for multistage units and similarly several casings in series will magnify greatly the effect of inaccurate data.

In conclusion a thorough and accurate knowledge of the process gas properties and contaminants, its quantity and the pressure requirements of the system will all contribute substantially to the selection of a centrifugal compressor which will adequately perform the required service with economy, flexibility, and reliability.

Can Specifications Improve Compressor Reliability?

Vendors can only bid on what you tell them you want. If it's reliability, specs must be detailed. These guidelines will help you with often overlooked items.

Royce N. Brown, Dow Chemical Co., Houston

A SPECIFICATION should outline to the bidding vendors just what it is you want to accomplish with the equipment and how reliable it must be. Compressor reliability can be achieved if it begins prior to vendor design and fabrication. Therefore, the beginning of a truly reliable piece of equipment has its beginning with the specification.

Basic data. Enough data must be supplied to rate the compressor. For example, the mass flows, inlet and outlet pressures, inlet temperatures, type of gas or gas physical constants need to be itemized. The basic data must be very clearly stated, complete to the extent necessary to achieve a common understanding between the owner and vendor.

Operations. Very few pieces of equipment operate at a single set of conditions. Thus, operating conditions must be divided into a set of normal conditions and those other than normal. The entire anticipated range of operating conditions should be defined either by range limits or alternate operating conditions. To establish guarantee points and name plate rating points, unique conditions within these parameters must be selected. After defining the normal operating conditions, consideration should be given to "off-design" conditions. For example, these would include abnormal molecular weight deviations as might be encountered in the startup of a machine on a gas other than the design gas. Another example might be encountered in a holding period of operation on the equipment, while part of a process is not in operation.

In describing abnormal, unusual or "off-design" operating conditions, even insignificant items should be mentioned, if for no other reason than to spur conversation at a later date. While these conditions may frighten the vendors at the time, it is better to point them out in the specifications than to use the "Oh, by the way" method to the service representative at startup. Examples of little potential vendor surprises are: HCl, H_2SO_4, or chlorides in the gas, even though they are in the ppm levels. Many times it is erroneously assumed that a vendor's knowledge of the process is more than your own, and therefore, the vendor should have been aware of these unusual conditions. The same reasoning can be applied to known potential process upsets. If the vendor has had an unfortunate experience with a process similar to yours and recognizes it in your process, he may well anticipate the problem. However, you should be much more aware of your potential upsets than the vendor. An example might be the temperature runaway potential in the gas being fed to a hot-gas expander. At least the specification could question the vendor on the degree of over-temperature tolerance his machine can handle. Another example is the potential of the sudden dead ending of a centrifugal or axial compressor during switching operations of batch-type reactors.

Some consideration of surge characteristics should be made, and unless your own plant design includes antisurge controls, definitely mention it in the specification. Possibly, the vendor should quote antisurge equipment with his offering. At a later date, the specification will serve as a good reminder to both the vendor and purchaser if some mention of compressor control is made. The details may be worked out at any convenient point, such as at the bid review or possibly later in a coordination meeting.

Another way to decrease your reliability is to not mention fouling in the specifications if process experience exists, not necessarily directly from operating centrifugal compressors but from smaller plant operation with reciprocating equipment. By mentioning a fouling potential in the specification, it may be possible for the vendor to include some form of washing in his offering. Alternatively, he may be able to allow additional head margin in his design. By working together, the period of time before having to clean a unit can be extended, but only if the service has been specified.

Materials. Past experience by the Operating Department provides a guide to materials specifications. It is foolish to assume that the vendor is completely knowledgeable about materials for your process. Stating the Operating Department's experience with materials or defining materials in the specifications, the vendor can use your experience in his design and not have to gain his experience the hard way at your expense (if you knew and didn't specify). However, you do not have to make the specification so rigorous that you prohibit vendor experience with materials. By proper wording of the specification, you can invite comments to reveal vendor experiences in materials. The main point is that by specifying a material other than what would normally be standard, a flag is raised to call the vendor's attention to a possible situation other than routine or standard. A large number of forced outages are traced to vendor design, but they really reflect a lack of application knowledge, which could have been prevented by purchaser communication in the specifications.

Hardware consists of anything from nuts and bolts to accessory items such as lube pumps, couplings, etc. Plant experience should definitely be applied to the specification of hardware items. Since process compressors are basically custom built, most vendors will accommodate special requirements.

The specification should respect the proprietary rights of a particular vendor. A patented or trade-name seal should not be specified on open bidding, if it is peculiar to one vendor. However, a particular item may not necessarily belong to that particular vendor alone if you, the customer, have contributed to the particular design and experience. Unless definite agreements were made prohibiting use on other vendor machines, in the interest of reliability, this particular design should be specified openly whenever it can be used in your best interest. Several examples follow. Generally, our experience has shown that the self-leveling, double-acting, tilting-pad, thrust bearings are very reliable when they are properly sized. While this may not be a standard item with many vendors, they will furnish it upon request if specified. An end seal you have field modified with a vendor to suit your particular need is your property, unless you have waived your possession. There may be an interstage seal on a compressor which performs well in your service; therefore, by specification the vendor would consider using this seal in your machine.

Couplings. It is impossible to specify a completely foolproof coupling, particularly by manufacturer and model number. Still, in the interest of reliability, the specification should cover the subject of couplings. This requires the use of plant feedback and a lot of research on the part of the specification writer to determine the latest developments in coupling design. Unfortunately, when a vendor furnishes a coupling, it often becomes just another outside purchased item which he in turn purchases with a minimum specification and strictly on low price. However, by writing a comprehensive coupling section into your specification you are making the vendor aware that you want the coupling to receive enough vendor attention to assure you that he is furnishing a reliable unit.

Gear units. When gear units are furnished by a compressor or driver vendor, they are usually purchased strictly on price with very little application know-how being contributed by the specifying vendor. The specifying vendor, in turn, expects the gear vendor to know everything there is to know about applying gears. Unfortunately, our experience has shown this not to be true. A comprehensive performance or specific gear specification will tend to emphasize your desire for gear reliability to the vendor furnishing this accessory item.

One useful technique in improving the reliability of accessory items is a specification requirement that all purchase type components must have purchaser's approval prior to vendor commitment to his supplier. This at least opens the door to some control on the type of item that the vendor will buy. Because of your specification he will now know he has to defend his choice of equipment and will, hopefully, exercise more care at the time of application engineering.

Lube and seal systems. Vendor management has rarely been particularly sympathetic to the cause of a reliable lube and seal system. Oil systems are relatively high cost items to the manufacturer. After all, the manufacturer is in the compressor business so why spend capital for facilities to build lube systems, as well as extra money for components? Fortunately, there seems to be some recent change in this attitude, however, the purchaser must still exercise great care in specifying lube and seal systems. The following is a partial list of items which must be considered:

- **Piping.** Specify materials—steel has numerous advantages. Specify welded construction and the minimum use of fittings. Tape type thread lubricants should be avoided.
- **Pumps.** Specify the type and number of pumps required. Minimum: two full-size pumps.
- **Driver.** If steam is available, a good choice is a steam turbine driver for the main pump and an electric motor for standby. Give the minimum steam condition. Give the expected voltage drop on the electric system if more than 10 percent is expected, especially on large systems.
- **Accumulator.** Required if slowed response on standby driver is expected.
- **Filter.** Specify twin filters with a noninterrupting transfer valve. Specify the degree of filtration (10 microns or lower—possibly even one micron in some applications).
- **Cooler.** Specify twin coolers with noninterrupting transfer valve. Specify materials of construction, utilities and fouling factor.
- **Reservoir.** Retention time should be 5-8 minutes. Specify a coating inside and out.
- **Environment.** Even bolts may require a specification to prevent excessive corrosion.

Specific design information may be requested in the specification to aid in later design reviews. An item that may require purchaser checking is the compressor's lateral and torsional criticals. Specification coverage of design requirements will assist in actually receiving the information.

Instrumentation. At specification time, the complete review of instrument requirements should be made on the part of the purchaser.

A clear understanding of the objective and purpose of equipment instrumentation must be developed by the user before specifying the instrument system. The system can be very basic, supervisory, or diagnostic. Within the scope of the specification, it is not necessary to be completely arbitrary. Allow for vendor participation. Either have him recommend a system or comment on your requirement and offer alternates. In the interest of reliability, compressor instrumentation can be best accomplished in the design stages, rather than added singly after the compressor is built.

Often plant safety requirements call for instrument connection sizes larger than a vendor would carry as a standard item. It is very sad to have an undersized lube system pressure connection cause a unit to shutdown when an adequately specified connection could have prevented it. Therefore, the specification must be comprehensive and cover the subject in depth.

Shipping preparation should be specified. A review of storage problems and environmental problems should dictate the degree of packaging required.

Installation. The installation environment should be reviewed so that any hazardous conditions are defined. Also the compressor location (indoor or outdoor) needs to be specified. Geographical location may be significant.

Basically, the specification should cover the quality of installation required. Part of the installation specification must come from the equipment supplier. The equipment vendor must provide enough detailed installation information so that any peculiar requirements of his equipment are specified. To these, the user's experience must be added.

Contractor specifications should be so specific that he will have his most skilled tradesmen assigned to the critical areas. The contractor will not know which are the critical areas unless these are described. Also, any special alignment techniques should be specified such as the use of optical alignment or other more sophisticated methods. Tolerances must be very carefully specified. The requirement for any special equipment has to be pointed out to the contractor. The equipment may come only partially assembled. Enough information and drawings must be supplied to permit the proper assembly and setting of clearances. The degree of inspection required should be described. The types of grout must be specified. Nothing should be assumed as to the contractor's ability.

Writing the specification. Up to this point there has been no mention of how you start to write a specification, or the basic organization thereof. Since most user companies have established formats, the foregoing comments are intended to be general enough to fit an existing format. Also, there are industry standards available such as API. Standards provide a good base to work from, but they must be supplemented, not only to answer the question posed in the standard, but also to custom fit your specification to the particular application.

Another useful approach is to attempt to get a broad review of your specification by Operations, Maintenance and Engineering. The specification must have an author who can write the basic document, and who later edits the draft into the scope and language of the final specification. Single authorship, with committee review, is preferred over committee authorship. Only in rare instances is vendor participation in specification authorship really justified.

Costs. Following these specification guidelines will probably result in higher compressor costs because they require the vendor to deviate from a so-called standard. It is, of course, within the vendor's commercial interest to prefer standards. Standards keep costs down, sometimes, but standards are minimum specifications with regard to reliability. Standard components may be used to make up a custom assembly which will aid in lowering first cost and also replacement parts. Oftentimes, the history available on standard parts may actually enhance reliability. Care must be taken, however, not to compromise too far. As an example, there is some advantage to using standard compressor wheels. However, standard compressor wheels may well lead to a rangeability compromise and must be carefully considered. The specification must control the vendor's choice of standards and require custom design, when necessary, to achieve the desired reliability and performance. A good specification that controls, both standards and custom design, can contribute to both the reliability of the unit and still maintain a reasonable first cost.

Design Specifications, Accessories

F. H. Rassman, Allis-Chalmers, Milwaukee, Wis.

DESIGN SPECIFICATIONS are those which in varying degrees set forth the manner in which the compressor is to be constructed. Commonly included in this type of specification are:

- Materials of construction and allowable stresses
- Type of bearings and allowable bearing loading
- Relationship of operating speed to torsional and lateral critical speeds
- Maximum allowable peripheral speeds
- Type of bolting—studs, tap bolts, thru-bolts, etc.
- Type of drain connections—flanged or threaded
- Detailed lubrication system requirements
- Test requirements
- Quality control requirements.

In some instances, the use of design specifications prohibits consideration of some manufacturers' standard line of equipment and therefore results in higher cost compressors. Whether or not additional quality and value is received may be difficult to evaluate. Compressor manufacturers do not as a whole ignore good practice in the absence of a detailed specification; they have too much at stake to do so. However, the user should specify design features when he has prior experience which dictates their necessity.

Industry specifications. The most widely used industry specification is the previously mentioned API Standard 617 "Centrifugal Compressors for General Refinery Service." Other specifications which are frequently referenced are NEMA SM-20 "Mechanical Drive Steam Turbines" for the standards on flange loading and piping strains, NEMA SM-33 "Gas Turbine Sound and Its Reduction," for the noise standards, AGMA Gear Specifications, ASME "Unfired Pressure Vessel Code" and various ASTM material specifications.

Operating conditions. Two of the more important items to be related in a compressor specification are the normal operating point of the compressor and the expected range of operation. There may in fact be several "normal operating points" required for certain processes. These can occur due to seasonal climatic variations, process load variations, etc. Proper selection of the compressor design point requires adequate knowledge of these factors, so that all the operating points can be achieved and the compressor can be designed to achieve maximum efficiency at the most frequent points of operation.

Multiple operating points can be achieved in several ways:

- Variable speed—as can be achieved by using a steam or gas turbine driver
- Suction throttling of the compressor
- Use of variable inlet guide vanes.

The latter two methods are predominantly used when a fixed speed driver such as a motor is used.

Power guarantees at off-design operating points are often requested. While they are necessary to evaluate the over-all operating economy of a compressor, because of the variables affecting the shape of the performance curve, a greater degree of uncertainty exists in predicting the performance at off-design points than at the design point. Performance testing to verify guarantee points is routinely done by most manufacturers.

Accessories. The entire rotating equipment train must be properly analyzed to insure freedom from problems due to lateral and torsional mode critical speeds, severe transient torsional stresses in the case of salient pole synchronous motor drives, rotor instability and thrust problems due to unaccounted for coupling tooth friction or lock-up, etc.

The axial forces generated by friction in gear type coupling teeth when connected rotors expand axially can be severe. API Standard 613, "High Speed Gear Units," requires a friction coefficient of 0.15 be considered. We conducted one test which indicated friction coefficients as high as 0.22. Other data in the literature indicate friction coefficients greater than 0.3. The limited data available imply that the actual coefficient is a function of the tooth shape and finish, relative loading, and misalignment. We presently are evaluating all systems assuming a 0.25 friction coefficient.

In addition to the rotating equipment train, it is, of course, necessary to provide a lubrication system which satisfies the requirements of the complete installation. The petroleum industry generally follows the API specification for lubrication systems which has been part of API 617, but has recently been revised and will be re-issued shortly as API Standard 614 "Lubrication and Seal Systems." Most compressor manufacturers also offer a general purpose lubrication system, which is less expensive than the API system, and of ample reliability for air compressor service. Instrumentation requirements, particularly bearing temperature sensors, shaft vibration detectors, shaft position indicators, etc., must be coordinated for all equipment in the compressor train.

Reliability. While there certainly are problems with compressors, our records indicate that compressor problems were completely overshadowed by gear problems during a three-year period. A ratio of approximately four gear problems to every compressor problem led to the generation of a stringent gear specification on our part. The problems encountered included pitch line runout, skewing of one gear element with respect to another, inadequate backlash, shaft failure, tooth failure due to inadequate lubrication, and bearing failures.

The practice followed by Esso that only compressors of similar rating successfully operating in a similar application be considered places severe restrictions on some manufacturers. If there are sufficient compressor vendors with the necessary experience for the particular application under consideration, this type of policy can be adhered to. However, there are a significant number of applications where the design integrity and past performance of the manufacturer in other applications must be respected.

Noise. U.S. legislation (Walsh-Healy Act) limiting environmental noise dictates at least a brief mention of the problems encountered with compressor systems. Predominantly mechanical or structure borne noise emanates from motors and gears. Unattenuated noise levels of 90 to 95 db are common for this type of equipment. Centrifugal compressors emit predominantly air borne noise. Noise levels vary drastically depending on type of construction, peripheral speeds, types of attenuation, etc. When specifying compressors, consideration must be given to all applicable federal, state and local codes, noise generated by other equipment already in the plant, and then reasonable noise limitations should be given in the specification. The vendor must then offer a suitable design and include necesary attenuation devices in his offering.

How to Select Materials for Centrifugal Compressors

Several metallurgical factors must be considered if a compressor is expected to survive under severe operating conditions. Here is a review of these items and the measures taken by manufacturers to achieve a long satisfactory machine life.

Joseph A. Cameron and **Frank M. Danowski, Jr.**, Elliott Co., Jeannette, Pa.

CONSIDERATION OF THE MATERIALS for use in rotating machinery must begin with careful study and appraisal of expected operating conditions. In the discussion which follows, attention is concentrated on centrifugal compressors, but similar considerations are required in other types of turbomachinery such as axial compressors and steam and gas turbines. In addition to the nominal conditions of operation, it is also necessary to take into account expected variations in these conditions. It is necessary to know whether variations are expected to be transient, or of long duration. Sometimes, machinery is purchased for intermittent operation, and this is usually the most severe kind of service.

Environmental factors which must be considered include:

- The operating stresses of rotating parts are determined by the design, size, and speed of rotation, and these stresses must be known for the obvious reason of avoiding selection of materials having insufficient strength.
- With regard to the casing and other stationary machine elements, the operating stress must also be known, and for these parts, the stress is determined primarily by size, design, and internal pressure.
- The gas which is to be handled by the compressor must be evaluated for aggressiveness with respect to the materials of construction. If high corrosion rates are expected with ordinary materials, alternate materials must be selected, even though substantial additional cost may be incurred.
- The expected temperature of operation must be determined. Temperature is a factor in all of the above three items. It affects the rate at which corrosion reactions proceed, affects the strength of material, and determines susceptibility to brittle failure.

PROPERTIES WHICH REQUIRE CONSIDERATION

Virtually all of the properties of materials must be considered in evaluating materials for use in centrifugal compressors. A list of the properties includes:

- Tensile properties at room temperature and actual operating temperature.
- Modulus of elasticity.
- Coefficient of thermal expansion.
- Susceptibility to brittle failure (toughness).
- Damping.
- Fatigue strength.
- Thermal conductivity.
- Specific heat.
- Hardenability.
- Weldability.
- Corrosion resistance.

Not all of these properties are important in every application, but various combinations of required properties must be evaluated depending on service requirements in each instance. A notable omission from the above list is long time high temperature properties such as creep or stress rupture. Only infrequently do centrifugal compressors operate at temperatures high enough to require consideration of these properties. They are much more commonly considered in turbines.

ASPECTS TO BE CONSIDERED

It would be impractical, in one discussion, to attempt to cover all of these aspects of centrifugal compressors involving metallurgical considerations. A few of those which are of most current interest have been selected:

1. Impeller fabrication.
2. Impeller materials.
3. Sulfide corrosion cracking.
4. Hydrogen embrittlement.
5. Brittle failure.
6. Dimensional stability.

IMPELLER FABRICATION

Types of impellers. The heart of a centrifugal compressor is the rotating element, and a great deal of attention is given to the fabrication of centrifugal compressor impellers. Fully shrouded impellers having solid back wall and cover, as shown in Figure 1, are the most common type in multi-stage centrifugal compressors. Impellers of this type present more difficult fabricating problems than the open or semi-closed type shown in Figure 2,

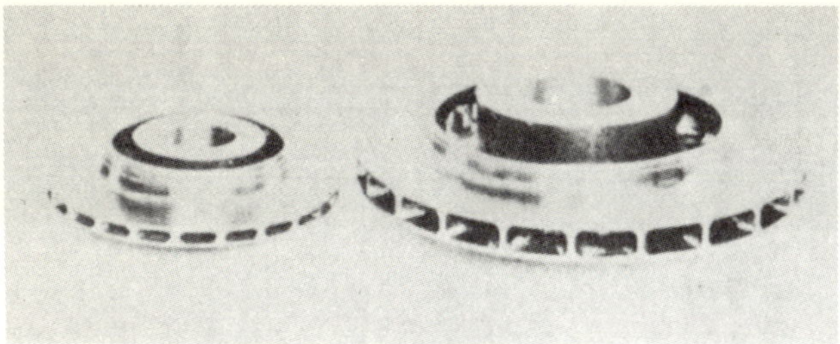

Fig. 1—Typical fully shrouded impellers

Fig. 2—Typical fabricated open impeller

Fig. 3—Methods of impeller fabrication

more commonly used in single stage machines. For this reason, attention is focused primarily on the fully closed impellers.

In most cases, centrifugal compressor impellers are fabricated from materials which are ordinarily considered difficult to weld; but, satisfactory techniques have been developed. Thousands of impellers have been fabricated successfully; in fact, the history of impeller welding goes back more than 30 years. There have been changes and improvements in materials and processes through the years. Probably the most significant advance was the advent of low hydrogen materials and processes. Prior to the development of low hydrogen welding materials, one of the most difficult problems was that of delayed cracking. Impellers that apparently were sound immediately after welding would occasionally be found to be seriously cracked after standing at room temperature for periods ranging from a few hours to a few days. Also, earlier impeller forgings were usually made from acid open hearth steel with relatively high phosphorus and sulfur contents. In more recent years this practice has been changed. Lower impurity basic electric furnace steels are now used, and weldability has been improved.

Methods of fabrication. Several methods of fabrication are shown in Figure 3. Number 1 in this figure illustrates construction in which blades are fillet welded to the hub and cover. This is the most common method of fabrication. Welding may be done by manual metal arc, gas shielded metallic arc, submerged arc, or firecracker procedures. Welds are readily inspected using magnetic particle methods on magnetic materials and liquid penetrant methods on non-magnetic materials. This method of fabrication has the advantages of being the least costly welding procedure and requiring the least capital equipment cost. The chief disadvantages are poor accessibility when the gas passage height is low, or the backward lean angle is high, and the intrusion on the aerodynamic passage of the fillet welds.

Number 2 is similar to Number 1 except that the welds are full penetration. This is a more costly welding procedure, which seems to offer few, if any, advantages. The same welding procedures may be used as in 1 above. Theoretically it would be possible to reduce the fillet size and thereby increase the area of the aerodynamic passage, but in practice this reduction is small.

Number 3 has a partial blade machined integral with the hub and cover and a butt weld down the middle. This procedure has been used successfully, but is more difficult than it appears, especially in blades having high backward lean angles, and there is some difficulty in achieving a satisfactory weld contour around the leading edge of the blades.

Number 4, slot welding, is used chiefly as an alternate to fillet welding in applications where the gas passage height is too shallow for accessibility, or the backward lean angle is too high to permit conventional fillet welding from the periphery of the impeller. This weld may be deposited by tungsten arc inert gas, metallic arc inert gas, manual metal arc, submerged arc welding, or a combination of these. As compared with fillet welding, there is a considerable increase in the amount of weld metal that must be deposited, and therefore, in welding time. Compensating for this to some extent is the lesser intrusion of the weld into the aerodynamic channel. As already mentioned, impellers that cannot be fillet welded can be fabricated in this manner. Impellers may be slot welded to either the hub or cover. On the component opposite to that slot welded, blades may be attached by fillet welding, or they may be cast or milled integrally with the hub or cover.

Number 5 shows a hub or cover electron beam welded to a blade. While a good deal of experimental work has been done, production experience with electron beam welding of impellers is limited. The chief problem has been extension of the interface between the blade and hub or cover from which a crack can propagate into the weld on cooling as shown in Figure 4. There is an operational problem in maintaining preheat during the time required to evacuate the welding chamber. Possibly this could be overcome using quartz lamps or a defocussed electron beam to preheat. Electron beam welds should preferably be stressed in tension, but the configuration of the part in this case produces a stress in shear. Inspection of these welds is very difficult. In addition, capital equipment cost is very high and utilization of the equipment is low.

Number 6 illustrates a construction without welding where the impeller is made from a one piece casting or machined from a solid forging, possibly by electrodischarge machining. In one piece castings, the tooling cost is very high to provide for varying channel heights and impeller cover eye openings. Electrodischarge machining is a relatively slow process, and produces what is known as a recast layer on the surface of the part, Figure 5. Studies have shown that this recast layer can reduce the fatigue strength of the part by as much as 60 percent unless it is removed by careful grinding.[2]

Number 7 shows various configurations of riveted blades which have been used. In comparison with welded construction, the riveted parts are less strong, and there is a potential source of fatigue cracking due to the stress concentrations around the rivet holes. Numbers 7a and 7b have rivet heads which protrude into the aerodynamic passage detracting from the cleanliness of this passage. If there is any dirt in the gas being compressed, it can hang up, and may cause balance or corrosion problems.

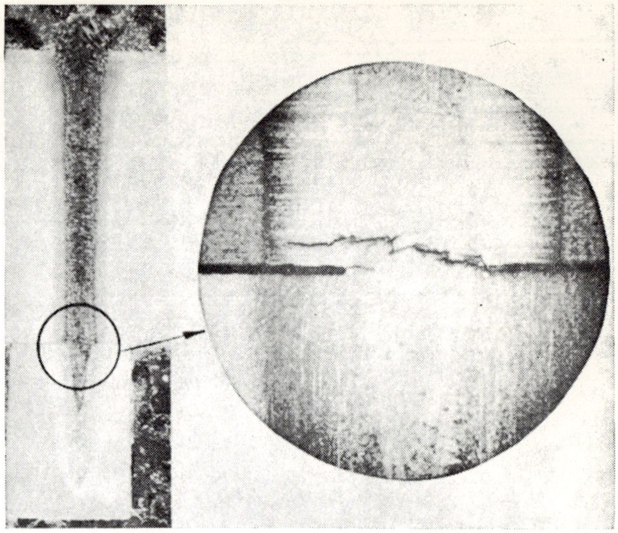

Fig. 4—Electron beam weld showing interface cracking

Fig. 5—Recast layer on electrodischarge machined surface

IMPELLER MATERIALS

Usual materials. Impeller disc and cover forgings are usually made from a low alloy steel, such as AISI 4140 or 4340. The final selection of the particular low allow grade for each application depends on the strength level desired and the size of the impeller. AISI 4140 is satisfactory for most applications; AISI 4340 is used for larger impellers or higher strengths because of greater hardenability and resistance to tempering. The greater resistance to tempering makes it possible to achieve higher yield strength while maintaining a relatively high tempering temperature with 4340 as compared to 4140. The usual minimum permissible tempering temperature is 1100° F, which makes possible a good balance between strength, toughness, and low internal stress.

The forgings are heat treated to the desired strength level, Table 1. After machining and welding, impellers are heat treated for the purpose of tempering the heat affected zone in the base metal adjacent to the welds. The base metal hardness varies with the yield strength required, but is frequently about Rockwell C-26. In the heat affected zone immediately after welding the hardness can be as high as Rockwell C-45 or 50. After tempering at 1100° F this value falls to Rockwell C-28 to 30. The use of more complex post weld heat treatments will be discussed later, but when a simple tempering treatment is employed, the temperature cannot be higher than 1100° F because this is the minimum tempering temperature for impeller hub and cover forgings. A post weld heat treatment at a temperature above the minimum tempering temperature for the forgings could result in a loss of strength in the hub or cover base material.

Special materials. Where more corrosion resistance is needed than is available from ordinary low alloy steels, such as in some chemical and petrochemical process gas compressors containing moisture and corrosive gases, the next step in improved resistance is usually AISI Type 410 stainless steel which contains about 12% chromium. Weldability of Type 410 is comparable to that of the low alloy steels.

A further improvement in corrosion resistance for more aggressive environments can be obtained through

TABLE 1—Typical properties of impeller materials

Material	Ultimate tensile strength	Yield strength 0.2%	Elongation %	Reduction in area, %	BHN
AISI 4140	100,000 psi min.	80,000- 90,000 psi	16 min.	50 min.	212-235
AISI 4140	115,000 psi min.	95,000-120,000 psi	16 min.	50 min.	269-321
AISI 4340	120,000 psi min.	105,000-120,000 psi	17 min.	43 min.	248-302
AISI 4340	125,000 psi min.	115,000-135,000 psi	16 min.	50 min.	269-321
AISI 410	100,000 psi min.	80,000- 90,000 psi	14 min.	40 min.	212-235
AISI 304	75,000 psi min.	30,000 psi min.	40 min.	50 min.	205 max.
AISI 316	75,000 psi min.	30,000 psi min.	40 min.	50 min.	217 max.
Armco 17-4PH	130,000 psi min.	105,000 psi min.	16 min.	50 min.	277-341
Monel K-500	130,000 psi min.	90,000 psi min.	20 min.	......	250 min.

the use of one of the precipitation hardening stainless steels such as Armco 17-4PH or 15-5PH. These grades, while appreciably more expensive, also have the advantages of further improved corrosion resistance as compared with Type 410, and much improved weldability. The precipitation hardening grades also have higher strength than Type 410 stainless steel, and are sometimes used for this reason where the corrosion resistance of Type 410 would be adequate.

The austenitic stainless steels, such as Type 304 and Type 316, Table 2, are occasionally used for very low tip speed impellers. The corrosion resistance is attractive in certain environments, but the yield strength of these materials is only about 30,000 psi imposing a rather severe limit on tip speed, and precluding their use for most applications. A further deterrent is the high coefficient of thermal expansion which creates difficulties with rotors built up by shrinking impellers on shafts. The coefficient of thermal expansion of the austenitic stainless steel impeller is about 50% greater than that of the usual low alloy steel shaft. If the temperature excursion during operation is sufficient, the differential growth may cause the impeller to come loose on the shaft. A heavy shrink fit cannot be used to offset this problem because of the low yield strength of the material.

Usually the temperature excursions in single stage compressors are such that the differential coefficient of expansion can be handled, and these materials have given considerably improved resistance to such gases as sulfur dioxide.

Another material which has been used for centrifugal compressor impellers is Monel K500. This material has been used chiefly for exposure to halogen gases in the moist condition, and the record is generally good, although corrosion rates sometimes are high enough to require occasional replacement. It has also been used because of its resistance to sparking in oxygen compressors. Some 20 years ago there were several failures in hydrogen reformer compressors, but these were attributed to improper heat treatment resulting in a high level of internal stress. Since that time many additional impellers have been fabricated using improved heat treating practices, and there have been no further difficulties. With Monel K500, as with the austenitic stainless steels, allowance must be made for the high coefficient of thermal expansion. The yield strength of Monel K500 is considerably higher than that of the austenitic stainless steels making it possible to use heavier shrink fits with Monel K500.

Compression of halogen gases, and especially chlorine, presents an interesting problem. If the chlorine is

TABLE 2—Dependence of failure on material yield strength in hydrogen and nitrogen atmosphere

	Average ultimate strength, psi[a]	Stress-rupture tests: Conditions for failure stress, psi[b]	Stress-rupture tests: Conditions for no-failure stress, psi[c]
Low strength, 4140			
80°F un-notched	135,000[d]		
80°F notched			
10,000 psi N_2	241,000 (4)		
6,000 psi H_2	207,000 (2)		
10,000 psi H_2	204,000 (3)	190,000 (30 sec) 175,000 (9 hr)	185,000 (8) 170,000 (5)
250°F notched			
10,000 psi N_2	221,000 (2)		
6,000 psi H_2	207,000 (2)		
10,000 psi H_2	185,000 (1)	185,000 (3min.)	175,000 (6)
High strength, 4140			
80°F un-notched	228,000 (2)		
80°F notched			
10,000 psi N_2	362,000 (2)		
2,000 psi H_2	135,000 (2)		
6,000 psi H_2	121,000 (4)		100,000 (5) 110,000 (3)
10,000 psi H_2	89,000 (3)		70,000 (4) 80,000 (4) 90,000 (3)
250°F notched			
10,000 psi N_2	274,000 (1)		
2,000 psi H_2	92,000 (2)		
6,000 psi H_2	96,000 (4)		70,000 (5) 80,000 (3)
10,000 psi H_2	82,000 (3)	65,000 (23 min)	
350°F notched			
10,000 psi H_2	103,000 (1)		

[a] Number of specimens shown in parentheses.
[b] Time shown in parentheses:
[c] Time in days shown in parentheses.
[d] Reported by the Bethlehem Steel Co.

truly dry, standard materials may be used quite satisfactorily provided that the temperature is kept below about 275° F.[3] In a few instances with wet gas, chlorine compressors have been built completely from unalloyed titanium. The impellers were of the open-inducer type. The casings were vacuum cast in graphite molds, and the casing volute casting had a cleaned weight of 1800 lbs. While titanium is exceptional for use with wet chlorine (see Figure 6), it cannot be used in dry chlorine, because in the event of a rub which would generate an elevated temperature, titanium is pyrophoric. This characteristic can be suppressed with moisture in the chlorine, and the generally accepted minimum moisture content is 0.015%.

Aluminum alloy compressor impellers have been used in great numbers, but in a limited range of applications. A steady stream of higher strength aluminum alloys is becoming available, and the applications may be expected to increase. Limitations on use of aluminum alloys have been (1) decrease in strength above about 300° F, and (2) difficulties in fabricating closed impellers from the high strength compositions. Integral cast closed impellers have been used extensively in such applications as diesel turbochargers refrigeration compres-

Fig. 6—Titanium castings for compressor handling wet chlorine

sors, and some centrifugal compressors on air. Aluminum alloys are avoided in impellers for oxygen service. In the event of a rub between the rotating and stationary parts causing the aluminum to reach a high temperature, there is a possibility of a thermit reaction with oxides on the surface of iron or steel parts.

Both aluminum and titanium are sometimes selected for centrifugal compressor impellers because of their low density, even in environments where corrosion will not occur. The lower density causes a shift in the critical speed of a rotor which may be very advantageous.

Coatings. Coatings for the prevention of erosion and corrosion of steel compressor impellers have been used, but only sparingly. A few machines have been built with a coating of electroless nickel, and while service to date has been rather brief, results seem to be satisfactory. Limited use has also been made of organic coatings such as one of the phenolic resins. The usual concern with coatings of this sort is durability. Most compressor users do not want to open the machine for recoating of parts more than once every two or three years, and it has not been possible to give assurance that coatings will last this long. In one application involving air containing finely divided water droplets and some sulfide dioxide, a phenolic coating has performed satisfactorily when renewed at intervals of about six months. In this case the machine is of the single-stage type where the impeller can be changed more readily than is the case with multi-stage compressors.

Open type impellers are much easier to coat than fully closed impellers. The difference is primarily in ease of cleaning. Most coatings require a white finish abrasive blast cleaning, which is difficult because of poor accessibility in fully closed parts. It is impossible in closed impellers of low passage height. Most coatings, and especially those requiring baking for curing, are very intolerant of discontinuities such as sharp corners or porosity in the surface being coated.

Aluminum alloy impellers in plant air compressors are almost universally anodized. Conventional anodizing is sometimes employed, but more commonly impellers are hard anodized to obtain a coating having greater thickness and hardness than conventional anodizing. This hard anodizing provides protection from both corrosion and erosion.

SULFIDE CRACKING

Background. In recent years there has been a great deal of interest in the possibility of sulfide cracking in centrifugal compressor impellers. There is extensive literature on sulfide cracking in oil well casing materials[4] going back some 20 years. Incidents involving centrifugal compressors have been rare, as reported by Kohut and McGuire[5] and Moller.[6] Nevertheless, the subject is, justifiably, one of serious concern because of the potential of a failure and loss of service of a vital link in the production chain of a chemical or petrochemical plant. Greer[7] presented an excellent paper describing the effects of more than a dozen variables on resistance to sulfide corrosion cracking. Other investigations dealing specifically with sulfide cracking in compressor impellers have been reported by Scheminger, Ebert, and Aul[8] and by Keller and Cameron.[9]

Essential features. While it is impractical to attempt to review all of the data in detail, the salient points may be summarized by noting that in order for sulfide cracking to occur it is necessary that the following conditions be fulfilled:

1. Hydrogen sulfide must be present.
2. Water must be present in the liquid state.
3. The pH must be acid.
4. A tensile stress must be present.
5. Material must be in a susceptible metallurgical condition.

When all of the above conditions are fulfilled, sulfide cracking may occur with the passage of time. Failure is not instantaneous.

Inhibition. It is frequently not possible to remove the hydrogen sulfide or moisture from the gas. Some interesting work has been reported on the prevention of sulfide cracking by the addition of inhibitors; but, at least as far as centrifugal compressors are concerned, the use of these inhibitors does not seem to be widely practiced. Since enough time will always eventually be accumulated in any operating machine, the only variable which is capable of adjustment seems to be the condition of the material.

Metallurgical condition. Numerous investigations have shown that the optimum microstructure for resistance to sulfide cracking is tempered martensite resulting from heat treatment by quenching and tempering. These studies have shown that low alloy steels having a maximum yield strength of 90,000 psi and a maximum hardness of Rockwell C-22 are not susceptible to sulfide cracking, even in the most aggressive environments. As the strength level increases above 90,000 psi, the threshold stress required to produce sulfide cracking may actually decrease in very severe environments. Warren and Beckman,[10] for example, have reported that with a yield strength of 100,000 psi the threshold stress required to produce sulfide cracking approached the yield strength, while at a yield strength of 140,000 psi the threshold stress dropped below 30,000 psi as illustrated in Figure 7. The test environment used by these authors involved a hydrogen sulfide water system at 104° F and 250 psi.

Effect of pH. Treseder and Swanson[12] have shown that the effect of pH is quite marked. For example, their experimental parameter increased from 4 at a pH equal to 2 to nearly 14 at a pH equal to 5 as shown in Figure 8. The importance of pH was also shown in the work of Keller and Cameron.[9] Among the tests conducted in this work was a series on AISI 4140 which was quenched and tempered after welding to a base metal yield strength of 126,000 psi. The test specimens were stressed to 80% of the yield strength and tests were conducted in triplicate. At pH 2.5 all three specimens failed, while at pH 4.2 none of three failed. A similar series of tests was conducted on AISI 4140 which was quenched and tempered before welding to a base metal yield strength of 83,000 psi with only a tempering treatment at 1100° F after welding. Again the test stress was 80% of the yield strength. At pH 2.5 three tests were conducted and all failed. At pH 4.2 there were eight failures in nine tests, but at pH 6.5 there were no failures in three tests. The test environment was room temperature water saturated with hydrogen sulfide, and the pH was controlled by addition of hydrochloric acid or ammonia water.

The acidity of the gas in a compressor is determined by condensing a sample of the gas and making a pH determination. When the concentration of hydrogen sulfide is sufficient to saturate the water, the pH value is about 4.3. If the hydrogen sulfide present is not sufficient to saturate the water, the pH will be higher. Further, the pH may be influenced by the presence of other constituents in the gas stream which are soluble in the water. Hydrogen chloride, for example, is capable of depressing the pH to values well below 4.3 and accelerating attack. Other possible constituents could reduce the acidity, raise the pH, and reduce severity of attack. For these complex reasons it is difficult, if not impossible, to specify a threshold concentration of hydrogen sulfide in a gas below which sulfide corrosion cracking will not occur.

Effect of temperature. A strong inverse relationship with temperature has been reported by Hudgins[13] with time to failure increasing markedly from 75° F to 150° F. Most centrifugal compressors with hydrogen sulfide in the gas operate at these temperatures or higher, particularly when the temperature rise through the machine is taken into account.

Service experience. While the limits of 90,000 psi yield strength and hardness of Rockwell C-22 were selected for very aggressive environments, experience suggests that the environment in centrifugal compressors is considerably less aggressive. There are many welded AISI 4140 and 4340 low alloy steel impellers in service which were given a simple tempering treatment at 1100° F after fabrication, and which have suffered no distress, although the environments are known to contain hydrogen sulfide.

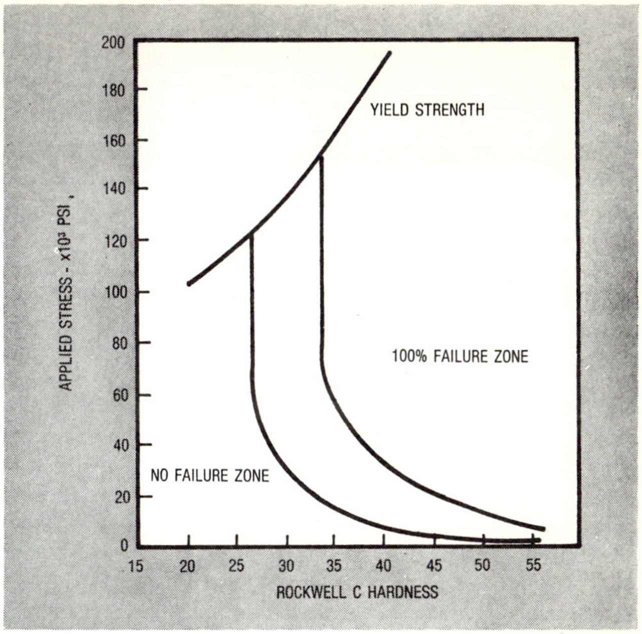

Fig. 7—Threshold stress for sulfide cracking as a function of hardness

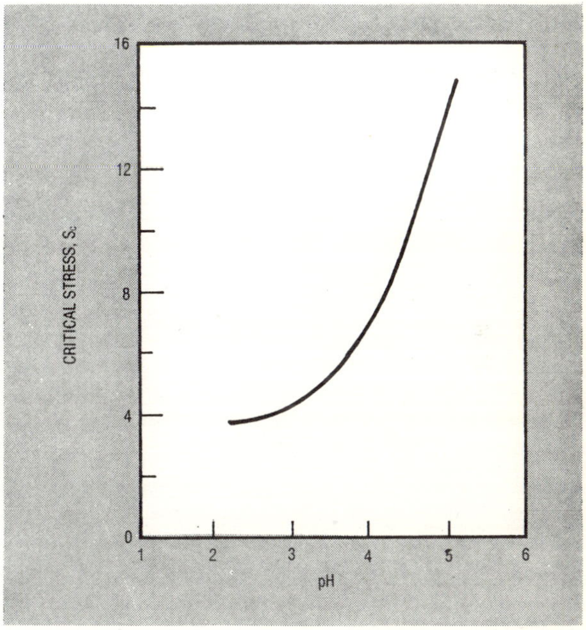

Fig. 8—Critical stress for sulfide cracking as a function of pH

Perhaps in these cases the operating temperature of the compressor is above the dew point of the gas so that liquid moisture is not present, or possibly the composition of the gas is such that the pH is greater than 7. A paper by Garwood[11] cited other cases where satisfactory service has been obtained in exposure to environments containing hydrogen sulfide at hardness values of Rockwell C-28.

In many of these cases the hardness of the base metal is approximately Rockwell C-26, and the hardness of the heat affected zone adjacent to the welds is Rockwell C 28-30. In one case, as reported by Moller,[6] in ethylene

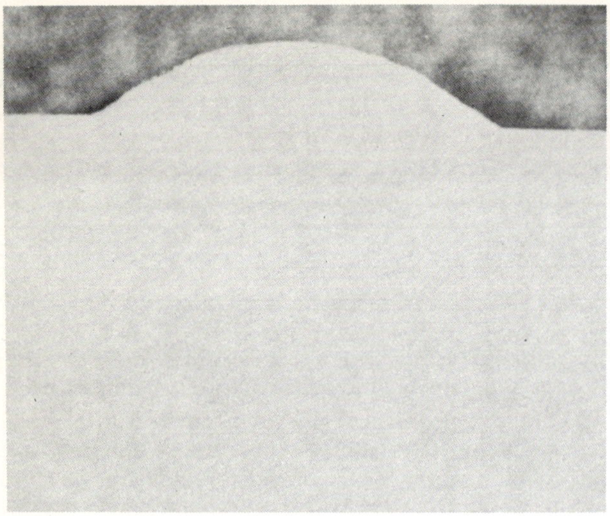

Fig. 9—AISI 4140 welded and tempered showing heat affected zone

Fig. 10—AISI 4140 quenched and tempered after welding showing no heat affected zone

off gas service there was cracking in the heat affected zone adjacent to some of the welds because of the omission of the 1100° F post weld heat treatment after repair welding. In this instance the hardness of the heat affected zone was Rockwell C 45-50. In the portions of the impellers which had not been repaired, and where the hardness of the heat affected zone was Rockwell C-30 or below, there was no cracking. It must be concluded that, in this instance, the severity of the environment was rather mild, and the common specification requirement of Rockwell C-22 maximum was not needed to prevent sulfide corrosion cracking.

Prevention. Compressor manufacturers are now equipped to perform a complete quench and temper heat treatment after fabrication by welding. When this is done, the heat affected zone disappears as is shown in Figures 9 and 10, and the impeller is of uniform hardness. With the use of such treatments, it appears safe to increase the maximum permissible yield strength up to at least 110,000 psi.[9] In evaluating the maximum permissible yield strength, it is instructive to take into account the actual operating stress in centrifugal compressor impellers and its relationship to the yield strength. This stress increases with the square of the speed of rotation. According to API 617, it is required that impellers be overspeed tested, prior to assembly of a compressor rotor, at 115% of the maximum permissible continuous speed. The maximum continuous speed is in turn 105% of the design speed. Thus at overspeed the actual speed is 121% of design, and the stress at overspeed is 145% of the stress at design speed. Assuming the stress during overspeed testing approaches the minimum specified yield strength of the material, the stress at design speed would be only 69% of the minimum yield strength. At maximum continuous speed rating the operating stress would be only slightly higher, still only 75% of the yield strength. These conservative stress values operate to reduce the risk of difficulty due to sulfide cracking.

The shaft of centrifugal compressors is made from a heat treated low alloy steel, usually the same grades, AISI 4140 and 4340, as in impellers. It is often required that the yield strength be over 90,000 psi to accommodate the maximum applied stress under the coupling. At this location the shaft is not exposed to the process gas. The operating stress in the portion of the shaft inside the casing where it is exposed to the process gas is quite low —often under 5000 psi. For this reason, and taking into account data such as that previously cited from Warren and Beckman, the yield strength is usually not restricted.

Stationary parts including the casing and diaphrams made from carbon steel and cast iron meet the maximum strength and hardness requirements for immunity to sulfide cracking with margin to spare. Barrel type compressors for very high pressure service are sometimes made from low alloy steel forgings heat treated to yield strengths above 90,000 psi and hardness over Rockwell C-22, but the operating environment is not one likely to give rise to sulfide corrosion cracking.

HYDROGEN ENVIRONMENT EMBRITTLEMENT

Types of hydrogen environment embrittlement. Another topic which has been of much interest in recent years is that of the effect of hydrogen in centrifugal compressors. The hydrogen problem takes three different forms which have been described by Jewett et al:[14]

1. *Hydrogen chemical reactions.* This has also been called hydrogen attack, and occurs at elevated temperatures. The minimum temperature is usually cited as about 425° F. Material selection for elimination of hydrogen attack in steel is well documented by the Nelson charts.[15] Hydrogen attack has not been a serious problem in centrifugal compressors. High temperature and high hydrogen concentrations have not often occurred simultaneously. When the temperature is high, the hydrogen content has been low and vice versa.

2. *Internal hydrogen embrittlement.* In this case delayed failure may occur, especially with high strength steels. Many examples are found in the manufacture of

hardened and electroplated parts such as springs, washers, aircraft landing gear struts, etc. Internal hydrogen was also the cause of flakes in forgings which were involved in the failure of several large turbine and generator rotor forgings some 20 years ago.[16, 17] The problem of flaking in large forgings has been eliminated with the use of vacuum degassed steel.

3. *Hydrogen environment embrittlement.* This is the aspect of the hydrogen problem which is of most significance in centrifugal compressors. As the name implies it occurs while the metal is stressed in hydrogen. Material exposed to high pressure hydrogen, but subsequently tested at room temperature does not exhibit this embrittlement.

Features of hydrogen environment embrittlement. There has been extensive literature developed on the last two forms of hydrogen embrittlement including papers by Elsea and Fletcher,[18] Walter and Chandler,[19] Beck et al,[20] Cavett and Van Ness[21] and Steinman, Van Ness and Ansell.[22] With all of the research, however, NASA[23] stated "There are considerable gaps in the data available; essentially, no thresholds have been established for specific influences such as pressure, temperature, and stress much less the influence of combined factors." To this list might be added the effect of variations in gas composition.

There are several important distinctions between the last two types of hydrogen embrittlement. Internal hydrogen embrittlement does not show up immediately after charging the specimen, but will cause delayed failure if the specimen is held under a modest stress for a period of time. Hydrogen environment embrittlement, conversely, causes immediate embrittlement. Specimens usually fail during or very shortly after loading or not at all. Hydrogen environment embrittlement failures start at the surface while internal hydrogen embrittlement cracks are initiated below the surface. Most of the time, failure in hydrogen environment embrittlement is in successive incubation periods. The incubation period before each step in crack formation is the time required for hydrogen to diffuse to the region of active crack progression.

Hydrogen environment embrittlement is sensitive to strain rate, because it requires the movement of hydrogen in the presence of a stress gradient. Conventional impact tests do not indicate embrittlement because the strain rate is high, and the time is too short to permit diffusion of hydrogen to the advancing crack tip. The effect is at a maximum at low strain rate, but the degree of embrittlement is independent of holding time under pressure even when the specimens are held under stress in high pressure (10,000 psi) hydrogen.

Categories of hydrogen environment embrittlement. Walter and Chandler[19] have separated metals into four categories according to the degree of embrittlement observed in high pressure hydrogen:

1. *Extreme embrittlement.* High strength steels and high strength nickel base alloys are in this category where embrittlement is characterized by a large decrease of notch strength and notched and unnotched ductility and some decrease in unnotched strength in 10,000 psi hydrogen. Metals in this category usually fail with one catastrophic crack which propagates into the specimen leaving a thin shear lip around the periphery except at the site of crack initiation.

2. *Severe embrittlement.* The majority of metals tested were in this category which includes most of the materials used in centrifugal compressors. Embrittlement is characterized by reduction of notch strength and notched and unnotched ductility, but little or no reduction of unnotched strength. Metals in this category usually fail with many surface cracks, some of which are quite deep.

3. *Slight embrittlement.* This category includes commercially pure titanium, copper, beryllium and the austenitic stainless steels having an unstable austenite structure. Embrittlement is characterized by a small decrease of notch strength and notch ductility. Metals in this category exhibit numerous small, shallow, blunt cracks, but low power magnification is required to see them.

4. *Negligible embrittlement.* Materials in this category include aluminum alloys, stable austenitic steels, and copper. They fail in the same manner in hydrogen and in air with no reduction in strength or ductility due to hydrogen environment embrittlement. Surface cracking is not observed.

Effect of variables. As has been indicated, the effects of variables on hydrogen environment embrittlement have not been as fully defined as is desirable, presumably due to the extreme difficulty and cost of the experimentation.

Hofmann and Rauls[24] have reported on the effect of temperature. In their work, embrittlement seemed to be most extreme near room temperature, decreasing as the temperature was raised or lowered. This work was on normalized low carbon steel. As shown in Table 2, other studies on AISI 4140 quenched and tempered to two different strength levels showed the same degree of embrittlement at 80° F and 250° F. The higher strength material, having a yield strength of 212,000 psi, exhibited a greater degree of embrittlement than other specimens with a yield strength of 127,000 psi.

Elastic properties, including the yield strength of a material, are the same in hydrogen as in air. Some plastic deformation is required to initiate hydrogen environment embrittlement. For this reason it has been observed that low alloy steels quenched and tempered to produce a tempered martensite structure are less prone to embrittlement in a hydrogen atmosphere than lower strength materials. In a tempered martensite structure, the ratio of yield strength to tensile strength is usually 0.85 or higher, while in lower strength materials the yield-tensile ratio may be as low as 0.5 because of the difference in heat treatment. Thus, deformation begins at a lower fraction of the tensile strength in the lower strength materials with the result that a greater depression in tensile strength due to hydrogen environment embrittlement is observed.

The effect of purity of the hydrogen is most important. As little as 1% oxygen has been reported to eliminate embrittlement in some alloy steels. Other impurities in the gas, which could combine with the metal to form a surface film presenting a barrier to the passage of hydrogen, would also be expected to be beneficial. Conversely, constituents in the gas stream which would be inert or which would attack the surface and prevent the formation of a barrier film would not have any beneficial effect, although possibly not a detrimental effect. If oxide films are developed, but are subsequently ruptured by plastic deformation, they would become ineffective in retarding hydrogen environment embrittlement. Some work has been done on protective coatings, but with limited success. The chief difficulties have been discontinuities or lack of adherence. Coatings are thin and could be damaged to the extent of destroying the protective effect by a momentary internal rub between rotating and stationary compressor parts, or by a foreign material passing through compressor in the gas stream.

The pressure of the hydrogen gas is an important variable, presumably at least in part as it affects the rapidity with which the gas can diffuse into an advancing crack. In higher pressures where the gas penetrates the crack more rapidly, conditions favoring the progress of hydrogen environment embrittlement at the crack tip are reached more rapidly. Consistent with this, it has been observed that fatigue strength is reduced in hydrogen environments, and that crack growth rates are increased over those observed in air.

Prevention. It has been indicated that an important variable in determining the degree of hydrogen environment embrittlement is the yield strength of the material. In many cases gas streams which are high in hydrogen also contain hydrogen sulfide and moisture. As was discussed in a previous section, it has, in recent years, become quite common to apply a maximum yield strength specification. Most commonly the yield strength has been limited to 90,000 psi maximum. Immunity to hydrogen environment embrittlement has been achieved simultaneously with immunity to sulfide corrosion cracking.

Most of the research on hydrogen environment embrittlement has been conducted on materials having a yield strength of 180,000 psi or higher. When notched specimens have been tested both strength and ductility have been substantially reduced. This embrittlement has been observed to decrease with decreasing strength.

It cannot be said that the limit has been precisely defined, but it appears that yield strengths up to about 120,000 are acceptable in centrifugal compressors without undue risk of embrittlement. This effect may be associated with the lesser fracture toughness of higher strength steels, and with the increased stress gradient at the base of a notch or crack in higher strength materials.

The relationship between the stress at operating speed and that during factory overspeed testing has been reviewed earlier. As in the case with sulfide cracking, this relationship provides a margin of safety in hydrogen environment embrittlement. During overspeed testing there may be plastic deformation and some blunting of stress concentrations. Such testing is commonly carried out in a partial vacuum. Subsequently, operation at a lower stress in a more severe environment is then less hazardous.

LOW TEMPERATURE OPERATION—BRITTLE FAILURE

Background. When compressors are required to operate at subzero temperatures, consideration must be given to the problem of brittle failure. As temperatures fall, all materials become stronger and less ductile, and some materials become increasingly susceptible to brittle failures. Those materials which have this susceptibility include most of the materials commonly employed in compressor construction.

In order for a brittle failure to be initiated, there must be stress exceeding the yield strength. While the nominal stress is always well below the yield strength, it is not possible to design and manufacture any engineering structure without stress concentrations, and it is from these sites that brittle failures propagate. Brittle failures, once initiated, can propagate at very much lower stresses.

Such failures propagate with the speed of sound, and this is the reason for reports of a loud noise accompanying brittle failures, such as those which occurred when welded ships broke in half during World War II. The problem, of course, was not with the welding, but it had not been encountered previously because brittle failures in individual ship plates frequently did not propagate across riveted joints. Individual plates were replaced when they cracked. In the case of welded ships, however, the path for propagation across the joint existed, and catastrophic failures were encountered.

Criteria for resistance to brittle fracture. There have been several changes in the criteria for susceptibility to brittle failure. In the work done immediately after World War II, the most commonly accepted criterion was the Charpy V notch impact tests using energy absorption to determine susceptibility. It gradually came to be realized that the previously used Charpy keyhole test was not as definitive as would be desired for reasons shown in Figure 11.

It may be seen that in the transition zone there is a spread in the Charpy keyhole values where the energy level obtained is unpredictable while the Charpy V notch test is more definitive. For this reason the Charpy keyhole test was superseded by the V notch test, still using energy absorption as a criterion.

Other tests have been used including the explosion bulge test, the drop weight test, and the drop weight tear test. These tests have the advantage of being more direct measurements of the susceptibility to brittle failure. They suffer, however, from the disadvantage of having associated with them one or more of such problems as requiring large size test specimens, expensive test setups, test specimens not readily obtainable from some forms of material, etc.

The ultimate in criteria for resistance to brittle fracture seems to be fracture mechanics. A great deal of

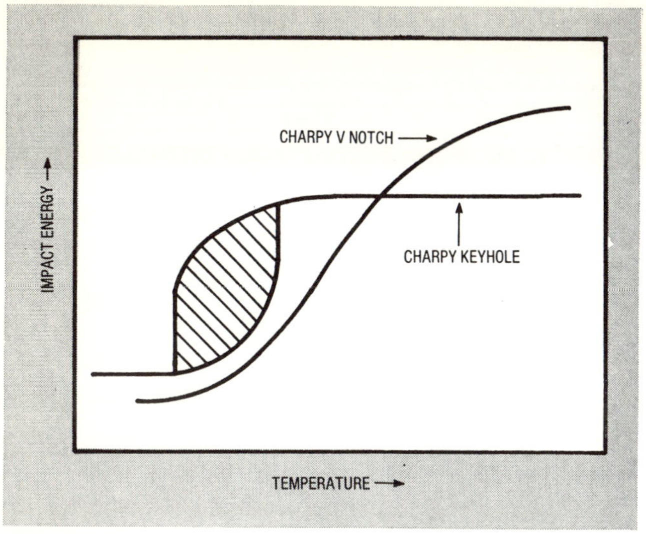

Fig. 11—Charpy keyhole and V notch impact energy as a function of temperature

work has been done in this area over the last 15 to 20 years, particularly with respect to ultra-high strength materials and certain non-ferrous alloys such as aluminum and titanium. At the present state of development, fracture mechanics is not fully applicable for the low alloy steels at the strength levels and the temperature levels commonly encountered in centrifugal compressors. The fracture mechanics approach is the best one for taking fully into account the fact that flaws exist in all materials and all structures. There is a finite size below which flaws will not propagate. The actual size is dependent on the strength of the material and the applied stress. It seems probable that ultimately fracture mechanics criteria will be developed, and will replace measurements presently in use.

Materials and fabricating procedures. Casing materials for low temperature service generally require only modest strength, and there are several grades of both wrought and cast products.

For fabricated casing material, fine grain carbon steels are available with satisfactory properties at temperatures to −50° F, 2¼% nickel steel plates will go to −75° F, and 3½% nickel steel plates may be purchased for service to −150° F. In the case of the 3½% nickel steel at −150° F, ordinary melting procedures are not capable of yielding satisfactory results in thicknesses over about four inches, but considerably improved results are obtained with electroslag melted material. The improvement is believed to be due to the substantially lower level of impurities. The properties of these materials at low temperatures are sometimes enhanced by accelerated cooling after the austenitizing treatment. This is particularly helpful in heavier thicknesses, say over three or four inches.

In order to take advantage of these materials, it has been necessary to develop fabricating techniques including, most especially, welding procedures. No great problem is encountered in meeting the requirements at temperatures down to −50° F using carbon steel filler metal for either manual metal arc or submerged arc welding. For temperatures below −50° F alloy steel filler metal is required and special fluxes are needed for submerged arc welding. These are available.

Fluxcore welding with products designated by the American Welding Society as E70T-X is attractive for use at ordinary temperatures because the deposition rate is higher than that of manual metal arc, and the technique may be used in positions where submerged arc welding is not applicable. Satisfactory low temperature impact results have not yet been obtained on fluxcore deposits consistently, but development along these lines is continuing.

Cast materials for casings are also available for service to −175° F using one of several low alloy nickel containing steels with 2¼% nickel being satisfactory to −100° F, 3½% nickel to −150° F, and 4½% nickel to −175° F. Below −175° F materials must be used which have an austenitic structure. Down to at least −320° F conventional cast austenitic stainless steels such as the cast equivalent to AISI Type 304 possess good fracture toughness. A modified 20% nickel cast iron has also been used for application at −260° F, and toughness values indicate that it would be acceptable to −320° F. While Type 304 is more difficult to cast, and more expensive, it also can be welded more readily. In conventional austenitic stainless steels such as Type 304, the weldability is good and the properties of the base metal, heat affected zone, and weld metal are all good at low temperatures. The modified 20% nickel cast iron has better foundry properties and requires fewer welds. There is more of a problem, however, with welding, and especially with the heat affected zone of the base metal. For this reason, welds in critical areas are not permitted, but non-critical locations are sometimes repaired.

In both materials, because of the austenitic structure, the coefficient of thermal expansion is about 50% greater than that of standard ferritic casings. This is not an insurmountable problem, but it must be taken into account, especially if the rotor is constructed of a material with a lower coefficient of expansion. Rotors for operation at −260° F have been fabricated from 9% nickel steel. This alloy has satisfactory resistance to brittle failure at least as low as −320° F. Its coefficient of expansion is similar to that of the low alloy steels, and thus much lower than that of austenitic casings. In the form of forgings and plates up to about two or three inches in thickness the alloy has attractive properties, but it is not available in the heavier thicknesses required for casing flanges. Castings of 9% nickel steel are not used because, while they can be made, they do not have attractive properties.

Materials for rotating parts. Extensive work at the Naval Research Laboratory by Pellini and associates showed that the energy absorption requirement for the prevention of brittle failure increased with the strength of the material. While a level of 12 or 15 ft.-lbs. in the Charpy V notch test is sufficient for ordinary mild steel, the energy absorption requirement might be 30 ft.-lbs. or more for quenched and tempered low alloy steel.[26] Recognizing this problem, Gross and associates at Lehigh[27] and at U. S. Steel[28] proposed a criterion using a direct measure of notch ductility.

They examined both the lateral contraction at the root of the notch and the lateral expansion on the compression side opposite the notch. They concluded that either would serve, but that lateral expansion was a great deal easier to measure than lateral contraction, and for this reason, expansion was selected. Gross et al proposed a criterion of 15 mils lateral expansion for a wide range of strength levels. This has since been incorporated in the ASME Boiler and Pressure Vessel Code, Section 8, Division 1 for materials having a tensile strength greater than 95,000 psi. There is, however, still some difference of opinion on this subject as indicated in a recent paper by Puzak and Lange[29] of the Naval Research Laboratory.

The Naval Research Laboratory work shows a superiority for other types of tests, especially the drop weight and drop weight tear tests. Unfortunately these tests require test specimens which are difficult or impractical to obtain from some forms of material.

In the case of rotating elements, AISI 4330 or 4340 can be heat treated to give acceptable values using lateral expansion, per ASME Boiler and Pressure Vessel Code, Section 8, Division 1, down to temperatures at least as low as −150° F with a yield strength of 90,000 psi or greater. At −100° F, the same impact test requirement can be met with a yield strength of about 105,000 psi.

Some years ago 3½% nickel steels such as AISI 2340 were used for this application. More recently AISI 4330, which is lower in nickel but higher in chromium and molybdenum, has become almost universal. The impact requirements can be met more readily with the 43XX series because of increased hardenability and better response to heat treatment.[30] Below −150° F recourse must be had to other materials such as as the 5% nickel steel recently developed by Armco under the tradename Cryonic 5 or the 8 or 9% nickel steels. As previously mentioned, 9% nickel steel has been used for some years quite successfully at temperatures down to −320° F in both rotating and stationary applications. For some applications aluminum alloys have also been used successfully. These temperatures are becoming increasingly of interest in connection with liquefied natural gas projects.

STABILITY

One of the most insidious problems that can occur in the manufacture and use of turbomachinery is dimensional instability. When instability is encountered during manufacture it may be impossible to achieve required close tolerances on stationary parts and concentricity on shafts. Long term instability which may occur in service can cause leakage at casing joints, or a rotor to go out of balance with resulting excessive vibration.

Casings. Instability leading to difficulty in manufacture can occur with centrifugal compressor casings, but in practice is not a major problem. Casings of either the cast or fabricated type are given a stress relief heat treatment during manufacture at about 1100° F, a temperature which insures a low level of internal stress. Generally there is more difficulty with distortion during machining of castings than fabricated casings.

The wall thickness of castings is tapered in order to achieve progressive solidification of the casting, with the result that the amount of metal removed in machining is not equally distributed. This tapered thickness is not encountered with fabricated casings. The amount of material removed in machining is more uniform, and usually less than with castings. Further, castings require substantial amounts of repair welding. Defects which are encountered during final machining, and which require welding are particularly troublesome.

Long term service instability of compressor casings is very seldom encountered, and for this reason literature in this area is non-existent. This type of instability is more commonly encountered with turbines because of the higher operating temperature, greater temperature excursions in service, and higher thermal gradients. In addition, particularly with modern high temperature, high pressure turbines, the wall thickness of the casing is greater. Several papers, including one by Reisinger and Scharp[31] have been published on the subject of steam turbine casing distortion and cracking due to thermal gradients. Precautions are taken to avoid this by the use of slower heating and special design features in the casing.

Other internal stationary parts in centrifugal compressors, such as diaphragms, are most commonly made from cast iron. Occasionally when higher strength is required, ductile iron or fabricated mild steel is used. In all cases the castings or fabrications are treated to produce a low level of internal stress, and difficulty with instability in service is virtually unknown.

Rotating parts. Rotating parts for centrifugal compressors are, likewise, carefully treated in order to insure a low level of internal stress. Shafts made from quenched and tempered alloy steels such as AISI 4140 or 4340 are tempered at a minimum temperature of 1100° F, and difficulty with achieving concentricity during manufacture is rarely encountered. As in the case of stationary parts, the problem is a good deal more serious with turbines, and, again, published literature on this problem generally deals with turbines. For turbine rotors with integral discs, a vertical heat treatment is recommended. Further, such rotor forgings are tested for thermal stability in a special test apparatus where the rotor may be rotated at a speed of about 2 rpm while being heated to a temperature of about 1000° F with the eccentricity being measured by the use of dial indicators having long extension arms reaching into the test furnace.

Possibly the most common cause of true instability is due to a deficiency in prior heat treatment with the result that the material on one side of the shaft has a coefficient of thermal expansion different from that on the other side by about 1%. This is sufficient to cause a significant deflection when the shaft is heated to 800-1000° F. This and other possible causes of instability have been discussed at length in two excellent papers by Barker and Jones[32] and Timo and Parent.[33]

ACKNOWLEDGMENT

Originally presented at the Second Turbomachinery Symposium, Texas A&M University, College Station, Texas, October 1973.

LITERATURE CITED

[1] Mishler, H. W., Monroe, R. E., and Rieppel, P. J., "Studies of Hot Cracking in High Strength Weld Metals." Welding Journal, Research Supplement, pp. 1s-7s (January, 1961)

[2] Kahles, John, "Electrical Discharge Machining." Metals Handbook, 3, pp. 227-233 (1967)

[3] Brown, M. H., DeLong, W. B., and Auld, J. R., "Corrosion by Chlorine and by Hydrogen Chloride at High Temperatures." Industrial and Engineering Chemistry (July, 1947)

[4] Vollmer, L. W., "The Behavior of Steels in Hydrogen Sulfide Environments." Corrosion, 14, pp. 324t-328t (July, 1958)

[5] Kohut, G. B., and McGuire, W. J., "Sulfide Stress Cracking Causes Failure of Compressor Components in Refinery Service." Materials Protection, pp. 17-22 (July, 1968)

[6] Moller, G. E., "Corrosion, Metallurgical, and Mechanical Experiences of Petroleum Refinery Compressors." NACE Task Group T-8-1 Interim Report (1968)

[7] Greer, J. B., "Factors Affecting the Sulfide Stress Cracking Performance of High Strength Steel." NACE Paper No. 55 (1973)

[8] Scheminger, W. J., Ebert, H. E., and Aul, E. L., "Resistance of Some Standard Compressor Materials to Hydrogen Sulfide Stress Corrosion Cracking." ASME Paper 71-Pet-25 (1971)

[9] Keller, H. F., and Cameron, J. A., "Laboratory Evaluation of Susceptibility to Sulfide Cracking." Carrier Corporation N.A.C.E. Corrosion 74, paper 99, Chicago Ill., March, 1974.

[10] Warren, D., and Beckman, G. W., "Sulfide Corrosion Cracking of High Strength Bolting Material." Corrosion, *13*, pp. 631t-646t (October, 1957)

[11] Garwood, G. L., "Material Selection for Downhole and Surface Equipment for Sour Gas Condensate Wells." NACE Paper No. 53 (1973)

[12] Treseder, R. S., and Swanson, T. M., "Factors in Sulfide Corrosion Cracking of High Strength Steels." Corrosion, *24*, pp, 31-37 (1968)

[13] Hudgins, C. M., "The Effect of Temperature on the Aqueous Sulfide Stress Cracking Behavior of an N80 Steel." NACE Canadian Western Regional Conference (1971)

[14] Jewett, R. P., Walter, R. J., Chandler, W. T., and Fromberg, R. P., "Hydrogen Environment Embrittlement of Metals." NASA CR2163 (March, 1973)

[15] Nelson, G. A., "Action of Hydrogen on Steel at High Temperature and High Pressure." Welding Research Council Bulletin 145, Section II (October, 1969)

[16] Emmert, H. D., "Investigation of Large Steam Turbine Spindle Failure." ASME Paper 55A172 (1955)

[17] Schabtach, C., Rankin, A. W., Fogleman, E. L., and Winne, D. H., "Report of the Investigation of Two Generator Rotor Fractures." ASME Paper 55A208 (1955)

[18] Elsea, A. R., and Fletcher, E. E., "The Problem of Hydrogen in Steel." DMIC Memorandum 180 (October 1, 1963)

[19] Walter, R. J., and Chandler, W. T., "Effects of High Pressure Hydrogen on Metals at Ambient Temperature." Final Report NASA Contract NAS19 (1969)

[20] Beck, W., Jankowsky, E. J., and Fischer, P., Hydrogen Stress Cracking of High Strength Steels." Naval Air Development Center Report MA-7140.

[21] Cavett, R. H., and Van Ness, H. C., "Embrittlement of Steel by High Pressure Hydrogen Gas." Welding Research Supplement, pp 316s-319s (July, 1963)

[22] Steinman, J. B., Van Ness, H. C., and Ansell, G. S., "The Effect of High Pressure Hydrogen Upon the Notch Tensile Strength and Fracture Mode of 4140 Steel." Welding Research Supplement pp 221s-224s (May, 1965)

[23] Anon, "Effects of Hydrogen on Metals." NASA Tech Brief 69-10372 (September, 1969)

[24] Hofmann, W., and Rauls, W., "Ductility of Steel Under the Influence of High Pressure Hydrogen." Welding Research Supplement, pp 225s-230s (May, 1965)

[25] Deegan, D. C., "Exposure Tests in High Pressure Hydrogen Gas." U.S. Steel internal report (September 19, 1967)

[26] Puzak, P. P., and Pellini, W. S., "Evaluation of the Significance of Charpy Tests for Quenched and Tempered Steels." Welding Research Supplement, pp. 275s-290s (June, 1956)

[27] Gross, J. H., and Stout, R. D., "Ductility and Energy, Relations in Charpy Tests of Structural Steels." Welding Research Supplement, pp. 151s-159s (April, 1957)

[28] Gross, J. H., "The Effect of Strength and Thickness on Notch Ductility." Welding Research Supplement, pp. 441s-453s (October, 1969)

[29] Puzak, P. P., and Lange, E. A., "Significance of Charpy V Test Parameters as Criteria for Quenched and Tempered Steels." Naval Research Laboratory Report 7483 (October, 1972)

[30] Schwartzbart, H., and Sheehan, J. P., "Impact Properties of Quenched and Tempered Alloy Steels." Armour Research Foundation of Illinois Institute of Technology—Project B011 (September, 1955)

[31] Reisinger, R. H., and Scharp, C. B., "Thermal Stress Protection in Starting and Loading Boiler-Turbine-Generator Combinations." ASME Paper 60PWR3 (1960)

[32] Barker, A., and Jones, F. W., "The Reversible Bending of Turbine Shafts with Temperature." Institution of Mechanical Engineers (London) (1955)

[32] Timo, D. P., and Parent, D. F., "Thermal Distortion of Turbine Rotors." ASME Paper 58A270 (1958)

Casing, Nozzles and Auxiliary Piping

R. C. Burns, Elliott Co., Jeannette, Pa.

Cast casings have been used for many years and experience has proven this type of construction to be reliable and satisfactory. In recent years the fabricated type casing has been used for many applications and it also has proved to be a satisfactory and reliable design. Its principal advantages are in better control of shipments and the elimination of the defects which frequently occur in a cast casing.

When defects occur in cast casings, they usually can be repaired satisfactorily. Unfortunately, it frequently happens that certain defects are not discovered until the casing is partially machined or leak tested hydrostatically. Repairs made at this time may require stress relief heat treatment and usually require remachining. All this requires additional time and frequently causes a delay in final shipment.

Costs of fabricated casings of the usual sizes are similar to those of cast steel. At the present time, costs are not competitive with cast iron. Therefore, if service conditions permit the use of cast iron, this will be the more economical construction.

Horizontally split vs barrel type. The horizontally split casing construction provides access to the rotor and internal parts when inspection or maintenance work is needed. If pipe connections must be located in the upper half, the piping should be designed so that the casing can be easily opened.

When high pressures are involved, particularly with low molecular weight gases, it becomes more difficult to maintain a leak-tight joint with a horizontally split casing. The discontinuities at the bearing bores at each end of the casing and the changes in flange contour around the inlet and discharge heads make it difficult to seal this type flange joint at the higher pressures. For such conditions the barrel-type casing with a vertical, circular joint is preferred.

The barrel-type casing provides greater flexibility of possible nozzle arrangements and the ability to disassemble it without disturbing the main pipe connections.

Casing sealants. Horizontally split casings require some kind of sealing compound on the joint. Some type of silicone rubber compound, either with or without a catalyst agent, is generally used today. Fiber-type gaskets or string gaskets are not recommended.

An O-ring gasket may be desirable, particularly for high-pressure applications. O-rings are used regularly in the vertical split of barrel-type casings. There is no reason why they should not be used on horizontal joints also. No sealant or O-ring can produce a tight joint if the flanges are not properly designed and accurately machined.

Nozzles. Nozzle connections usually are directed up or down depending on the particular installation and piping arrangement.

The discharge connection usually is taken off tangentially from the casing to match the gas flow leaving the last stage. This connection could be rotated to come off the side if necessary but this usually would require a special design and also a special pattern if a cast casing is used.

The inlet connection is larger than the discharge and it becomes very difficult to bring this off the side of a horizontally split casing and still obtain satisfactory gas-flow distribution into the first stage of the compressor. This arrangement has been used for refrigeration-type compressors but is rare for industrial or chemical process work.

Small nozzles for intermediate side loads or extraction usually are located in either the up or down position. Normally these can be located to the side also if desired.

Nozzles directed to the side from the upper half casing will interfere wth free access to the horizontal-joint bolts in that area. Therefore, such nozzles should be avoided if possible.

Auxiliary piping. The user should make his specifications for auxiliary piping as flexible as possible. The specifications should be clear and specific, and should recognize the special problems that may be involved with auxiliary pipe connections to the casing. One such problem occurs with cast-iron casings when threaded connections are not allowed. Frequently, threaded connections are necessary in locations where it just is not feasible to make a flanged connection. Seal welding or brazing threaded connections to cast iron may cause cracks and leaks which are worse than using a simple threaded connection. The specifications should allow some flexibility to meet special situations of this kind.

Piping loads and moments can create a problem of casing distortion and misalignment. The manufacturer must recognize that some pipe loads will occur and should design his equipment to withstand reasonable loads without harm. Some standards have been established in this regard. There are NEMA Standards, such as SM21 and SM22, which apply to mechanical-drive steam turbines. There is an API Code covering centrifugal compressors for general refinery service which is based on the NEMA turbine standards.

The API Code provides standards for pipe forces and moments which are acceptable for most compressors of normal design and which do not impose unreasonable hardships on the piping design. These piping loads may be excessive, however, for some special-purpose, lightweight, high-speed units.

D. F. Neale, Union Carbide Corp.,
South Charleston, W. Va.

USERS HAVE FREQUENTLY CONSIDERED engineering responsibilities complete after specifications and design details have been resolved with counterparts in the manufacturing organization. Expediters see that work proceeds on schedule, and inspectors see that specified non-destructive tests are performed.

Operating critical process compressors now demands continuing emphasis on quality. This assurance might be gained by intensified user inspection in the factory. This approach is costly, demands scarce talent, and may deteriorate shop labor relations. More importantly, it does not acknowledge the important premise that the manufacturer is best qualified by virtue of experience, access and personnel to maintain quality control in his operations.

One might then ask, "Why should the user be involved in shop quality control?" There are two reasons:

1. To ensure continuing emphasis on quality as the responsibility for the equipment progresses from the engineering design organization and becomes increasingly susceptible to the pressures of shop schedules, costs and labor influences, and

2. To ensure utilization of the user's operating and maintenance experience in critical areas.

We are attempting to secure these objectives through a Quality Assurance Program by requiring documentation of conformance with critical elements of our specifications and the manufacturer's engineering design requirements. These documentation requirements are briefly summarized in Table 1. Two levels of documentation are required. Under materials, for example, certification is required for shafts or main rotors, casing pressure resisting parts and critical bolting. Only records are required for elements such as seal parts, casing bolts and blades.

Any material that has not met the applicable specification, but is considered acceptable by the manufacturer, or any repair of pressure resisting or major dynamic parts must be approved by the user. The objective is to fulfill the second purpose of the program—the utilization of user knowledge and experience in critical areas or decisions.

Our inspectors have traditionally witnessed casing hydrostatic tests. In addition, the program requires documentation of other critical nondestructive tests when specified by the user or routinely performed by the manufacturer. Documentation of intermediate tests is not required when they are primarily to release the component for further manufacture.

We require documentation of critical dimensions, allowable deviations, and actual measurements of key surfaces. The emphasis on critical bolting is continued by requiring documentation of the class of fit of shaft and rotor through-bolt threads.

Balance and running tests. Requirements for documentation of dynamic balance and running tests are not particularly startling. Several observations may be made, however:

- Testing the machines with the prebalanced contact coupling hub in place is preferable. Unbalance of the assembly should not be corrected by modification of replaceable parts such as the coupling, thrust collar or balance piston.

TABLE 1—Manufacturer's shop quality assurance documentation requirements

1.0 Materials
- 1.1 Certified records of materials and heat treatment of shaft and main rotor, pressure resisting casing and critical bolting.
- 1.2 Record of specifications for seal parts, high strength bolts, blades and other parts specified by user.
- 1.3 Documentation of user approval of off-specification materials.

2.0 Tests—Records of:
- 2.1 Witnessed hydrostatic tests.
- 2.2 Other critical non-destructive tests, e.g., shaft and wheel X-ray or ultrasonic.

3.0 Repairs—Documentation of user's approval of repairs to:
- 3.1 Pressure resisting parts.
- 3.2 Major dynamic parts.

4.0 Machining and Fits—Documentation of basic dimensions, maximum and minimum acceptable deviations and actual dimensions of:
- 4.1 Shaft and rotor runout and interference fits.
- 4.2 Diameters and surface finishes of bearings and seals.
- 4.3 Fit of shaft and rotor through-bolt threads.

5.0 Balancing and Mechanical Runs—Certified records of:
- 5.1 Machine balance of rotor components and assembly.
- 5.2 Running test:
  - 5.2.1 Speed, time and ambient temperatures.
  - 5.2.2 Vibration, frequency levels and critical speeds.
  - 5.2.3 Bearing metal temperatures when available.
  - 5.2.4 Bearing and seal oil flows, pressures and temperatures.

• Documentation of critical speeds when they are observable is needed. Although there is some uncertainty in this regard, unbalance is not *forced* to identify criticals.

• Proximity probes are used extensively, especially on machines with relatively light rotors. We have worked with several manufacturers to develop radial and axial-probe holders which are adjustable and replaceable with the machine in operation. We prefer that these be installed and used during the test, but have not been consistently successful in this regard.

• Thermocouples or resistance temperature devices (RTD) are being used to measure bearing-metal temperatures. This data is recorded during running tests.

• Bearing and seal oil pressures, flows and temperatures, and seal leakages are recorded.

API standards demand compliance with part of the last requirement, but some manufacturers are slow to incorporate it into their standard shop tests. We are convinced that it is an especially worthwhile emphasis however, as poor bearing fits, improperly installed seals, and even improperly cast or drilled oil passages have been found. Both the user and supplier benefit when such deficiencies are identified while repairs can be made in the supplier's factory instead of in the field where conditions may be adverse.

Participation of Engineering machinery specialists in tests is being intensified. Their effectiveness has been diminished, however, because of several problems:

• Late receipt and resulting inadequate review of the manufacturer's test plans,

• Insufficiently definitive plans regarding test conditions, the types and calibration of instruments, data to be taken, and the sequence of the test,

• Late notification of test dates and frequent postponements, have caused tests to be missed or less qualified observers to be sent. Slippages may unavoidably occur, but this only emphasizes the need for improved coordination between the manufacturer's shop scheduling personnel and the contract administrators or customer's men,

• Test reports are notably bland, lacking explanation and interpretation. A comprehensive test report can constitute a valuable start toward the records which we consider essential to effective operation and maintenance.

Final inspection and assembly. Opinions differ regarding the optimum extent of compressor inspection following mechanical spin or performance tests. Many ask, "Why risk damage or error by unnecessarily opening an apparently satisfactory machine?"

Our experience seems to justify mechanical inspection following test. Bearings and seals are inspected for wiping, hot spots or other signs of distress. Horizontally-split, as opposed to barrel-type casings, are usually opened for inspection even though there were no indications of unexplained contact or vibration. Frequently, they must be opened anyway for separate shipment of rotors or for protective spraying of internals.

Table 2 lists data which should be recorded during final assembly after inspection. Included are journal and thrust-bearing clearances and percentage contact areas; axial and radial shaft, wheel and thrust-collar runouts; bearing, seal and labyrinth clearances; and rotor axial float.

These records accomplish a secondary but significant objective of the Quality Assurance Program. This is the accumulation of *as-built* data for reference during installation, failure analysis and maintenance planning. Maintenance personnel participated in defining the data requirements and see significant benefits in having it available.

Preparation for shipment. The variations in size, weight and geometry of compressors, and in the environments to which they are shipped, complicates the statement of ironclad rules for preparation for shipment (packaging). In general, however, machinery should be shipped as fully assembled as is compatible with the design characteristics of the machine, the method of shipment and the mode of field storage. Rotors, guide vanes, diaphragms, seals, etc., are generally better protected by the casing than by separate packaging. Rotors, of course, should be blocked to prevent rotation or oscillation on their bearings. Consideration should also be given to removal of rotors when they are overhung, extremely flexible, or when the casing is soleplate mounted and flexible.

The manufacturer is the best source of advice in this matter. If special handling, field storage, or field assembly of separate items are required, the manufacturer's packaging plans should state this information.

Overhung masses, such as intercoolers and oil coolers, and especially prominent or vulnerable instruments, valves, piping and control panels, should be removed and packaged separately. These packages should be attached to the main skid when an invulnerable location is available.

The vagueness of the foregoing recommendations and the high incidence of damage prior to erection, identifies the need for effective user packaging specifications. These

TABLE 2—Quality control assembly records

1.0 Bearings—Clearance and contact areas of:
- 1.1 Journal bearings.
- 1.2 Thrust bearings.

2.0 Seals—Clearances at:
- 2.1 Bearing housing.
- 2.2 Process fluid.
- 2.3 Wear rings and labyrinths at each impeller.

3.0 Rotor
- 3.1 Shaft dimensions at bearings and seals.
- 3.2 Axial and radial runout at:
  - 3.2.1 Wheels.
  - 3.2.2 Thrust collar.
  - 3.2.3 Shaft.
- 3.3 Axial rotor float:
  - 3.3.1 Shoes removed.
  - 3.3.2 Shoes installed (ref. to stops).

TABLE 3—Packaging and shipping specifications

1.0 Method of Shipment
- 1.1 Routing (land, sea) and conveyance.
- 1.2 Number of loading and unloading operations.
- 1.3 Special limitations of handling equipment or clearances.

2.0 Site Climatic Conditions
- 2.1 Temperature range, wind velocity and humidity.
- 2.2 Special conditions—sand, snow, fumes.

3.0 Site Storage Plans
- 3.1 Planned inspection on receipt.
- 3.2 Protection.
- 3.3 Maintenance—rotations, desiccant renewals, oil checks.

4.0 Closures
- 4.1 Flanges: Cover and gasket materials and thicknesses, number of bolts and seal wires.
- 4.2 Threaded connections.

5.0 Cleaning and Coatings
- 5.1 Internal casing, bearing and seal housings.
- 5.2 Flange facings.
- 5.3 Exposed machined surfaces.

6.0 Packaging and Crating
- 6.1 Overwrapping and sealing of machined surfaces.
- 6.2 "Boxing" of projections, crating of detached parts.
- 6.3 Skids.

7.0 Marking: Weights, lift points, rotor "blocks," and desiccants.

8.0 Special Handling and Inspection Requirements (by Manufacturer)

TABLE 4—The manufacturer's contribution to installation

1.0 Improve Communications on:
- 1.1 Shipment.
  - 1.1.1 Parts susceptible to damage.
  - 1.1.2 Adjustments susceptible to change.
- 1.2 Receipt at site.
  - 1.2.1 Recommended handling precautions.
  - 1.2.2 Inspection for damage.
  - 1.2.3 Claim procedures.
- 1.3 Storage—Before and after erection.
  - 1.3.1 "Compatible" coatings.
  - 1.3.2 Desiccant renewal.
  - 1.3.3 Rotor rotation.

2.0 Recognize the Importance of Installation Manuals by:
- 2.1 More timely issue.
- 2.2 Concise and relevant contents.
- 2.3 Make specific to machine; include:
  - 2.3.1 Custom features.
  - 2.3.2 Summary of tests.
  - 2.3.3 Special precautions.

specifications which contain information as shown in Table 3 provide for satisfactory protection in transit and constitute the first step in site custodial efforts. The importance of this consideration is emphasized by the fact that some machines may be placed on their foundations months or years before startup.

It is apparent that satisfactory packaging cannot be ensured by a general specification. Therefore, manufacturers should submit a packaging plan. This plan should summarize procedures and materials, and provide a sketch of timbers and containers used to comply with the specifications. This will ensure that the user's requirements are mutually understood and that appropriate expertise is applied to this important consideration.

Installation instructions. Installation is thought of as the user's responsibility. The manufacturer may increase his contribution in a number of ways, however. Some of these demand that installation manuals be upgraded. Their quality should reflect their frequent role of defining the means by which the user shall meet his obligation to install machinery in compliance with the manufacturer's recommendations.

- Manuals should be issued earlier. The desire of manufacturers to complete design, drawing and installation of all appurtenances to the machinery before issue is understandable. They should be issued at least eight weeks prior to equipment shipment however, to provide for review, modifications, distribution and training of erection, maintenance and operating personnel. Users are discouraged from suggesting modifications of the manuals, but it is sometimes necessary.

- Manuals should be concise and relevant. Whole catalogs of component suppliers are often included—without identification of which particular size or model applies. The possibility of error or delay is real and should be avoided.

- Manuals should include more specific information relating to the individual machine. Custom features such as vibration probe holders are frequently overlooked in the somewhat stereotyped manuals we often see. Also, a summary interpretation of the running test should be included. Any modifications or special precautions for erection or run-in which are developed or identified during the test should be cited.

- More comprehensive instructions should be provided for storage prior to, and following erection. These instructions should include:

a. Recommended frequency and materials for replacement of protective coatings and desiccants.

b. Frequency and extent of rotor rotation; recommended bearing lubrication during these rotations.

- Parts which are particularly susceptible to damage or adjustments which may change during shipment, should be identified for the guidance of the user's site inspectors. Late identification of damage results in costly delays.

We frequently request several copies of a single, large-scale, cross-sectional drawing of the machine. This drawing is invaluable for group training and maintenance planning. It is easier to procure such items if they are specified prior to issuance of the purchase order.

These and other recommendations are summarized on Table 4.

Field disassembly and run-in. Differences in the effectiveness of the manufacturer's quality control, and in our own field craftsmen and inspectors can significantly alter the optimum extent of field disassembly and inspection.

In general, we believe the craft skills and the facilities available in the manufacturer's shop surpass those in the

TABLE 5—Contents of field handling and installation guidelines

1.0 Objectives and Scope
- 1.1 Guidelines; not instructions.
- 1.2 Only authority for general exceptions to manufacturer's instructions.

2.0 Receipt and Unloading
- 2.1 Lifting: Use of lugs, respect for shaft extensions and other projections.
- 2.2 Maintenance of integrity of packaging; damage reports.

3.0 Intermediate Storage
- 3.1 Area preparation, soil treatment, grading.
- 3.2 Rotor rotation, oil levels.
- 3.3 Records.

4.0 Placement on Foundation
- 4.1 Anchor bolt templates and cleanliness of sleeves.
- 4.2 Foundation laitance removal.
- 4.3 Soleplate and shim quality.

5.0 Erection
- 5.1 Early cleaning and commissioning of lube system.
- 5.2 Grouting, emphasis on: selection of materials, machine base cleanliness, protection, preliminary alignment, wedge removal, bolt tension, and dowels.

6.0 Piping
- 6.1 Precautions on removal of flange covers, fit up of piping anchors and guides, and pipe cleaning.
- 6.2 Suction screens installation and removal.
- 6.3 Lube and seal oil system.
  - 6.6.1 Initial flush—bearing bypass.
  - 6.6.2 Final check.

field—particularly in grass-roots construction projects. Accordingly, we prefer to reduce field work by achieving the highest practical degree of confidence that the machine is properly assembled, cleaned and protected before leaving the shop. Field examination may then be limited to items such as bearings, seals and lube systems. Our previously cited Q/A requirements are an important step in this direction.

The manufacturer's installation instructions are key documents for achieving a successful installation. The site engineer and manufacturer's erector should review the instructions in detail with the craftsmen. A plan of attack, with predefined inspection points and records, should be developed. Any changes or exceptions to the manufacturer's instructions should, of course, be cleared through the erector. Serious oversight and unnecessary compromise of the guarantee can be avoided.

Supplemental guidelines for field handling and installation have been developed. These will alert field personnel to critical areas and achieve the highest practical level of consistency compatible with the varying instructions provided by machinery manufacturers. Subjects covered are shown in Table 5.

Several points from these guidelines should be reemphasized:

1. Manufacturer's instructions should be reviewed in depth.

2. Conflicts or contradictions must be resolved jointly between the user's engineers and the manufacturer's erector.

TABLE 6—Erector responsibilities

1.0 To Inspect:
- 1.1 Condition on receipt at site.
- 1.2 Condition on opening.
- 1.3 Final alignment.
- 1.4 Completed installation.

2.0 To Advise and Instruct on:
- 2.1 Handling and storage procedures.
- 2.2 Erection and alignment methods.
- 2.3 Commissioning and startup procedures.
- 2.4 Mechanical and hydraulic performance problems.
- 2.5 Needed spare parts, special tools or instruments.

3.0 To Monitor and Report on:
- 3.1 Equipment physical damage.
- 3.2 Craft competence.
- 3.3 Unsatisfactory mechanical procedures.
- 3.4 Unsatisfactory operating conditions, controls or practices.

4.0 To Approve:
- 4.1 Work in progress.
- 4.2 The completed installation.

5.0 To provide an effective and authoritative communication channel between the manufacturer and the user on mechanical (as opposed to commercial) considerations.

TABLE 7—Erection date

1.0 Machinery Mounting
- 1.1 Grouting and shim materials.
- 1.2 Anchor bolt torque values, if specified.

2.0 Internal Clearances and Fits
- 2.1 Journal and thrust bearing clearances.
- 2.2 Seal clearance where possible.

3.0 Alignment
- 3.1 Bearing pedestal fit and alignment.
- 3.2 Coupling face angularity.
- 3.3 Coupling rim alignment.

4.0 Rotor
- 4.1 Rotor number.
- 4.2 Slow roll run-out.
- 4.3 Initial vibration (in operation).

5.0 Piping
- 5.1 Spring support deflection.
- 5.2 Cold spring (x, y, z direction).

6.0 Gears
- 6.1 Journal and thrust bearing clearance.
- 6.2 Tooth contact.
- 6.3 Tangential and axial tooth backlash.

7.0 Motor
- 7.1 Rotor/stator air gap.
- 7.2 No-load vibration.
- 7.3 De-energized coast down vibration.

3. Protection of the machine is not relaxed during erection.

4. Complete logs are maintained.

I have suggested ways by which the manufacturer may improve his communications and advices to the user. This leads to the role of the manufacturer's erector.

Primary responsibilities of the erector are listed in Table 6. It is apparent that his role is important and demanding. The following qualifications and preparations will enable him to fulfill it satisfactorily.

- The erector should be knowledgeable regarding the machinery involved, its manufacturing and test history, sub-suppliers, and contents of the manuals. This requires the manufacturer to assign his erector early and to thoroughly brief him before he reports on site. The erector should not be indoctrinated at the expense of construction time and personnel.

• The erector should be experienced and highly qualified mechanically, should understand labor relations and should be skilled in human relationships. Good communications and mutual respect between the erector and site personnel are essential.

Our specifications now require that a resumé of several probable manufacturer's erectors be submitted for review. This information, and the recommendations of our machinery specialists, will be forwarded to site management.

• Timing of the erector's services varies with the critical process requirements and complexity of the machine. Opening the machine is avoided until it is ready for alignment. The service may then be held to two visits: preliminary alignment and grouting (including piping fit-up); and final alignment, oil system flush, and run-in.

The erector can play a key role in maintaining the integrity of the guarantee. This is especially true in cases where machinery has been in place for long periods before startup. Then the erector is frequently called upon to visit, inspect and certify *as new* condition prior to startup.

Erection and run-in data. I have previously cited certain data which have been collected through our Quality Assurance Program. This data, of course, constitutes a record of critical dimensions, clearances, etc., as the machine left the shop.

Table 7 shows data which should be logged prior to commissioning. Some of this appears rather commonplace —but may not be after several years and turnover of personnel.

Specify Shop Testing to Improve Compressor Reliability

Douglas F. Neale, Union Carbide Corp., Chemicals and Plastics, South Charleston, W.Va.

Compressors purchased under API Standard 617[1] will receive the mechanical running test defined by the Standard. Although the user may specify modifications or additions to this test, his major choice is limited to the additional purchase of a shop thermodynamic performance test. Under special circumstances, a field performance test may be chosen in preference to the shop test.

MECHANICAL RUNNING TESTS

A compressor mechanical running test is a rotor spin test which evaluates the dynamic characteristics of the rotor-bearing system and the hydraulic integrity of the contract shaft seals. Capacity or efficiency of the compressor is normally not even roughly checked.

The mechanical running test defined by API-617 calls for operation of the compressor at speeds from 0 to 110 percent of the maximum continuous speed with a four-hour run at the maximum continuous speed.

Rotor vibration is recorded throughout the compressor's operating speed. While operating at maximum continuous speed a vibration amplitude sweep is made through a frequency range of 0.25 times synchronous to twice vane passing frequency. When specified, frequency, peak-to-peak displacement, critical speed and phase shift are measured and recorded. Spare rotors are supposed to be run in exactly the same way.

When design considerations permit, the mechanical test is run with the shaft seals in place. Seal performance is checked with approximate design differential pressures across the seals. Manufacturer's exceptions to the Standard as well as special user additions require considerations which are discussed in the following sections.

Test plan. User specifications should require the manufacturer's detailed test plans for review prior to the test with all exceptions to the API Standard highlighted. A good plan includes a description of the test set-up and instrumentation, test procedures, test conditions, an outline of the proposed report and report forms.

Shaft seals. A typical exception to API-617 is the condition under which shaft seals are checked. To stay within shop driver power capability, the four-hour, maximum continuous speed mechanical running test may be performed with an evacuated nitrogen or helium atmosphere. Casing pressure during the run is too low to permit the required full differential pressure seal leakage test. Seal leakage is checked later with the compressor at full pressure but running at a sometimes arbitrarily selected reduced speed.

Failure to fully simulate seal design operating conditions may help explain why seal performance during plant startup may not duplicate the apparently satisfactory test performance. Therefore, a two-phase seal test is recommended (and cited in the manufacturer's test plan) when API-617 requirements cannot be satisfied in a single test. The first check should be made at *design maximum continuous speed* and the maximum seal differential pressure permitted by driver power limitations, and the second check at seal *design differential pressure* and the maximum speed permitted by the driver.

Standard 617 requires bearing and seal oil temperatures and pressures together with seal oil in-leakage be measured and reported. Our specifications add the documentation of bearing oil flows. We have not previously been concerned about seal out-leakage. During a recent test, however, the capacity of the shop oil system exceeded the compressor design seal-oil requirement. Seal oil overflowed the shop seal-oil drainage system and led to identification of excessive oil out-leakage because of a seal design error. If the excess leakage had gone undetected, a serious delay in startup would have resulted as the leakage exceeded the contract seal oil pump capacity.

Gas temperature rise. The evacuated helium tests, frequently undertaken to reduce shop driver power requirements eliminate the development of meaningful head and temperature rise across the compressor. When flow is permitted, however, as in a throttled suction air run, forewarning of a mechanical or aerodynamic deficiency may be measured by evaluating air temperature rise. Although a crude indicator, much higher than predicted rise may indicate a mechanical deficiency (excessive labyrinth clearance; excessive internal recycle caused by a casing or diaphragm casting error) or the need for a comprehensive thermodynamic performance test.

Thrust bearing alignment and runout. If minimized during mechanical running tests, thermodynamic load subordinates the important consideration of thrust bearing alignment and runout under load.

If a thrust bearing housing is misbored, rotor axial forces during the mechanical test can be insufficient to "seat" the thrust collar against the misaligned bearing. Radial deflection and vibration may be observed later during operation under load. By imposing varying pressure differentials across the compressor and mapping the axial and radial vibration and phase relationships, this deficiency will likely be detected during the mechanical test. Rotor axial vibration may also indicate thrust collar runout, easily checked by instrumenting an axial position proximity probe to read out in the vibration mode during the running test.

Lube/Seal oil system. Accepted criteria for testing compressor train lube and seal oil systems is API Standard 614[2]. The Standard provides for user option on contract versus the manufacturer's shop lube system during the mechanical running test. Some users prefer to use the contract system to provide an incentive for thorough lube system cleanup, confirm operational compatibility with the machinery train and avoid costly field identification and correction of oil system problems. Costs, delivery requirements and unavailability of shop space or a pump driver power source are among the considerations which may persuade one to yield on this requirement.

Other mechanical test requirements. Other steps toward maximizing the assurances gained from mechanical spin tests include:

- Recording of radial vibration and phase relationship from end to end of the compressor and of the train, if a tandem test.
- Recording of bearing metal temperatures using the contract RTD's or thermocouples.
- Recording of test train coast-downtime of occasional value, but easily recorded.
- Verification of variable inlet guide vane operation and calibration, if applicable.
- Use of contract coupling hubs with equivalent half weight of spacers imposed—on both ends of the compressor if it is a drive-through type. Shop couplings are frequently a poor simulation of the heavy, long spacer—length couplings commonly used today.
- Always run the spare rotor.

THERMODYNAMIC PERFORMANCE TESTS

The most common performance error in centrifugal compressors is in design head. When the driver is a steam turbine, hot gas expander, or two-shaft gas turbine, a head error can usually be compensated by a speed change. The speed change has very little, if any, effect on driver efficiency in the speed range that would accommodate most compressor head errors. If the driver is an electric motor, excess compressor head must be throttled away. This is a direct loss of power. If the head is low, design capacity may not be attainable.

If the driver is a single shaft gas turbine, it may not be possible to compensate for high head by speed reduction. Even with gas turbine air bleed modulation, the turbine power output may decline faster with decreased speed than the compressor power requirement, and a reduction in compressor capacity may be required. If the head is low, specified head can probably be attained by increasing the speed within the normal + five percent margin. An increase in gas turbine speed will directly increase the mass flow of air and significantly increase fuel cost.

A recent experience illustrates the value of a performance test on a fixed speed compressor. Changes made to reduce the head of a 17,000 hp air compressor, found to be six percent high, resulted in a decrease of 550 hp. Annual power savings were estimated at about six times the cost of the test.

A specially designed "fix" was less than 25 percent effective in correcting the head of a large air compressor found to be nine percent low.

Fortunately, a duplicate compressor with spare rotor was nearing completion at that time. The manufacturer used combinations of impellers from the two rotors to make four trial fixes and performance tests before design head was attained. Had this trial and error sequence been conducted in the field using impellers that were fabricated as needed, the fix might have taken years.

Other incentives for thermodynamic performance tests include:

- Strict test requirements encourage greater emphasis on design and manufacturing quality control.
- Imposition of aerodynamic loads achives mechanical verifications beyond those normally achieved during no-load mechanical testing.
- High energy costs of large compressor trains create justifiable concern about even fractional percentage points of thermodynamic efficiency.

SHOP PERFORMANCE TESTING

The very complete and authoritative PTC-10 Power Test Codes[3] do not eliminate the need for user review and acceptance of test plans. Options available under the Code, exceptions taken by manufacturers and the complexity of properly simulating design point operating conditions for multi-stage, multi-sidestream compressors make detailed review and mutual agreement on certain key points essential. Some of these points are discussed in following sections.

Class of test. Class I[3] tests are run with the specified gas at the speed, inlet pressure, inlet temperature and cooling for which the compressor is designed. Class II[3] tests (calculation on the basis of perfect gas laws generally permissible), and Class III[3] tests (calculated on basis of "real gas" equations) are those in which the specified gas and/or operating conditions cannot be duplicated on the test stand.

Class I tests are nearly always preferred but may not be practical. For example, when the specified gases are not available, power and cooling required for compression exceeds shop capabilities and safety considerations (e.g., toxicity, flammability, oil in the air of a high pressure closed loop air test) are unacceptable. Class I tests are not used. As a result API-617 permits Classes II and III tests. The manufacturer and user should reach an agreement on the class, the test gas and the basis for thermodynamic matching and determination of gas properties to be used in calculating test results.

Open versus closed loop. An open test has the advantage of low cost, satisfactory safety and well-established physical properties of air to calculate test results. For substitute gases, however, a closed loop test is necessary. A closed loop test is potentially more accurate, as the ability to vary gas mixture and loop conditions may permit the very important parameters of design and test volume ratios and machine Mach numbers to be more closely matched. The closed loop also permits the elevation of suction pressures and temperatures which may be required for a Class I test.

Selection of test gas. An open test may satisfy the requirements of known physical properties, safety and thermodynamic matching when the specified gas is air (e.g., in a Class I test), a hydrogen-hydrocarbon mixture, natural gas, or other light hydrocarbon. Helium and nitrogen are commonly used in closed loop tests on light gas machines. With heavy specified gases, such as propane and chlorine, the equivalent test speed to satisfy design volume ratios and Mach number with air typically exceeds design speed. To avoid excessive wheel stresses, heavier gases such as CO_2, a fluorocarbon refrigerant, or a fluorocarbon-nitrogen mixture may be used.

Test data acquisition and accuracy. The Test Code[3] does not establish tolerances on the specified performance of compressors or, except in limited instances, define permissible error in test data. The Code does categorize tests by class and defines facilities and practices expected to provide reliable results from those tests.

The Code is rigorous regarding the location of test points, and minimum number of readings, stability and duration of test run for each point.

The Code requires that "Initial calibration of the instruments involved shall be made immediately prior to the test. Recalibration after the test shall be made on those instruments of primary importance which are liable to variation as a result of use." The type of instruments to be used as standards and for test data acquisition is also defined in the Code.

Test suction pressure. It is common practice to air test at low power by reducing suction pressure and exhausting to the atmosphere. Despite some uncertainty in the correction to be used for significant differences between design and actual Reynolds numbers and friction factors, an uncorrected error is usually in the user's favor. For example, arbitrarily select an air compressor (pressure ratio equals 3.0). It will show over twice the error in head, from a given absolute error in suction pressure measurement, when tested as an exhauster than when tested as an atmospheric air compressor.

Shaft power determination. The Power Test Codes permit direct determination of compressor shaft power input by torque meter or reaction mounted drivers, or indirect determination by heat balance, heat input to a loop cooler or by measurement of electrical input to a drive motor. The most common approach is an abbreviation of the heat balance method wherein the gas horsepower" is calculated from the enthalpy rise of the gas. This method ignores bearing friction, shaft seal leakage and casing heat loss.

Currently available optical-type torque meters built into shop test-coupling spacer tubes are proving to be reliable and capable of retaining calibration. Properly instrumented strain gages with a transmitter system provide a very acceptable alternative and may make it easier to use the contract coupling hub and simulate coupling spacer half-weight. Any torque measurement approach, however, gives little insight into the nature of a thermodynamic performance deficiency if it exists.

System mechanical losses. PTC-10 specifies that "the heat equivalent of the mechanical losses (bearings) shall be determined from the temperature rise of the lubricating oil" and that the "heat equivalent of the . . . friction loss in mechanical seals, if used, shall be determined and included in the mechanical losses." A review of eight manufacturer's test reports showed wide variations in practice. In several reports mechanical losses were ignored; only "gas horsepower" and sometimes "hydraulic efficiency" (a term not defined in the Code) were reported. In other reports mechanical losses were calculated from "standard" data or an oil heat balance. Loss values varied from one to over four percent of total shaft horsepower and bore no consistent relationship to compressor running speed or size. Mechanical losses, evidently, may be significant in relationship to the four percent horsepower guarantee of API-617. The basis for determination of mechanical losses should, therefore, be clearly established prior to the test and resultant values included in the test report.

FIELD PERFORMANCE TESTS

One or more of the following situations may require a field performance test:

- A tight construction schedule may not provide time for shop testing.
- Compressor performance is not critical during early operation.
- Thermodynamic performance of an existing compressor appears to be deteriorating, possibly from fouling, erosion or loss of design clearances.
- Power consumption of the compressor train has become more critical.
- Process conditions or gas composition have been or will be changed.

The following field test precautions can help:

- Plan in advance. When possible, design piping and control systems to accommodate the test.
- Consult with and include the manufacturer. Especially agree on a test procedure and the basis for calculation of gas thermodynamic properties.
- Carefully review PTC-10 test requirements regarding instrumentation, test conditions and allowable fluctuation of readings during the test. For critical parameters (e.g., suction and discharge pressures and temperatures, and gas flow and analyses), improve or calibrate instrumentation to the point that agreement can be reached with the manufacturer on the validity of results.

• Arrange with Operating personnel to reduce production if necessary to achieve fully stable performance. Provide sufficient assistance and/or automatic logging to permit essentially simultaneous readings of all points.
• Confirm stable operation of the entire compression cycle and associated process before proceeding. Record at least three sets of data at each point (and preferably more as designated by the Code.)
• Work closely with Operating personnel to identify and log data which, although remote from the test cycle, may assist in the evaluation of inconsistent test data.

Despite careful preparation, field tests can be limited by:

• Piping straight runs less than specified in PTC-10.
• It may not be possible to attain the design point while remaining within plant operating constraints. Control valves may not be installed to permit discharge throttling; safety considerations may prohibit suction throttling below atmospheric.
• Plant operating flexibility is insufficient to reach surge and overload.
• Special provisions necessary for shaft torque measurements are not readily applied in the plant.

LITERATURE CITED

[1] API Standard 617, "Centrifugal Compressors for General Refining Services," Third Edition, October 1973.
[2] API Standard 614, "Lubrication, Shaft-Sealing, and Control Oil Systems For Special Purpose Applications," September 1973.
[3] "Power Test Codes," ASME PTC-10, 1965.

ACKNOWLEDGMENT

Originally presented as part of a panel discussion on centrifugal compressor testing ASME Petroleum Mechanical Engineering Conference, Tulsa, Okla., Sept., 1975.

Electric Motor Drivers for Centrifugal Compressors

Electric motors are easy to apply as compressor drivers. With their high efficiency they fit today's need to save energy. Follow these guidelines for a reliable and successful application.

J. C. Moore, General Electric Co., Schenectady, N.Y.

In the past, the decision to use electric motors for centrifugal compressor drives has been made only after all other possibilities to use either a steam or gas turbine have been exhausted—or so it has seemed to motor suppliers. This picture may, however, be changing as a result of the increase in fuel costs, particularly petroleum. In addition to the considerations of heat balance and investment cost, a third factor—the availability of feed stock—must be considered. Why burn a part of the needed hydrocarbons when electric power generated from coal, hydro or nuclear energy can be purchased with a resultant saving of feed stock to make more product? Obviously, with these alternatives this should not occur—particularly when electric motors of high efficiency and reliability are available to drive almost any configuration of centrifugal compressor.

CORRECT MOTOR SIZE SELECTION

Nothing is more useless than an installed motor that is too small to start the machine to which it is coupled. There is no excuse for this problem because the widespread use of computers by machine designers makes it practical to predict accurately brake horsepower requirements of the compressor. It is easy to avoid this problem because motors are available in standard ratings in larger horsepower sizes in average increments of about 15%. Therefore, a practice of selecting the standard motor rating that matches or exceeds the brake horsepower of the driven machine will result in 0 to 15% margin.

The practice of adding 10% to the calculated brake horsepower and then using the next larger standard horsepower motor will result in margins of 10 to 25%.

With this wide selection of standard ratings, it is easy to find a good match between driven machine size and motor rating without the use of continuous overload capabilities.

Past practice has been to use motors with an overload capability of 115%. This can reduce investment cost if the machine is to be operated at 115% load. The motor life, however, will be shortened because temperature rise at 115% load is 90°C. Conversely, the life of the same motor will be increased if this motor is operated at 100% load or approximately 70°C rise.

CORRECT MOTOR TORQUE SELECTION

Torque selection would be easier if a simple match between full-load brake horsepower and motor rating were the whole story. To get the benefits of the motor's simplicity, the plant designer must understand its peculiarities. When a motor is starting, it draws four to seven times its full-load current. This depresses the voltage, particularly on large motors, and results in a reduction of motor starting torque by the square of the voltage drop. If voltage drops to 80%, the torque drops to 64% $[(80\%)^2 = 64\%]$. This will be no problem if the designer has planned for this condition and matched the motor speed torque curve at reduced voltage to that required by the compressor.

It is good practice to apply a motor with torque capability at the voltage which the system drops to when the motor is being started that is at least 10 percent greater than the torque required by the driven machine at all points between zero and full speed. See Fig. 1. A motor with special torque characteristics, a stiffer power system or a method of unloading the driven machine may be required. On some very large motors, all three of these may be required to ensure a drive that will operate.

MOTOR VOLTAGE AND STARTING METHOD

Selection of proper motor voltage is usually determined by the available electrical system and its size. The most frequently used voltage for 500-4,000-hp motors is 4,160

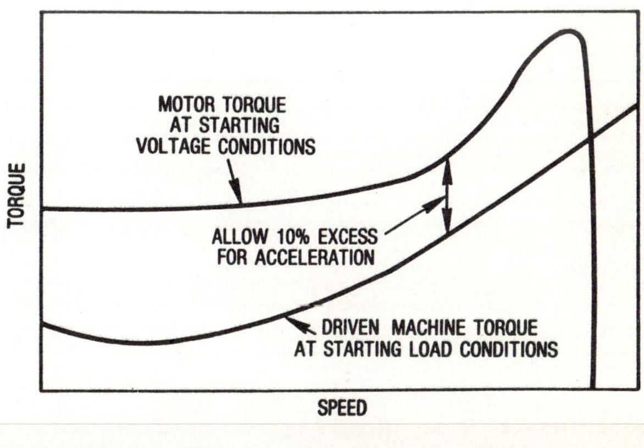

Fig. 1—Typical motor vs. driven machine speed-torque curves.

volts. On large electrical systems, motors 4,000 hp and larger may have to be rated 6,600 or 13,200 volts because of available switchgear.

When motors are started, they draw high inrush current that depresses line voltage. The amount of voltage drop depends upon the size of the electrical system. Standards usually indicate this voltage dip should be less than 10 percent; however, there are many good systems where the motor driving a compressor is the largest in the plant and where voltage dip is 20 to 30 percent or more.

As pointed out previously, motor torque varies as the square of the applied voltage. This reduced available torque usually requires that motor-driven compressors be started with inlet valves or guide vanes closed. Most compressors will operate under these conditions for a short time (usually 60 to 120 seconds) without damage from heating or surge until the motor reaches full speed.

In many cases, compressor breakaway torque is 15 to 20 percent and as the compressor comes up to speed, it evacuates itself until the torque at full speed is 15 to 20 percent. These curves typically are as shown in Fig. 2.

With low starting torque requirements, it is possible on large ratings to design special low-inrush low-torque motors that will cause less voltage drop when starting.

The three commonly used methods of reduced-voltage starting in use at present are: reactor start, auto-transformer start and unit transformer start. The most popular of the three is the unit (or captive) transformer start because it does not require extra switching, can be designed to meet individual system requirements simply by selecting the transformer impedance and can permit use of medium or small motors on 13,800-volt systems without penalizing motor characteristics. Reduced voltage starting is never a drive requirement, but is a requirement to prevent or limit electrical system disturbance.

WK² OF THE COMPRESSOR

Energy required to accelerate the inertia of the driven machine shows up as heat in the motor rotor as the machine comes up to speed in exactly the same way the kinetic energy of a moving automobile is converted to heat in the brake linings as the automobile is stopped. Frequent stops or too heavily loaded stops of an automobile will cause brake linings to overheat and fail. In a similar fashion, starting the motor too frequently or starting a driven machine with a load inertia larger than the motor was designed for, will cause motor rotor winding to overheat and fail. Here the analogy stops for, if the motor is properly matched to the load inertia, the rotor winding will last indefinitely. Industry standards have been established for the load WK^2 capability of the standard motor. These values are high enough for most centrifugal compressor drives. Many high-speed compressors require step-up gears. The WK^2 at the motor shaft is proportional to the square of the gear ratio. For this reason, special motor characteristics may be required.

MOTOR SPEED AND TYPE

Motor speeds. Centrifugal compressors of higher speeds are being used more frequently than ever before. The top motor speed available is 3,600 rpm on a 60-Hertz system.

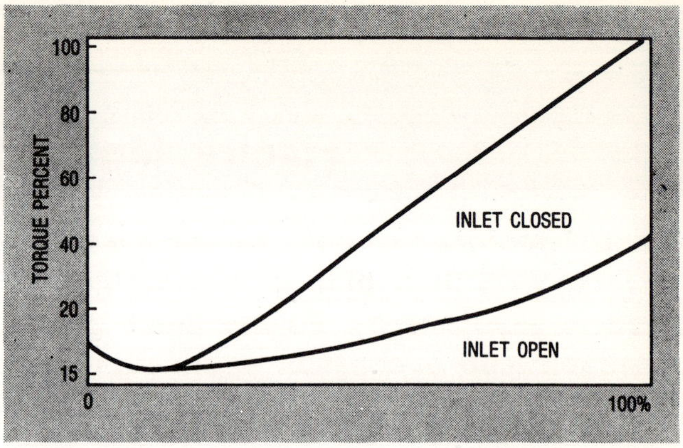

Fig. 2—Typical compressor speed-torque curve.

Fig. 3—Four 45,000 hp wound rotor induction motors connected in tandem.

To meet requirement speeds higher than 3,600 rpm, it is necessary to use step-up gears.

Introduction of a step-up gear permits a normal induction motor speed of 1,800 rpm. For 500 to 20,000-hp drives, 1,800 rpm is the lowest first-cost induction motor. Similarly, the lowest first cost synchronous motor speed is 1,200 rpm in sizes from 5,000 hp and up. There is really no size limit when building a motor. Four 45,000-hp motors (Fig. 3) coupled in tandem to drive a 180,000-hp axial-flow compressor at Ames Aeronautical Laboratory, Moffet Field, Calif., have been operating for many years.

Motor types. There are three types of motors available for compressor drives:

1. Induction
2. Synchronous
3. Wound rotor induction

The induction motor is the first choice for drives from 500 to 5,000 hp because it has one insulated stator winding and one uninsulated, shorted rotor winding. For 1,200, 1,800 and 3,600-rpm compressors, no speed increaser is required. The 1,800-rpm motor is the least expensive of the three speeds. Therefore, this speed is usually selected for higher speeds using step-up gears.

The synchronous motor is usually the primary choice for very large ratings because of price when power factor and efficiency are prime considerations. These motors have three windings: one insulated winding on the stator and two windings on the rotor—one insulated and one uninsulated amortisseur winding that is used for starting only.

The insulated synchronous motor rotor winding requires a DC power supply that is usually supplied from a brushless exciter mounted on the motor shaft.

Synchronous motors are available at the same speeds as induction motors. However, in practice only the 1,200-rpm speed is used. Step-up gears are always used to match compressor speed requirements. Control and operation of synchronous motors require more devices, but these are well understood and are very reliable. The system is such that the operator controls the machine in the same way he controls an induction motor.

Synchronous drives are more efficient, can operate at 1.0 pf or 0.8 pf, and are lower in first cost in the larger horsepower sizes. Recent changes in motor prices have altered specific relationships of synchronous versus induction motor prices, but a general relationship continues to exist as shown below:

For geared drives	Lowest first cost Induction 1,800 rpm	 Synchronous 1,200 rpm
1. 500-15,000 hp	X	
2. 6,000-50,000 hp		X

The overlap between 6,000 hp and 15,000 hp is intentional because specific drive characteristics and requirements will change the choice.

For very large drives above 30,000 hp, use of non-self-starting synchronous motors at 3,600 rpm is possible. These are synchronous generators that must be brought up to speed with a starting turbine or motor and synchronized with the system and operated as a motor. Obviously, it is a requirement on these drives that the compressor be unloaded to 15 to 20 percent of full load to keep starting equipment size reasonable.

Wound-rotor induction motors have not been used for compressor drives because of their cost and high losses when resistors are used for speed control. One exception is very large drives used in wind tunnels where there is no other way to control the compressor output except by speed control.

Increased cost of energy and availability of thyristors in large power sizes are causing a new interest in the wound-rotor motor because instead of discarding slip power as heat at lower than synchronous speed, this power is "recycled" by solid-state rectifier inverter techniques and put back into the power system (see Fig. 4). On applications where speed control is required (and particularly where load cycle requires operation at reduced speed for long periods), this drive results in relatively high efficiency over the entire speed range.

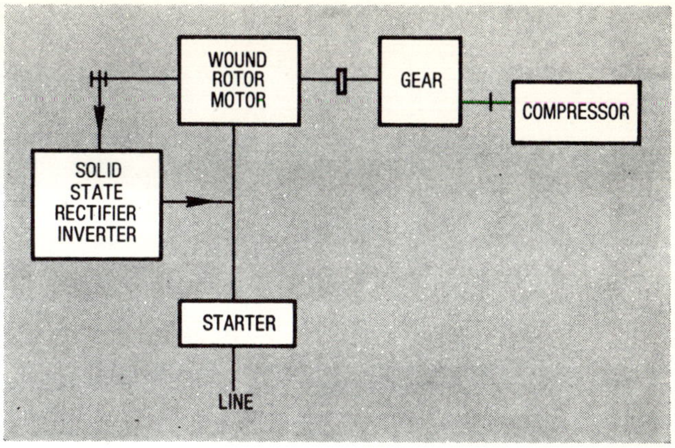

Fig. 4—Slip frequency is recycled for high efficiency at reduced speed.

MOTOR ENCLOSURES

Motor winding life can be substantially increased by the proper choice of enclosure.

- Open dripproof—for use in indoor, relatively clean locations.

- Weather Protected II (WP II)—for use in outdoor locations. This enclosure can easily be modified to include filters when used in extremely dirty locations.

- Totally Enclosed, Water-Air-Cooled (TEWAC)—for use in outdoor or indoor, clean or dirty locations. Since heat from motor losses is taken away by water, motors having this enclosure are well suited for installation in small rooms.

- Totally Enclosed, Fan-Cooled (TEFC)—These machines are arranged so that heat is transferred from the motor to the ambient by utilizing an air-to-air heat exchanger either mounted on the motor frame or built as a part of the motor with fins cast into the end-shields or frames.

- Hazardous locations—This requirement must be met by pressurizing the motor enclosure with clean air or inert gas. Explosion-proof motors are not generally available above 500 hp.

INSULATION SYSTEMS

The insulation system used on most motors in the 500- to 20,000-hp size is rated Class B and is based on mica and epoxy. Mica has long been referred to as the standard for dielectric capability, particularly in the area of corona resistance. Mica is applied to the coils, and then by use of various vacuum pressure methods (see Fig. 5), the coils are impregnated with epoxy which is highly resistant to moisture. Epoxy is inert to most chemicals, has excellent mechanical characteristics, and because of the impregnation process, serves to enhance mechanical strength of the

Fig. 5—Complete stator being placed in vacuum pressure impregnation tank.

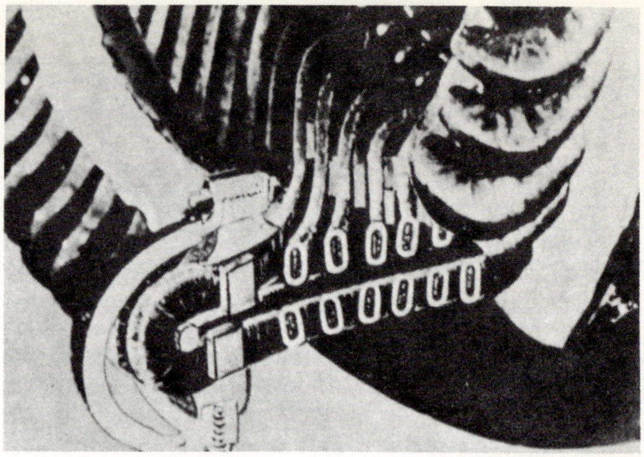

Fig. 6—Bracing system secures end turns to limit coil movement during starting.

winding by completely filling it and bonding it together. The mechanical strength of the epoxy mica system is great. To it have been added bracing systems like the system shown in Fig. 6, utilizing polyester glass materials to mold a support system designed to hold motor end turns practically immobile during high-current starting conditions. These systems virtually eliminate coil movement and chafing between turns which have been a primary cause of winding failure in the past.

Class B insulation, normally furnished with NEMA standard 100 percent rated motors, operating at 80°C rise by resistance in a 40°C ambient, has a design life of approximately 15 to 20 years, which matches normal plant life. Because the rate of an oxidation reaction is doubled for each 8-12°C rise in temperature, the life of a motor insulating system is cut in half for each 8-12°C the total temperature of the winding is increased. Therefore, another way of applying motors more conservatively is to specify a lower than standard temperature rise.

Because of torque limitations discussed previously, in practice lower temperature rise can better be secured by specifying an oversize motor. This procedure does not penalize motor operating characteristics, power factor and efficiencies, if no lower than ¾ load operation is contemplated. Most motors will operate at approximately 55°C rise at 75 percent load. For temperature rise at loads in between 75 percent and full load, the curve, Fig. 7, is typical.

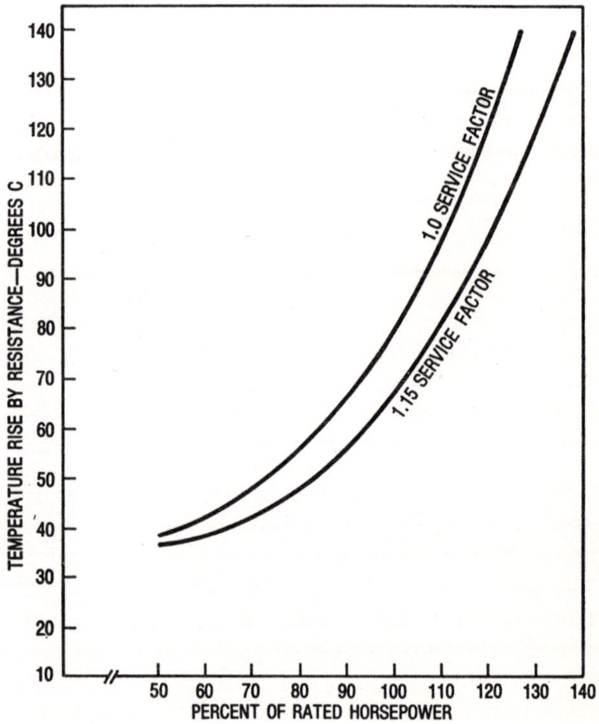

Fig. 7—Typical temperature rise vs. percent of rated horsepower curve.

CONTROL AND PROTECTIVE EQUIPMENT

Functions provided by the controls should include those listed in Table 1. In addition, the following will give the motor greater protection and should be used on large and critical drives:

Temperature detectors (stator winding). Temperature detectors in the stator windings are available on all motors being considered. With relays installed in the control, the signal from these detectors can be made to sound an alarm when there is an abnormal change in motor

TABLE 1

Function	Ind. motor	Syn. motor
1. Stall protection	X	X
2. Running overload protection	X	X
3. Protection against single-phase operation	X	X
4. Protection against failure to synchronize	..	X
5. Protection against loss of excitation	..	X
6. Protection against operation out of synchronism	..	X

winding temperature. This signal can be used as an indication of motor distress so that the operator can check for overload, clogged filters or other obstructions in the ventilating passages to the motor. A second relay can be arranged to shut the machine down when operating temperature exceeds the design limit by a nominal amount (approximately 10°C).

Bearing temperature detectors. Bearing temperature detectors or relays are available that will sense bearing temperature on sleeve bearings and should be used on large drives to minimize damage. Bearing temperatures usually do not change unless a loss of oil film (caused by foreign material in the oil, loss of oil supply, change in alignment, etc.) has caused bearing damage. These devices are available for all motors and should be used on all drives where downtime must be held to a minimum. High-temperature alarm and high-temperature cutoff are available and should be set at a nominal amount above normal operating temperature and at the maximum temperature permited by the design, respectively. These devices may not give sufficient warning to saving the bearings, but journal and rotor damage is prevented.

Lightning and surge protection. Lightning arrestors and surge protection should be installed on all motors rated 1,500 hp and larger or 4,000 volts and higher. It may be desirable to protect even smaller machines on critical drives. Lightning arrestors can be located in the control or at the main plant substation if there is no exposed line between this point and the motor. Surge protection must be located at the motor terminals. These surge capacitors slope off the voltage surges caused by switching other equipment in the same fashion that surge tanks or flow restricting devices soften hydraulic surges caused by valving in a hydraulic system. Analogous to the hydraulic system, it does little good to locate the surge tank very far from where the surge can cause damage. Over-size terminal boxes with lightning arrestors and surge protection capacitors are optional accessory items for large motors.

Space heaters. Space heaters should be installed on motors located out-of-doors or indoors where changes in ambient temperature, when the motor is shut down, can cause accumulation of moisture by the dew point phenomena. These heaters simply maintain motor temperature slightly above ambient so that dew will not form during temperature changes. Corrosion of mechanical parts and moisture damage to insulation is thereby minimized or eliminated.

Differential protection. Damage caused by arcing when a motor fails can be minimized by use of differential protection. This is accomplished by comparing the current in each phase of motor winding to sense any current unbalance caused by winding turn insulation failure. An alternate and less expensive system, called a ground sensor relay, is frequently used on critical drives. This device senses a line current unbalance caused by a fault to ground. Both systems are very sensitive and will limit motor punching damage from arcing to ground.

MECHANICAL CONSIDERATIONS OF MOTOR-GEAR-COMPRESSOR ARRANGEMENT

Most considerations discussed thus far are well understood by electrical engineers. However, in many cases, mechanical coupling of the motor, gear and compressor is often overlooked. The three most important factors to be considered in a discussion of mechanical coupling are torsional analysis, vibration and coupling selection. Not withstanding their importance, it is possible to deal with them in only a cursory fashion in the following sections.

TORSIONAL ANALYSIS

Mechanical consideration of the motor-gear-compressor arrangement must be carefully designed as stated by Thames and Heard.[1]

This article develops methods of calculating torsional performance of a motor, gear compressor system by the use of analog computers. However, many motor-compressor drives are put together without this type of analysis—with surprising success. It has been this author's experience that when a torsional analysis is made, either shaft sizes of the gear or motor, coupling flexibility, and occasionally the size of the gear is changed to improve torsional performance.

CONTROL OF VIBRATION AND BALANCE

General. Excessive vibration is undesirable because it imposes unnecessary forces and stresses on equipment, its foundation and nearby equipment. In addition, vibration is often the source of undesirable air-borne noise.

Excessive vibration may be produced by:

1. Excessive vibratory forces, resulting in the generation of forced vibrations of unacceptable levels

2. Resonant amplification, resulting in the generation of unacceptable vibration levels by normal vibratory forces

3. A combination of Items 1 and 2.

The vibration level experienced by an installed machine is a function of many variables. Some of these variables are controlled by the motor manufacturer; some are functions of the foundation, driven equipment, alignment, etc., and are obviously beyond the control and responsibility of the motor manufacturer. Responsibility for satisfactory mechanical operation of the motor-driven compressor system generally lies with the compressor builder or the user who is putting the system together.

System vibration at resonance. A motor and the machines it drives form a system which has many resonant frequencies. Vibration that will result from impressing periodic or vibratory forces on the system will be influenced by the relationship between impressed forcing frequencies and resonant frequencies. The desirability of preventing the coincidence of a forcing frequency with a resonant frequency is obvious. It is desirable to have the resonant frequency 25 percent above or below the forcing frequency.

Motors are designed and built so that resonant frequencies of motor components by themselves, or as an assembly, do not approach any known forcing frequency. It is possible, however, that the combination of the mass and stiffness of a motor with the mass and stiffness of a base or foundation will result in a combined resonant frequency which lies too close to a primary forcing frequency, resulting in excessive vibration. The problem of resonant amplification should be one of the concerns of those assembling and/or installing the motor-driven compressor system.

SOURCES OF MOTOR VIBRATION

Mechanical unbalance of rotor. All rotating bodies have a certain amount of unbalance and also have a point defined as residual unbalance. This point represents the practical lower limit to which balance can be refined. As standard procedure, rotors of new motors are balanced to generally accepted limits for most applications. The degree of balance a given motor has is generally defined by the amount of vibration that will be produced under certain standard test conditions. Vibration produced under the conditions is not only a function of the actual amount of unbalance in the rotor, but is also a function of the mass of the motor, the stiffness of the supports on which the motor is mounted, etc. This system is used because most people are not concerned with the number of ounce-inches unbalance the rotor of a motor may have, but rather with vibration this unbalance will produce at some measurable point on the motor.

Because, in most cases, it is not practical for a motor manufacturer to duplicate actual motor mounting conditions. NEMA (National Electrical Manufacturers Association)[2] has set up standard mounting and test procedures. For these tests, the motor is mounted on soft rubber mounts, which results in a system resonant frequency well below running speed frequency of the motor. The purpose of this mounting is to decouple the motor from foundation effects so that vibration produced by rotor unbalance, magnetic forces, etc., can be measured without inconsistent outside influences.

Factory limits for vibration produced by a motor when tested in the previously described manner, are given in Table 2.

TABLE 2—Factory-test vibration limits (in mils peak-to-peak vibration)

Motor speed (rpm)	NEMA standard	Special
3,000 and up	1.0	0.5
1,500 to 2,999	2.0	1.0
1,499 to 1,000	2.5	1.0

Note: Large 3,600-rpm motors usually require special balance.

It should be recognized that the vibration amplitude limits given in Table 2 apply to a given set of conditions of mounting, temperature, etc., for factory tests and are not necessarily the limits which can be expected under actual operating conditions. Vibration amplitudes that are produced under actual operating conditions may be more or less, depending upon a variety of factors.

Magnetic forces. Inherent in the design of electric motors are magnetic forces acting across the air gap between the rotor and stator. These forces produce the useful work performed by the motor. However, they also produce side effects that can result in undesirable vibration and noise. Magnetic flux across the air gap produces a steady force which tries to pull the rotor and stator together; the flux also produces a varying force which tries to vibrate the rotor and stator at a frequency of twice the frequency exciting the motor. These forces are fairly well neutralized by centering the rotor in the stator to produce a uniform air gap and are minimized by design and connection of the stator windings. However, the forces cannot be eliminated; those remaining are responsible for normal motor vibration and noise at twice-line frequency and the resultant harmonics.

These forces can be transmitted from the motor to the driven equipment, foundation and adjacent structures, and resonant amplification in these parts can result in excessive vibration and noise. In addition, when a motor is installed, it can be subjected to conditions that may increase twice-line frequency vibrations by increasing the amount of vibratory force and/or changing resonant frequencies, either of which may result in resonant amplification. Vibratory forces may be increased by running the motor at higher than rated voltage. Mechanical misalignment can cause shaft deflections that produce air gap dissymmetries with the resultant increase in unbalanced magnetic forces.

Resonant frequencies of major motor components can be changed by characteristics of the foundation on which the motor is installed, by type of coupling used to connect it to the driven equipment or by differential thermal expansion.

When designing a motor-driven compressor installation, the coincidence of system resonant frequencies with various forcing frequencies present in the system should be avoided. Primary forcing frequencies originating in the motor are:

- Running-speed frequency
- Line (power supply) frequency
- Twice line frequency.

SOURCES OF SYSTEM VIBRATION

Vibration of a motor-driven system is not only a result of vibratory forces produced by the motor and driven equipment, but also the dynamic response of the system. This system includes bases, foundation, piping and all associated parts and structures. The system, then, must include all components participating in the vibrations.

Primary vibratory forces associated with motor operation have been discussed, and it is beyond the scope of this article to discuss vibratory forces generated by the wide variety of compressors. Each type of compressor may produce a particular set of vibratory forces and they act on the system and create vibratory responses.

The manner and precision with which a motor is coupled to the driven equipment can greatly influence the amount of vibration produced by the motor and system. Proper selection of couplings for the particular application is important, and precise alignment between shafts is necessary to maintain vibration at acceptable levels.

The amount of vibration various types of machinery can safely be subjected to without incurring undue hazard of premature failure, safety, etc., varies greatly. This is determined not only by the type of machine, size, speed, etc., but also by the critical nature of equipment application and what other consequences may occur due to a forced shutdown or premature failure. Generally, it is up to the user to determine standards of performance.

Customers who specify vibration limits for motors generally do so in terms of mils vibration produced at the bearing housings with the motor operating at rated line voltage and frequency on standard NEMA mounts. This appears to be the most practical approach to the problem because it recognizes that the manufacturer cannot duplicate actual installation characteristics for test purposes; it also recognizes that there are other possible sources of motor vibration in addition to rotor unbalance.

COUPLINGS—CHOICE IS IMPORTANT

There are many types of couplings in use, but most can be classified as either rigid or flexible.

Flexible couplings are the type most often used with induction and synchronous motors. For compressor drives, the flexible coupling comes in many forms: gear type, flexible-disc type, pin and bushing type, spring-grid type and rubber-biscuit type.

Experience has shown there is considerable misuse of the flexible coupling. Many users have the mistaken idea that flexible couplings eliminate the need for precision line-up of equipment. This is not true! The function of flexible couplings is mainly to compensate for some minor misalignments which might occur during operation due, for example, to change in relative foot heights as motor and driven equipment heat to operating temperature.

A typical coupling arrangement is illustrated in Fig. 8. It is good practice to have a "centering" or thrust-carrying bearing as an integral part of the equipment, which determines, within fairly close limits (0.015 inch), the position of the rotating element of the driven equipment (fixed shaft in Fig. 8). If the coupling shown in this arrangement were of the rigid type, the position of the motor shaft would be set relative to its own bearings except for the axial growth of the shafts due to temperature changes or due to a change in position of the "centering" bearing. Reasonable motor bearing clearances ("A" and "B") would then protect the motor bearing from having slingers come in contact with the babbitted end of the bearing.

Flexible couplings of the disc type and/or rubber-biscuit type also limit axial movement of the shafts sufficiently to prevent motor bearing damage with normal bearing clearance.

Many designs of flexible couplings (gear type, pin and bushing type, spring-grid type) permit relative motion between the shafts. Unless this movement ("C" + "C") is limited to less than the motor bearing clearances ("A" + "B"), it is possible for shaft slingers to come in contact with the motor bearing. Axial thrust may then be transmitted through the coupling to the bearing shoulder. Experience has shown that in many applications, sufficient axial thrust to cause bearing damage may be transmitted through the flexible coupling. The limited end-float coupling is designed with typical clearances as shown in the table, Fig. 8, to prevent shaft slingers from contacting motor bearing positioning surfaces.

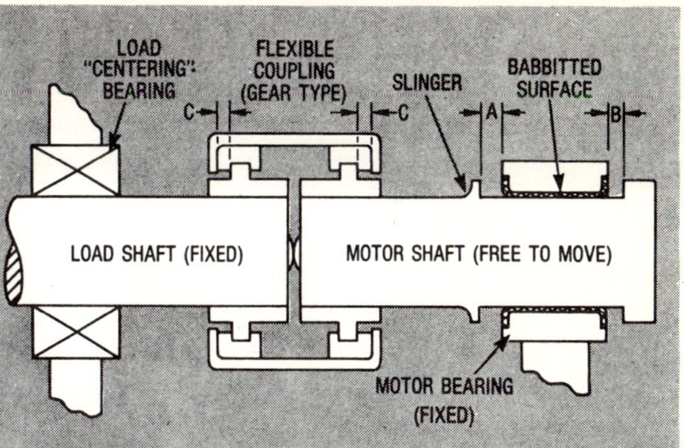

MOTOR SHAFT POSITION	MOTOR END FLOAT (INCH)	A	B	C
SHAFT ENDS TOGETHER	1/4	11/64	5/64	3/64
	1/2	11/32	5/32	3/32
SHAFT ENDS APART	1/4	5/64	11/64	0
	1/2	5/32	11/32	0

Fig. 8—Recommended clearances with limited end-float flexible coupling.

Axial thrust may be due to:

1. Gravity forces, because the rotational axis is not level
2. Forces produced within the coupling due to misalignment wear
3. Axial expansion of the shaft system and displacement of the "centering" bearing.

Regardless of how well-designed the coupling is, when it is operating under load, load torque will produce sufficient friction between coupling elements so that a fairly large axial force may be required to change axial spacing of coupling halves. This friction depends upon a number of factors such as misalignment, amount of wear on the elements and cleanliness of the assembly. Thus, a loaded coupling tends to be an axially "locked" device and will transmit thrusts produced by items (1), (2) or (3) mentioned previously.

LITERATURE CITED

[1] Thames, P. B. and Heard, T. C., "Torsional vibration in synchronous motor-geared-compressor drives," AIEE Paper 59-657 presented at the AIEE Eastern District Meeting, Baltimore, Md., May 19-21, 1959.
[3] NEMA Standards MGI-12.06.

ACKNOWLEDGMENT

This paper was originally presented at the Third Turbomachinery Symposium, Texas A&M University, College Station, Texas, October 1974.

Compressor Seal Fundamentals

Centrifugal compressors are greatly limited by the capability of shaft seals. Here is a description of the various seal types and their performance in typical process compressor service.

W. E. Nelson,
Amoco Texas Refining Co., Texas City, Texas

EVERY CENTRIFUGAL compressor must have some means of limiting or eliminating gas leakage along the shaft where it comes through the casing. This is accomplished using various types of shaft seals. Seals are usually arranged so that intake pressure must be resisted, but it may be substantially above atmospheric pressure. Centrifugal compressors are greatly limited by the capabilities of shaft seals. This article will describe various shaft seals and their performance in typical process compressor service.

API classifications. Shaft seals may be divided into the following categories by API Standard 617, "Centrifugal Compressors for General Refinery Services."[1]

- Labyrinth
- Restrictive carbon rings
- Mechanical (contact) type
- Liquid film or floating bushing type
- Liquid film-type with pumping bushing.

Some seals, mainly those used in refrigeration machines, are not classified by API. The first two seal categories are usually operated dry, when casing pressure is less than 200 psig, the outer terminal exhausts to atmosphere, the gases handled have low flammability and are non-toxic in nature. The last three categories require seal oil consoles either separately or as part of the lube system and are used in flammable and toxic gas service. None of these is a perfect seal. They are all controlled leakage devices. That is, they must allow a certain amount of process gas, seal gas or fluid, to leak past the seal points or faces to avoid contact or to permit lubrication and cooling of the faces.

Balancing pistons. A single flow compressor is unbalanced; it has inlet pressure on one side and discharge pressure on the other. Means must be provided to balance the force exerted toward the inlet side. The usual method is to install a balancing drum at the discharge end of the assembly as shown in Fig. 1. It includes a labyrinth seal to break down the compressor discharge to suction pressure.

Discharge pressure is applied to the inboard end of the balancing piston and the outlet end is piped to inlet. The differential force thus acts in an opposite direction to the forces acting on the impellers. The piston can be sized to overly or partially balance at normal conditions, but is supplemented by an axial thrust bearing that maintains proper rotor position and takes the unbalanced load.

This arrangement also permits equalizing the sealing chamber for the high-pressure end with a suction seal chamber by means of a balance line. Both shaft seals are then limited to sealing against suction pressure.

Double suction and "back-to-back" impellers (those mounted so their thrusts oppose each other) normally do not require a balancing drum. It is usually more economical to apply a balancing drum or piston whenever possible.[3]

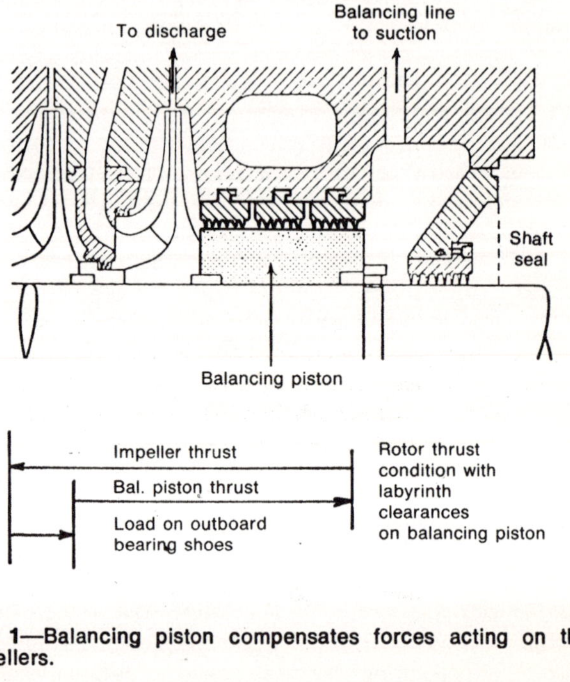

Fig. 1—Balancing piston compensates forces acting on the impellers.

Labyrinth seals are the oldest type seals, simple to build and relatively trouble-free (Fig. 2). Leakage rates are relatively high (1-2 percent of compressor flow on small units). They are used universally at interstage seal points, at the shaft ends of air blowers and compressors, and in other applications when a small loss of process gas or seal gas (air, nitrogen, etc.) may be tolerated. Leakage across a labyrinth seal depends on the number of teeth, diameter of the packing ring, clearance and pressure to be sealed. There is always leakage and care must be taken that this leakage creates no hazard. This seal can be used for a dirty gas.

The normal operating limits are about 200 psi maximum or 20 psi per inch of axial length for a straight pass labyrinth seal. This works out to be above 5 psi per tooth.

The clearance usually provided for turbomachinery shaft seals, using labyrinth blading, is maximum bearing clearance plus one mil per inch of diameter of labyrinth. Clearance is considered excessive when it exceeds two mils per inch of diameter plus the bearing clearance. Labyrinth seals on casings operating at pressures in excess of 50 psig have 8-20 blades. Lower pressure casings contain three to six blades. Straight labyrinth seals are used on an interstage shaft, on suction eye seals and on low-pressure outside seals. Staggered and tapered diameters under the labyrinths can reduce the leakage as much as 40 percent.

Many successful design variations are in use. Multiple labyrinth lands or teeth may be part of a soft stationary ring (bronze, aluminum) dovetailed into casing grooves, or the teeth may be of hard material (steel) rotating with the shaft and running against a stationary sleeve of softer material. In either case, occasional touching should increase the clearance without further harm. The second version with rotating hard teeth is popular in Europe and is gaining favor in the United States; it permits running in with almost zero clearance, digging its own clearance grooves and maintaining low leakage rates. The soft mating material (smooth or stepped in diameter) may be bronze, aluminum, a lead-tin alloy, soft cast iron, carbon-graphite or (for high temperature) stainless steel honeycomb. When the operating temperature exceeds 250° F, monel or stainless steel strips are used.[4]

A stepped tooth or staggered labyrinth design reduces leakage by over 40 percent compared to the straight pass type. The use of honeycomb labyrinths offers even better control of leakage rates (up to 60 percent reduction of a straight pass type). Honeycomb seals operate at approximately one-half the radial clearance of conventional labyrinth seals. The honeycomb structure is composed of stainless steel foil about 10 mils thick. Hexagonal-shaped cells make a reinforced structure that provides a larger number of effective throttling points (see Fig. 3). In addition, stainless steel honeycomb retains its strength at temperature and pressure levels which would cause weakening of an aluminum labyrinth.

It is possible to alter the labyrinth seal for many gas applications. The packing can be exposed to full discharge pressure and a port or lantern to bleed off leakage through the first section is connected to intake. Another sectionalizing port may be used to inject another gas at slightly higher than intake pressure. This prevents process gas leaking out, but permits injection gas leakage both to atmosphere and to intake. The operation may be reversed and the port connected to an ejector, maintaining a pressure below atmospheric. In this case, the process gas and atmospheric air would leak to the port and be removed.

These improvements are not attractive in these days of higher energy costs because even under the best of conditions the leakage rate is too high and the energy costs to operate the improvement are also costly.

Other problems also occur when using this type seal. Thermal growth of the mating surfaces must be carefully

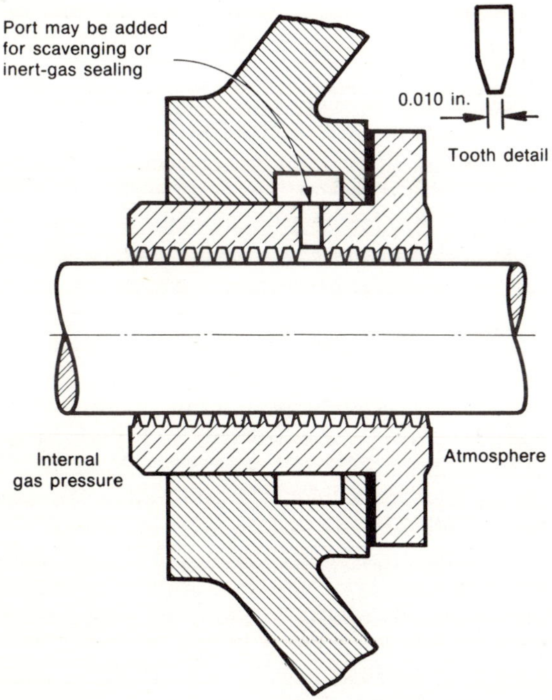

Fig. 2—Labyrinth seals are the oldest type and relatively trouble-free.[1]

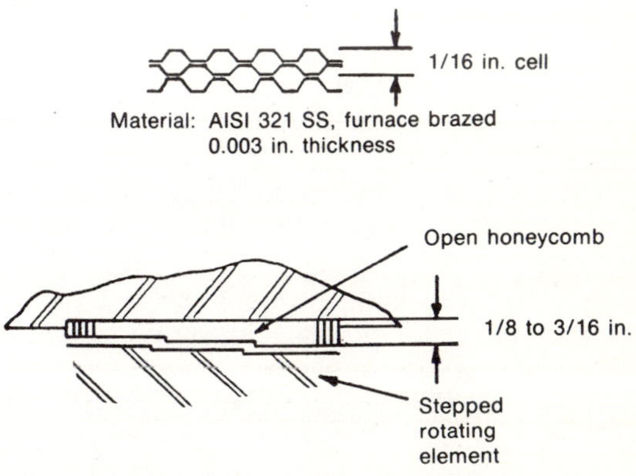

Fig. 3—Schematic of honeycomb labyrinth seal used to control leakage rates.

controlled to maintain correct clearances under all operating conditions.

Restrictive carbon ring seals operate on the principle of capillary flow through a close clearance with the sealing fluid, usually a gas or vapor. Multiple stationary carbon rings, often segmented and held together by a garter spring (as on steam turbines) are used on certain process compressors (see Fig. 4). If the seal gas contains some humidity to lubricate the carbon rings, radial clearances between the ring and shaft or shaft sleeve may be very close and leakage rates substantially lower than those of labyrinth seals of similar space requirements may be achieved. The seal may be operated dry (as in the labyrinth type), with a sealing liquid (as in the mechanical type) or with a buffer gas. An exception is where lube-oil is applied to carbon-ring seals by drop-feed lubricators. This is done to reduce the frictional heat of the higher pressure seal contacts and where the lube oil is not a nuisance.

General operating limits are about 200 psi maximum or 25 psi per *active* ring. The basic restrictions and problems cited for the labyrinth seal also apply to this type. In addition, the carbon rings can shatter easily. Its basic advantage lies in reduced leakage if proper clearances are maintained (about two mils total at operating pressures and temperatures). This seal can only be used in a relatively clean gas service.

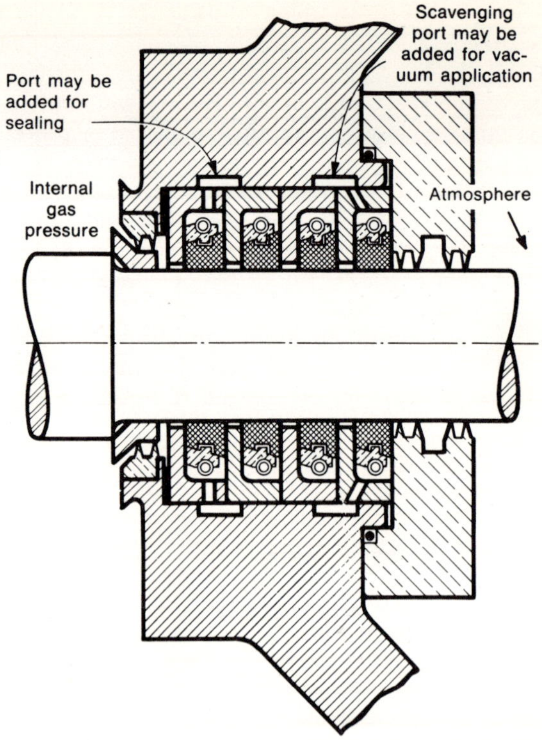

Fig. 4—Restrictive carbon ring seals used in relatively clean gas service.[1]

Liquid-film or bushing seal. The shaft seals considered thus far were concerned with relatively low pressure compressors operating at discharges of less than 200 psig, 250° F and the seals operated dry. An oil film or bushing seal is, basically, simply a close-clearance bushing surrounding the shaft. Sealing is accomplished by the sealing face between the end of the bushing and its housing, and by the restriction of the small clearance area between the bushing and the shaft. An oil-film shaft seal consists of two bushings which are pressurized with 2 to 4 gpm of seal oil at a midpoint about 5-15 psi greater than the gas chamber pressure. The inner bushing facing the gas is shorter than the outer bushing. Tight clearances limit leakage to a few gallons per day.[2] The bushings are usually made of babbitt-lined steel ¾ to 2 inches long. With 0.004 to 0.006 inches diametral shaft clearance, the bushings are floating—that is, they may follow shaft movements. At high pressures they can act as bearings and can cause instabilities or affect the rotor dynamics unless adequate precaution and care are taken in their design. Seal oil pressure is maintained at 5 to 15 psi above inboard gas pressure by controllers or by pressurized overhead tanks. Seal oil is allowed to flow in two directions: through an inboard bushing or a face seal to a high-pressure drain area and out through a trap or float valve (sour oil), and through an outboard bushing to an atmospheric drain (sweet oil). The inboard seal oil will absorb some process gas in the drain area. If the gas is sour, this trap oil must be treated or reclaimed; therefore, the inboard oil flow must be kept down to a few gallons per day. The bulk of the oil circulation passes through the outer bushing which is somewhat longer than the inner bushing to break down the greater pressure drop. The entire assembly is sealed into the casing with suitable O-rings and gaskets to avoid gas leakage (Fig. 5). The seal oil system with its controls and safety provisions is complicated and costly, but the seal, if properly applied, is free of wear and very reliable.

Sealing limits are about 3,000 psi for a basic two-ring design (one low differential, one high differential), and up to 5,000 psi for three to four ring seals (one low differential, two or three high differential).

The leakage rate will vary *directly* with the differential pressure and wetted perimeter: with the *cube* of the clearance and with the *square* of the eccentricity and will vary *inversely* with oil viscosity, temperature and length. Shear work done on the sealing fluid during its passage through the bushing raises its temperature.[6]

Clearance between the bushing and shaft, and the length of the bushing, must be selected to obtain minimum leakage without exceeding fluid temperature limitations. Flatness, parallelism and surface finish of the mating seal faces must also be carefully controlled to obtain maximum seal effectiveness. Axial clearance of the bushing rings and the housing is critical and is completely ignored by most manufacturers. There should be about 15 mils clearance per bushing ring between the ring and the housing. Where the rings are mounted back-to-back, there will be 25 to 35 mils clearance total for the seal.[5]

In higher pressure seals, more than one outboard (i.e., high differential) seal ring may be used. Generally, it is

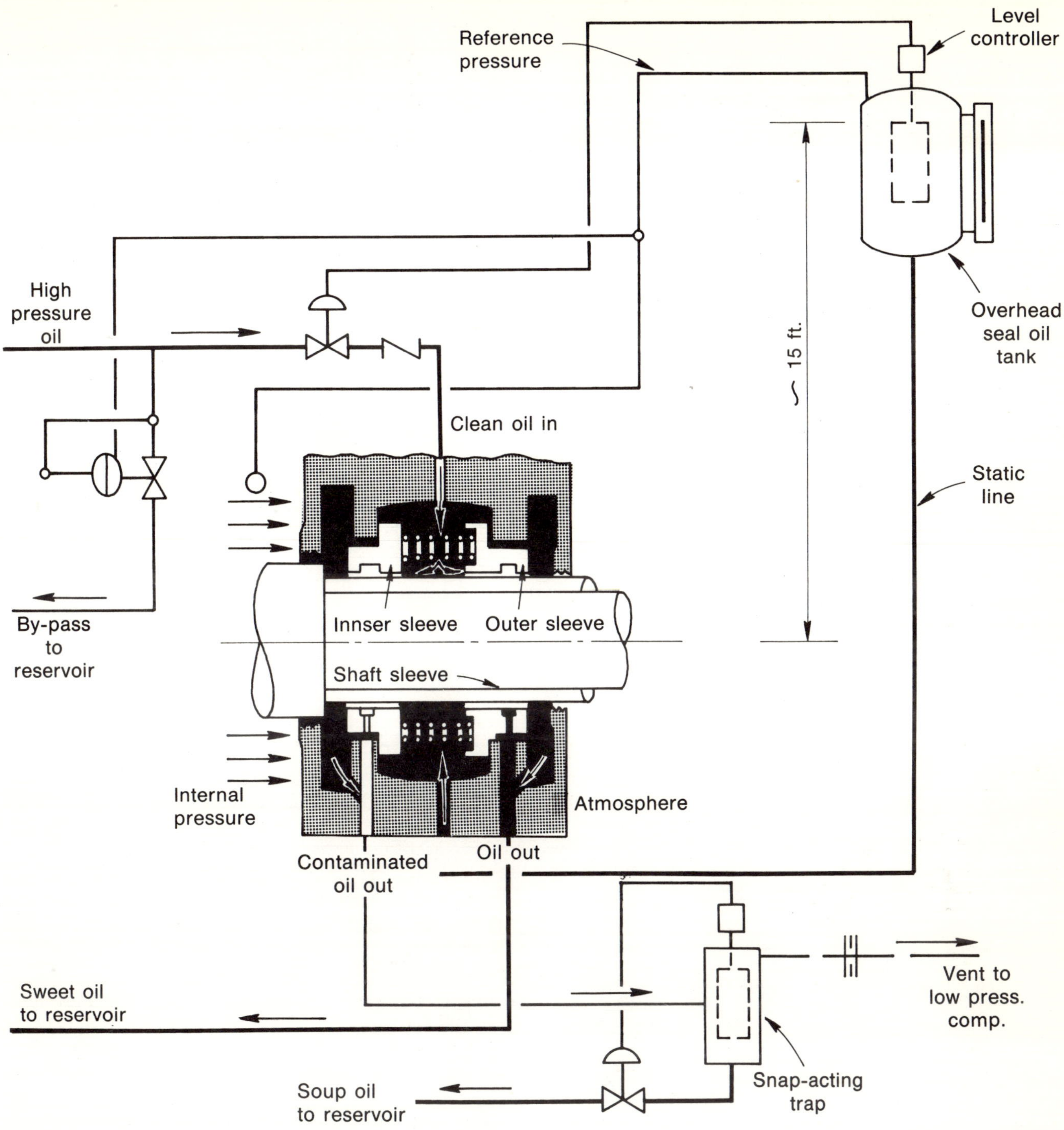

Fig. 5—Oil film type seal system for high pressure compressors.

desirable to use a single ring because the inboard ring of the two operates with up to 80 percent of the total pressure drop across it. The outer ring with the lower differential causes lubrication and cooling problems that can shorten the life of one or both rings.

Since all the oil returned to the sweet oil system passes through the bushing, the oil temperature will be higher than a bearing where only part of the oil passes through the tight bearing clearances.

Seal oil systems are covered by API Standard 614.

Mechanical contact seal. Difference between this seal and the oil-film seal are: the clearance is reduced to zero, the sealing faces are normal to the shaft instead of longitudinal and normal and the seal requires a greater differential pressure (20 to 50 psig) above the internal casing pressure.

The features of the seal are shown in Fig. 6. Unattached floating carbon rings are presumed to run at half shaft speed. Mating surfaces are lapped to two microns. Inner gland oil leakage is less than the oil-film seal; the oil

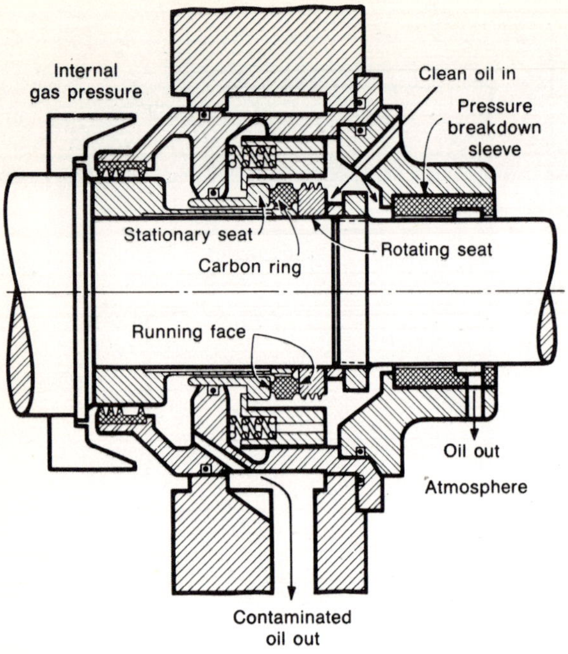

Fig. 6—Clearance is reduced to zero in the mechanical contact seal.[1]

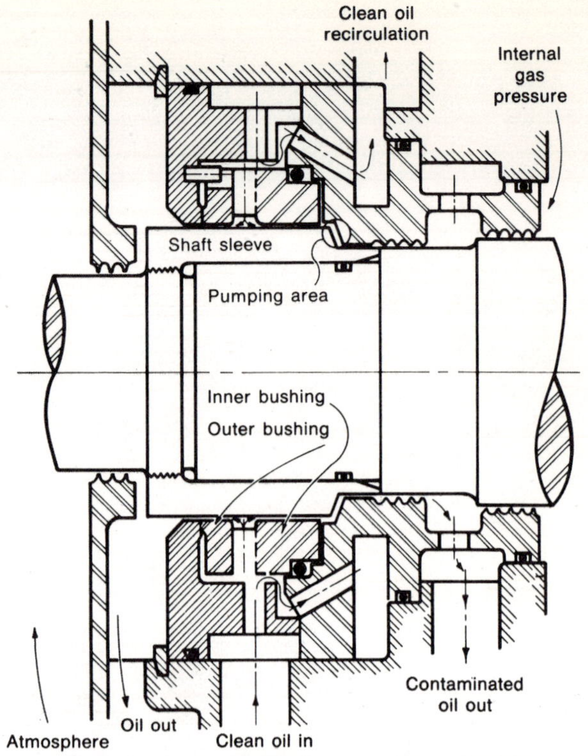

Fig. 7—Liquid film type seal with pumping bushing.[1]

supply system is less complex and it has an automatic closing device. The seal will contain the internal gas pressure when there is a loss of seal-oil pressure or at shutdown. Small auxiliary pistons are activated by internal casing pressure which maintains a contact between the stationary seat and the carbon ring to prevent gas escaping to atmosphere. Higher differential pressure on the seal oil permits higher compressor discharge pressure surges without gas blowing past the seals. An overhead tank is not required unless holding time is advisable to eliminate the need for immediate compressor shutdown on loss of the seal oil pumps.[2]

Operating limits of the contact type seal are about 500 psi. The fragile nature of the carbon ring makes the seal hazardous to use in process compressors. Other disadvantages are a result of the complex assembly which makes them hard to install correctly. Deterioration of the contact faces with operating time causes the sour oil leakage rate to increase which can cause a shutdown at inopportune times.

Effectiveness of this seal depends on keeping mating faces on the collar and seal ring parallel and flat. The collar is made of steel, specially heat-treated and lapped to a surface flatness of 15 (10^{-6}) inches. The seal ring is made of a refined grade of carbon, shrunk within a steel shroud. The ring is contoured geometrically so that dimensional stability is ensured and surface flatness preserved for long life. Secondary seals are rubber (O-rings). If the seal is operated with a seal oil system, face pressures may be several pounds per square inch. If the seal faces must run "dry" (cooled and film lubricated by a seal gas) face pressures are kept very low and carefully

TABLE 1—Leakage comparison of various seals[6]

Dry Types	Gas Leakage Index
Straight Pass Labyrinth	100
Staggered Labyrinth	56
Honeycomb Labyrinth	40
Restrictive Ring	20
Mechanical Contact (Running Dry)	2
Oil Types (Oil Loss)	
Mechanical Contact 6¾ in. Face Diameter 30 psi Differential 5,000 rpm	0.03 gal./Hr.
Liquid Film or Bushing 5½ in. Bore Diameter 0.007 in. Clearance 5,000 rpm 60°F Oil Rise	1.75 gal./hr. or 55 Times the contact type

controlled with light springs and pressure-compensated face designs. In either case, microscopic film lubrication must be established at running speeds to avoid rapid wear. Consequently even the face seals show some leakage, but at a far lower rate than bushing or labyrinth seals.

Sealing at standstill is practically perfect. This fact is one of the greatest assets of face seals. They also can use a combined lube and seal oil system.

Liquid film type with pumping bushing. This is a patented variation of the oil-film seal which uses the dynamic pumping action of the inner bushing. Instead of the straight bushing a conical sleeve and bushing or a radially stepped arrangement is substituted which has rather large clearances over the shaft. The inner seal oil-flow is impeded by the dynamic pumping action of the cone or radial steps acting as an impeller (see Fig. 7).

Note that seal oil flows through the seal oil supply area at all times which is different from the liquid film bushing. This seal is limited to sealing pressures of about 3,000 psi maximum. The oil leakage rate is less than the liquid film or bushing type but is not as low as the contact or mechanical type.

Leakage comparison. The relative effectiveness of various type seals can be seen in Table 1. This table has been modified by the author to include honeycomb labyrinths that were not covered by reference source.

LITERATURE CITED

[1] API Standard 617, 3rd Edition, "Centrifugal Compressors for General Refinery Services."
[2] Scheel, Lyman F., *Gas Machinery*, Gulf Publishing Co., Houston.
[3] Gibbs, Charles W., "Compressed Air and Gas Data," Ingersoll-Rand, New York.
[4] Fulleman, John, *Advances in Petroleum Chemistry and Refining*, Chapter 8, Wiley and Sons, New York.
[5] Elliott, W. B., Amoco, personal experiences and suggestions
[6] Gartmann, Hans, *DeLaval Engineering Handbook*, McGraw-Hill.

Compressor Lube and Seal Systems

When reliability is important in single-train compressor installations, specifications beyond those described in API Standard 614 must be considered. Here are the important lube and seal-oil system points to review.

R. E. Ruckstuhl, Humble Oil & Refining Co., Baton Rouge, La.

MINIMUM REQUIREMENTS for lube and seal-oil systems for large compressors are covered in a new API Standard 614 to be issued soon. Additional requirements should be considered to improve the reliability of single-train compressor installations designed for long, uninterrupted operation. The most important considerations are discussed in this article.

Seal buffer gas. Oil lubricated seals usually operate for longer periods, at low leakage rates, than non-oil-lubricated seals. However, combining a seal-oil system with a lube system creates several potential problem areas, particularly if the process gas is dirty. The best solution is to use a clean buffer gas in the seal area to keep the dirty gas away from the seal and seal oil.

Pumps. Three lube-oil pumps (one operating and two automatic-start spares) have been used on many new projects for greater reliability than using only one spare. The actual cost of the third pump is not great if included in the original inquiry, particularly if a third pump would have been purchased for spare parts. If two separate sources of electrical power can be provided, the normally operating pump and one spare can be motor driven. Turbine driven pumps are very expensive to operate because of poor efficiency of small steam turbines. If, however, there is a surplus of available steam, the turbine-driven pump can be the normally operating pump. If utilities are unreliable, a shaft-driven pump should be considered.

Accumulators should be provided in both the seal-oil and lube-oil systems. Seal-oil pressure must be maintained to keep process gas in the compressor if a complete utility failure stops all seal-oil pumps. A 10-minute seal-oil holdup time is desirable to allow time for corrective action. A shorter holdup time may be used for contact seals that provide an automatic pressure shutoff after coast-down. A separate accumulator is used to keep lube-oil pressure above the low oil pressure trip while the auxiliary pump is being started. Fifteen seconds holdup is probably the largest lube-oil accumulator that can be justified. The normal accumulator is not effective for lubricating bearings during coast-down after a complete utility failure. However, an overhead standpipe that will allow a small gravity lube-oil flow for a short coast-down period can be provided at very low cost.

Oil conditioner. Lube-oil systems associated with steam turbines tend to have moisture accumulation in the oil. An oil conditioner should be specified for these systems. It should be required to remove water from the oil on a continuous, or regular intermittent basis.

Piping. Stainless steel piping should be specified for the critical sections of the lube and seal-oil piping between the filters and the bearings, seals and control devices.

Incremental capacity of lube and seal-oil coolers, filters, reservoirs, pumps, piping, etc., costs very little if included in the original specifications, and can improve reliability of the auxiliary systems. As an example, an acoustical enclosure around a gear may require additional oil that may not have been included in the original oil system. Or an oil whirl problem may require lower oil temperatures than the oil coolers were originally designed to provide. Or cooling tower water may sometimes be warmer than design.

Bearing temperature instrumentation. High temperature alarms should be specified to warn of bearing failures. The sensing points should be in the bearing metal. A dial thermometer should be placed in each oil drain as a check on the thermocouples. Do not expect the oil-drain temperature to react as fast, or be nearly as accurate an indication of problems, as the bearing-metal thermocouple.

Oil temperature control. A temperature control instrument should be provided to maintain a constant lube-oil temperature. It may control the water to the lube coolers, or bypass some of the lube oil around the coolers. Once you depend on an instrument for this control, a high-temperature alarm should be provided in the oil line from the coolers to warn of controller malfunction.

Trip devices. Capability of checking trip devices, without stopping the compressor, should be specified for machines that cannot be shut down at least every two years for these checks. Specifying this feature and designing adequate safeguards to prevent both accidental trips and accidental loss of protection requires an extra effort. There should be a signal light to show when the protective devices are in the test position. It must be reset to the normal operating position after a test.

Annunciators should be specified to have a first-out feature to show which alarm signal occurred first, and which trip actually shut down the machines. The first-out feature should be designed so the acknowledge button will silence the annunciator horn without losing the first-out indication light. The annunciator may be local, or in the control house with the process instruments for which the operator is responsible. If local, one or more alarms should be brought to the control house to warn the operator of trouble at the compressor. However, someone must go to the local annunciator to determine the cause of the problem. If the annunciator is in the control house, the operator can immediately identify the problem and determine how quickly corrective action must be taken. For example, alarms for low level in the lube-oil reservoir, or a dirty lube-oil filter, do not require as rapid corrective action as major vibration.

ACKNOWLEDGMENT

This article is a portion of presentation made by the author to a panel session at the AIChE's South Texas Section, Annual Technical Meeting, Oct. 21, 1971, on "Reliability Considerations When Specifying Compressors, Drivers and Auxiliaries."

Test Compressor Performance in the Shop

Compression equipment must perform as expected. Failure to do so can result in many startup and operation difficulties. Here's what can be done to avoid problems before the machinery is installed.

Royce N. Brown, Dow Chemical U.S.A., Houston

RECENTLY there has been much written about the mechanical reliability of compressor trains. There is no question as to the importance of a high degree of mechanical reliability since it generally results in a "go" or "no go" situation. Good performance attainability is generally not as consequential, because partial load operation may be possible with a deficient compressor. A fact not generally realized is that the subject of performance is a requirement for reliability! For a machine to be truly reliable it must not only run well mechanically but must perform at 100 percent of its capacity whenever called upon. It is then very desirable to test prove performance prior to receiving a machine, or as an alternative, to field test shortly after installation prior to entering production.

ASME TEST CODE

The basis for code testing is the ASME Power Test Code PTC 10-1965 Compressors and Exhausters.[1] Several specific points made in the Code were intended as guiding principles, yet are often misunderstood. The following facts must be considered:

- The Code establishes the rules for a test, including the definitions of Code or Non-Code.
- The Code is not a textbook on testing.
- The Code has to assume that the gas properties of the gases involved are known. It recognizes that this is not always the case, but must place the burden of knowledge of gas on the test's participating parties.
- The Code establishes a basis on which to agree or disagree. Ultimately the final test procedure and methods must be agreed on by the purchaser and vendor.

The Code attempts to categorize testing and thereby establish an inherent degree of accuracy. These categories are based on methods of test and methods of analysis.

The Code establishes three classes of tests. Class I includes all tests made on the specified gas (whether treated as perfect or real) at the speed, inlet pressure, inlet temperature and cooling (if applicable) conditions for which the compressor is designed and is intended to operate, i.e., an air machine or a gas loop test. Classes II and III include all tests where the specified gas and/or the specified operating conditions cannot be met. Class II and Class III basically differ only in method of analysis of data and computation of results. The Class II test may use perfect gas laws in the calculation while Class III must use the more complex "real gas" equations. An example of a Class II test might be a suction throttled air compressor. An example of Class III test might be a CO_2 loop test of a hydrocarbon compressor. Fig. 1 shows code allowable departure from specified design parameters for Class II and III tests.

Another facet of the Code is the establishment of instrumentation for the compressor test. The type, number of points and locations on the compressor are defined and apply to all classes of Code tests. The Code is specific about the number of readings per point and the minimum duration of the run. Calculation methods are also supplied using perfect gas thermodynamic relations for Class II tests and real gas relations for Class III. Class I inherently requires little correction; but when corrections are needed, either perfect or real gas relations can be used

Variable	Symbol	Range of test values limits – % of design value	
		min.	max.
Volume ratio	qi/qd	95	105
Capacity-speed ratio	qi/N	96	104
Machine mach number	Mm		
0 to 0.8		50	105
Above 0.8		95	105
Machine Reynolds number where the design value is	Re		
Below 200,000 centrifugal		90	105
Above 200,000 centrifugal		10*	200
Below 100,000 axial compressor		90	105
Above 100,000 axial compressor		10**	200

Mechanical losses shall not exceed 10% of the total shaft power input at test conditions.

*Minimum allowable test Machine Reynolds number is 180,000.
**Minimum allowable test Machine Reynolds number is 90,000.

Fig. 1—Allowable departure from specified design parameters for Class II and Class III tests.

depending on the nature of the gas. The Code draws heavily on work presented by Schultz[6] using methods directly from his paper. Calculation methods will be further discussed later in the paper.

In actual practice very few tests are run as "True Code," or by Code definition, as Code tests. Each vendor uses some form of deviation to either speed up the test or to fit his facilities. Most of these shortcuts are not serious—in fact they contribute to keeping the cost of testing reasonable. The user must understand fully the Code requirements and where these deviations occur. An example of a nonserious deviation would be the waiver of the time limit per point, i.e., the time required for settling out. This may be based on stability of the data only, rather than data stability and a minimum time. Another deviation generally taken, which the user must evaluate, is the gage calibration procedure immediately prior to and after the test. The use of manometers keeps this requirement to a minimum. There are quite a number of the above examples, too numerous to mention, none of which generally is serious. The user should realize, however, in permitting these deviations he may not have a true code test.

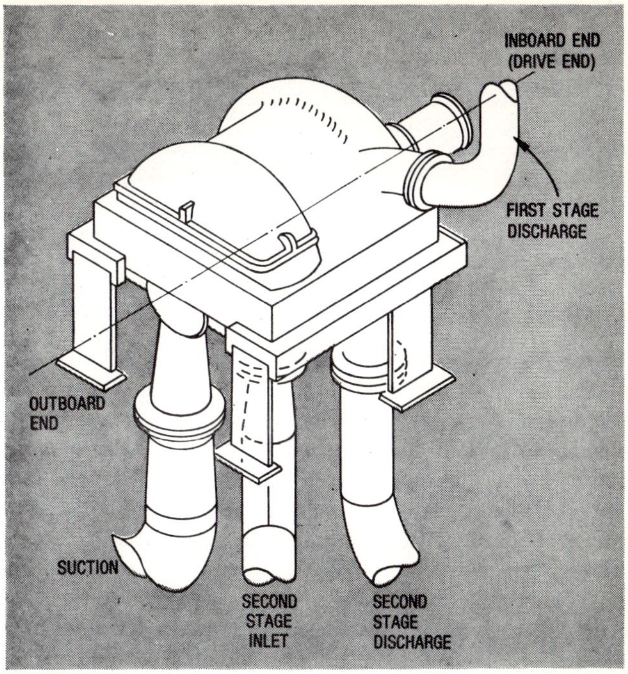

Fig. 2—Typical multi-stage centrifugal compressor.

Testing the untestables. Compressors, which fall outside of Code limits due to the nature of the gas, vendor shop test limits, or machine speed limits, may still be tested and useful information obtained for the user.

The first and most obvious question is how and to what degree one evaluates or judges the test. The most obvious reply is "it depends." On this seemingly vague note, let us explore some of the possibilities.

On a simple "once through" compressor, the gas properties may dictate a test speed higher than the physical capability of the unit. For a Class II or III test, the Code requires an equivalent test speed higher than rated speed so that the wheels are operated at the design volume ratio. When this is not possible the machine cannot be Code tested. In the case of the "once through" compressor with two or maybe three impellers, a predicted air curve may be developed. The air curve is derived by using the basic wheel characteristics and calculating a wheel-to-wheel rematch of the compressor on air. The shape of the curve will not be truly maintained. However, the design flow area can be reasonably explored to check capacity and head. It is reasonable to assume if the predicted air curve is reproduced, that the wheels did perform as designed and the compressor should rematch on design gas and produce expected flow and head. Power requirements are somewhat less reliable, but a reasonable measure of efficiency may be made.

For more complex compressor configurations with multiple inlets and/or outlets (Fig. 2), even more judgments must be made. The unit may have to be tested one section at a time. A different test speed may be required for each section. Another method of testing requires the removal of some impellers. While tedious and expensive, it might well be justified. In some cases, particularly if more than one section is tested at one time, additional instruments may be required. Instruments may have to be installed internally within the compressor. The internal instruments are required to separate the individual section performance from the whole. This test normally uses air as the test gas so the vendor must prepare air-expected curves for use in test evaluation.

When testing requires significant interpretation a judgment must be made as to whether or not the nature of the test will develop good temperature rise data. It may be necessary to install total temperature probes at the OD of the last impeller prior to the flow leaving the casing. When properly installed this type measurement provides a much more reliable reading of true temperature rise and therefore gives a more accurate basis for compressor efficiency and input power required. Generally the vendors resist the specification of internal instruments. Therefore, good understanding of users' requirements and objectives is essential. If costs are realistically evaluated, air equivalent testing will cost less than loop testing.

Loop testing. A closed-loop test may be necessary to performance test within the Code. There is more inherent accuracy in the loop test than in the air test as volume ratio matching can be more closely achieved. The loop test has several limitations that may not be obvious. Loop tests are generally expensive and time consuming. They may become so complicated with complex compressor configurations that they become impractical to set up. Finally, the number of gases available for shop-loop testing is quite limited. Air which is quite available for normal testing is not particularly suitable for loop testing. Fig. 3 shows a typical shop test loop arrangement.

In a loop test, air and oil come into contact causing the danger of an explosion. Extreme caution must be used.

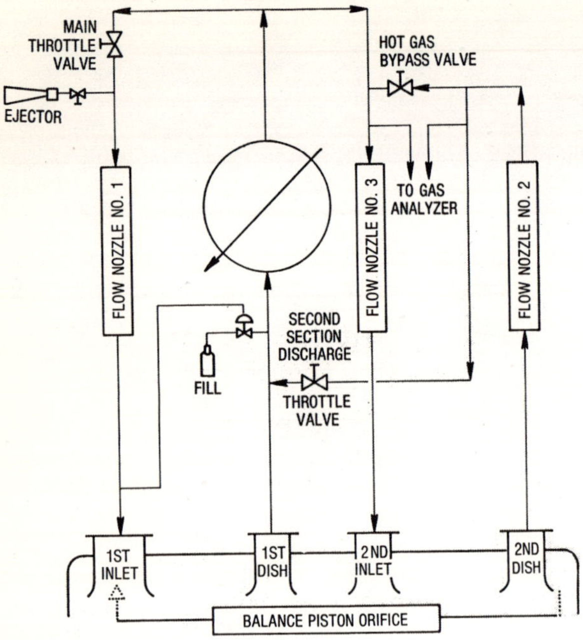

Fig. 3—Shop test loop arrangement.

All combustible, toxic and other gases where safety is involved are disqualified. This narrows the field considerably. The gas cost factor makes the problem even more difficult. The problem of known gas properties adds the final complication. The limitations just covered, together with the cost factor, help to provide the incentive to get a useful test from an open air test even if the instrumentation is more complex.

Test correlation. Earlier mention was made of gases and their correlation calculation. A point was made that the Code could not assume the role of being the final authority on gases. Do not expect the vendor to be all-knowing in the areas of gas properties. The problem of gas properties is the responsibility of the user. Much data is published on gases which together with high-speed computers make the job of defining gas properties somewhat easier today than it was at the time the Code was written. This is not to say that all properties of all gases are as well defined as they might be. Gas mixtures in particular are always a problem.

A few suggestions come to mind. Also some useful references are listed at the end of the article. The BWR[2,3] equation works well with many of the hydrocarbons. The Martin Ho[3,4] equation works well with refrigerants and chlorine. Once gas properties are established, correlation methods suggested by the Code provide results that are meaningful to the ultimate user of the compressor.

ACKNOWLEDGMENT

Originally presented at the Third Symposium on Compressor Train Reliability, Manufacturing Chemists Association, April 24, 1973, Chicago.

LITERATURE CITED

1 ASME—Power Test Code, Compressors and Exhausters PTC 10—1965.
2 Benedict, M., Webb, G. B. and Rubin, L. C., *Chemical Engineering Progress*, Vol. 47, No. 8, p. 419, August 1951.
3 Cooper, H. W. and Goldfrank, J. C., *Hydrocarbon Processing*, Vol. 46, No. 12, p. 141, December 1967.
4 Martin, J. J., Kapoor, R. M. and DeNevers, Noel, *AIChE Journal*, Vol. 5, No. 2, p. 159, June 1959.
5 Reid, R. C. and Sherwood, T. K., "The Properties of Gases and Liquids," McGraw-Hill, New York, 1966.
6 Schultz, J. M., ASME, Transactions, *Journal of Engineering for Power*, p. 69, January 1962.

Part II.

Centrifugal Compressors—Application/Performance

Best Approach to Compressor Performance

When a centrifugal compressor is specified for nonideal gases and the manufacturer's data is based on air, how do you predict design and off-design performance? This calculation procedure solves the problem.

Stephen A. Shain
Shell Development Co., Emeryville, Calif.

THE SPECIFICATION of a centrifugal compressor and prediction of design and off-design performance are parts of a problem that continually arise in process design studies. To make these calculations less difficult, a computer program was written which handles very general situations including multistage compression, inter- and intra-stage leakage, multicomponent gas streams, departures from perfect gas behavior, intercooling when necessary, and allowances for manufacturers' data on the basis of air to permit design for hydrocarbons and other vapors.

The equations used for the individual compression stages of the compressor are summarized in the sections which follow. The assumptions and limitations implied in the general efficiency and head coefficient correlations used, are discussed. The equations given are more general than those usually appearing in the literature.[1,4,6,8,9,12]

COMPRESSOR PERFORMANCE EQUATIONS

Outlet Temperature and Pressure. The temperature and pressure at the outlet of the compression stage are related to the hydraulic efficiency (polytropic efficiency) and head coefficient by the following set of equations:

$$T_2 = T_1 + \frac{(0.0012854)}{(60)^2 C_p}\left[\frac{1}{\varepsilon} + \left(\frac{\partial \ln Z}{\partial \ln T}\right)_p\right]\frac{\pi^2 D^2 N^2 \Psi}{2g_c} \tag{1}$$

$$P_2 = P_1\left(\frac{T_2}{T_1}\right)^m \tag{2}$$

where

$$m = \frac{1}{\dfrac{ZR}{MC_p}\left[\dfrac{1}{\varepsilon} + \left(\dfrac{\partial \ln Z}{\partial \ln T}\right)_p\right]} \tag{3}$$

The efficiency and head coefficient* are defined by

$$\varepsilon = \frac{(0.0012854)\, W_p}{H_2 - H_1} \tag{4}$$

$$\Psi = \frac{2\,(60)^2 g_c W_p}{\pi^2 D^2 N^2} \tag{5}$$

where the polytropic head, W_P, is defined

$$Wp \equiv (144)\int \frac{dP}{\rho} \quad \text{(Polytropic path)} \tag{6}$$

The properties appearing in Equations (1) and (3) are to be averaged over the compression path. The temperature equation, Equation (1), requires an efficiency which is an appropriate average over the path of the compression, but the pressure equation is only valid for a constant efficiency path.

Dimensionless Numbers. Dimensional analysis of the compressible fluid flow equations[7] indicate those variables and groups of variables which are important in specifying the performance of a centrifugal compression stage. Applicable equations are the equation of continuity, the momentum equation, the energy balance, and the equation of state. These equations must be solved subject to specified boundary conditions.

The solution to the equations is controlled by the magnitude of the following dimensionless groups:

The Grashof number

$$N_{Gr} \equiv \frac{g\beta \Delta T L^3}{\nu^2} \tag{7}$$

The Reynolds number

$$N_{Re} \equiv \frac{UL}{\nu} \tag{8}$$

The Prandtl number

$$N_{Pr} \equiv \frac{C_p \mu}{\lambda} \tag{9}$$

The Eckert number

$$N_{Ek} \equiv \frac{(0.0012854)\, U^2}{g_c C_p \Delta T} \tag{10}$$

If the reference temperature, ΔT, in the energy balance equation is taken to be the absolute temperature, T, the Eckert number is related to the Mach number

* The definition of the head coefficient, Equation (5), involves a factor of 2 which is omitted in some sources.

$$N_{Ek} = \frac{(k-1)\, N_{Ma}}{\left[1 + \left(\frac{\partial \ln Z}{\partial \ln T}\right)_p\right]^2} \quad (11)$$

In the above list of dimensionless groups, the Eckert number (Equation 10), may be replaced by

The Mach number

$$N_{Ma} \equiv \frac{U}{V_s} \quad (12)$$

The speed of sound, appearing in the denominator of Equation (12), is the speed at which a small amplitude disturbance will propogate through the gas in the absence of absorption losses, i.e., for an adiabatic reversible or isentropic process. The speed of sound is given by the relation:

$$V_s = \sqrt{144\, g_c \left(\frac{\partial P}{\partial \rho}\right)_S} \quad (12a)$$

The relations of thermodynamics may be used to transform this equation to a form which allows direct use of the equation of state:

$$V_s = \sqrt{144\, g_c\, k \left(\frac{\partial P}{\partial \rho}\right)_T} \quad (12b)$$

The final relationship for the speed of sound in a nonideal gas is:

$$V_s = \sqrt{\frac{144\, gc\, kZRT}{M\left[1 - \left(\frac{\partial \ln Z}{\partial \ln P}\right)_T\right]}} \quad (12c)$$

The calculation makes use of Equation (12c) evaluated at the inlet of a compression stage.

In the computer program, the characteristic velocity (U) is taken to be the impeller tip speed. In addition to the above groups, the equation of state and the boundary conditions are implicit in determining the solution.

Two compression stages will have the same solution in terms of appropriate dimensionless variables provided certain side conditions are satisfied. First of all, the boundary conditions in the two situations must be identical. Second, both compression stages must be geometrically similar, that is, the ratios of all dimensions must be identical, and third, the equation of state for the two gases undergoing compression must be identical.

Thermodynamic Variables. Solutions to gas motion equations are generally given in values of gas thermodynamic properties as functions of spatial variables in the impeller and diffuser. However, the solutions of interest are certain functions of these thermodynamic variables rather than their complete spatial variation. In particular, the solution may be thought of as the ratio of useful work to the total work or total enthalpy rise, and the ratio of useful work or pressure head to the kinetic head of the impeller. This first variable is the polytropic head divided by the actual work input and is referred to as the hydraulic efficiency. The second variable is the head coefficient. These variables have been previously defined by Equations (4) and (5). These ratios will depend on the boundary conditions and in particular on the ratio of the fluid velocity to the tip velocity of the impeller. This ratio is the volumetric flow coefficient (ϕ)

$$\phi \equiv \frac{Q}{\pi DNA} \quad (13)$$

and the formal relations for the efficiency and head coefficient and the flow coefficients are:

$$\varepsilon = \varepsilon\,(\phi; N_{Gr}, N_{Re}, N_{Pr}, N_{Ma}; \text{ equation of state parameters, geometric variables})$$

$$\Psi = \Psi\,(\phi; N_{Gr}, N_{Re}, N_{Pn}, N_{Ma}; \text{ equation of state parameters, geometric variables})$$

The volumetric flow used in defining the flow coefficient may be an inlet flow, an outlet flow, or any average of inlet or outlet flows. Several alternate approaches have been reported in the literature.[2,3,4,9,11] Better correlations appear to result from a use of intermediate flows, but correlations in this form are not always available from the manufacturers. The example given is based on inlet volumetric flows.

For impellers which are geometrically similar, the area (A) is uniquely related to the impeller diameter (D). However, the value of A has been left free so that it may be adjusted to improve the efficiency and head coefficient correlations or to put them in a convenient form, particularly if geometric similitude is not maintained. For example, the flow coefficient may be interpreted as the ratio of the volumetric flow to the manufacturers' design volumetric flow if a nominal area (A) is taken to be

$$A = \frac{Q\text{ (design point)}}{\pi DN} \quad (14)$$

This form is often convenient since the manufacturers' correlations are commonly in terms of percentage flows relative to the design point, i.e., Equation (14) fixes $\phi = 1$ at the design point. A similar definition of a nominal diameter (D) may be obtained from Equation (5) for $\Psi = 1$ at the design point.

Efficiency and Head Coefficient Correlation. When the heat loss to the surroundings is negligibly small, the Prandtl number may be dropped from the solution.[7] If the Reynolds number is large ($N_{Re}^2 >> N_{Gr}$), natural convection is no longer of importance; thus the Grashof number may be eliminated from the solution.[7] If the Mach number is smaller than 1, that is, if the speed of sound is not exceeded, at all points along the compression path the performance of the compression stage is insensitive to the Mach number.[9,12,13] For Reynolds numbers greater than 100,000, as is usual, the results are insensitive to the Reynolds number.[2,3,6]

The mechanical design of different impellers for a given frame or frames is not generally based on the principle of geometric similitude. Impellers for the same frame designed for varying volumetric flow capacities will typically have the same impeller diameter with a varying channel width or area of flow. The assumption of geometric similitude is then not strictly applicable. An assumption implied in the example calculation is that a single correlation for head coefficient and for efficiency applies for all of the impellers available in the computation. This assumption will be strictly valid only if the impellers are geometrically similar. However, it is necessary to make this assumption in many cases since more detailed information on the performance of the individual impellers is not always available.

The form of the correlations which is suggested for use reduces to:

$$\varepsilon = \varepsilon(\phi)$$
$$\Psi = \Psi(\phi) \qquad (15)$$

An example correlation of head coefficient as a function of volume coefficient is presented based on the data of C. A. Macaluso[4] for a single stage compressor. Since the actual physical dimensions of the compressor were not included as part of the data, the results are correlated only in terms of ratios of the appropriate quantities. That is, the head coefficient is replaced by the dimensional ratio (in enthalpy units)

$$\Psi \sim \frac{H}{N^2} \quad \left[\frac{\text{Btu/lb mass}}{(\text{rpm})^2}\right] \qquad (16)$$

and the volume coefficient (based on suction volume) is replaced by the dimensional ratio

$$\phi \sim \frac{Q_1}{N}\,\frac{cmf}{rpm} \qquad (17)$$

The results for a variety of speeds ranging from 5,400 rpm to 8,250 rpm and volumetric flows of 2,000 cfm to 4,500 cfm are presented in Figure 1. While there is scatter of the results, a single correlation can be presented with an accuracy of 5-10 percent. A point of particular interest is that the results are for compression of different gases (molecular weights of 120.9, 137.4, and 170.9). The surge limit or minimum value of the volumetric flow coefficient and the choking limit or maximum value of the volumetric flow coefficient are reduced to single values of the volumetric flow coefficient when presented on the basis of Figure 1.

The above results and others from the literature[9,11] suggest that correlations for both head coefficient and efficiency as functions of volumetric flow coefficient may be used for a variety of nonideal gases provided the appropriate dimensionless forms are retained. In the absence of information on the particular gas being compressed, manufacturers' information based on air can be used in the computation even though exact similitude would not be obtained.

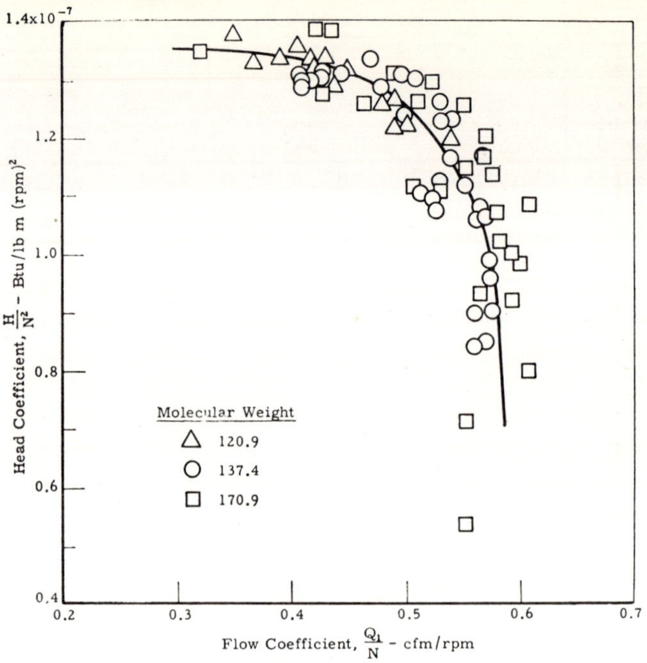

Fig. 1—Head correlations for a variety of speeds.

Compressor Performance Calculations. The performance calculation for a compressor or series of compressors is accomplished on a wheel-by-wheel basis. The performance of each impeller or compression stage is considered in sequence by application of the general set of performance equations, Equations (1), (2), and (3). These equations allow calculation of the temperature and pressure at the outlet of a compression stage provided that the efficiency and head coefficient are known.

The specific definitions of the coefficients and dimensionless groups used have been presented as Equations (4) and (5). The efficiency and head coefficient are represented as functions of a single variable, the volumetric flow coefficient, Equation (15). The correlation of efficiency and head coefficient with volumetric flow coefficient is often a satisfactory representation of the performance of not only a specific impeller, but also of other impellers of similar design but which are scaled for use at different volumetric flows. This correlation of efficiency and head coefficient provides a convenient condensation of information on the performance of a compression stage for a variety of operating conditions.

There are limitations inherent in this approach, and some of these have been discussed above. The calculation procedure is sufficiently flexible so that additional information, when available, can be included.

Outline of Program Logic. The program logic may be schematically outlined as follows:

1. Read in input data.

2. Calculate recycle (leakage) flows.

[If the compressor calculations are incorporated with other process calculations, recycle flows (not necessarily leakage, as in this example) may depend on the com-

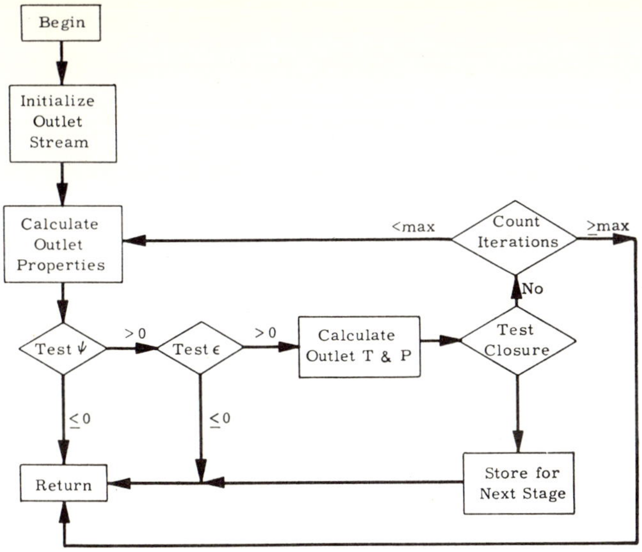

Fig. 2—Block diagram of compressor stage calculations.

pressor discharge conditions. In that case an over-all material balance may require iterative application of the entire compressor calculation.]

3. Calculate physical properties for the suction.
4. Initialize stage discharge physical properties.
5. Calculate appropriate average flow coefficient.
6. Calculate head coefficient and efficiency.
7. Calculate new discharge temperature and pressure.
8. Compare current and previous discharge temperature. If closure to within a specified tolerance is not achieved, return to (5). If closure is achieved, calculate the next stage starting with (4). After the last stage, continue with (9).
9. Print out final results.

Figure 2 is a simplified diagram of the subroutine which calculates the performance of a single stage.

Use of the Computer Program. The input data required by the program can be divided into the following categories:

Component data—consisting of a specification for vapor components in the form appropriate to the subroutines which calculate the compressibility factor, Z, and other physical properties required for the computations.

The equation of state used in the example is based on a modified version of the Martin and Hou equation of state[5,10]

$$p = \frac{RT}{(v-b)} + \frac{A_2 + B_2T + C_2e^{-KT/T_c}}{(v-b)^2} + \frac{A_3 + B_3T + C_3e^{-KT/T_c}}{(v-b)^3} + \frac{A_4}{(v-b)^4} + \frac{B_5T}{(v-b)^5}$$

Impeller correlation data—specifying a correlation of efficiency and head coefficient of the impeller or impellers with the volumetric flow coefficient. (The correlation may be based on inlet flow, outlet flow, or any average of inlet and outlet flows from a compression stage. Variations in the efficiency and head coefficient with Mach number and Reynolds number may be included, if desired.)

Compressor sizing data—consisting of a list of impeller scale parameters or physical dimensions to be used in the correlations for efficiency and head coefficient. (These usually are a characteristic impeller area and diameter, but other variables which reflect the capacity of the impeller may be used.)

Stream data—consisting of a specification of the inlet stream temperature, pressure, and molar flows for each component.

Miscellaneous specifications—including the rotational speed for each frame, and several bounds and increments which must be specified.

The output from the program is a reiteration of the input data and a statement of the calculated properties for the inlet stream, i.e., the density and the volumetric flow. A listing of parameters and properties for the outlet stream of each compression stage is presented. These include the stream number, the frame number and impeller number, the nominal diameter and area for the impeller, the rotational speed, the temperature, pressure, density, molecular weight, the volumetric flow, and arithmetic average value of the specific heat ratio, the value of the exponent which defines the compression path (given by Equation 3), the value of the average flow coefficient, the Mach number, Reynolds number, head coefficient, efficiency, the head and horsepower. When a stage consists of intercooling, the temperature and pressure of the stream is presented, and, if phase separation occurs, molar flows of both the vapor and the liquid stream are presented.

In addition to the calculated properties, a warning flag, FLAG, is indicated for each compression stage. Values of FLAG are used to indicate the course of the computation and signal any difficulties which may have been encountered during the computation. The warnings which may be indicated by FLAG are:

- The Mach number exceeds the critical Mach number.
- The average value of the flow coefficient is less than a minimum value.
- The average value of the flow coefficient is larger than a maximum value.

- The number of iterations for the calculation at this stage has exceeded the maximum number of iterations, i.e., the calculation has failed to converge.
- The calculated head coefficient is negative (may result from inappropriate extrapolation of correlation).
- The calculated efficiency is negative (as in head coefficient above).
- Intercooling has resulted in condensation.
- The temperature has exceeded the maximum allowable temperature.

Example Calculation. An example is presented on the application of the above procedure using a large scale digital computer. Calculations are made for a three-stage propylene refrigeration compressor. The shape of the head-capacity and the efficiency-capacity relations was taken to be the same for all three wheels, with the flow coefficient based on inlet flows. The head-capacity relation for the second stage was used as supplied by the manufacturer. The shape of the curves for the first and third stages differ by 2-5 percent at the extremes in the range of 60-140 percent design capacity. An efficiency-capacity relation was not available for the wheels in this machine. The manufacturer was able to supply an efficiency-capacity curve for a similar machine at a specific speed which was close to the design condition at the second stage. This curve was used to calculate an efficiency-capacity relation for the range in capacity of from 60-140 percent of the design points. Again, the wheels may differ by approximately 5 percent at the extremes of the range. The flows in the recycle (leakage) streams were taken as 1.5 percent at each stage and 6.5 percent from the discharge of stage 3 to the suction of stage 1. Figure 3 is a schematic diagram indicating the interstage flow, stream 12, and the intrastage flows, streams 8, 9, and 10.

The results of the computations are summarized in Table 1 and are compared with a performance test.

The discharge pressures of 312-276 psia and power requirements of 2949-3107 hp are somewhat above the 275-200 psia and 2,400-2800 hp ranges from the original design specifications. However, the calculated values are close to those from the actual performance tests for this machine. The calculated discharge pressures (third stage) are within 1-3 percent of the pressures actually achieved with even closer agreement for the brake horsepower. The agreement is excellent considering that manufacturers' head and efficiency data for air has been used to calculate performance for propylene under conditions where it is a nonideal gas.

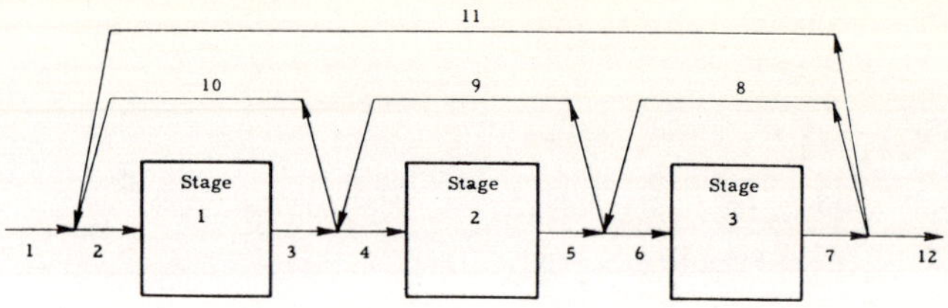

Fig. 3—Flow diagram for a three-stage compressor.

Why This Approach? The approach presented here is intended as a thermodynamically consistent method for performance calculations when nonideal gas phases are important. It is hoped that the presentation will encourage publication of further experimental results which will allow evaluation of this approach and outline more clearly its limitations. The author would like to emphasize to those concerned with design, manufacture, and performance of such compressors that methods and computer capabilities for evaluation of thermodynamic properties of fluids are becoming generally available and should extend the usefulness of many design methods. The physical properties programs[10] used here have been made available to the Physical Properties Project Subcommittee of the American Institute of Chemical Engineers.

ACKNOWLEDGMENTS

The author acknowledges the assistance of Mr. C. V. Sternling who supplied the programs and correlations used for estimation of physical properties, and of Mrs. D. J. Albert and Mr. E. L. Necoechea who programed the calculations for the computer.

NOMENCLATURE

A	nominal area, sq. ft.
C_p	heat capacity at constant pressure, Btu/lb. mass-°R
D	nominal diameter, ft.
g	local acceleration of gravity, ft./sec. sq.
g_c	gravitational constant, 32.16 lb. mass-ft./lb. force-sec. sq.
H	enthalpy, Btu/lb. mass
k	heat capacity ratio, C_p/C_v
L	characteristic dimension, ft.
m	exponent, defined by Equation (3)
M	molecular weight, lb. mass/lb. mole

Table 1—Example Calculations Compared With Performance Test

Flow[a] Lb.-Mole/Hr.	RPM	SUCTION Temp., °F	SUCTION Press., psia	SUCTION cfm[b]	DISCHARGE (3rd Stage) Temp., °F	DISCHARGE (3rd Stage) Press., psia	DISCHARGE (3rd Stage) cfm	DISCHARGE (3rd Stage) Brake, hp[c]	PERFORMANCE TEST Press., psia	PERFORMANCE TEST Brake, hp
3250	8950	78	80	3563	231	312	1093	2949	308	2820
3500	8950	78	80	3837	225	304	1197	3025	298	2870
3750	8950	78	80	4111	218	292	1325	3078	286	2930
4000	8950	78	80	4385	211	276	1487	3107	268	2980

a–100 Percent Propylene. b–For feed flow, temperature, and pressure. c–1.02 x Compression hp.

N	rotational speed, rpm
N_{Ek}	Eckert number, defined by Equation (10)
N_{Gr}	Grashof number, defined by Equation (7)
N_{Ma}	Mach number, defined by Equation (12)
N_{Pr}	Prandtl number, defined by Equation (9)
N_{Re}	Reynolds number, defined by Equation (8)
P	pressure, psia
Q	volumetric flow, cmf
R	gas constant, equal to 1.987 Btu/lb. mole-°R, or 10.73 psia-cu. ft./lb. mole-°R
S	entropy, Btu/lb. mass-°R
T	temperature, °R
U	characteristic velocity, ft./sec.
V_s	velocity of sound, ft./sec.
W_p	polytropic head, ft.-lb. force/lb. mass
Z	compressibility factor

Greek Symbols

β	coefficient of thermal expansion, vol fract/°R
ε	polytropic efficiency, defined by Equation (4)
λ	thermal conductivity, Btu/sec.-ft.-°R
μ	viscosity, lb. mass/ft.-sec.
ν	kinematic viscosity, sq. ft./sec.
π	constant, equal to 3.14159
ρ	density, lb. mass/cu. ft.
ϕ	volumetric flow coefficient, defined by Equation (13)
Ψ	polytropic head coefficient, defined by Equation (5)

Subscripts

1	refers to compression stage inlet
2	refers to compression stage outlet

LITERATURE CITED

[1] Church, D. M., Chem. Eng., 69, No. 5, 127 (Mar. 5, 1962); 69, No. 6. 167 (Mar. 19, 1962); and 69, No. 7, 117 (April 2, 1962).

[2] Davis, H., Kottas, H., and Moody, A. M. G., Trans, ASME, 73, 499 (1951).

[3] Davis, H., Trans. ASME, preprint No. 56-A-122 (1956).

[4] Macaluso, C. A., ASME preprint No. 60-HYD-15 (1960).

[5] Martin, J. J. and Hou, Yu-Chun, *AIChE Journal*, 1, 142 (1955).

[6] O'Neill, P. P. and Wickli, H. E., Trans. ASME, B84, 248 (1962).

[7] Schlichting, H., "Boundary Layer Theory," Pergamon Press, New York. 1st edition (1955), p. 248-253.

[8] Schultz, J. M., J. Eng. Power, Trans. ASME, A84, 69 (1962).

[9] Sheets, H. E., Trans. ASME, preprint No. 51-SA-19 (1951).

[10] Sternling, C. V., private communication.

[11] Wiesner, F. J., Jr., Trans. ASME preprint No. 60-HYD-17 (1960).

[12] York, R., Jr., *Ind. and Eng. Chem.*, 34, 535 (1942).

How Compressor Components Affect Performance

One component influences the next like a chain reaction affecting over-all compressor performance. Effects can be seen when components are designed for one gas and the machine operates with another.

Sulaiman Sayyed, Delaval-Stork V.O.F., Hengelo, Netherlands

MYTHS AND misunderstandings concerning centrifugal compressor components confuse many engineers. This article will describe the operating principles of compressor components and their effect on over-all performance. Such knowledge can help insure the correct selection of commonly used hardware and to better understand the potentials and limitations of the selected machine itself.

A centrifugal compressor primarily consists of a series of compression stages (Fig. 1) where compression is achieved by applying inertial forces to a gas. Gas enters the compressor body through an inlet nozzle. As it passes through a series of rotating blades or an impeller, those forces are transmitted by adding velocity energy through the acceleration of the gas. Therefore, impeller design becomes the single most important element of the whole machine.

After the impeller, the gas enters into a diffuser where it decelerates and the velocity energy is transformed into pressure energy. After the diffuser, the gas passes through a volute or a collector where velocities are further reduced before the gas is discharged to the process through a discharge nozzle in a single-stage compressor or carried over to the next stage in a multi-stage compressor. Therefore, a centrifugal compressor is often called a dynamic machine as all the compression is achieved by continuous dynamic action of blades and channels. Poor design or the mismatch of any channels has a direct influence on the over-all performance characteristics of the machine.

A realistic way to understand a centrifugal compressor is to categorize each compression stage into two basic groups: rotating and stationary parts.

ROTATING PARTS

The rotor, in simple form, consists of one or a series of impellers mounted on a shaft. Impellers can either be shrunk-on or keyed or a combination of both depending upon the operating speed and the prevailing stress levels the amount of shrink that can be achieved without kinking the shaft.

Centrifugal impellers can be divided into the following categories according to the blade design configurations:

a) Backward leaning blades—where the outlet angle (β_2) is less than 90 degrees

b) Radial blades—where the outlet angle (β_2) is 90 degrees

c) Forward leaning blades—where the outlet angle (β_2) is greater than 90 degrees.

See Fig. 2.

Blade design is limited to the first two types (Fig. 2) in industry because of their constantly rising head characteristics from design to surge. Forward leaning blades

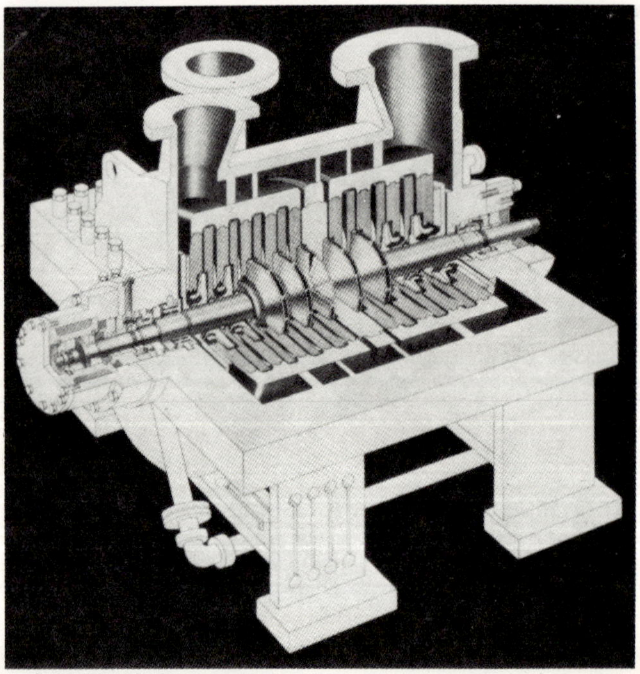

Fig. 1—Multi-stage compressor.

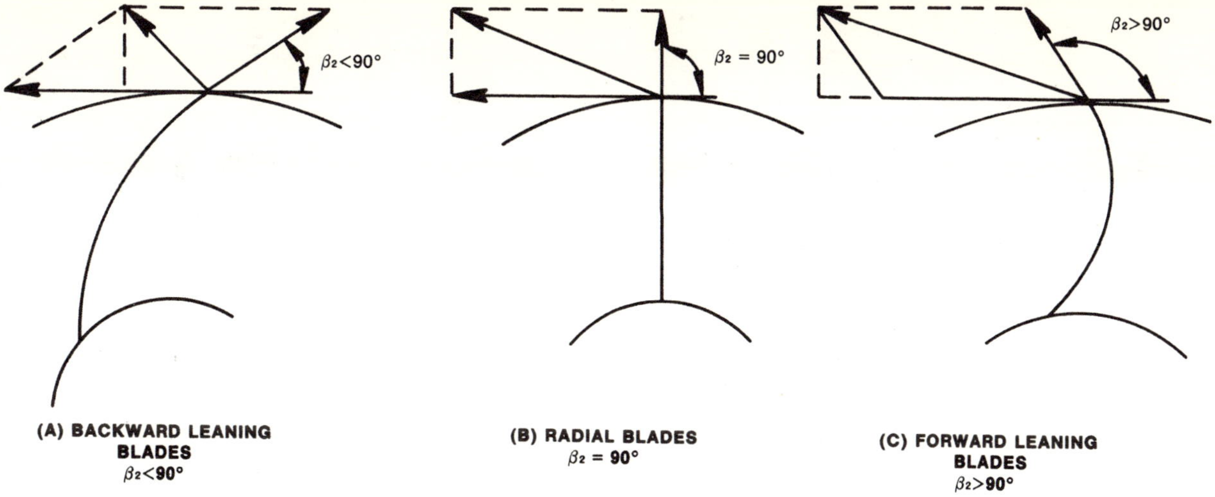

Fig. 2—Blade design configurations.

are not desirable because of their inherent quality of unstable performance; that is, the head increases and decreases for increasing flow. The present discussion will therefore be limited to only backward leaning impellers.

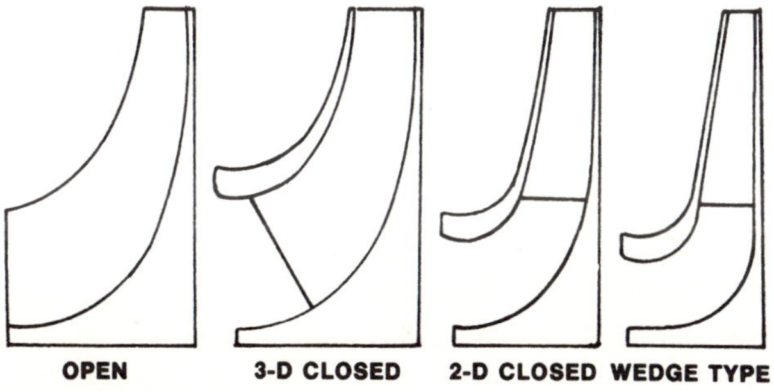

Fig. 3—Types of impellers.

Fig. 4—Shroudless impeller.

Impeller types. Impellers fall into two main groups: shrouded and shroudless, or more commonly known as closed and open impellers (see Fig. 3).

Shroudless or open impellers usually have the blades extending inward almost to the full length of the hub (see Fig. 4). Thereby the blades are designed in line with the approaching flow and further reduce the relative velocity inlet mach number. The major advantage of this type is in having lower stresses than shrouded impellers for the same flow coefficient and at the same operating speed, as the weight of the shroud does not have to be carried by the impeller. Therefore, open impellers can be operated at higher speeds without exceeding material stress limitations.

Higher operating speeds permit a higher compression ratio. Thus, open impellers have the capability of developing higher head at greater flows than enclosed impellers with the same flow coefficient and blade design. Offsetting this advantage, however, are the problems of both a design and cost oriented nature associated with the need to maintain minimum clearances at close tolerances between the rotating blades and adjacent stationary diaphragm wall. These clearances must be small enough to obtain the desired performance but with enough space to prevent rubbing between the two surfaces caused by a possible rotor shift. In addition, the blade design often requires special care in blending various sections of the blades from inlet to discharge.

Shrouded or closed impellers can be divided into the following types:

- **Three dimensional impellers.** The blade or vane extends farther down towards the inlet than two dimensional impellers but not as far as open impellers (see Fig. 5). It is part of a simply drawn geometrical shape of a cone or a cylinder tilted at an angle to the surface of the paper. Major advantage: a close match of flow and blade angle is achieved with very little incidence

Fig. 5—Three dimensional shrouded impeller.

Fig. 6—Two dimensional shrouded impeller.

along the impeller inlet. Results: well matched inlet conditions which contribute to improved over-all machine performance.

• **Two dimensional impellers** (see Fig. 6). The blade consists of part of a cylinder or a combination of cylinders perpendicular to the surface of the hub and shroud. Impeller design is comparatively simple. A mean inlet velocity is assumed which results in some incidence both at the impeller hub and shroud. This compromise on incidence angles can adversely affect performance, especially on impellers in the high-flow range. However, in the lower flow range the effect of incidence change across the blade inlet becomes insignificant as the impeller inlet is rather small.

The simplified design and construction makes the use of this type impeller of benefit in the lower flow range.

• **Wedge type impellers** are also closed two dimensional impellers especially designed for very low flows. With these impellers care is taken to provide as square a channel as possible by increasing the blade thickness from inlet to discharge, forming a wedge shape (see Fig. 7). This type of design increases the equivalent hydraulic radius at the impeller discharge, minimizing channel losses in the impeller thus resulting in higher efficiency

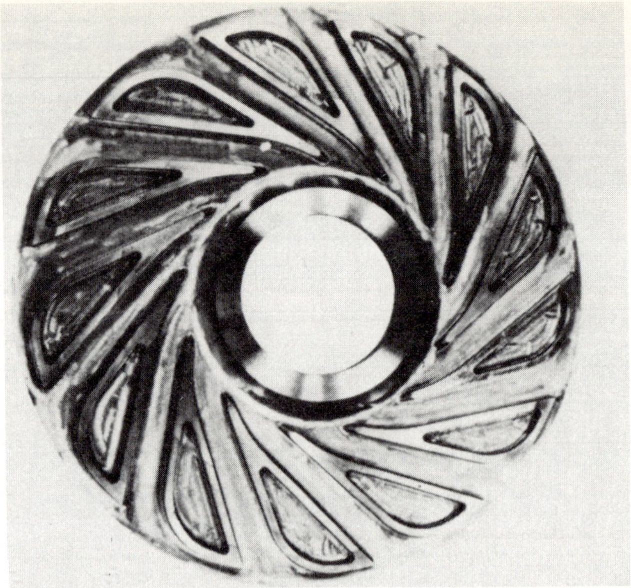

Fig. 7—Wedge type impeller without cover plate.

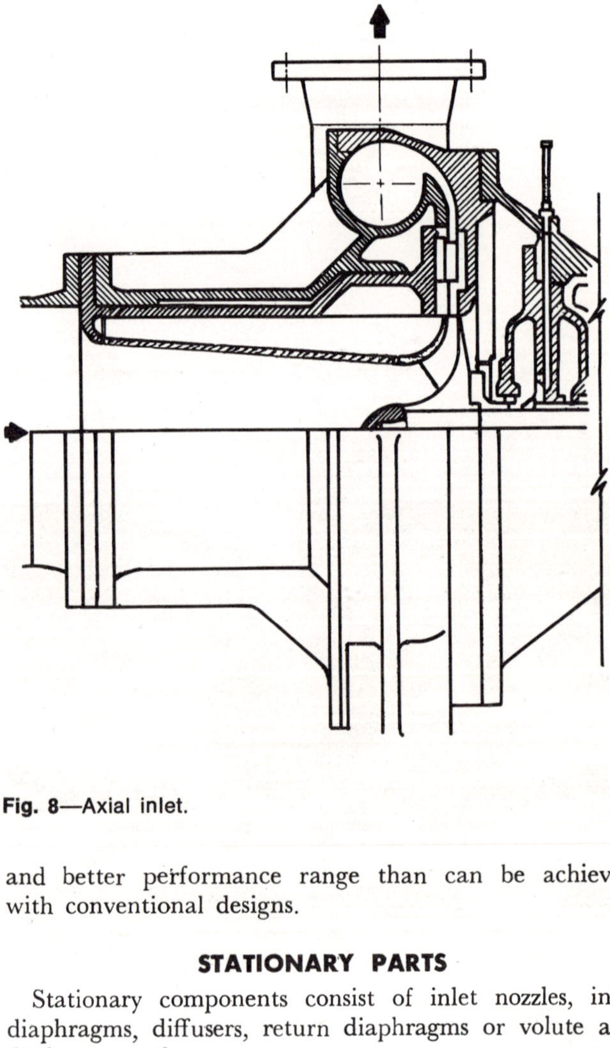

Fig. 8—Axial inlet.

and better performance range than can be achieved with conventional designs.

STATIONARY PARTS

Stationary components consist of inlet nozzles, inlet diaphragms, diffusers, return diaphragms or volute and discharge nozzles.

Inlet nozzle. Functions of the inlet passage are to accelerate the fluid from the inlet flange to the impeller

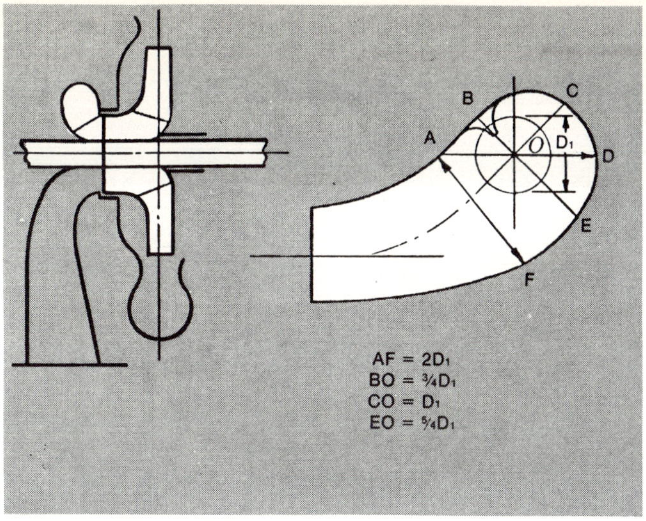

Fig. 9—Volute inlet.

Fig. 10—Inlet diaphragm.

eye with minimum losses and to provide a uniform velocity profile. Since the flow path between the inlet nozzle and the impeller eye is short and inlet nozzle flow velocities are relatively low, friction losses are very small. However, inlet nozzle design has an important bearing on the velocity distribution immediately ahead of the impeller which affects impeller efficiency.

Basically an axisymmetric inlet flange combined with a simple convergent nozzle is the most efficient way to lead the flow. The two basic designs of inlet nozzles can be grouped as follows:

- **Axial inlet.** An axial, tapered flow path which gradually reduces toward the impeller eye is definitely the best, especially for single stage machines, as it assures uniform velocity distribution to the impeller inlet (see Fig. 8.)

- **Volute inlet** is used in multistage compressors as a practical necessity because space limitations usually prevent the use of a long axial inlet to the first stage impeller. For low specific speeds this type of inlet is quite efficient, but for high specific speeds the detrimental effects of a sharp 90-degree turn immediately in front of the impeller becomes more pronounced. The velocity distribution is not uniform and flow tends to cling towards the inner radius. This non-uniformity of flow should be taken into account in the design of the flow passage allowing enough axial distance in front of the impeller to minimize its adverse effect on performance (see Fig. 9).

Inlet diaphragms are designed to lead the flow towards the eye of the impeller in a very smooth and systematic way without creating too drastic a change between the shroud and hub velocities (Fig. 10). The flow pattern in a centrifugal compressor is very complex and depends upon many factors. Primary flow in the direction of the passage centerline is directly influenced by the passage geometry, Coriolis force, viscous effects and indirectly by the secondary flow pattern. Secondary flow is generally caused by the forces from passage curvature and Coriolis effect acting on the gradients of primary flow. Viscous flow and the presence of a boundary layer on the passage walls also contribute to the secondary flow. This flow is quite common in open impellers and could be of quite a different pattern than in shrouded impellers. Viscous losses in impellers are therefore associated with boundary layer, and the growth of this boundary layer depends on velocity variation along flow surfaces. To arrive at the exact velocity distribution leading to an optimum performance is quite difficult but certain fundamental guidelines must be followed with boundary layer considerations, impeller blade loading and the ratio of relative velocities.

Inlet guide vanes. Sometimes inlet guide vanes are used as an extension to the inlet diaphragm. Guide vanes may be of the stationary or adjustable (movable) type. The three typical configurations and their relative position to the impeller inlet are pre-rotation, radial and counter-rotation.

With a pre-rotation setting of an inlet guide vane, the head and capacity of the stage is reduced with a slight improvement in stage efficiency due to reduced relative inlet velocity.

Using a counter-rotation setting, a slight increase in head and capacity occurs with a drop in stage efficiency and stable range.

The most efficient configuration is the radial inlet guide vane setting. Using this setting, a smooth, accelerating inward flow takes place towards the eye of the impeller.

Diffusers. Centrifugal compressors are usually fitted with vaned or vaneless diffusers (see Fig. 11). The influence of the diffuser on compressor performance cannot be overemphasized. It is in this area where an efficient transition of kinetic energy to static pressure takes place. Therefore to meet the desired objectives of high pressure, high efficiencies and a large margin between surge and design point, it is essential that utmost attention be given to diffuser design.

Vaneless diffusers commonly used, are parallel wall types because of economic and design considerations, though some experimental work has been done on divergent type diffusers. The use of parallel-walled, vaneless diffusers is preferred where operating conditions require considerable flow changes or where deposits from compressing dirty gases are expected. In the vaneless diffuser, fluid follows a logarithmic spiral. In a vaned diffuser, the vanes are used to guide the flow deviating from the logarithmic path to achieve faster diffusion and to reduce the amount of friction losses.

Vaned diffusers may be very simple, as in a conical design where the number of blades and their thickness distribution are determined in such a way that the blade channel has an area distribution equivalent to that of a cone or two-dimensional diffuser; or a very extensive and expensive design using three-dimensional blades with splitter vanes. Generally, a small vaneless area is left before starting the blades. This vaneless area helps reduce the circumferential pressure gradient at the impeller tip and reduces the mach number at the inlet to the vaned diffuser. To achieve proper diffusion, care must be taken to have a good ratio between the opening and the length of the channel. Performance improvement using a vaned diffuser can be as high as 3 to 4 percent in comparison to a vaneless diffuser but the over-all high velocities and high mach numbers restrict the compressor's envelope of operation causing higher noise and vibration levels.

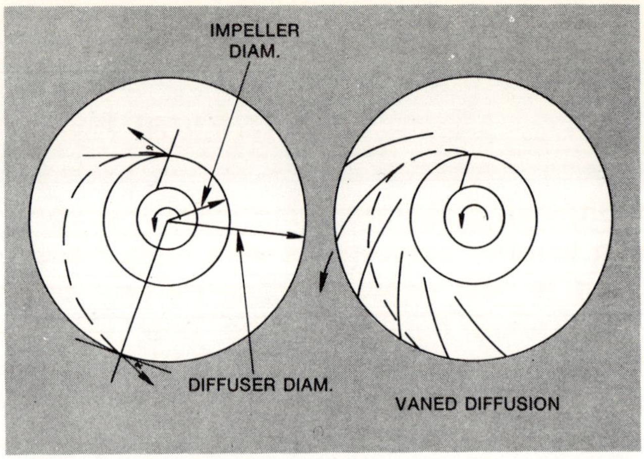

Fig. 11—Vaneless and vaned diffuser.

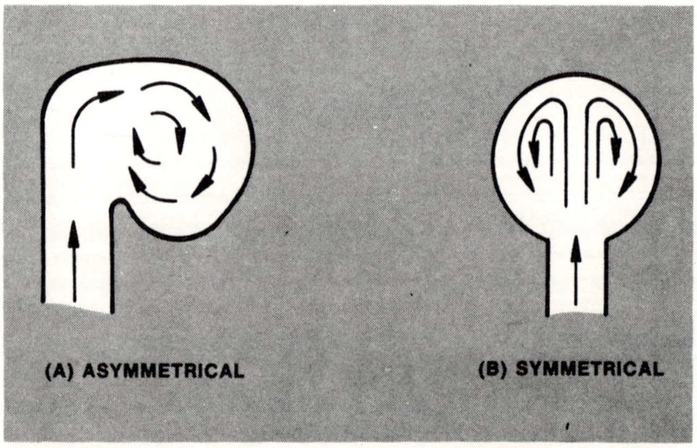

Fig. 12—Volute.

Crossover and return channel. Little is known about the design of return channels, nor is much experimental data available to support the existing philosophy where the flow is merely dumped in the return passage. Some work has been done using a more logical approach whereby about 20 percent deceleration takes place in the return vane throat with respect to the diffuser exit, followed by a more gradual diffusion in the return channel to a level which would provide a slight deceleration to the inlet of the next stage. Testing this approach in an actual compressor was done with considerable success in improving over-all performance from earlier designs.

Volute, or collector as it is generally known, is another area where little experimentation has been done to improve performance. Its hydraulic characteristics are determined only on theoretical considerations. The velocity distribution across any of its sections is not uniform, and is further complicated by frictional drag. The radial outward component of the absolute velocity causes a spiral motion which is outwards in the middle but inwards near the volute walls (see Fig. 12).

Therefore, in many new designs one sees asymmetrical rather than symmetrical volutes.

Volutes are generally designed with the philosophy of having constant average velocity for all sections.

An outlet nozzle is always placed directly after the volute with a nozzle area equal to the outlet area of the volute. It has been established experimentally that for a nozzle inlet, a conical element of 6 to 12-degree included angle is the most efficient for velocity conversion. Although such an included angle is ideal, physical limitations often make it impossible to have the desired nozzle length.

Even though the impeller or rotor provides the only means of adding energy to the gas and the stationary components convert velocity energy into pressure energy, the influence of one on the other must not be underestimated. The influence of one component on the next acts like a chain reaction affecting the over-all performance. This effect can be seen quite clearly when the components are designed for a particular gas and the machine is operated with a different gas. Furthermore, the greater the number of stages the more pronounced is the effect.

The combined effect is generally presented in the form of an over-all performance curve of the stage or the machine. Basically the characteristics of the curve are the same regardless of how they are expressed. If the components are well matched and well designed aerodynamically, one can be very sure that the machine will perform as expected.

Centrifugal Hypercompressors for Ethylene Service

Traditionally, reciprocating compressors have been used to achieve the very high pressures required in low density polyethylene processes. Larger plant capacities and new technological developments make centrifugal compressors a possible candidate for this service. Here's what must be considered if centrifugal compressors are to be used.

R. A. Strub, Sulzer Brothers, Ltd., and
C. Matile, Burckhardt Engineering Works, Ltd.

REGULAR industrial production of low-density polyethylene (ldPE) started about a quarter of a century ago and spread very quickly over the world. Although different processes are used, the flow diagrams are, in their main features, all similar. The two main categories are:

- Vessel reactor processes, generally working under pressures of 1,500-2,500 bar
- Tubular reactor processes, with a pressure at reactor inlet ranging between 2,000 and 3,000 bar, or more
- The last category usually reaches higher conversion grades but both need recycle such that the hypercompressor capacity is about 4-6 times the fresh gas consumption. Power consumption of the hypercompressor also is about 4-6 times that of the primary compressor.

Expansion of ldPE-production and consumption slowed slightly but is still expected to double within about seven years, compared with the past four to five years. Single-stream capacities increased more than 10-fold within 20 years, from the range of 4-10,000 t/year to about 50-120,000 t/year today. Presently, this tendency shows no signs of weakening or of reversing.

From the point of view of technology, design and manufacture, nothing would make it impossible to increase the size of the plant constitutive elements to two or three times that of today. For reactors and high-pressure pipings it is basically a question of production equipment. Though the largest hypercompressors and extruders actually made are already fairly large ones, it would be possible to produce still larger ones and, if necessary, to place two or three units in parallel.

However, use of huge reciprocating compressors would be increasingly unattractive for the following reasons:

- High weights for installation and maintenance
- Heavy foundations and large space requirements
- High inertia forces and moments
- Generation of pressure pulsations and of mechanical vibrations in the whole system, with the need of damping devices and of very strong fixations.

Therefore, major manufacturers of hypercompressors began some years ago to study the possibility of using centrifugal compressors instead of reciprocating for gas compression duties of ldPE-production. If it were possible, the inconveniences mentioned would be avoided. The following facts may be considered as settled:

- The lower limit of practical feasibility for a hyperturbocompressor lies at a gas capacity of 80-100 t/h.
- Three intercoolings will be necessary (instead of only one with reciprocating units), in the case of usual water or air cooling.
- Most of the main design features and calculation methods may be derived from the combined experiences in the fields of high-pressure centrifugal compressors for ammonia or methanol-synthesis, and large high-pressure boiler feed pumps.

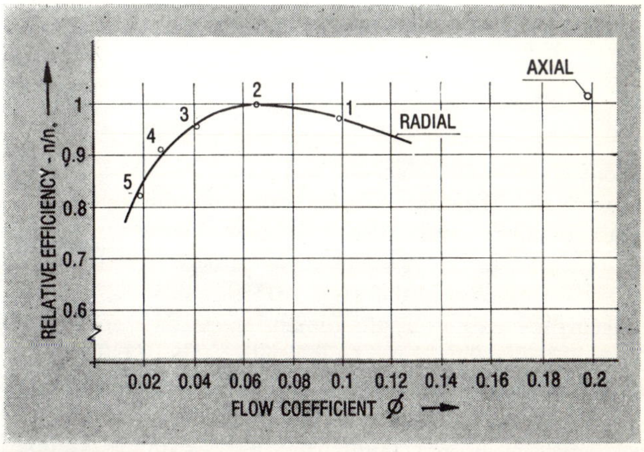

Fig. 1—Relative efficiency as a function of flow coefficient for radial stages.

- Real, specific problems are directly bound to polyethylene production and can only be solved by production plant tests. (These problems are chiefly the risk of fouling or plugging by low polymers or PE-dust and the possible thermodynamical consequences.)

- Because the primary compressors' capacity is only 25-35 percent that of the hypercompressor, the threshold of feasibility would lie still higher than for the corresponding hyper.

Following is a summary of the actual state of experiences and development of the centrifugal hypercompressor, with special consideration for a 125 t/h-unit for the compression of ethylene, working between 245 and 2,450 bar.

LAYOUT AND PRINCIPAL DIMENSIONS OF COMPRESSORS

Before defining the layout of the compressor installation, the influence of the various parameters which can greatly impair machine efficiency must be evaluated. According to the relative value of various losses, design and layout philosophy of the installation may be largely influenced.

Fig. 1 shows the relative efficiency of a given impeller size as a function of the flow coefficient. One can see that stage efficiency drops markedly in the neighborhood of ϕ-values of 0.03 and lower. For this reason, recent projects endeavored to maintain the ϕ-values above 0.03.

This leads to speeds in the range of 30,000-40,000 rpm for which no commercial steam turbine is available for direct drive. Use of speed increasers must be taken into account, and gear losses should not be too large, since it would minimize the advantage of the high-speed compressor.

Previous studies based on 100 t/h capacity and on the assumption of direct drive showed that speed had to be limited to about 18,000 rpm to find a suitable steam turbine. This was only possible with a back pressure turbine or one with limited partial condensation. Due to this speed limitation, the impeller diameter of about 160 mm would lead to very small ϕ-values (0.0125-0.0175) and a stage efficiency of about 64 percent. Later, a capacity of 125 t/h was selected and was considered as a minimum for decent efficiency.

The following values have been established:

Mass flow (125 t/h) .... 34.8 kg/s
Inlet pressure .... 245 bar
Inlet temperature .... 40° C
Discharge pressure .... 2,450 bar

Four compressors placed in series each having six stages were selected. Fewer machines with a greater number of stages are not recommended. Machines of eight stages lead to too slender rotors. A cooler is placed between each compressor casing to cool the gas to 40° C.

According to Fig. 1, the dimensionless flow coefficient of the impeller should not be lower than about 0.03 which leads to an impeller diameter of 100 mm rotating at 36,500 rpm. Efficiency values for these smaller impellers are corrected to take into account the proper Reynolds Number and the surface roughness. With a higher degree of polish, absolute efficiency of 77 percent for the first stage impeller and 74 percent for the last stage could be reached. However, these values are optimistic and efficiencies of 75 percent and 73 percent are selected for the first and the last stage of the compressor train.

TABLE 1—Ethylene compression stages

Compressor casing	Pressure p (bar)	Temperature T (°C)	Specific volume v (m³/kg 10³)	Flow coefficient ϕ (—)	Width of impeller at O.D. b (mm)	Efficiency η (%)	Power N_i (mw)
I inlet	245	40	2.53	0.046	5.3	75	4.99
outlet	715	89	2.23	0.041	4.6		
II inlet	715	40	2.05	0.038	4.5	74	5.06
outlet	1250	77	1.95	0.036	4.1		
III inlet	1250	40	1.86	0.034	3.9	73.5	5.10
outlet	1820	73	1.81	0.033	3.7		
IV inlet	1820	40	1.75	0.032	3.7	73	5.13
outlet	2450	70	1.71	0.031	3.6		
Total						73.8	20.28

The various stations along the compression are defined by Table 1. To simplify the presentation, the following assumptions have been made:

- No pressure drop between the compressors

- Same isentropic head of 107.5 kJ/kg in each one of the four compressors.

It must be noticed that the last assumption (same isentropic head) has very similar power consumption in all compressors, and gas outlet temperatures decreasing with the pressure level. Both effects are desirable. In a final design, of course, other considerations may lead to a slightly different division of pressures.

Efficiencies and power values given in Table 1 consider only the labyrinth leakage of the impeller. The following losses are excluded:

- Seal friction

- Seal oil supply

- Bearing friction

- Ethylene discharge to lower pressures.

Oil losses at shaft seals. Because no tests at pressure levels and differential pressures prevailing in the installation are available, calculation of oil losses is based on tests carried out at lower pressures, corresponding to synthesis gas compressors. For the conversion of the losses at higher pressures, we have to consider the influence of the modification of viscosity with pressure.[1]

Oil flow should be proportional to the third power of the radial clearance S for constant viscosity. However, oil temperature increases with smaller radial clearances and consequently oil losses are greater. The influence of the seal length L is a linear function, although the losses should be somewhat greater than inversely proportional to the seal length due to a larger temperature increase. For a given peripheral speed and clearance the loss has been assumed to be proportional to the diameter D.

Based on the above philosophy, the flows referred to the selected geometry of the oil seals have been estimated.

We shall assume the maximum oil losses as follows:

At seal	245	715	1,250	1,820	2,450 (bar)
Oil loss	43	96	145	193	243 (cm³/s)
Power loss	1.5	9	25	47	80 (kw)

The power loss can be calculated from:

$$N = \frac{(Q_2)\ (\Delta P_2)}{\eta} \times 10^{-3}$$

Where: N = Power loss, kw
Q_2 = Oil loss, m^3/s
ΔP_2 = Differential pressure at seal, N/m^2
η = Pump efficiency = 0.75.

Shaft seals can only work properly when the oil is neither dissolved nor diluted by the gas. This eliminates most of the natural mineral oils which are all more or less dissolved by ethylene, especially at higher pressures. However, several synthetic fluids have been developed as lubricants for the reciprocating hypercompressors used in the ldPE-industry which resist this action of the gas and could be used as sealing fluids.

Ethylene gas losses. Besides losses at the impeller seals already considered in the stage efficiency, a certain amount of gas losses occurs at the balancing piston of the compressors or, if the gas pressure is reduced to the level of sealing oil pressure, at the shaft seal. Based on tests with air and neglecting the compressibility, which is very small, the losses can be calculated. For the compressor design dealt with here, the average lengths of the seals are of 50 mm with a diameter of about 70 mm. A radial clearance of 0.2 mm is considered practical with this type of machine.

Calculations indicate that the power losses due to the ethylene gas at the balancing pistons are relatively great and that it is more economical to compress the sealing oil to a higher pressure level to meet the ethylene pressure than to reduce the ethylene pressure to a lower level.

A pressure drop of the ethylene gas greater than the ΔP generated in one casing would lead to a relatively high temperature increase of the gas and a danger of polymerization. On the other hand, the compression of oil at high pressure level is mechanically difficult. A cascade system could be used in which the pump working at higher pressure level could be placed with its drive in a pressure vessel maintained under intermediate pressures. In this case, normal pumping equipment could be used. Another possibility would be to use balanced double-flow rotary screw pumps, if necessary, connected in series in case the total pressure increases were difficult to reach in a single casing. Of course, slow running piston pumps could be used. The pump would be similar to a small ethylene piston compressor.

From all these possibilities, the double-flow balanced screw-type pump concept seems to be the most attractive one.

Friction losses of seals and bearings. Seals can be considered as supplementary radial bearings. Friction losses can be estimated on the basis of the test results since the ΔT_{oil} is known. For the tested ring, a value of 2 kw is obtained. Since the oil seal for the compressor is about four times longer, a loss of about 8 kw is assumed per seal or 16 kw for two seals, i.e., about 0.3 percent of the 5.5 mw of one compressor unit.

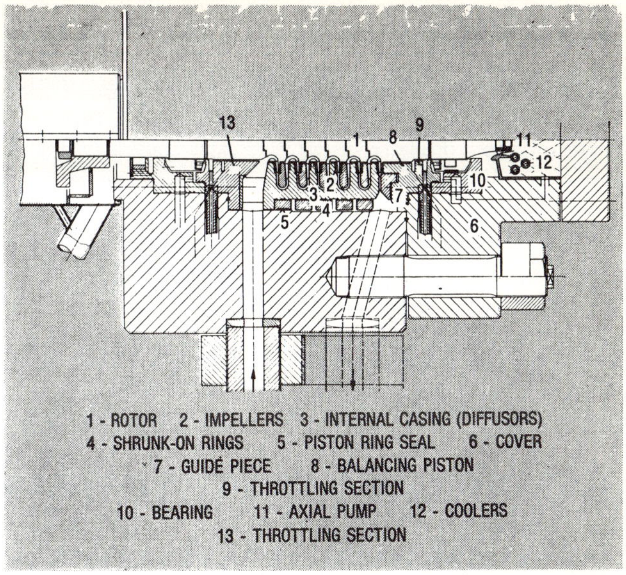

Fig. 2—Design of a radial hypercompressor for ethylene service.

Radial bearing losses can be assumed to be 0.15 percent or 8 kw for each machine since the bearings are narrower than the two sealing rings.

An estimation of power loss generated by the axial bearing is necessary. As shown in the literature,[3] power dissipated by an axial bearing greatly depends on the type of lubrication. It is clear that in this case only a directed lubrication is permissible. A power saving up to 50 percent can be achieved. For instance, at 36,500 rpm a bearing lubricated by flooded lubrication absorbs about 170 kw. The same type of tilting-pad thrust bearing would absorb about 60 kw when fitted with directed lubrication.

A solution is to use a thrust transmission gear with rigid coupling. This permits a greater bearing load while eliminating gear coupling friction due to axial expansion of the shaft. A system capable of absorbing 5 tons of axial thrust per machine is necessary for this type of equipment.

Compressor design. The design principle can be seen from Fig. 2. Because ethylene losses are relatively high when compared to the pumping power of the sealing oil, it is advantageous to try to seal the shaft at the corresponding working pressure. An ethylene pressure reduction to the suction pressure of the preceding machine at the shaft seal of compressor IV fitted with a balancing piston, for example, increases the ethylene losses by about 2 × 310 = 620 kw (two shaft ends) and reduces the oil pumping power by only 140 − 75 = 65 kw. Another reason for not doing this is because ethylene temperature would be too high at the end of the throttling.

Because losses due to ethylene leakage at the balancing piston are important, try to eliminate at least one balancing piston on one casing by axially connecting two machines so that their axial thrust is opposed.

In addition, try to avoid all ethylene and sealing oil leakage to lower pressures or to atmosphere by closing the casing at the nondriving shaft end. This is feasible if the respective shaft ends are submitted to exactly the same sealing oil pressure, regardless of operating conditions.

In the present design, the study of one compressor was based on the above principle. The last casing, submitted to the delivery pressure of 2,450 bar, is shown in Fig. 2. The rotor (1) is built up by the impellers themselves (2), which are joined together by electronic welding. Tests were carried out for gas turbine rotors, and the results were very satisfactory. On a rotor of about 400 mm diameter and a total length of 396 mm, comprising three discs welded together, a maximum eccentricity of only 0.04 mm was noticed. Due to this remarkable small deformation, the pieces may be finished prior to welding. Tests showed that full penetration of the weld is reached. This type of impeller assembly eliminates internal hysteresis of the rotor, while increasing its rigidity and strength for torque transmission. Impellers are made with a special gold brazing technique.

The internal casing (3) is split horizontally and both halves are held together by a series of rings (4) shrunk around the casing. A good symmetry is therefore achieved, avoiding flanges and bolts.

This internal casing is submitted externally to discharge pressure, pressing both halves together tightly. A piston ring seal (5) avoids bypass of gas from the discharge side to the suction. Seals between the cover and the casing could be a delta gasket or, as is the case for the other static seals, by O-rings with nonextrusion washers. The cover (6) is bolted on the casing and prestressing is accomplished by the usual method reverting to heating rods and thermal expansion of the bolts. During fabrication the complete external casing is submitted to autofrettage which eliminates subsequent peak stresses by local yielding.

The internal casing is free to expand axially and is guided in a piece (7) fitted in the cover. The opposite end is fixed by screws to the integral cover of the external casing.

The nondriving end of the shaft has a balancing piston (8), followed by a short throttling section (9) through which a small amount of gas is continuously withdrawn against the oil, providing for a sufficient sealing gas stream. Intensity of this gas stream is adjusted by external throttling and measured by a suitable device. Since this gas is polluted by oil mist, it can be cleaned and sent back to the process stream.

In the described cross-section, the nondriving end of the shaft is under full suction pressure. In this case, a combination of seal bearing (10) has been selected,[4] the necessary flow of lubricating oil being maintained by an axial pump (11) fitted on the shaft. Necessary coolers (12) are also integrated in the design. This combined bearing is fitted with tilting pads and supported by resilient elements to insure maximum damping capacity. The combined seal bearing was tested at peripheral speeds up to 120 m/s and 180 bar sealing differential pressure.

On the driving end a conventional oil seal and a separate bearing are provided. Here again, a small quantity of sealing gas is withdrawn against the oil. In this design the shaft has a throttling section (13) insuring that the pressure selected at the seal corresponds to suction pressure of the compressor placed before this machine, and this simplifies the sealing oil system. If such is not the case, corresponding throttling is eliminated.

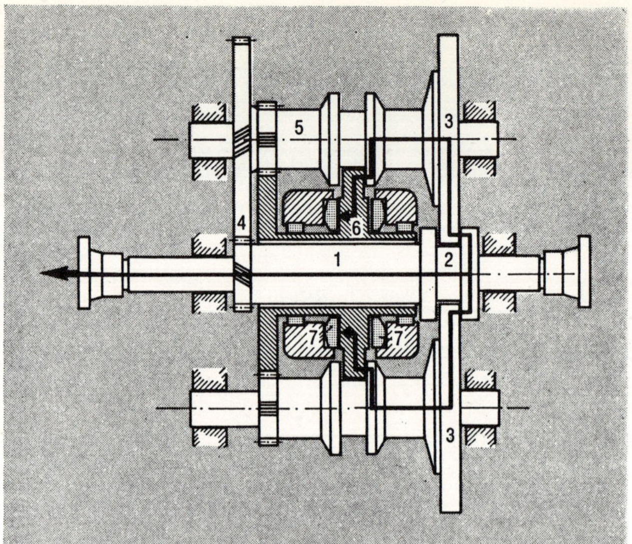

Fig. 3—Thrust block.

Radial bearings are of the tilting pads type insuring a better running stability. In this proposal, no axial bearing has been incorporated in the design, since either a thrust block will be used or a thrust transmission through the driving gear to a low-speed axial bearing is incorporated in the gear.

Principle of the thrust block is shown in Fig. 3. The driving shaft (1) is transmitting its thrust through the interlocked mating pieces (2) and (3) to an axial bearing (6) placed concentrically around the shaft. Gears (4) and (5) are synchronizing gears.

The thrust block was tested at 18,000 rpm under an axial load transmission of 15 tons, which is much more than the capacity of an axial bearing fitted directly on the compressor shaft which could hardly reach 5 tons. Furthermore, the power losses are much smaller with a thrust block. They are about 150 kw. Such a thrust block could be sufficient for at least two casings coupled rigidly to each other. This compares to a loss of 100 kw per bearing with directed lubrication which is about half of the power dissipated by normal lubrication. However, directed lubrication is not easy to achieve.

DISCUSSION OF THE COMPRESSOR DESIGN WITH RESPECT TO KNOW-HOW

It is clear that these machines are rather different from what was used so far. For this reason it is worthwhile to make a few comparisons with past experience on turbomachines having a high degree of power concentration, like large boiler feed pumps.

Balancing system. One could ask the question whether another balancing system should be used as in the case for

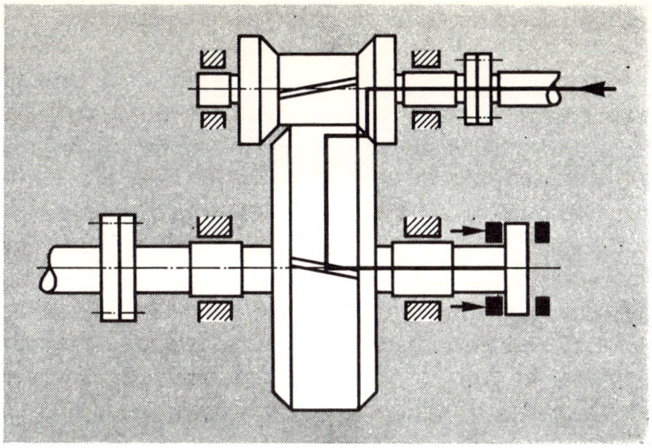

Fig. 4—Principle of thrust transmission through gears.

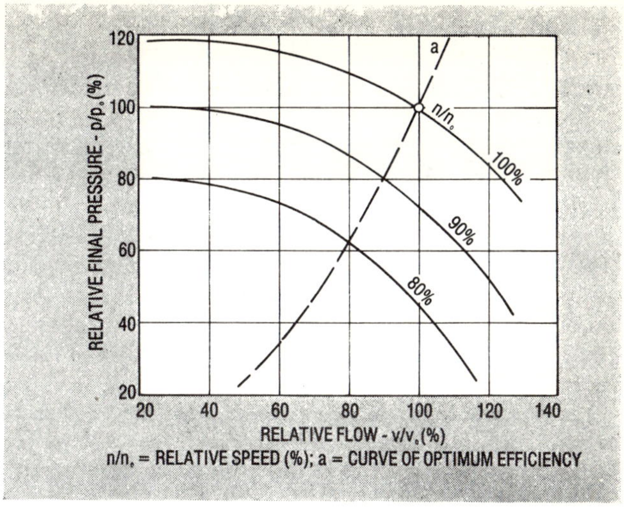

Fig. 5—Compressor pressure-flow characteristics.

boiler feed pumps. A study[7] shows that relative leakage losses are smallest for a balancing disc system. However, additional friction losses of the balancing disc show clearly that this system has no advantage over a balancing piston for machines having less than seven to eight stages. The balancing disc system must, therefore, be rejected.

Seals. As said previously, experience at low speeds with seals working at 2,500 bar pressure differential is available. No difficulty should be experienced with the sealing problem. The supply of sufficient oil at pressure level as high as 1,250 bar or more could give some difficulties, but we are confident that either balanced rotary pumps or slow running piston pumps can be used successfully.

Gear. Using a gear makes the problem of finding a suitable turbine easier, but the gear problem is probably difficult to solve.

The speed of a 12-mw condensation turbine able to drive two compressors will be in the range 10,000 to 15,000 rpm. A single-stage normal gear is not sufficient for this duty because of too high loads on teeth and bearings. In principle two solutions are possible:

- Two 6-mw gears, one for each compressor
- One 12-mw gear, with double action on the pinion.

Turbine with a drive at both ends. In this case the rotors of each compressor have two shaft seals. Residual thrust can be transmitted through the gear to a slow-running axial bearing. Here again, a simple 6-mw gear would not be safe enough. The best solution would be a double-path gear for each compressor. The loss would be about 300 kw for each gear.

Turbine with one shaft-end driving two compressors coupled in parallel. In this case a simple gear is too heavily loaded. A double path or a gear with double action on the pinion on each side of the turbine shaft could be used. However, these solutions lead to great design complications and are not advisable.

Turbine with one shaft driving two compressors. In this case, the gear can be single-stage gear with two compressors in opposition with only two atmospheric shaft seals. A gear with double action on the pinion must be used. If the compressors are in line, the group must have four atmospheric shaft seals. The gear will be of the double path type.

Opposed compressors insure the best possible balance of axial forces; the remaining ones must be taken over by axial bearings or by a thrust block. Total losses due to gear and axial bearings are about 840 kw. These relatively big losses, based on previous experience, are due to the high circumferential velocity. They could perhaps be slightly reduced by careful studies and trials.

Fig. 4 shows the principle of thrust transmission to the slow-running gear. High-speed gears based on this principle can be derived from the many gears actually in service.

Characteristics. Pressure-flow characteristics are given in Fig. 5. As can be seen, a certain amount of the flow variation can be realized by speed variation. The characteristics are similar to those of high-pressure boiler feed pumps. It is, therefore, possible that surging will be greatly damped out as is the case for water pumps.

SUMMARY OF THE PRESENT DEVELOPMENT'S RESULTS

Efficiency.

- Summing up total losses including gears, sealing oil and ethylene throttling, an over-all efficiency of about 65 percent can be estimated.
- The version with high-speed compressors is a necessity if an acceptable efficiency must be reached, however, at the cost of a more complicated installation. Compared to a solution for about 18,000 rpm leading to direct drive, an improvement in efficiency of 5-10 points or a saving of 10-15 percent in power consumption can be estimated.

• Even the high-speed solution consumes about 25 percent more power than slow-runnig reciprocating compressors now in use.

Design and operation of the compressors.

• Realization of the 36,500-rpm machines does not offer great technical difficulties.

• High-pressure oil seals should not give rise to mechanical difficulties.

• A certain amount of capacity regulation can be obtained by speed variation. The same method will be used to adapt the compressors to the different levels of delivery pressure required.

• Problems created by the fouling or plugging of the clearances and narrow channels by polymers carried by the gas can only be solved by extensive tests in ldPE-production plants under normal operating conditions.

Drivers.

• Steam turbines with back pressure or limited partial condensation can be found for 16,000-18,000 rpm up to the required power of 11-12 mw for each group of two compressors.

• Condensation turbines would have lower speeds between 10,000 and 14,000 rpm which makes the use of a speed increaser unavoidable.

• Gas turbines of this power have still much lower speeds in the range of 5,000-6,000 rpm. Used as drivers, they would make much more complicated and expensive gears necessary.

Transmission.

• For this equipment it is advisable to eliminate all gear coupling and especially between gear and compressors. In this case, thrust due to teeth friction is also eliminated.

• According to this, the axial thrust must be transmitted to the low-speed gear forming the fix point of the shaft assembly. When doing so, the thrust bearing has a greater capacity and the operating safety is enhanced.

• The most loaded and weakest point in the layout having a speed increaser to reach 36,000 rpm is certainly the gear itself.

• Use of a gear with thrust transmission to a slow-running axial bearing is a better solution than using a high-speed thrust bearing on each casing. Use of an automatic oil balancing piston located inside the casing would induce considerable friction losses.

Layout.

• The layout with two opposed compressors offers inherent advantages, simplifying the oil sealing system. The driving gear, however, is more loaded and complicated.

• Within the range of capacities which can be contemplated for the near future, the space requirement of a centrifugal compression set with its driver would be

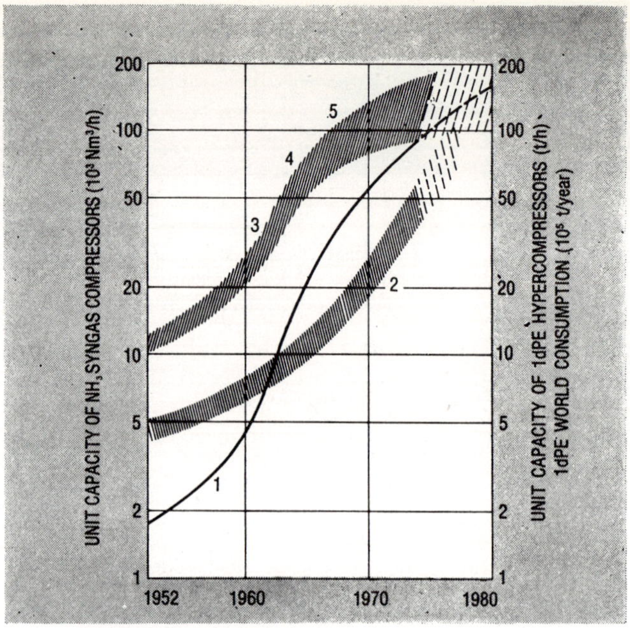

1. WORLD CONSUMPTION OF ldPE
2. LARGE HYPERCOMPRESSORS IN OPERATION FOR ldPE PRODUCTION
3. LARGE NH_3 SYNGAS COMPRESSOR UNITS: RECIPROCATING COMPRESSORS UP TO ABOUT 50,000 Nm^3/h, CENTRIFUGAL COMPRESSORS ABOVE

Fig. 6—Evolution of ldPE consumption and large hypercompressors compared with syngas compressors.

smaller than for a reciprocating compressor with motor but nevertheless at least half the space. On the other hand, the heat exchangers would need much more space so that the total required area would be about the same.

Conclusions. Centrifugal compressors for ethylene with delivery pressures of 2,000-2,500 bar are feasible for capacities of more than 100 t/h. They have several advantages compared with the reciprocating units. On the other hand, they have a certain number of disadvantages.

It is difficult to establish a clear comparison of costs of the two solutions because the economic conditions are different from one country to the other, especially regarding the price of the different kinds of energy. For units of the size which can be contemplated today (100-125 t/h), it appears that the investment costs for the compression set would be of the same order for the reciprocating as for the centrifugal unit (if it were readily developed). This estimation includes the compressor with complete drive, the intercoolers, the foundation and the erection. It does not include the boiler plant.

Therefore, it seems doubtful whether the centrifugal hypercompressor may be attractive in this range of capacities. Another consideration must be taken into account. The corresponding primary compressors could not be designed as centrifugal ones, in spite of their much lower delivery pressure, since their capacities are still too small (30-40 t/h). It is to be feared that a combined plant with a reciprocating primary and a centrifugal hyper would not be a satisfactory solution since the primary must be stopped from time to time for replacement of valves or piston rings.

The situation would be very different for still larger capacities, for instance, double the ones considered in our present studies. A centrifugal hypercompressor with a capacity of 200-250 t/h and a power consumption of about 40 mw would have an improved efficiency. As this output could be reached only by means of two or more reciprocating compressors in parallel, all other aspects (space requirement, investment, maintenance costs) would be in favor of the centrifugal unit. Furthermore, the primary compressor could probably also be built as centrifugal.

Thus, the final choice depends chiefly on future development of ldPE consumption and production on one side and on the development of large single-stream production plants on the other side. Fig. 6 shows the general features of the evolution over the years. If no basic modifications take place in the polymerization processes and in the market conditions, a further increase of the ldPE consumption may be expected approximately according to the dotted line.

LITERATURE CITED

1. Strub, R. A., "Solution de quelques problemes peu usuels de mecanique poses par l'industrie chimique." *Bulletin Technique de la Suisse romande* No. 21, 1963, pp. 351-359.
2. Bara, P., Durchflussbeiwert u fur Labyrinthdichtungen. Report No. 1324, Sulzer Brothers Ltd., Research and Development Department, Sept. 27, 1962.
3. Leopard, A. J., "Direct lubrication for tilting pad thrust bearings," *Journal of Tribology*, November 1970 (Glacer Metal Co.).
4. Strub, R. A., Stuffing box of rotating shaft (combination of shaft seal with radial bearing), U.S. Patent 3756673 (Sept. 4, 1973). Application July 9, 1971 (Switzerland), Brown Boveri Sulzer Turbomachinery Ltd., Zurich.
5. Strub, R. A. Ausgleichsysteme des Axialschubes in Pumpen. Report Sulzer, W., No. 21313, Oct. 25, 1954.
6. Matile, C., "Industrial Reciprocating Compressors for Very High Pressures," *Sulzer Technical Review*, Vol. 53 (1971), No. 2, pp. 97-104.

Part III.

Centrifugal Compressors—Instrumentation and Control Systems

How to Instrument Centrifugal Compressors

Unless capacity control is provided for centrifugal compressors, the system may become unstable, causing surging and upsets in the compressor and process.

Richard E. Daze, The M. W. Kellogg Co., New York

THE PERFORMANCE CHARACTERISTICS of the centrifugal compressor are such that unless adequate and proper capacity control is provided, through instrumentation, the compressor circuit may become unstable, causing surging and upsets in the compressor and process. The objective therefore of any centrifugal compressor control system is to achieve smooth capacity regulation and to prevent the compressor from surging, even though the process flow drops below the surge limit of the compressor.

This article will describe some of the methods used to control the capacity of centrifugal compressors and illustrate typical instrumentation for several compressor services.

System Characteristics. To determine the optimum compressor control and instrumentation requirements for a given application, thé performance characteristics of the compressor and of the connected system must be established Basically the centrifugal compressor is a variable capacity, nearly constant pressure ratio machine. In contrast, the reciprocating compressor is generally a constant capacity, variable pressure ratio machine. The operating point of the compressor is always determined by the intersection of the compressor pressure vs. capacity curves and the system pressure vs. capacity curve.

There are three types of system characteristic curves which are normally encountered. They are:

- **All flow resistance** as in gas pipe lines and approached in recycle circuits,
- **Constant pressure** as in systems employing back-pressure regulator or other flow sensitive throttling device (Condensing systems, such as propane or ammonia refrigeration are usually of the constant pressure type),
- **Combination of the above** which apply to the more general cases, where part of the total pressure drop is constant, and part is due to flow resistance.

The various types of system curves are shown together with a typical compressor performance curve in Figure 1.

Variable Speed Control. Speed variation is a simple and effective method of controlling the capacity of a centrifugal compressor. It has many advantages. Variable speed can provide an infinite "family" of compressor head-capacity curves over the driver speed range. Figure 2 shows that capacity control by speed variation is most efficient as it requires the least part load horsepower compared to other control methods.

From the curves in Figure 1, it is seen that the effect of speed on the capacity of the compressor varies with the type of system resistance. If the system pressure is constant, a small speed change will effect a large capacity change as shown by increment "A". On the other hand, the same speed change in a flow resistance circuit would produce a smaller capacity change as shown by increment "C".

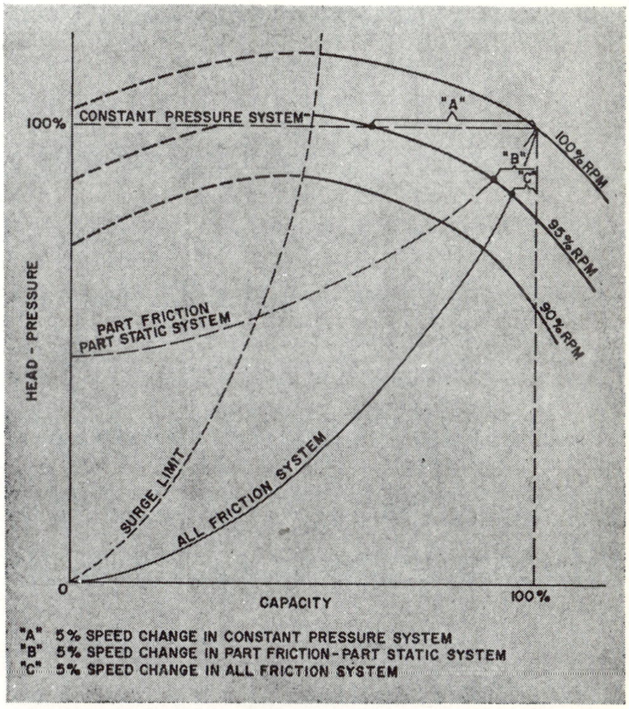

Fig. 1—The effect of compressor speed on system flow. A 5-percent decrease in speed of a compressor in a constant pressure system will effect a much larger reduction in capacity than a 5-percent decrease of compressor speed in a friction circuit.

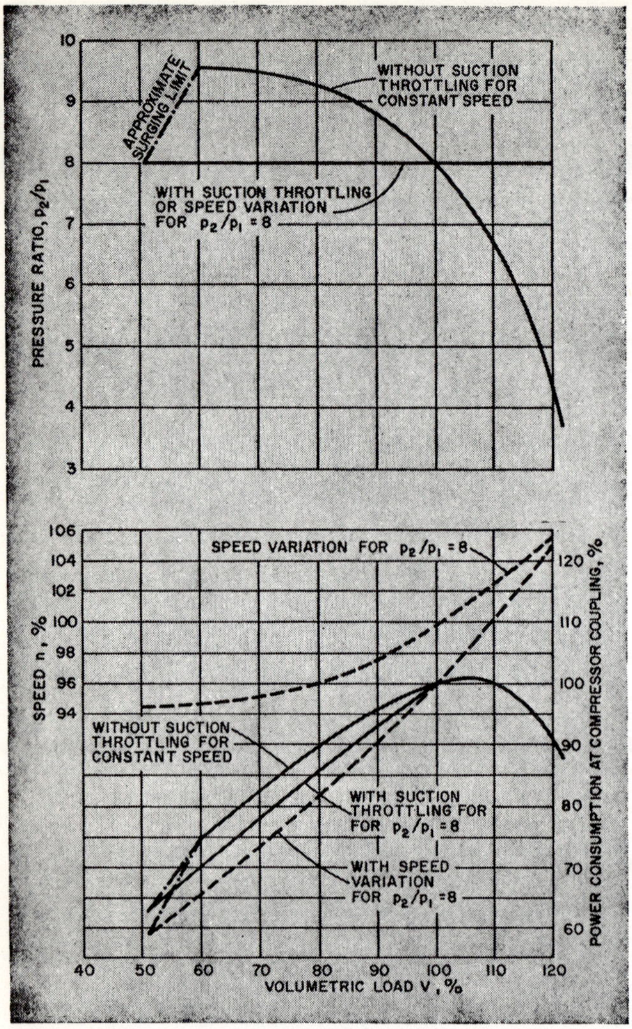

Fig. 2—Capacity control by speed variation.

The steam turbine is particularly suitable for variable speed duty. The turbine governor is usually provided with an air head so that a 3-15 psig instrument air signal from the process controller can be applied. The air signal resets the governor to maintain a new set speed. With the turbine under governor control, a change in speed caused by a change in steam conditions will be corrected by the governor when it senses a speed change. NEMA has defined the control characteristics of turbine governors and Table 1 gives the specification for various class governors. For most compressor control applications in process services, a NEMA Class "C" turbine governor is suitable.

Constant Speed Capacity Control. Many compressor installations today are based on constant speed because:

1. Driver speed is constant—such as with electric motors.
2. Limited speed variation—such as may occur with single shaft gas turbines.
3. Multiple compressor services require constant speed.

TABLE 1—Speed Governor Standards

NEMA* Class	Adjustable Speed Range (%)	Maximum Steady State Speed Regulation (%)	Maximum Speed Variation Plus or Minus (%)	Maximum Speed Rise (%)	Trip Speed (% Above Rated Speed)
A	10	10	0.75	13	15
	20	10	0.75	13	15
	30	10	0.75	13	15
	50	10	0.75	13	15
	65	10	0.75	13	15
B	10	6	0.50	7	10
	20	6	0.50	7	10
	30	6	0.50	7	10
	50	6	0.50	7	10
	65	6	0.50	7	10
	80	6	0.50	7	10
C	10	4	0.25	7	10
	20	4	0.25	7	10
	30	4	0.25	7	10
	50	4	0.25	7	10
	65	4	0.25	7	10
	80	4	0.25	7	10
D	30	0.50	0.25	7	10
	50	0.50	0.25	7	10
	65	0.50	0.25	7	10
	80	0.50	0.25	7	10
	85	0.50	0.25	7	10
	90	0.50	0.25	7	10

Oftentimes, several compressor casings in different process services are arranged for drive-through by a single driver. In the case of a gas turbine drive, this is done to match the compressor loads to the available gas turbine horsepower, since there are large horsepower increments between gas turbine ratings. With the multiple service arrangements, a change in speed would affect more than one compressor service, which may be undesirable. Hence, with multiple compressor arrangements, the driver—whether steam or gas turbine—is usually run at constant speed.

To affect capacity control under constant speed conditions several methods are available, namely:

- Discharge throttling
- Suction throttling
- Variable inlet guide vanes

Discharge Throttling. Throttling of the compressor discharge will cause the compressor to follow the pressure-capacity curve as shown on the typical performance curve in Figure 1. This method is not commonly used as it is inefficient and wastes power.

Suction throttling is simple and more efficient than discharge throttling because:

- Less pressure has to be throttled on the suction to accomplish the same result as discharge throttling. If a compressor has a compression ratio of say, 8, a throttling of 1 psi on the suction side is equal to throttling 8 psi on the discharge side. By throttling the suction, the actual CFM to the compressor is being increased even though the up stream CFM or weight flow is reduced. This allows the compressor to operate further out on its head capacity curve—closer to its maximum efficiency point—so that a much smaller amount of head or pressure must be throttled than with discharge throttling. Since less head or pressure is throttled, less power is wasted. The difference in the power requirements between discharge throttling and suction throttling is illustrated in Figure 2.

- The increased CFM to the compressor because of suction throttling also has the effect of lowering the surge point of the compressor system when compared to the upstream flow. Figure 2 illustrates the reduction of the surge point by suction throttling in contrast to discharge throttling. Suction throttling is accomplished by providing a butterfly valve or suction damper in the inlet piping to the compressor. These devices are usually operated by means of an air diaphragm and equipped with a clutch and handwheel for manual control.

Variable inlet guide vanes have a two-fold action:

1. The suction pressure is reduced (as in suction throttling) thereby creating a greater inlet CFM.

2. The flow is directed through adjustable vanes, so that pressure energy is efficiently converted to velocity energy. This velocity energy can, via adjustment of the vanes, be directed to cause prerotation of the gas, with or against compressor rotation. By adjusting the gas rotation relative to the impeller rotation, the first-stage impeller can be effectively loaded or unloaded.

Although variable inlet guide vanes can be put at the inlet of all stages of the compressor, it is more common to have only the first stage provided with this feature. The effect of the variable first stage flow performance has a significant effect on the performance of the subsequent stages so that the over-all compressor performance will be markedly changed. The basic effect of variable inlet guide vanes is to alter the impeller velocity diagrams. This can affect a change in impeller performance with a minimum of throttling, therefore, it is a more efficient method than suction throttling.

By using variable inlet guide vanes, the efficiency will be improved over a wider operating range. However, the peak efficiency may be slightly lowered because of fluid drag even when the inlet guide vanes are wide open.

Should the gas be dirty, there may be a tendency for the vanes to stick if any solids or deposits build up.

The variable inlet guide vanes are relatively expensive compressor features. They are justified only if the compressor is expected to operate at part loads for considerable periods of time and the resulting savings in driver power costs are significant.

Any of the above methods can be used for variation of the compressor capacity, in accordance with the process demands. However, in every case a minimum flow condition will be reached which may produce surging of the compressor.

Surge. It is an inherent characteristic of the centrifugal compressor that its performance becomes unstable at some minimum flow point. This is mainly caused by the drooping head characteristic between the highest pressure point and shut-off as shown by the dotted portion of the head-capacity curve in Figure 1. The surge point depends on such factors as the number of stages, the relative wheel loading, blade angle, method of control, etc. Compressors with few wheels—say three or four—may have this surge point at about 50 percent of design flow while compressors with many stages—say eight to ten—may have their surge point at about 85 percent of design flow. The greater the number of stages, the higher the surge point.

Since the turn-down range of centrifugal compressors is from 15 percent to 50 percent, it is essential that the compressor installation be provided with a manual or automatic flow controlled minimum bypass system to prevent the compressor circuit from going into surge.

With a flow controller measuring total compressor flow, the controller can be set at a predetermined minimum—say, the estimated surge volume plus 5 or 10 percent—so that when the process flow falls below the minimum set point, the kickback or recycle valve will open, returning gas to the compressor suction and thereby assuring a continuous minimum flow to the compressor inlet. Since the gas has been heated by the compression it is important that the bypassed gas be cooled to approximately the normal inlet gas temperature by means of a process cooler, a suction cooler, a special bypass cooler or by liquid injection into the bypass stream. If the gas were not cooled, a continuous temperature build-up would occur as the hot discharge gas mixes with the suction gas. Any increase over the normal gas inlet temperature will cause the compression ratio put up by the compressor to fall off.

In the more complex installations, the compressor may have one or more inlet or extraction side streams. If the side stream flow is of such a magnitude that its failure would cause the compressor to surge, then the side streams must be flow controlled to insure that the predetermined minimum flow will always be entering the compressor. Cooling of the recycled gas to the side stream inlets is often necessary.

It is important to have the bypass line take off between the compressor outlet and the discharge check valve. This arrangement will allow the compressor to operate within its own recycle loop even though the compressor discharge pressure is insufficient to lift the check valve.

On compressors with multiple connections, each side stream should be examined for the effect that boil-offs from evaporators might have on a compressor that has been shut down. If a flow of gas would cause reverse rotation of the compressor, the side stream should be provided with a check valve.

TYPICAL APPLICATIONS

Outlined above are some of the basic principles of compressor control; however, each specific application has its own peculiarities. Many of the applications such as air blowers for FCCU service, gas recovery, reformer recycle, and refrigeration cycles that are encountered quite frequently, lend themselves to establishing a basis of minimum instrumentation. Some of the more common arrangements are described on the next page.

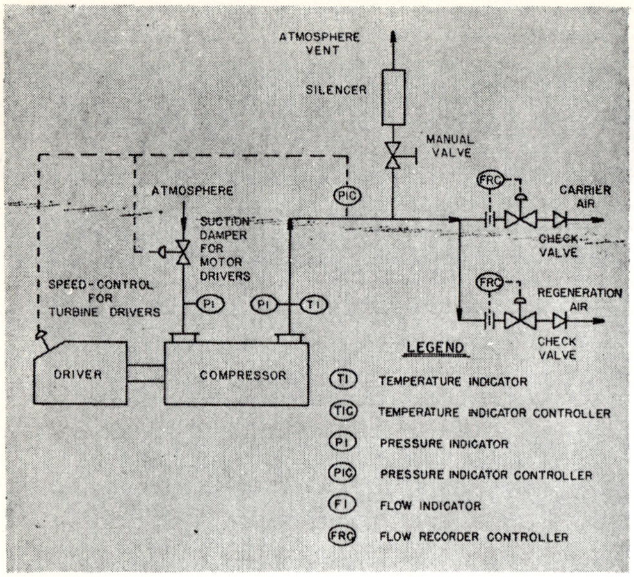

Fig. 3—Air blower for fluid catalytic cracking unit.

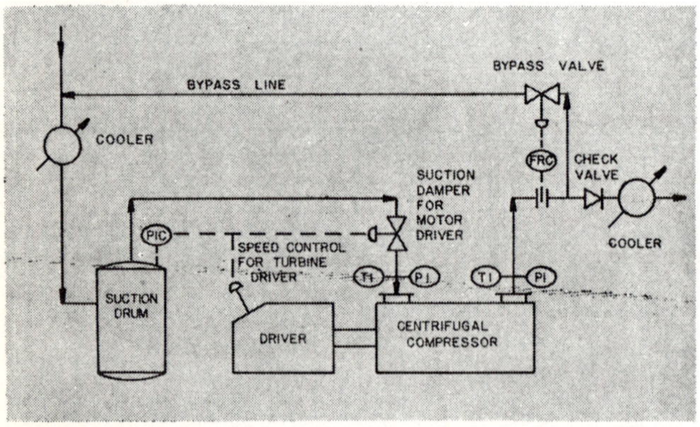

Fig. 4—Single case for gas recovery compressor in FCC unit.

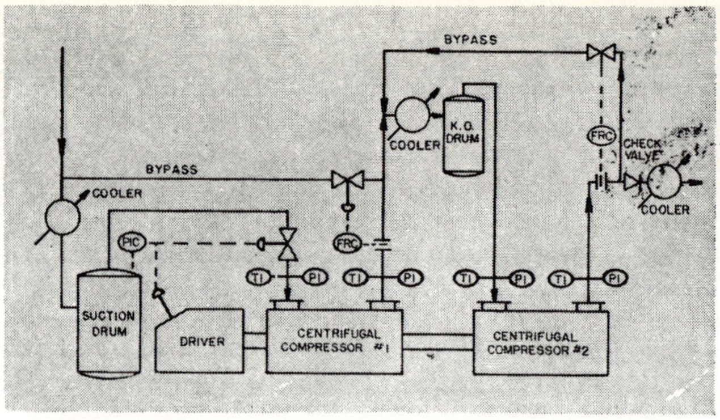

Fig. 5—Two-case gas recovery compressor for FCCU.

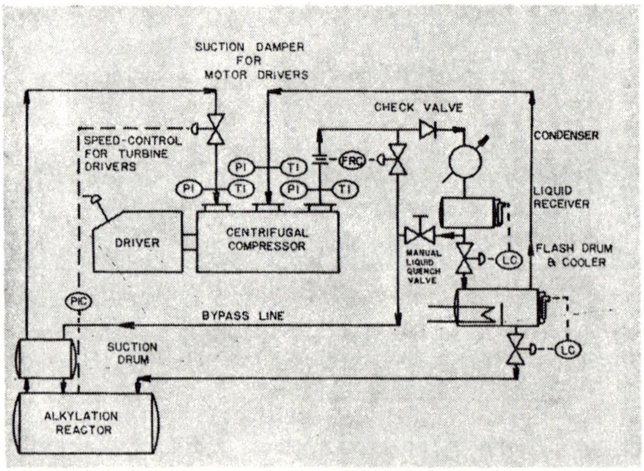

Fig. 6—Alkylation refrigeration compressor.

Air Blower For FCC Unit. The air blower in a fluid catalytic cracking unit operates against a relatively constant discharge pressure set by the pressure in the catalyst regeneration system. If both the carrier air and regeneration air are supplied by a single air blower, the air requirements for each service must be individually flow controlled and the compressor discharge pressure controlled as shown in Figure 3. If all the air from the blower is used at a single pressure level then the compressor speed or suction damper may be placed on flow control. Since the operating conditions on the air blower are relatively constant, the minimum flow bypass valve is usually manually operated.

Gas Recovery Compressor of FCC Unit. The gas recovery compressor of an FCC unit operates against an essentially constant discharge pressure system. Figure 4 shows the instrumentation for a typical installation. A pressure controller on the suction drum actuates the compressor control system to maintain constant suction pressure. If the flow drops below a predetermined minimum, the flow controller will automatically open the bypass valve insuring sufficient compressor flow to stay in the stable region. It should be noted, in Figure 4, that the hot discharge gas is cooled by means of a heat exchanger on the suction side of the system. If the discharge cooler had been used to cool the gas, some of the heavy ends of the gas would condense out. Hence, a lighter mol weight gas would be returned to the compressor suction. The pressure ratio of the compressor would be reduced if the molecular weight of the inlet gas was lower than the design molecular weight. Figure 5 shows a two-case arrangement driven by a single driver. Each case is provided with its own minimum flow kickback to insure there will be no significant change in molecular weight of the bypassed gas.

Refrigeration Compressor For Alkylation Unit. The refrigeration compressor in an alkylation unit operates against a constant discharge pressure as set by the liquefication temperature of the gas in the condenser. A side stream or economizer connection is oftentimes provided on the compressor to reduce the horsepower requirements. The economizer usually "floats" on the flash drum at some intermediate pressure. Figure 6 shows a typical alkylation compressor installation. The temperature in the reactor is kept constant by maintaining a constant pressure. A pressure controller actuates the compressor control system to maintain a constant pressure to the centrifugal compressor. If the flow falls below a predetermined minimum, the flow controller will cause the

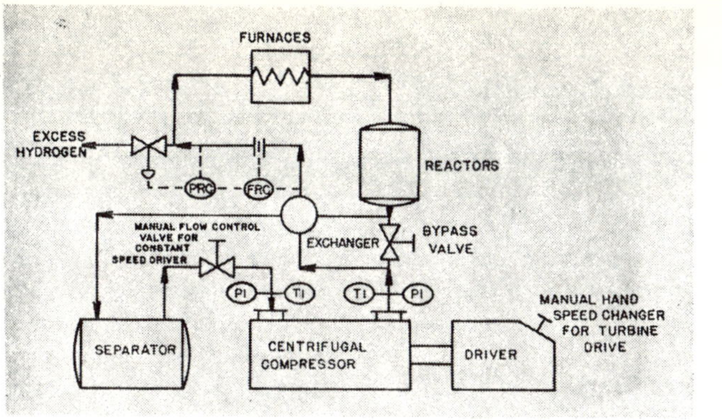

Fig. 7—Reformer recycle gas compressor.

bypass valve to open insuring sufficient compressor flow. The discharge gas recycled back to suction may be cooled by the addition of a liquid refrigerant quench. The side stream flow, while of a significant quantity, is not usually so large that its failure would cause the compressor to surge. No special instrumentation is provided for the side stream flow.

Reformer Recycle Compressor. The reformer recycle compressor operates in essentially a flow resistant circuit. That is, the pressure drop through the furnaces, exchangers, piping and reactors is almost all resistance. Referring to Figure 1, we can see that a compressor operating in an all resistance circuit is not likely to surge since the system curve is always to the right of the compressor surge curve. This characteristic makes the recycle compressor circuit inherently stable and no flow controller is normally required for minimum flow protection. The flow is usually manually set either by fixing the speed in the case of a variable speed drive or by manually adjusting the flow control valves in the case of a constant speed drive. During the course of a run as the catalyst activity decreases, periodic adjustments are made to either the speed setting or to the manual flow control valves. Figure 7 illustrates a typical reformer recycle installation. A manually operated bypass valve is provided for use at startup.

Refrigeration System. One of the more complex compressor applications is the low temperature refrigeration circuit. Figure 8 illustrates a typical refrigeration installation. Since the gas is condensed, the compressor operates essentially at a constant discharge pressure. Low level evaporator temperatures are held constant by maintaining constant suction pressure in the suction drum. A pressure controller on the suction drum actuates the compressor control system. In most cases, the sizes of the side stream flows are of the same order of magnitude as the suction streams. In order to maintain a good flow balance to all compressor impellers, the minimum flow controllers are installed in the suction side streams to the compressor. These flow controllers actuate bypass valves to supply gas flow to whatever stream has fallen below its predetermined minimum. However, the bypassed gas is at a high temperature level relative to the gas being compressed and unless it is cooled, the warm

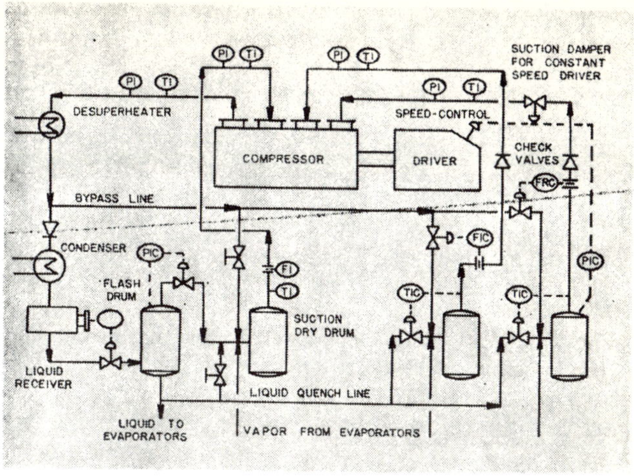

Fig. 8—Refrigeration system.

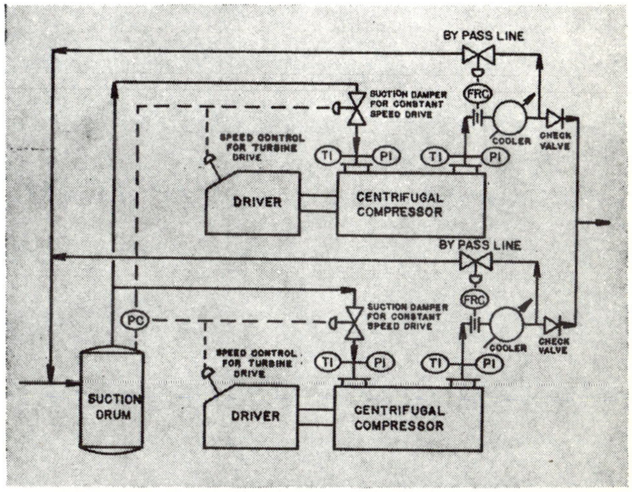

Fig. 9—Parallel operation of centrifugal compressors.

gas to the compressor suction will adversely reduce its pressure ratio.

Liquid refrigerant quench is used to maintain the gas temperature near the design gas inlet temperatures. At the compressor suction, the liquid quench is usually temperature controlled to maintain a relatively constant gas temperature. At successively higher inlet stages, whether the streams are flow and/or temperature controlled will depend upon the relative size of the side load compared to the flow of the main stream. Sufficient flow and temperature controllers must be put on automatic control to keep the compressor out of surge during normal operation. Additional manual flow and temperature controls may be provided to allow for starting up or other unusual operating conditions.

Parallel Operation. Centrifugal compressors operating in parallel are oftentimes used when:

- The process flow quantities are very large, indicating that more than one compressor is required to fulfill the process operating condition.

- Continuity of plant operation is of the utmost importance.
- Plant expansion requires adding compression facilities.

Oftentimes in revamping units, the compression capacity is increased by the addition of a centrifugal compressor. If the added compressor has a larger or smaller capacity than the original machine, its performance characteristics will obviously be different. Even duplicate machines may have different performance characteristics.

To insure stable operation and to prevent the effects of one machine surging to upset the rest of the system, minimum flow controllers must be provided for each individual machine. Figure 9 shows a typical arrangement. This arrangement provides for individual bypass lines and allows each unit to be started up separately, recycling gas within its own closed loop before being placed in parallel operation with the other compressor. If the flow to either of the compressors drops below its predetermined minimum, the bypass for that particular compressor will open keeping the machine out of surge and not upsetting the other compressors.

The above illustrations only indicate the minimum process instrumentation to insure stable operation. In addition, other instrumentation is normally provided, such as high liquid level alarm and shutdown on suction drums, high discharge gas temperature alarms and shutdown, low lube and seal oil pressure alarms and shutdowns and others, depending upon the installation. Sufficient instrumentation has been indicated on the schematic illustrations to run a field performance check for comparison with the compressor manufacturer's predicted performance, if a gas analysis is available.

Summary. While the control and instrumentation of each application must be studied individually, there are certain basic elements common to most installations which must be considered:

- The system characteristics must be established. It should be remembered that the operating point of the compressor is always determined by the intersection of the system pressure vs. capacity curve and the compressor pressure vs. capacity curve.
- Sufficient automatic flow control equipment must be provided to keep the compressor out of surge during normal operation. In addition, manual controls may be required for startup, shutdown or abnormal operating conditions.
- Recycled gas must be cooled if it is returned to the compressor inlet. Care must be taken to insure that cooling of the gas does not change its characteristics (i.e., mol weight).
- The bypass line must be taken off at a point between the compressor and the discharge check valve. With this arrangement the compressor can be recycling gas without having to lower the system pressure to a value lower than the compressor discharge pressure.
- Pressure, temperature, and flow indication points should be provided at each compressor inlet or outlet connection so that the data obtained from these instruments together with a gas analysis can be used to check compressor performance.

BIBLIOGRAPHY

NEMA Publication SM-20.
Escher Wyss News Reprint, #22009, Fig. 17, Page 12.

For Easier Compressor Control

A control system for multi-stage compression is designed and then tested with computer simulation.

A. Eli Nisenfeld and **Roger Miyasaki,** Olin Chemicals, Stanford, Conn.; **Tom Liem,** Aramco, Ltd., Saudi Arabia, and **J. M. Eskes,** Bechtel, Inc., San Francisco (All were associated with Bechtel, Inc. at the time this article was written.)

Computer simulation successfully integrates both process design and operational control of propane refrigeration. The simulation method applies to a $400-million liquified natural gas and sulfur plant proposed for Das Island in the Persian Gulf.

Significant insight into design requirements is given by the simulation technique, and leads to several critical recommendations. For one, compressor speed manipulation is unsatisfactory for suction pressure control because of oscillatory response and long settling time. This performance can cause overall instability because the propane process can interact with the feed gas and multi-component refrigeration systems to propagate effects.

Compressor control. Preliminary specifications stated turn-down on the compressors must be accomplished entirely by the control system—without relying on flaring and relief techniques. Further, the feasibility of any suggested system for this function had to be demonstrated before design could be approved.

An analog simulation of the processes and control system for both design and demonstration purposes was proposed by the engineer-constructor, Bechtel, Inc. The

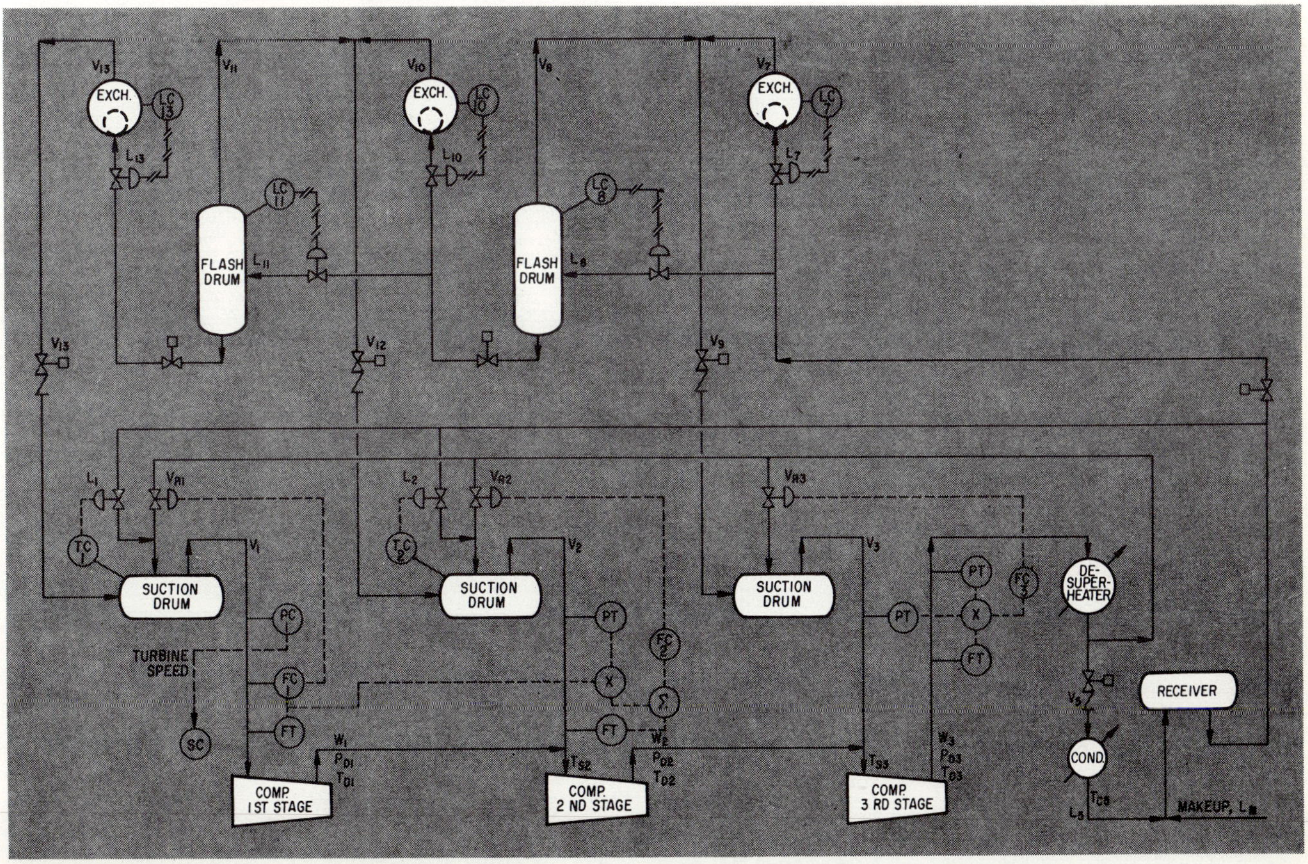

Fig. 1—How the three-stage compressor is controlled.

client agreed the technique was highly practical and considered the intended simulations to be significant factors in awarding the contract.

A propane refrigeration process, as shown in Fig. 1, was among four systems developed using simulation techniques. The process equipment includes a three-stage compressor with associated suction drums, propane receiver, propane flash drums, process exchangers, propane desuperheater, propane condenser and condensate receiver.

Assumptions used to simplify the model and produce equations for these elements are outlined in Table 1. For analysis of system trip conditions and to assure desired accuracy over a broad range of pressures, simulations were run also on a model excluding flash drums and vaporizers as shown in Fig. 2.

For simulation, individual performance curves for the compressor stages are approximated as linear relationships between head coefficients [(head)/(speed)^2] and flow coefficients [(inlet flow)/(speed)^2]. Parameters of the curves are determined from compressor manufacturer data for physical constants for propane. Eight differential equations describe material balances and valve operations for each stage. In addition, four algebraic equations represent flows between stages and six first-order differential equations account for energy balances.

The original intent was to perform analog simulation using two 10-V Electronic Associates Inc. Model 681 computers with 120 amplifiers each, and one Model TR-48 computer with 60 amplifiers. Seven Fisher Controls Co. TL-101 proportional-plus-integral controllers were to be wired directly to the computers to control three flows, two temperatures and one pressure, as well as simulate the compressor governor. Yet when the detailed equations were written, the model was too large for implementation on the available analog computers.

Rather than expand the system, it was decided to interface an existing Fisher Controls dc^2 process computer to the analog equipment, to share a portion of the real-time simulation load. The analog and digital computation were used concurrently for process modeling. The digital computer did not perform any control functions as it had when a similar combination was used to test design and control of a distillation column.[1]

Simulation. Equations describing the compressor stages, interstage interactions and material balances were implemented on the analog computer and run in real time. Time constants in these expressions were typically about 3 seconds.

Algebraic and differential equations representing the energy balances were programmed on the digital system. These were static, or had time constants typically an order of magnitude longer than those programmed on the analog computer. The equations tended to be coupled and algebraically complex.

Analog inputs to the digital program included 2 temperatures, 7 vapor flows, 3 gas flows, 4 densities and 2 masses. Analog outputs were 5 temperatures and 2 liquid flows. The simulation also required initial data parameters including 4 vessel volumes, 3 vaporizer temperatures, a make-up temperature and flow, 3 heat transfer ratios, 1 cooling water temperature, a total propane condenser area and volume, a propane condenser liquid density and 2 liquid temperature coefficient ratios.

TABLE 1—Assumptions to simplify simulation

Compressor stages
- Vendor-supplied curves adequately characterize performance
- No accumulation occurs
- Compressor-turbine moment of inertia can be lumped into a macro-model
- Fractional losses in the compressor-turbine system are proportional to the square of speed

Propane suction drums
- Liquid recycle is completely vaporized
- Heat of vaporization and specific heats are constant
- Molecular weight and compressibility are constant
- Perfect mixing occurs

Propane receiver and propane flash drums
- Vapor-liquid equilibrium exists at all times
- Vapor hold-up can be neglected compared to liquid hold-up
- Liquid density is constant
- Heat of vaporization and specific heats are constant
- Perfect mixing occurs

Heat exchangers
- Vapor-liquid equilibrium exists at all times
- Vapor hold-up is negligible compared to liquid hold-up
- Liquid density is constant
- Heat of vaporization and specific heats are constant
- Heat load is constant
- Perfect mixing occurs
- Parallel exchanges can be lumped for analysis

Propane desuperheater
- Perfect mixing occurs
- Molecular weight is constant
- Compressibility factor is constant
- Heat load is constant
- Specific heat is constant
- No condensation occurs

Propane condenser and liquid line to receiver
- Molecular weight is constant
- Compressibility factor is constant
- Condensing rate is proportional to heat transfer rate
- Pressure-temperature equilibrium exists at all times in the vapor space
- Heat of vaporization is constant
- Overall heat transfer coefficient is constant
- Water temperature is constant
- Condensate density is constant
- Condensate specific heat is constant
- Exposed surface area and liquid level are linear functions of liquid hold-up
- Perfect mixing occurs in both vapor and liquid phases

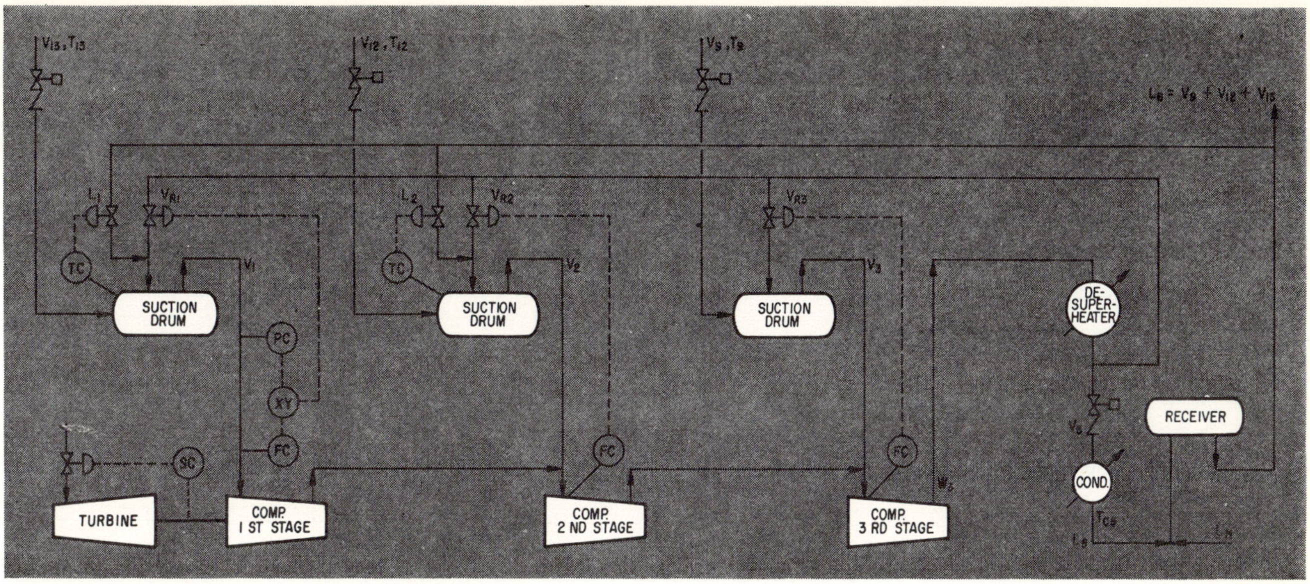

Fig. 2—Flash drums and vaporizers are excluded to increase accuracy over a broad range of pressures.

Logs were produced of all digital computer inputs and outputs. In addition 14 channels of strip chart recording were used to monitor analog computer signals, and trend recorders were linked with the process controller inputs and outputs.

The computational requirements imposed on the digital computer were reasonably strict owing to the complexity of the equations and the need for real-time operation. The clock on the system was accordingly modified to operate with 0.01-second timing, rather than the standard 1-second increments provided for process control. The computer was able to handle the increased load effectively, and ran through the set of calculations about seven times per second. Digital computer system performance was considered outstanding, under operating conditions more severe than those for which it was normally intended.

There are advantages which come from using a process computer as part of an analog simulation. For example, interfacing is facilitated because the digital input and output channels are typically designed to accommodate 1 to 5-V process signals, which are directly compatible with standard 0 to 10 or 0 to 100-V analog computer levels using simple scaling and biasing amplifiers. In addition, computer reporting capabilities are useful since all simulation variables appearing as digital system inputs and outputs are available for logging.

The particular digital computer system had a number of features which were also beneficial in the simulation application. The free-running front end make it possible to do all input and output scanning with no special programming. Moreover, this hardware-oriented interface required no software overhead, and therefore operated faster for the same accuracy and precision than would have been possible using multiplexers. This factor is significant because process simulation requires computation in an open-loop mode.

For logging the digital computer offered a "busy print" capability which allowed values to be buffered for output on a time-available basis. This permitted results to be obtained without slowing the simulation.

The programming language, pc^2, used with the computer system gave another advantage. The language was easy to learn, and had arithmetic capabilities that were powerful enough to write the necessary equations efficiently. Further, the compiler was resident in a read-only memory. Therefore, changes could be made on-line using a standard language instruction. This was useful for modifying equations during program development and for modifying constants and parameters while simulations were running.

Simulation effectiveness. This method of design forces engineers to examine the process, identify actual parameters and define all measurement and control points. Running the simulation then reveals interactions and operating constraints which might otherwise not become apparent before hardware implementation. Simulation of the propane refrigeration process and control system showed the effectiveness of proposed strategies in providing stability, controllability and surge protection under normal variation and emergency upset conditions.

When flow signals were compensated only by pressure, control system performance was generally unsatisfactory. However improvements were seen when the signals were temperature-compensated as well, to reflect true volumetric flow at the suction conditions of each compressor stage. With flow fully compensated, the control system was able to handle total first stage block valve failures and 60 percent step reductions in vaporized loads.

It was accordingly suggested that suction pressure control or anti-surge control be selected to manipulate the first stage recycle. This would involve selecting the greater of suction pressure or first stage flow, with the latter fully compensated for pressure and temperature. Under these conditions limiting time constants of 5 to 15 seconds were recommended for closure of the first stage suction block valves upon failure. This range was found to be bounded by surge for faster closures and limit-cycle oscillation for slower closures. Similar recommendations were made for operation of the recycle valves upon various step reductions in vaporizer load, and for actions to be taken in the event of second and third stage block valve failures.

Simulation of blocked discharge and system trip indicated that for satisfactory performance, the amount of liquid quench flow vaporized must be accounted for and vented. It appeared that the quench valves could be tripped closed in either of these cases. However additional study is considered necessary, since the lack of cooling could result in temperature rises, which could be amplified through the compressor stages and result in a higher temperature recycle.

The cost of simulation represented less than three percent of the overall design cost. The same process simulated entirely on a digital computer, using a program environment such as CSMP, would have cost three times as much. The economies associated with the analog simulation result because of the close relationship which can be achieved between the control system designer and his engineering tools.

LITERATURE CITED

[1] Nisenfeld, A. E., Miyasaki, R., Lynch, R., and Connover, J., "Using dedicated computers for on-line process control," *Instruments and Control Systems*, November and December 1972.

How to Control Centrifugal Compressors

Use performance curves to determine best control method for your particular applications.

E. A. Tezekjian, Elliott Company,
Mountainside, N. J.

THE TREND of modern day refinery and petrochemical processes dictates, for economic reasons, the replacing of positive displacement type compressors with centrifugal compressors. These dynamic compressors are essentially a constant pressure, variable volume machine (as contrasted to a positive displacement compressor being a constant-volume, variable-pressure machine) and must be regulated in almost all applications. Before control can intelligently be discussed, the operating characteristics must be understood.

Constant Speed Performance Curves. The most universal centrifugal compressor performance map would plot capacity versus head and versus efficiency. This discussion will concern itself with polytropic compression and, therefore, polytropic head and efficiency.

Figure 1 depicts a typical centrifugal compressor performance map operating at some fixed speed. A typical specific performance map plots capacity in terms of inlet cfm (cubic feet per minute) on the absicissa and brake horsepower and discharge pressure on the ordinate. The map is usually based on constant inlet conditions, such as pressure, temperature, MW and K value of the gas.

In Figure 1, the shape of the head curve is shown for a particular impeller design. Aerodynamic design of centrifugal compressors is beyond this discussion; however, it should be pointed out that centrifugal compressor impellers can be designed to some extent to achieve more nearly the sloping characteristics of that required to suit the systems requirements. For practical reasons, manufacturers usually will offer standardized impeller designs with standardized frame sizes.

A given centrifugal compressor design "knows" only its inlet volume and its speed. At a given speed it will develop a particular head for a given flow throughout. Consider the inlet cfm equation:

Equation (1)

$$Q_1 = WV_1 = \frac{W}{\rho_1} = \frac{WZ_1RT_1}{P_1\,144}$$

Where Q_1 = Inlet volume, cfm

W = Total mass thruput, #/min.

V_1 = Specific volume at inlet conditions, $ft^3/\#$

ρ_1 = Density at inlet conditions, $\#/ft^3$

Z_1 = Inlet compressibility factor

R = Universal gas constant

$$\frac{ft - lb}{lb - {}^\circ R}\left(R = \frac{1545}{\text{Mole Wt}}\right)$$

T_1 = Inlet temperature, absolute, °R (°F + 460)

P_1 = Inlet pressure, absolute, psia

From Equation (1) it can be seen that it is possible to change completely the design inlet gas conditions and still maintain design inlet cfm. For all practical purposes (within the scope of this discussion), the

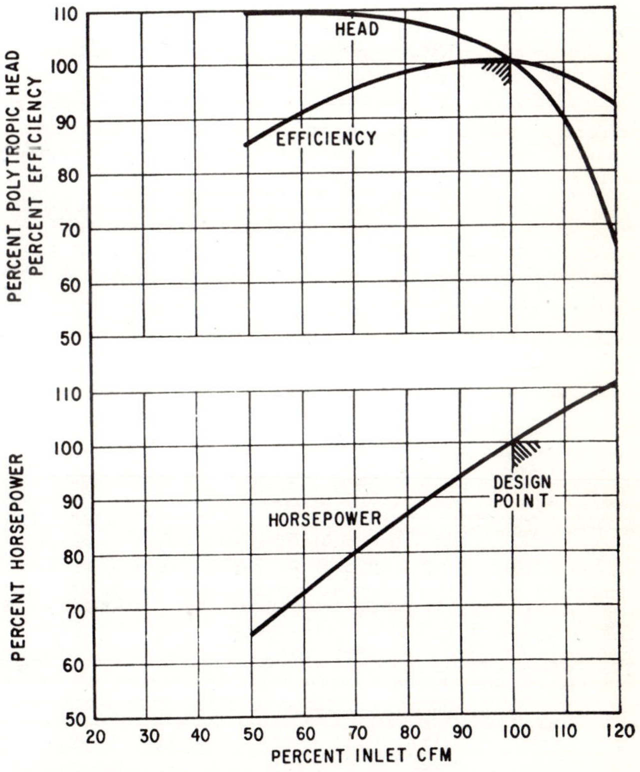

FIGURE 1—Typical constant speed performance curves for a centrifugal compressor.

polytropic head and efficiency of a given single stage centrifugal compressor remains the same even though the design gas and gas conditions have changed, as long as design inlet cfm and speed is obtained.

Effect of Changing Gas Conditions. Consider the polytropic head equation:

Equation (2)

$$H_p = Z_m R T_1 \left(\frac{K\eta_p}{K-1}\right)\left[\left(\frac{P_2}{P_1}\right)^{\frac{K-1}{\eta_p K}} - 1\right]$$

Where H_p = Polytropic head, $\frac{\text{Ft}-\text{lb}}{\text{lb}}$

Z_m = Mean compressibility factor

R = Universal gas constant,

$\frac{\text{Ft}-\text{lb}}{\text{lb}-{}^\circ\text{R}}\left(R = \frac{1545}{\text{Mole Wt.}}\right)$

T_1 = Inlet temperature, absolute, °R

K = Specific heat ratio, c_p/c_v

P_1 = Inlet pressure, absolute, psia

P_2 = Discharge pressure, absolute, psia

η_p = Polytropic efficiency

Since a given compressor design operating at design speed will produce design head when passing design inlet cfm, the discharge conditions, pressure and temperature will be determined solely by the thermodynamic properties of the gas being compressed. To illustrate this further, Figure 2 shows the characteristic curve for a given centrifugal compressor operating at constant speed but under varying inlet conditions.

Curve AB represents the characteristic for a centrifugal compressor designed to handle air at an inlet pressure of 14.5 psia, inlet temperature of 100° F, molecular weight of 29.0 (dry air), and K value of 1.40. This unit would develop 100 percent of design discharge pressure at an inlet capacity of 100 percent design cfm.

If now, the intake air temperature is decreased from 100° F to 40° F, all other conditions remaining the same, the discharge pressure would increase to 106 percent of design. This can be seen by the head equation (2) previously discussed. The horsepower required under these conditions would be more than design by the ratio of the absolute temperature $\left(\frac{460+100}{460+40}\right)$, since by decreasing temperature we increase density and, therefore, mass flow for a given volume rate (refer to Equation 1 and 3).

Equation (3)

$$GHP = \frac{(W)(H_p)}{(33000)(\eta_p)}$$

Where GHP = Gas Horsepower

W = Total Mass Flow, lbs/Min.

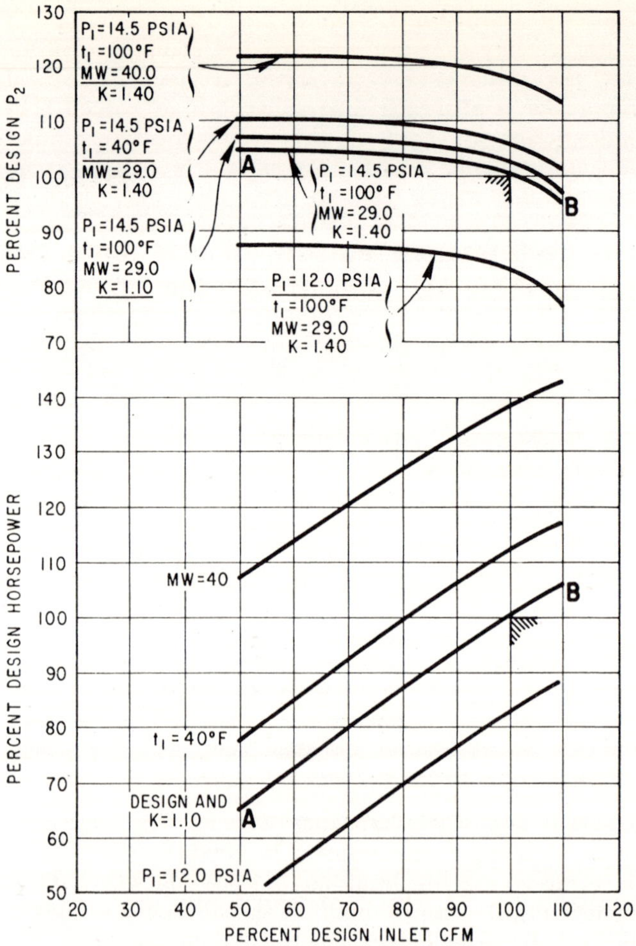

FIGURE 2—Effect of changing gas conditions on centrifugal compressor operating at constant speed.

H_p = Polytropic Head, Ft. lb/lb

η_p = Polytropic Efficiency

By lowering the inlet pressure from 14.5 psia to 12.0 psia, all other conditions remaining constant, the discharge pressure would drop to 83 percent of design—by the exact percentage the inlet pressure dropped, since the developed pressure ratio has not changed. The horsepower again would be reduced by the ratio of the inlet pressure to that of design $\left(\frac{12.0}{14.5}\right)$ since, in this case, the inlet density has been reduced and, therefore, mass flow for a given volume rate, (refer to Equation 1).

By changing the composition of the gas such that the molecular weight increases from 29.0 to 40.0, all other conditions remaining the same, the discharge pressure would increase to 118 percent of design. In this case, the horsepower would increase from design horsepower by the ratio of the molecular weights (40/29).

Lastly, if the ratio of specific heat value, K, is decreased from 1.40 to 1.10, all other conditions remaining the same, the discharge pressure will increase to 102 percent of design with no change whatsoever in compressor horsepower.

Capacity Limitations. There are definite limitations of the stability range of a centrifugal compressor. Limiting the minimum capacity of a given centrifugal compressor is a phenomenon called "surge" which normally occurs at about 50 percent of the design inlet capacity at design speed. This extremely complex phenomenon is probably still one of the most difficult problems in the field of fluid dynamics. To provide a simplified explanation, consider the single stage compressor operating at constant speed, discharging through a throttling valve. By throttling on the discharge valve, we increase the system resistance and, therefore, the head required by the compressor to overcome this resistance. As we continue to throttle this valve, less flow will be capable of flowing through the compressor.

This continues up to the point of maximum head capability of the compressor. At flows below this "surge" limit, the compressor head characteristic curve takes a reverse slope, resulting in decreased head capability. At this condition, the system back pressure exceeds that capable of the compressor delivery, causing a momentary backflow condition. At this time, however, the back pressure has been lowered, enabling the unit to again be capable of delivering flow higher than the flow at which the surge began. If the obstruction to flow downstream of the compressor is unchanged (i.e., same discharge valve position), operation follows back along the head characteristic curve until the peak head delivery is reached again. This cyclic action is what the industry properly calls "surge". To operate at flows below the surge flow, requires controls which will be discussed later.

While the stability range of a centrifugal compressor is commonly indicated from the rated point to the surge point, the unit can operate stably to the right of the rated point. The greater the load demand on a centrifugal compressor, the greater the "fall-off" in delivered pressure. The upper limit of capacity is determined by the phenomenon of "stonewall". Stonewall occurs when the velocity of the gas approached its sonic velocity somewhere in the compressor, usually at the impeller inlet. Shock waves result which restrict the flow, causing a "choking" effect—rapid fall off in discharge pressure for a slight increase in volume throughput. Stonewall is usually not a problem when compressing air and lighter gases; however, in compressing gases heavier than air, the problem becomes more prevalent as the molecular weight increases.

In discussing the operating range of centrifugal compressors, limitation to single stage compressors was purposely made for simplicity. As the number of stages increases, performance maps tend to show a more sloping curve with lesser stable range than shown in Figure 1 or Figure 2, as dictated by the particular application.

Variable Speed Performance Curves. A typical variable speed performance map is shown in Figure 3. With variable speed, the compressor easily can deliver constant capacity at variable pressure, variable capacity at constant pressure, or a combination variable capacity and variable pressure.

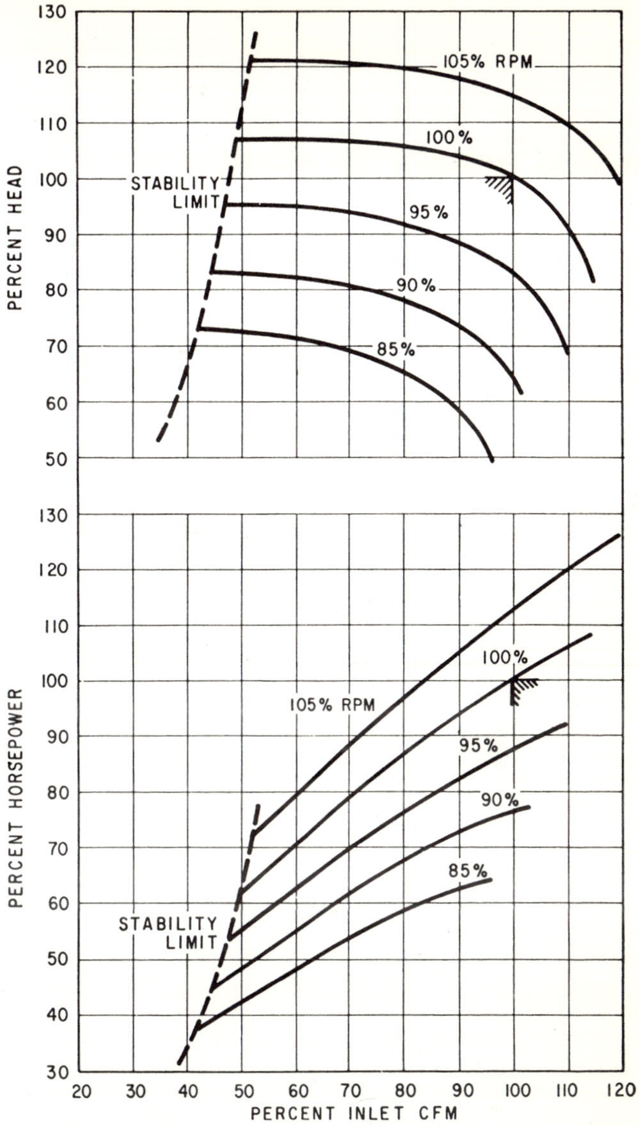

FIGURE 3—Typical variable speed performance curves for centrifugal compressor.

Basically, the performance of the centrifugal compressor, at speeds other than design, are such that the capacity will vary directly as the speed, the head developed as the square of the speed, and the required horsepower as the cube of the speed. As the speed devaites from the design speed, the error of these rules, known as the Affinity Laws, or Fan-Laws, increases.

By varying speed, the centrifugal compressor will meet any load and pressure condition demanded by the process within the operating limits of the compressor and the driver. It normally accomplishes this as efficiently as possible, since only that head required by the process is developed by the compressor. This compares to the essentially constant head developed by the constant speed compressor.

Factors Affecting Decision of Control. For the majority of centrifugal compressor applications, some form of regulation is required. The type of control used depends first on the compressor driver.

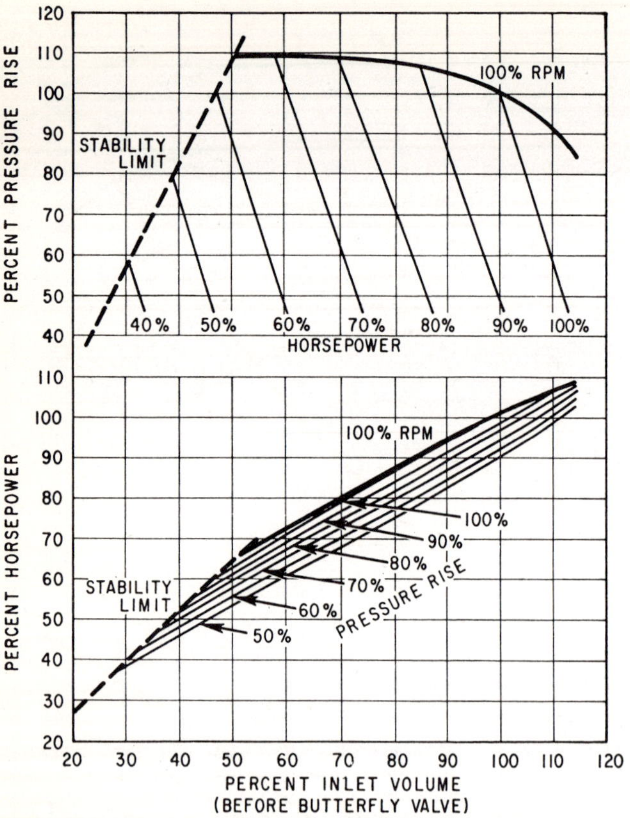

FIGURE 4—Typical constant speed performance curves. Compare the slope of the constant horsepower lines with Figure 5.

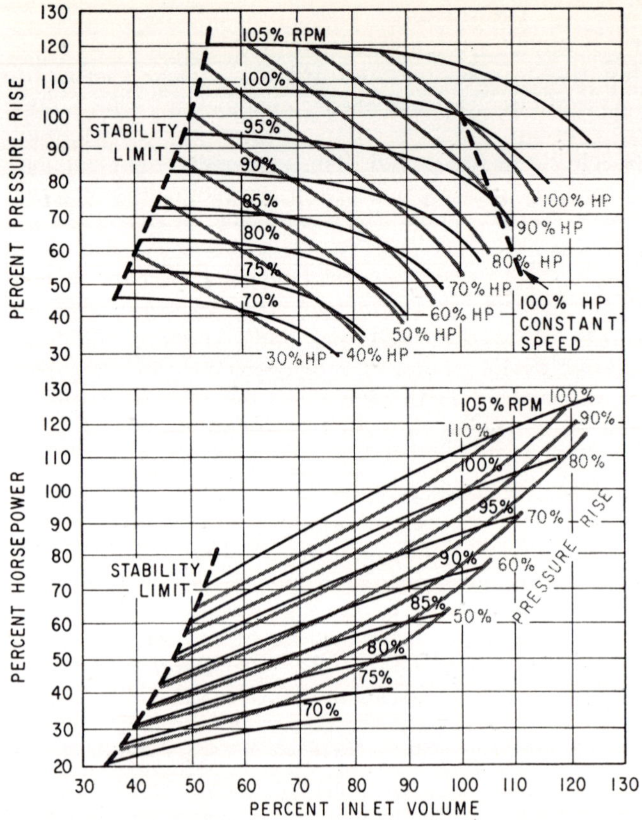

FIGURE 5—Typical variable speed performance curves. Compare the slope of the constant horsepower lines with Figure 4.

For turbine driven compressors, normal control is accomplished by varying the speed. This method of control permits a wide range of operation in a relatively efficient manner. Speed control is more efficient than throttling the flow at constant compressor speed since, by artifically creating resistance, an unrecoverable loss in power results. This can be seen by comparison of Figure 4 and 5 which shows constant speed and variable speed performance curves respectively. Compare the slope of the constant horsepower lines.

For motor driven compressors, the control can become more intricate, especially for the usual constant-speed type motors. For this type driver, means of obtaining control are:

1. Use of a hydraulic or electric coupling between motor and compressor to obtain speed variation. This is not a popular method of controlling speed because of severe coupling efficiency penalties throughout the range of speed operation.

2. Use of butterfly valve at the compressor inlet or at compressor discharge. Throttling at the suction is preferred, since by so doing, the gas density is decreased, thus meaning less mass flow for a given inlet cfm. In other words, throttling at the discharge does not take advantage of the head nor density reduction obtained by inlet throttling which artificially increases the compressor cfm toward the rated point, resulting in lowered horsepower requirements.

3. Use of adjustable inlet guide vanes which adjust the characteristic curve of the centrifugal compressor. The adjustable inlet guide vanes are most effective for conventional compressors developing less than 30,000 feet head or a multistage compressor with three stages or less. Power savings by the use of adjustable inlet guide vanes at the inlet of a compressor over suction throttling can approximate 10 percent for a single stage compressor, 5 percent for a two-stage compressor, and 3 percent for a three-stage compressor. In a sense, the adjustable inlet guide vanes change the aerodynamic design of the first impeller. It does this by prewhirling the gas entering the wheel in the direction of its rotation, thereby developing less head than at design. (An analogy to this "unloading" of the first wheel can be more easily understood when considering a sled being pushed downhill rather than on level ground). This is accomplished with a decrease in design efficiency. However, the ratio of the head to efficiency decreases from its design value, resulting in lowered horsepower.

Although not used commercially in the United States, adjustable inlet guide vanes are available at the inlet of every stage in conventional multistage compressors by some European manufacturers. Economics justify this expensive and sometimes complicated feature in most European countries because of high utility rates.

4. Use of a power wheel at the inlet of the compressor upstream of the first stage impeller. The gas to be compressed expands through a set of movable guide vanes and is directed upon the turbine blades. The power wheel or turbine wheel is theoretically designed to more effectively handle part load conditions than adjustable inlet guide vanes by itself. However, the power wheel has disadvantages, in particular, to the inherent parasitic losses at the wheel at design conditions resulting in lowered efficiency. This control method is not popular in the United States.

5. Use of adjustable diffuser vanes. Like the multiple adjustable inlet guide vanes, the adjustable diffuser vanes are not commercially used in the United States for the same reasons mentioned. It is important to note that devices such as these, when used with gases that are gummy, corrosive, or erosive, lead to heavy maintenance problems.

6. Use of a wound-rotor induction motor obtaining speed variation by varying the resistance in the rotor or "secondary" circuit This is an expensive and relatively inefficient drive and, therefore, not normally used for industrial centrifugal compressor drives.

7. Use of a direct-current motor obtaining speed variation by varying the field current by means of a rheostat. Again, this is a relatively expensive and inefficient drive, having commercial DC availability problems and, therefore, not normally used for centrifugal compressor drives.

Summarizing, the head-capacity control methods most commonly used domestically are:

- Speed control
- Inlet throttling
- Adjustable inlet guide vanes

The other controls mentioned are seldom seen in the United States. European high cost of power and fuel relative to labor and material there justify the more elaborate types of control.

With limitation to the three common types of control, typical forms of regulation can be studied.

System Characteristics. Unfortunately, from a user's viewpoint, manufacturers are sometimes unjustifiably blamed for providing a compressor design that does not function in the process as expected. In many instances, experience proves that the system resistance was not properly understood by the user in specifying the design and operating conditions of the compressor. It is extremely important, therefore, that the system resistance or characteristics be fully known before discussing or recommending a control system.

Figure 6 shows a plot of pressure versus capacity for a constant speed compressor. Shown on this plot are three different types of system characteristics. A compressor operating against a fixed head or pressure would have a system characteristic defined by AB. A close-coupled packaged refrigeration compressor such as used in alkylation units would follow a system resembling AB. In contrast to the forementioned system, a compressor discharging into a large system through a long run of pipe would have a system characteristic that follows Curve AD. In this case, all the delivered pressure by the compressor is used to overcome pipe line friction. A natural gas pipe line compressor is an example of this type of system. Most commonly, there are systems which have essentially fixed top pressure, except for some pipe friction in the system. A catcracker air compressor would typify a system represented by Curve AC.

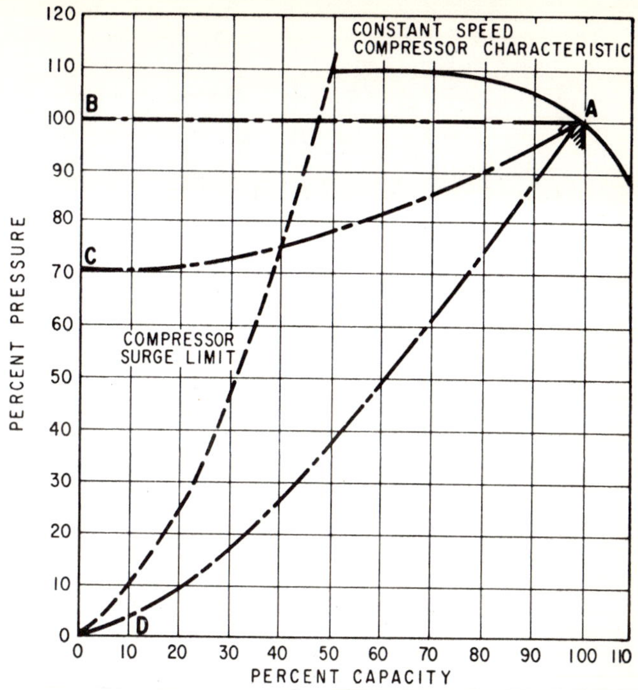

FIGURE 6—Pressure vs. capacity for constant speed compressor.

Note in Figure 6 that the compressor head-capacity characteristic curve follows none of the system characteristics. Variation of the compressor output to meet the demand of the system, therefore, requires controls to regulate the volume, pressure, or a combination. In all of the controls, either a high pressure oil system or a source of high pressure air will be required for the operation of a servomotor as the final power medium. Rather than discuss specific details of control mechanisms, schematic diagrams will be used along with a performance curve to illustrate the following types of control:

- Constant pressure control
- Constant pressure control—parallel operation
- Contant weight flow control
- Constant weight flow control—series operation
- Anti-surge control

Constant Pressure Control. Figure 7 shows a constant discharge pressure control system for a turbine-

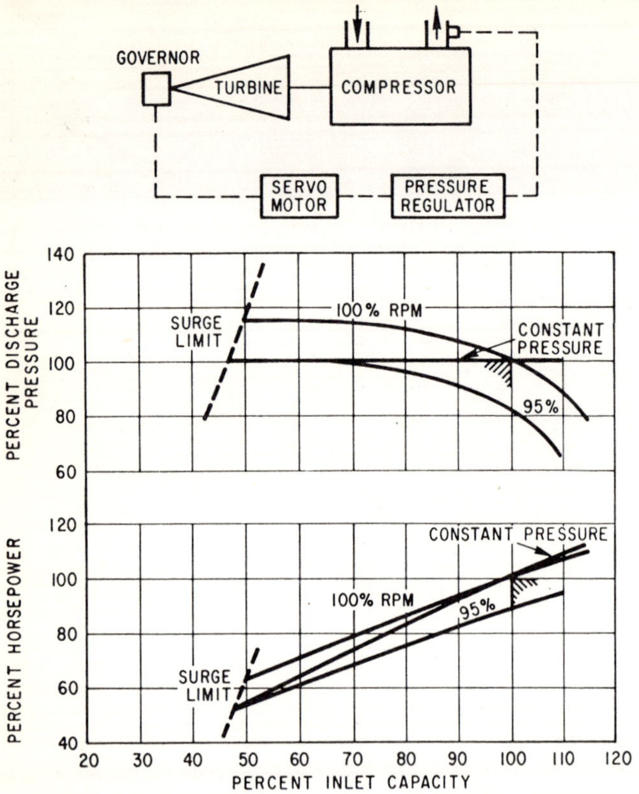

FIGURE 7—Constant discharge pressure control system for turbine driven compressor.

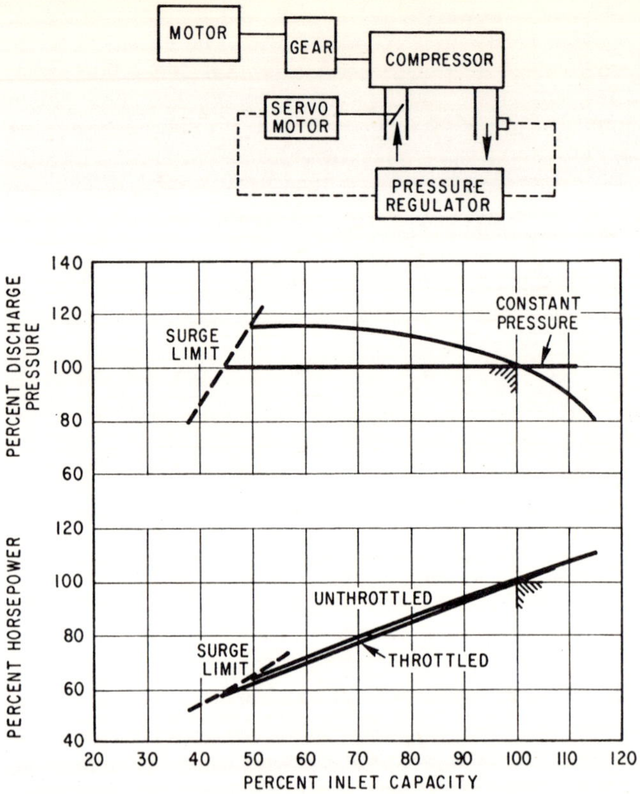

FIGURE 8—Constant discharge pressure control system for motor driven compressor with suction throttling.

driven centrifugal compressor. The pressure regulator may be air operated or hydraulically operated to position the servomotor. The servomotor in turn is connected to the turbine speed-governing system, or more simply, to the steam valve at the turbine inlet. Constant discharge pressure is maintained by varying the speed of the turbine. Note the performance curve in Figure 7. In this case, the compressor is able to develop constant head by lowering its speed. If there is a 10 percent head rise from design to surge, the approximate minimum speed to maintain constant discharge pressure will be $N_{des}\sqrt{\frac{100}{110}}$ or, roughly 0.95 N_{des}, as determined by the Affinity Laws previously mentioned.

Figure 8 shows a constant discharge pressure control system for a motor-driven centrifugal compressor. In this case, the servomotor positions a butterfly valve located at the compressor inlet. The discharge pressure is held constant by positioning the inlet butterfly valve, thus, throttling out the excess pressure ratio developed at flows lower than design. Note the performance curve in Figure 8. The performance map shows the savings in horsepower by throttling at the inlet of the compressor, as compared to throttling at the discharge.

If the servomotor in Figure 8 were to position adjustable inlet guide vanes rather than an inlet damper, the performance curve would take the shape as shown in Figure 9. For comparative purposes, this curve superimposes the curves of Figures 7 and 8. Note the horsepower savings over suction throttling. In contrast, note the horsepower advantage that speed control has over this method of control. Specific percentages are not repeated here because of lack of true meaning. In general, the greater the pressure ratio and number of compression stages, the more pronounced the difference will be between speed control and the other methods

Parallel Operation. Controlling two or more compressors operating in parallel and having identical characteristics would be relatively simple. The system described in Figures 7 and 8 would still apply, except only one pressure regulator would be required for both units. The two or more servomotors obtain a hydraulic or pneumatic impulse signal from the pressure regulator; and, in the case of the turbine driven compressor, would control the turbine speed. In the case of the motor-driven compressor, the servomotors would position the inlet butterfly valves or a set of adjustable inlet guide vanes. In any case, check valves must be installed at the discharge of each compressor to prevent any back flow of fluid to overcome any slight unbalance in the characteristics of the two compressors.

More complicated, but more likely, would be the system involving two or more compressors with dissimilar, or similar but not identical, characteristics. Figure 10 shows a combined performance map for two compressors which have dissimilar operating characteristics. To maintain constant discharge pressure, one

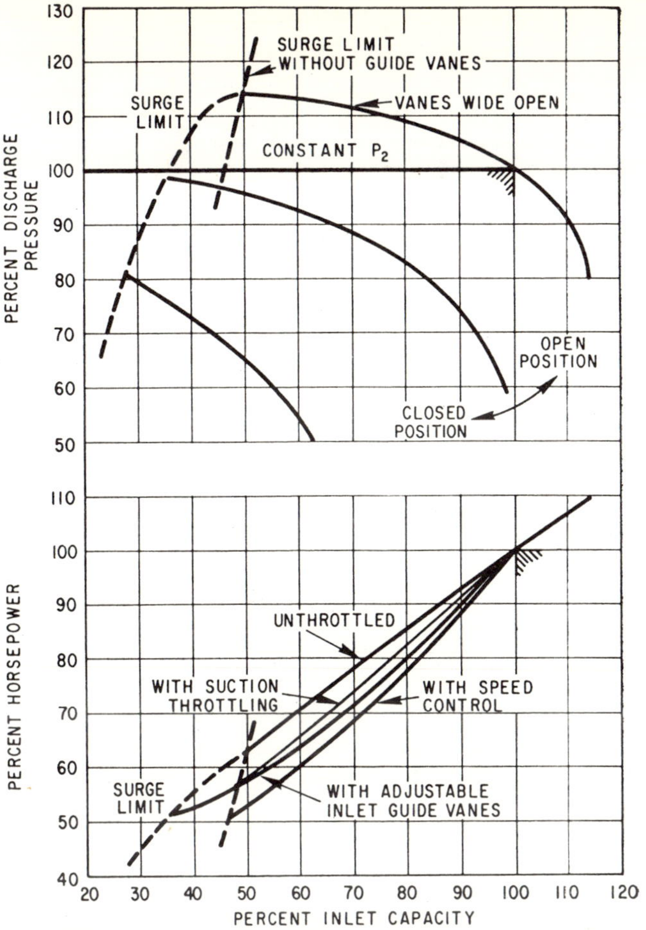

FIGURE 9—Constant discharge pressure control system for motor driven compressor with adjustable inlet guide vanes.

compressor would be operating at flow differently than the one in parallel with it. As a result, the control system would have to include a separate flow controller for each compressor unit. Figure 10 shows this system schematically for two motor driven compressors.

Constant Weight Flow Control. In the case of constant weight flow systems, for a turbine driven compressor, a servomotor actuated by a flow regulator would maintain constant weight flow by varying the speed of the turbine.

In most systems involving variable system resistance (Curve AC, Figure 6) constant weight flow control of some means is used. Figure 11 shows this control system whereby inlet pressure and temperature will vary over some known range. In this system, pressure and temperature compensation is included to adjust the signal transmitted by the flow regulator for true constant weight flow control.

The performance curve shown in Figure 11 requires no explanation.

For a motor driven compressor, the system schematically would be similar to Figure 8 with constant weight flow being maintained by positioning an inlet butterfly valve.

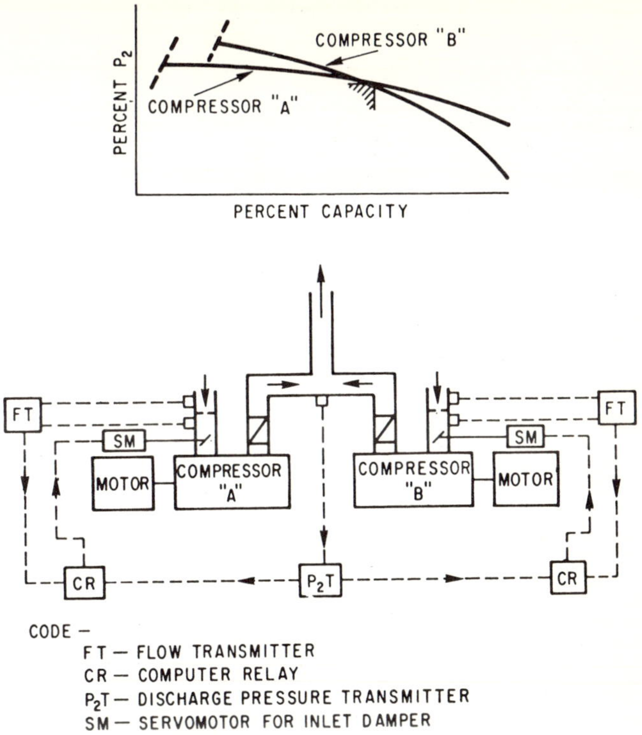

FIGURE 10— Constant discharge pressure control system for two compressors operating in parallel having dissimilar operating characteristics.

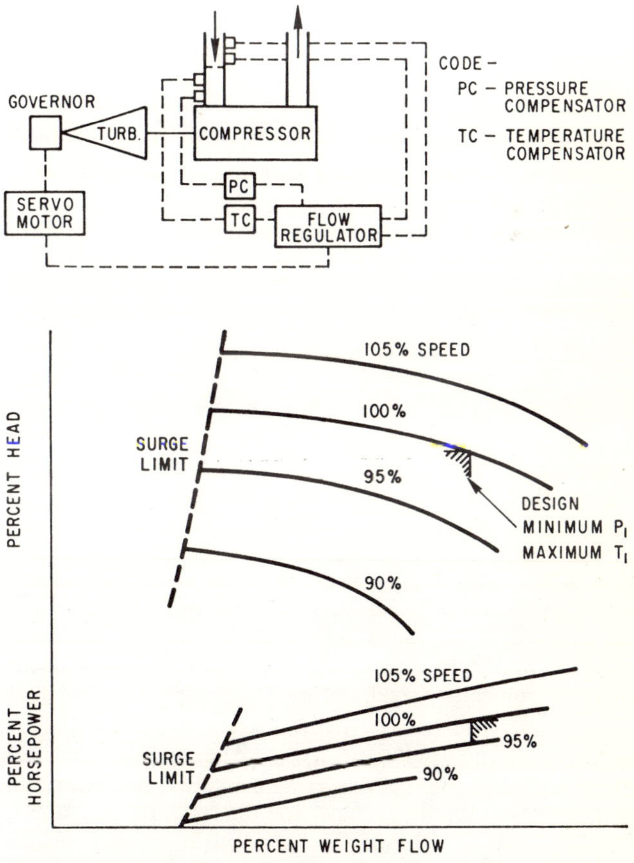

FIGURE 11—Constant weight flow control system for turbine driven compressor.

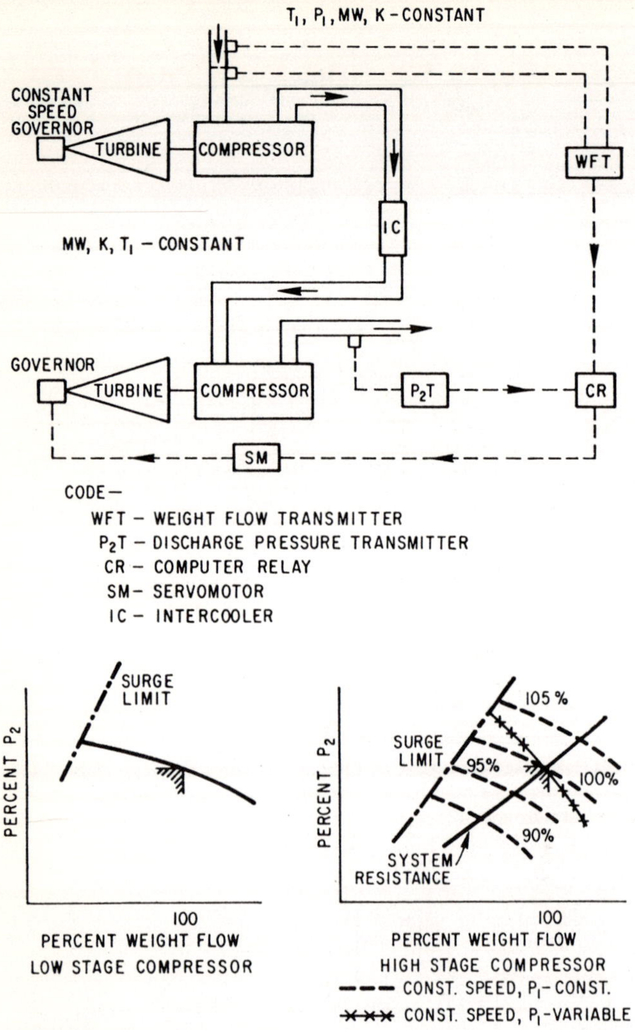

FIGURE 12—Constant weight flow control system for two compressors operating in series.

Series Operation. If, in Figure 11, an additional compressor body were added in the compression system by means of direct coupling to the first compressor body, the control system would not normally change in any way. That is, as long as one drive controls speed of both compressors, the system can be treated as a single body control problem. (Note: Surge control is not discussed here for simplicity: however, it should be pointed out that by-pass lines should be installed around each compressor, mainly to facilitate start-up).

If, however, each compressor body were driven by a separate drive, such as shown in Figure 12, a simple solution would be to operate the low stage compressor unit at constant speed, letting the discharge pressure rise or fall from design point, in accordance with the compressor's characteristic curve. The high stage compressor would then be speed controlled to maintain constant weight flow. Due to the system resistance, the combination of final discharge pressure and weight flow can indicate the operating speed, as shown on the performance map in Figure 12.

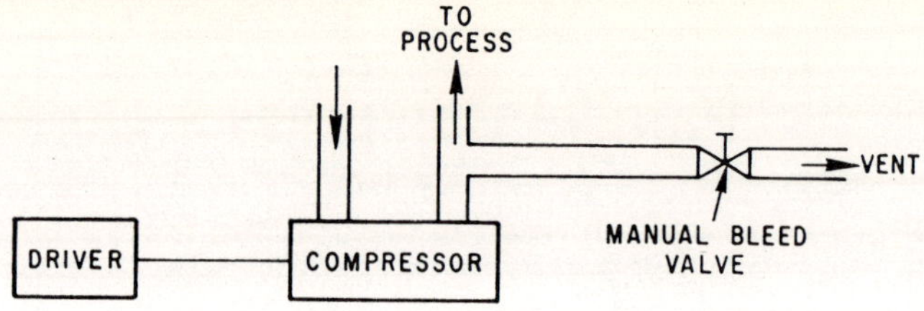

FIGURE 13—Manual surge control system.

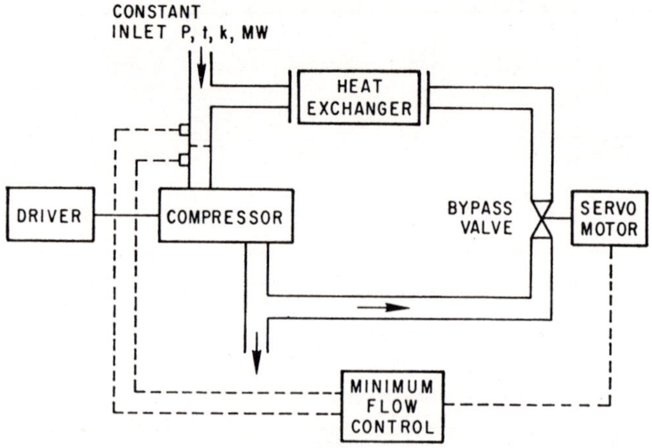

FIGURE 14—Automatic anti-surge control with recirculating bypass.

Anti-Surge Control. In some compressor applications, operation is practically always at design capacity. In this case, the surge control would merely consist of a manual bleed valve at the discharge of the compressor. In the case of compressing air, a non-toxic or inexpensive gas, the system which is the simplest possible is shown schematically in Figure 13.

More practically, however, operation at other than design condition would normally require some form of automatic anti-surge control. Consider the control essentially a minimum flow regulator which, through a servomotor, operates the surge valve as required to maintain stable operation. Again for compressing air, a non-toxic or inexpensive gas, the surge valve can vent to atmosphere. However, in many applications, the gas is expensive or toxic and, therefore, it becomes desirable to re-circulate the flow necessary to insure stable flow through the compressor.

Figure 14 shows a system in which automatic surge control is applied to a compressor handling an expensive gas. Note the bypass cooler which is required to remove the heat of compression before mixing with the process gas at the inlet. This prevents compressor performance fall-off which would occur should the inlet temperature rise above design value.

Centrifugal compressors can easily be controlled because of their inherent performance characteristics. If you keep in mind the thermodynamic factors affecting compressor performance, a control system can be designed to fit the process in which the compressor functions.

Parallel Compressor Control . . . What Should Be Considered

Process design information is collected here without mechanical details to make process studies easier when compressor sizing is required.

A. E. Nisenfeld and **C. H. Cho,**
Fisher Controls Co., Marshalltown, Iowa

CONTROL OF most compressor systems was not a problem until the energy crisis; control of systems with parallel compressors has always been a problem.

Control system engineers are recognizing that every process or unit operation is entitled to some design attention of its own, no matter how many times an apparently identical system has been implemented. This is true when we design control systems for centrifugal compressors and doubly so when these compressors are operating in parallel.

There are many parallel compressor applications in processing industries. Most common, perhaps, are refrigeration compressors in ethylene and propylene plants. Frequently three or more compressors are in parallel providing suction to a second pair of compressors in parallel. This configuration (Fig. 1) illustrates control problems which are uniquely associated with parallel compressors. In each set of compressors, one is a spare in standby having automatic startup controls.

Consequences. Bringing one compressor up to speed when it must pump against head established by a parallel compressor(s) is difficult and dangerous. If the "coming-up" compressor is tied into the discharge manifold too early, it can be thrown into very violent surge. If one compressor is failing or being shut down, bringing up the "new compressor" too late will cause the process to be starved for gas. Tying in too early can also cause reverse rotation of the compressor going down.

Perhaps the most serious problem, with due respect to the uniqueness of surge control requirements for parallel compressors, is balancing the load among compressors. The authors have seen perhaps 35 or 40 different parallel compressor systems and have yet to see two compressors in any of these trains with identical performances. At the same speeds, these compressors pump different quantities of gas and have different surge and stonewall points. (Stonewall is that phenomenon where a compressor can no longer pump the quantity of gas fed to it—sometimes called "choke.") Without proper balancing controls, the more efficient compressor often carries more than its share of the load and the less efficient compressor operates with its recycle valve (spillback) wide open, because the efficient compressor is taking all the feed.

From Fig. 1 we can readily see why surge control is difficult. Spilling back to the suction of one parallel compressor obviously affects the suction conditions and quantities of gas at other compressors.

Before control system design, process requirements and compressor operating characteristics must be understood. Process requirements determine the controlled variables; compressor characteristics dictate the strategy for capacity regulation.

Objectives of control. The primary objective of a control system is to meet and maintain system design specifications for thruput and discharge pressure in response to process demand variations. The system must provide compressor surge protection and orderly startup and shutdown. For parallel compressors the control system should be able to bring another compressor on line in parallel with one or more operating compressors.

Compressor control can be divided into two major functions:

Control related to process demand (capacity). Compressors are generally used for a process requiring a variable quantity of gas at constant pressure. As flow demand changes, control systems should respond to this change and maintain thruput by one or more of the following:

- Varying inlet pressure.
- Varying speed.
- Varying inlet vane angle.
- Varying spillback flow.

Varying inlet pressure is not the most economical because when we drop inlet pressure, the compressor uses more energy to reach discharge pressure. However, capacity of a centrifugal compressor is often controlled by placing a control valve in the suction line. With this control valve, compressor inlet pressure can be changed to obtain a new operating curve for compressor capacity regulation with a constant discharge pressure. Fig. 2 shows two sets of compressor curves for two different inlet pressure conditions. To obtain the change in flow rate

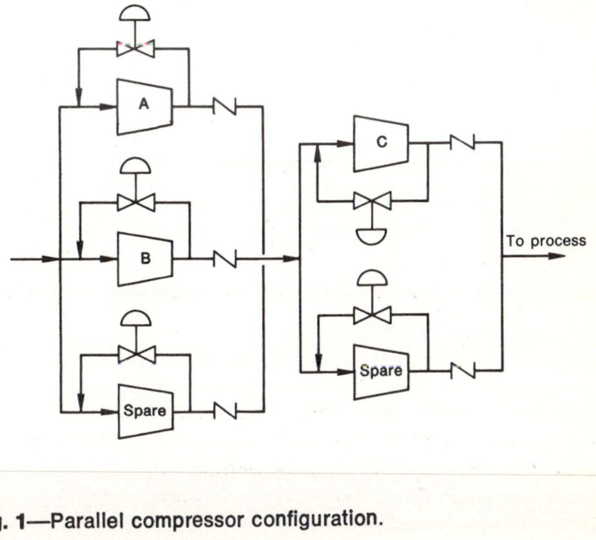

Fig. 1—Parallel compressor configuration.

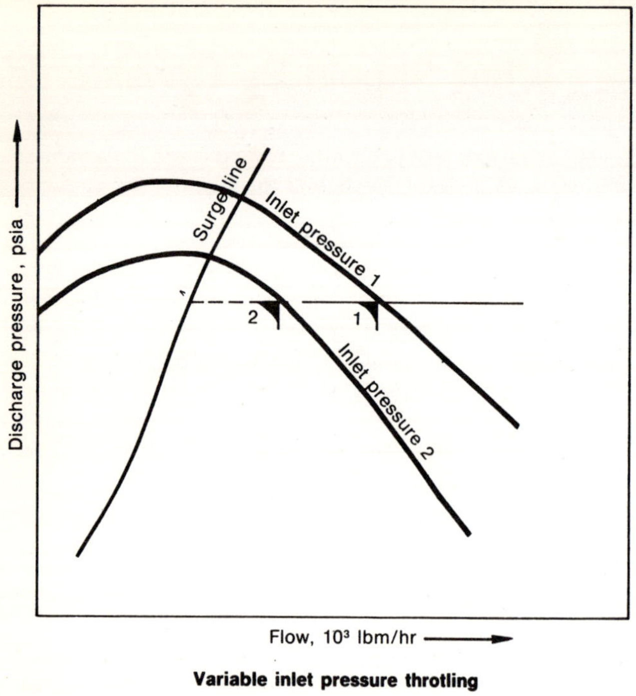

Fig. 2—Variable inlet pressure throttling.

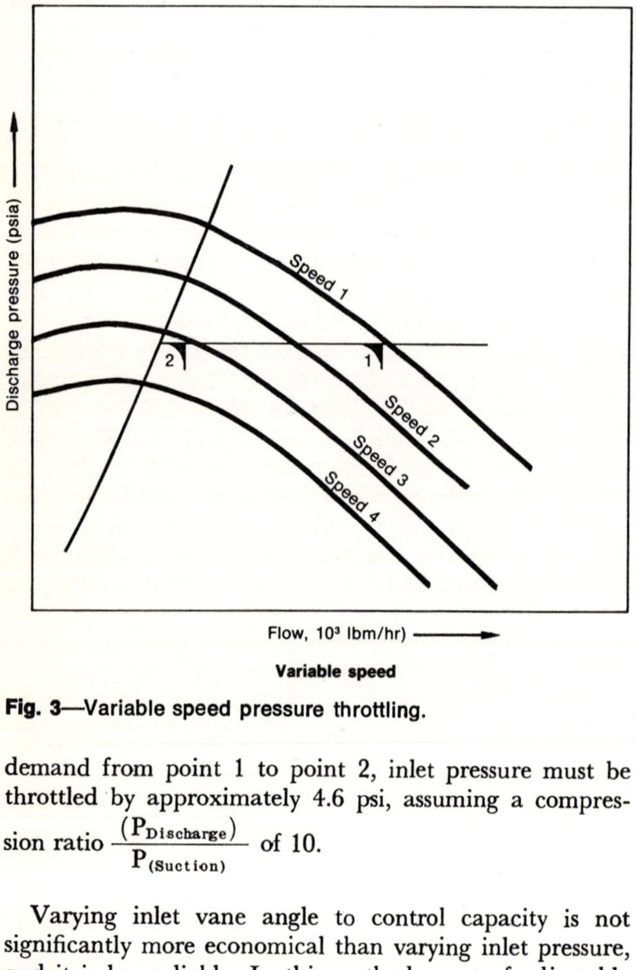

Fig. 3—Variable speed pressure throttling.

demand from point 1 to point 2, inlet pressure must be throttled by approximately 4.6 psi, assuming a compression ratio $\frac{(P_{Discharge})}{P_{(Suction)}}$ of 10.

Varying inlet vane angle to control capacity is not significantly more economical than varying inlet pressure, and it is less reliable. In this method, a set of adjustable guide vanes on the inlet to the compressor stage is used to pre-rotate or counter-rotate the gas stream with respect to impeller rotation. Therefore, this method unloads or loads the stage to lower or raise discharge head, at a constant speed and flow. Advantages of this method over varying inlet pressure are:

► The variable inlet vane can lower or raise (bi-directional) discharge pressure.

► It provides capacity regulation with less power consumption since it doesn't directly throttle flow.

Varying the speed of a compressor is the most economic means of capacity control, but it is not practiced as often as it could because of mechanical problems with speed transmitters and governors. Since the pressure ratio developed by a centrifugal compressor is related to impeller tip speed, variation of discharge pressure may be plotted as a percent of design speed. For example, Fig. 3 shows that a two-to-one flow turn down ratio can be achieved by varying maximum design speed by 20 percent. This method is also energy effective since power input is reduced by the square of the speed. One disadvantage is the sensitivity of capacity to speed change as shown in Fig. 3.

Varying spillback (recycle) flow is frequently used because it seems to be the safest. However, it is costly because gas is compressed and then dropped back to suction pressure. At a given speed, a compressor will pump a fixed quantity of gas against a fixed head. If a spillback valve is opened, net forward flow of gas will be reduced; but total flow through the compressor will remain constant and above the surge point.

Control related to surge protection. The purpose of an anti-surge control system is to protect the compressor and process from surge. It may be a separate system which overrides normal capacity controls or it may be part of the capacity control system.

The common anti-surge systems (Fig. 4) include the following:

Minimum flow, probably the simplest form of surge protection, uses a predetermined minimum flow as setpoint for the surge protection controller. For this system to work properly, the compressor must be operating at constant speed with constant gas properties; suction pressure and temperature must be virtually constant.

Minimum flow with automatic speed control provides surge protection by manipulating compressor speed to hold a minimum flow.

Flow-speed system uses compressor speed to determine the setpoint of the spillback flow controller. Therefore, the surge point is a function of compressor speed.

Guide vane position resets minimum flow controller. In a constant speed compressor, inlet guide vanes are often used for flow regulation. The surge point is a function of guide vane position as well as compressor speed. Guide vane position can be used as a remote setpoint to a spillback flow controller to compensate for change in surge point.

Flow-ΔP control is the best of the widely used protection systems because it is independent of compressor speed and suction conditions. The locus of the surge line and set point of the anti-surge controller, is determined by the following equation:

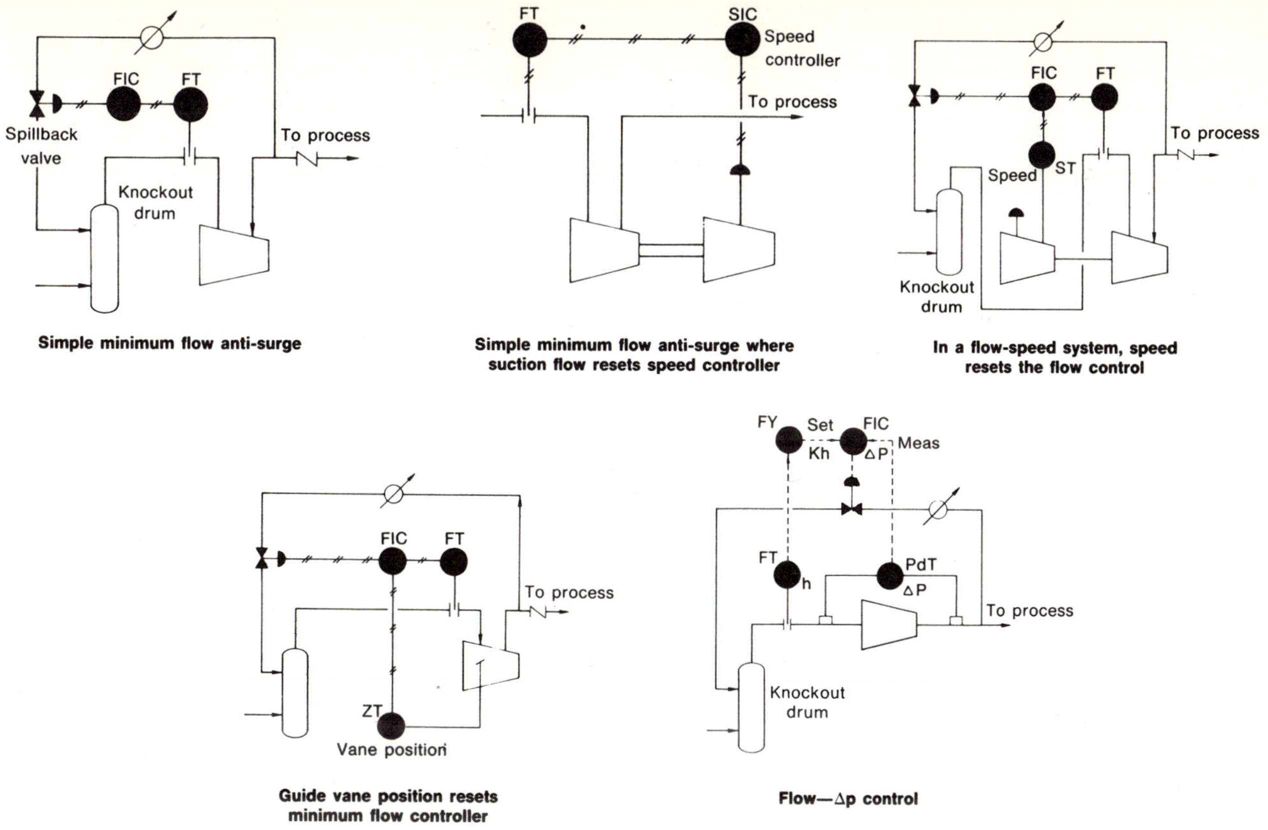

Fig. 4—Common anti-surge systems.

$$\Delta P = kh + b$$

where:

ΔP = Compressor discharge pressure-compressor suction pressure

h = Differential pressure across an orifice in the compressor suction

k = Proportionality constant

b = Bias to match control line origin with surge line origin.

Peculiarities of parallel compressors. The best way to design controls for parallel compressors is by simulation using a math model of the compressor and the process. The type of simulation required is a function of system complexity and nature of probable disturbances. For a simpler system, simulation can be by manual, graphic methods using compressor maps with simple process system curves. Large hybrid computer systems are required for very complex parallel systems.

A two compressor system, both machines operating continuously with 60 percent of total maximum flow as their design base, is the most straightforward example of parallel compressors. Let's examine specifics of load allocation control for this system.

There are two approaches: balance the load equally, or operate at maximum over-all efficiency. The difficulty of load balancing can be seen in Fig. 5 where the performance curve of one machine at a given speed is superimposed on the performance curve of the other at the same speed.

Surge points are different; while machine A operates safely at inlet flow B1, the surge protection system of machine B will be in operation. Both machines cannot pump flow B1 at discharge pressure P_{D2} at the same speed. Machine A will have to operate at a higher speed for equal load distribution.

But, if machine B is recycling gas to stay out of surge, it will not pump flow B1 forward at P_{D2} if operated on the indicated speed line. It too will have to be sped up. Since the surge point is a function of speed, among other things, it is possible there is no speed at which machine B can be operated to produce a net flow of B1 cubic feet/minute at a pressure of P_{D2} atmospheres.

One solution is to have machine A pump some quantity between A1 and B1, say A1 + 0.3 (B1-A1). Machine B would then pump B1 + 0.7 (B1-A1). The sum of the individual productions is 2B1. A control system was designed (Fig. 6) to solve the problem. Only the capacity controls on one machine are shown; no anti-surge controls are shown.

The demand controller is shown as a manually set FIC-1, but it can be a pressure controller, somewhere in the process, setting FIC-2 through characterizer f (N) or through FIC-1.

The characterizer compensates for the speed difference between the two machines. This control system was designed and calibrated by trial and error and has been in operation since 1970.

Efficiency became an important consideration when tests showed that the compressor system was using about 36 percent more steam than intended. The "safe" solution implemented maximized production through the less energy-efficient machine. Furthermore, depending on load,

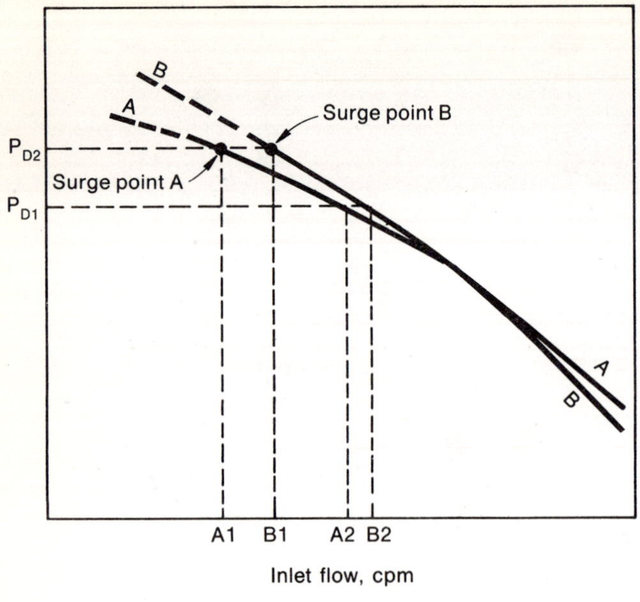

Superimposed performance curves

Fig. 5—Superimposed performance curves.

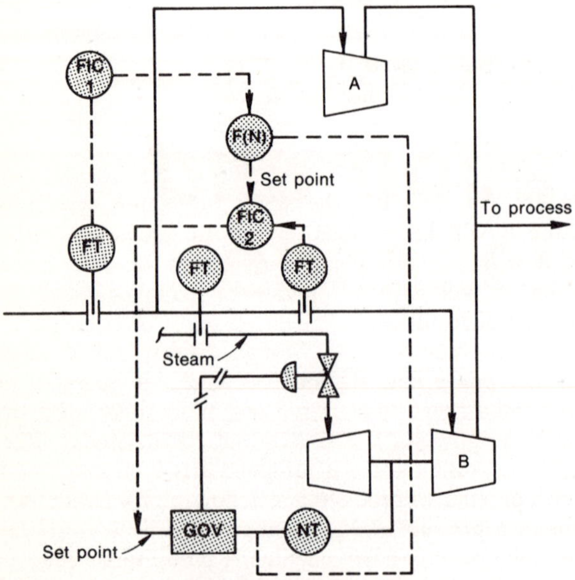

Load balancing control system

Fig. 6—Load balancing control system.

it was more economical to maximize thruput through A one time and B another. Occasionally one machine had to be run full out while the other had a slight amount of recycle.

In an effort to design an energy efficient system, manual graphic simulation was tried, without success. The system was finally simulated using a small, hybrid computer, using control strategy developed on a digital computer. Analog controller implementation was possible but impractical because it was too complex and unreliable.

Anti-surge considerations. Fortunately, effect of recycle flow on suction conditions in this system was not great. If spillback from one machine changed the common suction pressure significantly or rapidly, the simple control system of Fig. 14 would have trouble. When this occurs capacity control must be re-designed to counter this interaction. This can mean a complete, on-line change in control strategy.

Why on-line? Because economics require a strategy that cannot be executed when one machine is near its surge point. The near-to-surge strategy is uneconomic. Control strategies to cover different sets of operating conditions are desirable. The correct strategy can be selected on-line automatically. Such systems have been designed and implemented with a digital computer.

Automatic start-up. As previously explained, another important consideration in operating parallel compressors, is starting up a spare. Discharge pressure of machine A may be high enough to keep the check valve in machine B's discharge closed while B is trying to start up. If this check valve is located downstream of the anti-surge spillback, the anti-surge system may open the spillback valve to keep the compressor from surging. However, in order to force the check valve open and allow machine B to come on stream machine B's discharge pressure must be raised by closing the spillback valve. Will the machine surge? Probably.

This situation has to be handled by a carefully synchronized reduction of machine A P_D and an increase of machine B P_D. This has been done with analog controls, but it is difficult. Digital execution is recommended.

Suppose a check valve is placed inside the spillback take off? This is often done when there may be large volumes of gas in the spillback loop. A large volume of gas trapped inside the check valve could cause reverse rotation during compressor trip out. Compressor B would surge continuously unless the spillback valve opened and passed gas from the other machine.

At best, we are back to the problem of forcing the check valve to open. At worst the check valve will oscillate with the spillback loop. Once this cycle starts, nothing can be done except to take the spare down and try again.

There have been some successful attempts at automating on-line, spare start-up using relay logic or programmable logic controllers (PLC's) in combination with conventional analog controls. Although some of these systems have worked, today we see their implementation on digital computers.

These problems are not serious in every parallel compressor system. If they were, we would have a more complete answer by now.

There is no universal technique for designing compressor control systems, especially for parallel compressor systems. We cannot provide cookbook solutions but want to make you aware of potential problems.

The authors strongly recommend that control systems for parallel compressors be designed by dynamic simulation. We further recommend that these simulations be done on a hybrid computer.[1,2,3] Cost of simulation will be paid back many times by improved efficiency and safety of the resultant control system.

LITERATURE CITED

[1] Nisenfeld, A. E. and Schultz, H. M., Analog Simulation Aids in Control System Design.
[2] Nisenfeld, A. E. and Marks, A., Dynamic Simulation Prevents Start-Up Problems.
[3] Nisenfeld, A. E., et al, "Better Compressor Controls," *Hydrocarbon Processing*, April 1975.

ACKNOWLEDGMENT
Based on material copyrighted by and reprinted with the permission of the Instrument Society of America.

Part IV.

Centrifugal Compressors—Operation and Maintenance

Start up Considerations for Closed-Loop Overhung Centrifugal Compressors

A centrifugal compressor should be purged and pressurized so as to avoid damage to the thrust bearing

Bruno Bacchelli and **Ettore Bianco,** SNAM Progetti SpA, Milan, Italy

A CENTRIFUGAL COMPRESSOR with overhang impeller generally should not be started while it is under pressure. The thrust bearing in a compressor has a very low load carrying capacity or no capacity at all when the compressor is stopped. The loading capacity of the thrust bearing increases as the compressor speed increases.

On the other hand, if the pressurizing rate during startup is limited so as not to damage the thrust bearing, surging may occur. Therefore, pressurizing should be at some compromising rate—not too slow nor too fast.

Observations will be reported here to help establish the proper pressurizing rate and to give sufficiently detailed explanation of the phenomena connected with the various operation phases and dimensions of the various valves.

Startup steps. When a compressor is accelerated quickly, in the case of electric motor drives, pressurizing precautions usually can be neglected. In fact, the compressor can often be started while under pressure. However, it is essential to pressurize slowly when the compressor driver is of the slow acceleration type as in turbine drives. The following startup steps are used (Fig. 1):

Step 1: With intake valve V_1 and discharge valve V_2 closed, open startup bypass valve V_4 and vent valve V_5 to purge the compressor with gas.

Step 2: Close V_5 to pressurize the compressor.

Step 3: Open valve V_1 and V_2 to attain operating conditions.

The timing for this startup sequence depends on compressor acceleration characteristics and the size of valves V_4 and V_5. In all cases, V_5 should be larger than V_4.

The size of V_4 should be selected carefully. It should be small enough to prevent a too-rapid pressurizing of the compressor so that the load capacity of the thrust bearing is always higher than the thrust force. On the other hand, V_4 should be large enough to attain pressurizing with sufficiently high flow rates so that surging does not occur.

To signal when pressurization is complete it is advisable to use a differential pressure switch to measure P_1-P_2. A standard pressure switch to measure P_2 would give unreliable results since P_1 usually varies during startups. The value of P_1-P_2 should be small when it corresponds to complete pressurization of the compressor. A smaller differential pressure gives a higher compressor speed after completion of pressurization—all other conditions being the same.

WHEN V_4 IS LARGE

Purging (Step 1). As long as the vent valve V_5 is open, compressor flow is steady at Q_A corresponding to the flow

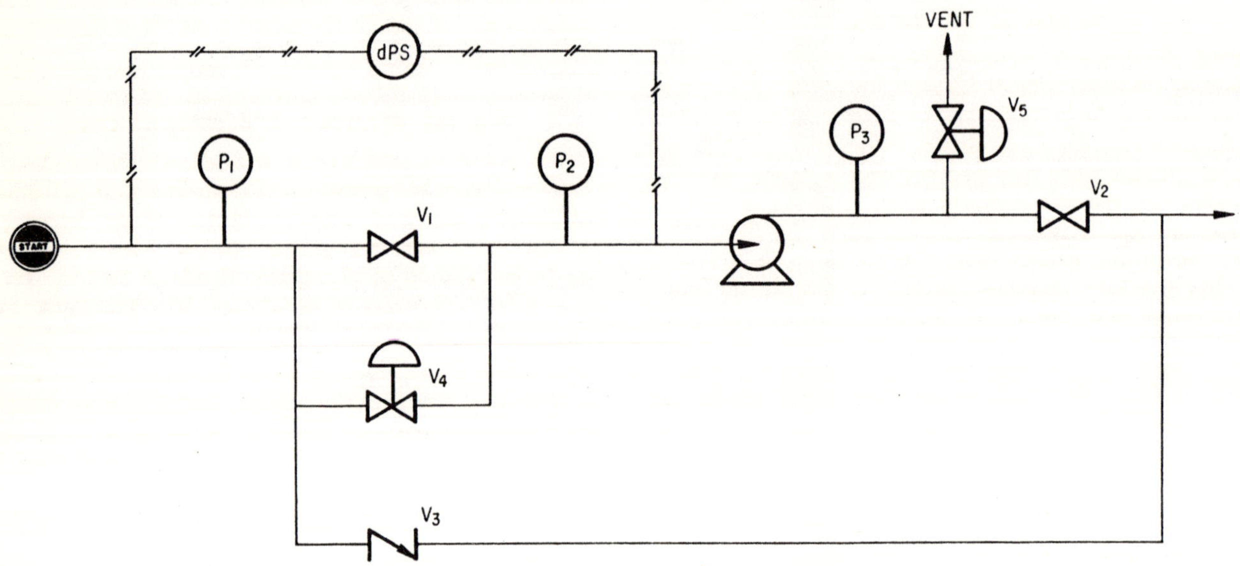

Fig. 1—The sizes of the bypass valve and the vent valve are important to proper startup.

determined by the pressure drop P_1-P_{atm} through valve V_4. Under these conditions the compressor starts and begins to accelerate. The vent valve V_5 should be closed as soon as the compressor reaches a speed such that the load carrying capacity of the thrust bearing is greater than the thrust force. This speed will be designated as n_1 (usually in the range 100-1,000 RPM).

Pressurizing (Step 2). With the vent valve V_5 closed, P_2 is approximately equal to P_3. The compressor flow begins to decrease rapidly from Q_A as the gas begins to pressurize the compressor stator and piping downstream of the compressor.

Fast acceleration of the compressor, or a large downstream volume to be filled, is represented in Fig. 2 by a decrease in compressor flow from Q_A to Q_3. The compressor can reach a speed of n_3. As long as the flow is determined by the pressure drop through V_4, compressor flow stays equal to Q_3 although the compressor speed increases to n_X. Then compressor flow decreases as shown by the path to point X in Fig. 2.

Slow acceleration of the compressor, or a small downstream volume to be filled, is represented in Fig. 2 by a decrease in compressor flow from Q_A to Q_2. When the compressor reaches a speed of n_2, the compressor flow is no longer determined solely by the pressure drop across V_4. Therefore, compressor flow continues to decrease as the compressor speed increases as indicated by the path to point Y in Fig. 2.

Pressures P_2 and P_3 increase approximately together as long as the flow rate through the compressor is higher than the flow at which the compression ratio is equal to one for the existing compressor speed. Thereafter, P_3 exceeds P_2 by an amount dependent on the compression ratio.

Surge capacity is reached at higher compressor speeds if the downstream volume is greater or compressor acceleration is faster—provided the size of valve V_4 and all other conditions are the same.

WHEN V_4 IS SMALL

Purging (Step 1). As long as the vent valve V_5 is open, compressor flow is steady at Q_B and the compressor starts and accelerates. As long as the compressor speed is lower than n_B (see Fig. 2), P_2 is almost equal to P_3. When the speed is greater than n_B, P_3 exceeds P_2 by an amount dependent on the compression ratio.

When the speed reaches n_1, vent valve V_5 should be closed. As in the earlier case, the speed n_1 is a speed such that the load carrying capacity of the thrust bearing is greater than the thrust force.

Pressurizing (Step 2). With the vent valve V_5 closed, the gas flow into the system is steady and determined by the pressure drop across V_4. However, the compressor flow rapidly decreases to only a fraction of the flow through V_4 and then remains constant as long as V_4 is controlling. Thereafter, compressor flow decreases and compressor speed increases as represented by the path to point Z in Fig. 2.

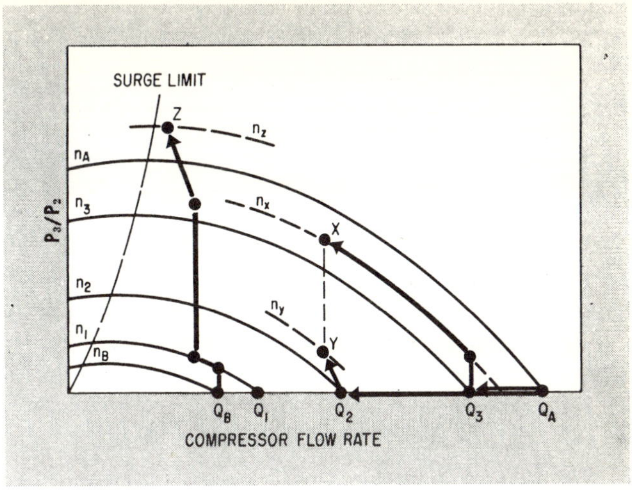

Fig. 2—Compressor speed after pressurizing depends on initial startup flow.

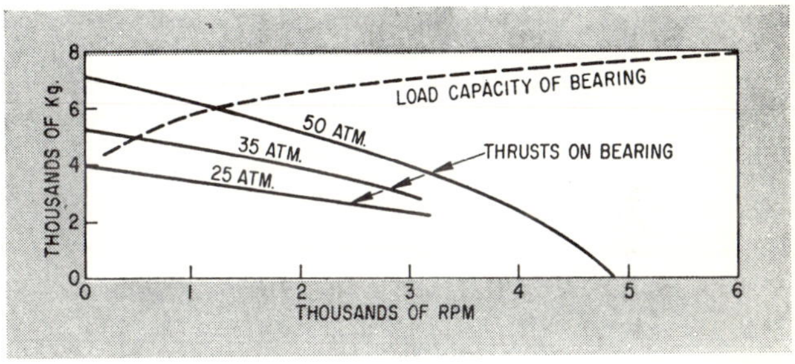

Fig. 3—Compressor speed should be fast enough for bearing load capacity to exceed thrust.

A larger V_4 will make the compressor pressure increase quicker. On the other hand, a smaller V_4 will give a lower purging flow Q_B. A smaller vent valve V_5 will give a higher purging pressure P_2 before the compressor starts. Also there will be a shorter time in which V_4 will be controlling after V_5 is closed. Furthermore, the compressor speed at which the surge capacity is reached will be lower.

THRUST BEARING CONSIDERATIONS

The thrust exerted by the compressor in the bearing is a function of gas pressure as shown in Fig. 3. Offsetting this thrust, the load capacity of the thrust bearing is greater at higher compressor speeds. This means the compressor should be pressurized slowly so that the thrust caused by gas pressure is always less than the load capacity.

Sizing of V_4 is a compromise. In summary, a smaller V_4 will give lower pressurization and a slower increase in thrust in the bearing. However, V_4 must be big enough to complete pressurization of the compressor before the compressor reaches the surge condition. These guidelines do not apply to compressors with shaft crossing both sides of the case where starting can be under pressure with the compressor accelerated either quickly or slowly.

How to Improve Compressor Operation and Maintenance

Centrifugal compressors can be designed and built with operation and ease of maintenance in mind. Here's what to .consider when specifying a new compressor.

Hugh M. Davis, Delaval Turbine Division, Trenton, N.J.

THE PROCESS INDUSTRY has experienced tremendous growth during the past two decades. The growth of single line and continuous process plants and the increasing use of automation have demonstrated the importance of component reliability. Machinery users are now demanding dependable performance, simplicity of operation and ease of maintenance. The suppliers of centrifugal compressors have been forced to review their designs and sometimes to design new equipment to satisfy these users' demands. This paper discusses the ways in which a centrifugal compressor can be built and used to satisfy a customer's operating and maintenance requirements.

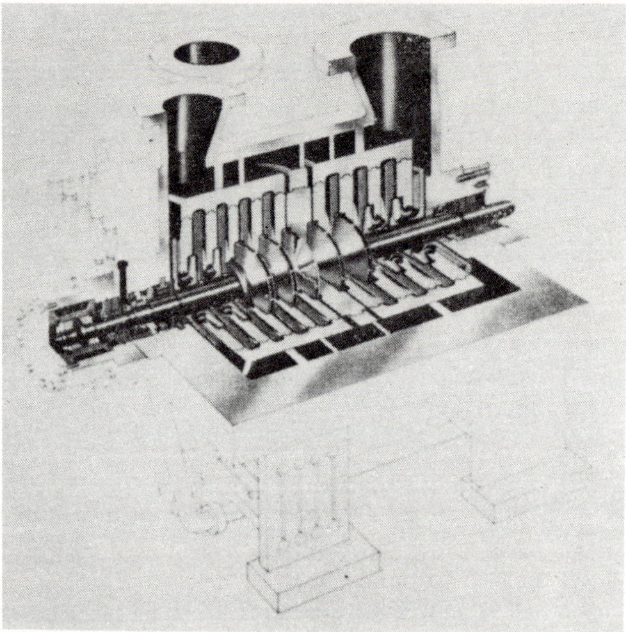

Fig. 1—Multi-stage fabricated case compressor.

Typical multistage compressor design. A typical multistage centrifugal compressor, designed to meet a particular customer's needs, is shown in Figure 1. This machine consists of 9 impellers in series and is designed to compress 4000 cfm of gas from an inlet pressure of 25 psi to a discharge pressure of 425 psi. Each impeller imparts velocity (kinetic) energy to the gas being compressed. This velocity energy is converted into increased pressure in the diffuser passage. The cross-over passage and the return guide vanes lead the gas to the next impeller where the compression is continued. The volume of the gas stream is reduced as it is compressed and each stage is designed to accept a successively smaller flow.

Range of application. Fig. 2 is a curve which shows the limits of applications for centrifugal compressors in terms of flow and speed. The speed is limited by the stresses in the impellers. The small flow, high speed compressors have the same working stress levels as the large flow, low speed machines. The compressor applications in the low flow range are almost entirely driven by motors and speed increasing gears. The compressors in the mid-range of flows are driven by motor-gears, steam turbines, and some gas turbines. The large, high flow compressors are practically all driven by steam turbines.

The size and operating speed of a centrifugal compressor have a direct effect on the operation and the maintaining of the compressor.

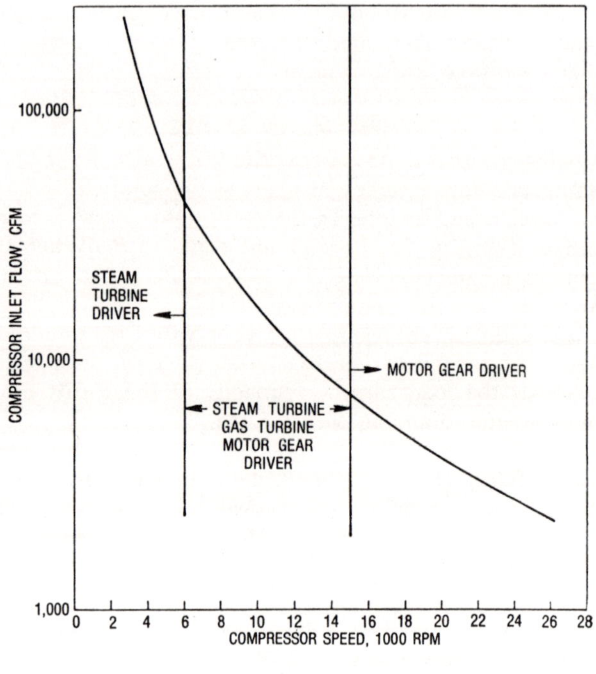

Fig. 2—Application chart for centrifugal compressors.

Fig. 3—Four-stage barrel compressor.

A small machine such as the one shown in Fig. 3, besides having high rotating speeds, is usually high pressure as well. Shaft alignment is more critical since the shaft ends are small and external pipe forces must receive proper attention to ensure good operation of the equipment. Maintenance is easier due to the light weight of the compressor components, but the clearances of the internal seals and bearings must be watched more closely due to their small physical size.

A large compressor, such as the one shown in Fig. 4, is more difficult to work on due to its size and weight. The clearances between rotating and stationary parts is less critical but the shaft alignment is more difficult to obtain, although more liberal tolerances are acceptable, because the components are not easy to move and special lifting facilities are required. The foundation for these large machines is also of special concern. Unless the supports are designed, constructed, and maintained properly the machinery may never achieve trouble-free operation.

Rotor dynamics. Modern process compressors are built in accordance with the API specification 617.[1] One important item defined by this specification is the natural frequencies of the rotor. These natural frequencies must not occur in the variable speed range of the compressor. The dynamics of a rotor can be studied with the help of the computer and the effect on the rotor of operative unbalance due to build-up or misalignment can be evaluated. These rotor unbalances will load the bearings. Computer analysis allows the engineer to predict these bearing loadings and to design a dependable maintenance free machine.

Fig. 5 shows the calculated and measured rotor re-

Fig. 4—Large turbine-driven centrifugal compressor.

sponse curves for an eight-stage compressor rotor. The measured values were obtained first during the mechanical test of the compressor. Although the vibration level was less than 0.7 mils and the bearing forces were below the design dynamic load limit, the steepness of the vibration curve near the maximum operating speed was understandably cause for concern. The rotor was modified and retested.

These results are shown in Fig. 6. The shop test shows quite low vibration amplitudes and low bearing loading. The compressor, once in actual service, will become unbalanced due to build-up on the rotor. The difference between the calculated and measured response curves shows that this compressor will be tolerant to considerable rotor deposits before it will have to be cleaned. Lightly loaded journal bearings, such as those used in compressors, can be unstable at high speeds, and a number of solutions to this problem have been used. The tilting pad bearing is widely used in compressors. Each shoe tilts independently to maintain its load carrying hydrodynamic pressure wedge. Extensive service in many types of compressors have proven the dependability of this bearing.

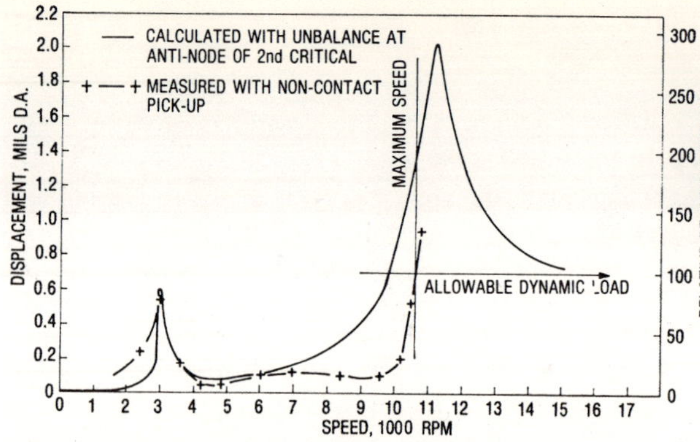

Fig. 5—Eight-stage rotor response curve.

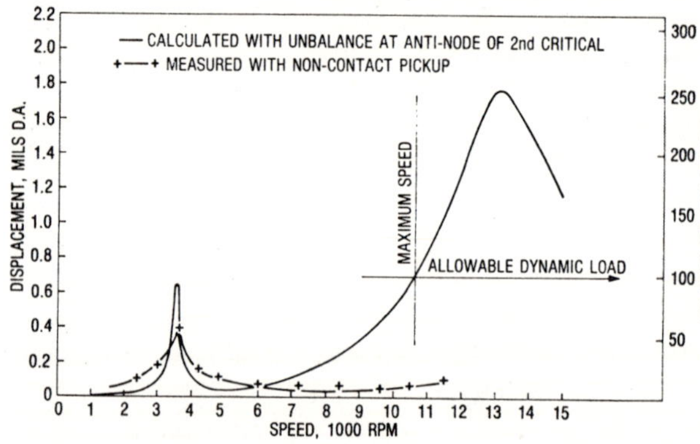

Fig. 6—Rotor response curve for modified eight-stage rotor.

Off-design operation. Most compressor users take trouble-free mechanical operation for granted, but they are concerned with compressor performance. Figure 7 shows a typical compressor performance curve. Uncomplicated and trouble-free operation can be expected in the stable performance region to the right of the surge line. Surging, or unstable operation, can occur in any centrifugal compressor when the inlet flow is reduced to approximately 60% of the design inlet flow or lower. Compressors that produce large pressure ratios, ratio of inlet pressure to discharge pressure tend to have more violent surges. When the compressor is operated repeatedly or for prolonged periods of time in surge the pressure forces can damage the internals of the machine. For those applications where frequent surge operation can be expected the compressor internals should be made of steel, instead of the more common cast iron material. When the compressor is operated in surge continuously it absorbs approximately 40% of the rated horsepower, however the flow thru-put is greatly reduced and under some conditions stops completely. The power required to drive the compressor in surge is therefore largely converted to heat. This causes excessive temperature build-up inside the compressor and in a matter of seconds, if the conditions are severe enough, can melt the soft labyrinth seals which control the internal leakage. The compressor performance suffers once the seals are damaged and the machine must be opened and the seals replaced to restore it to the original condition. Excessive temperatures in a compressor having a balance drum labyrinth seal made of a soft material with a low melting temperature can melt the seal. This will upset the rotor thrust balance and overload the thrust bearing. When the thrust bearing fails the rotor will shift axially and the impellers will rub against the stationary parts causing further damage. To help avoid these operating problems the compressor can be provided with a high temperature balance drum seal made from compressed metal fibers that will withstand several times the normal operating discharge temperatures.

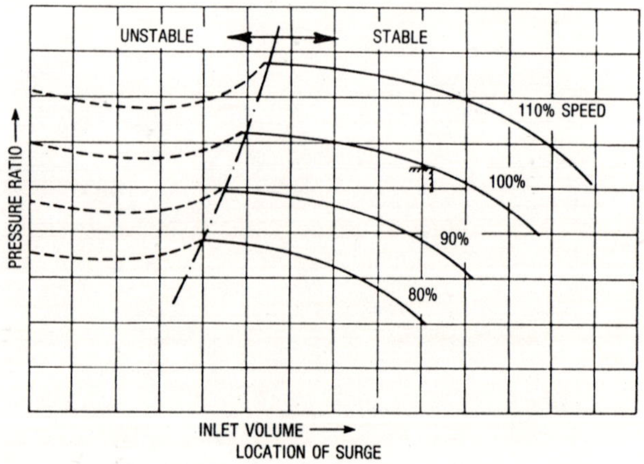

Fig. 7—Compressor performance curve.

Another safety feature that can be employed is a high temperature switch located in the balance drum leakage pipe. It is wrong to locate this switch in the discharge pipe. In this location the switch does not protect the compressor since there is not sufficient discharge flow to carry the heat to the switch when the compressor is operated in surge. The balance drum leakage pipe is the correct location for this protective device. There is leakage flow in this pipe even when the compressor is being operated completely shut-off.

Normal maintenance items. It is reasonable to expect trouble-free operation for periods up to two or three years between internal inspections for heavy duty commercial machinery. However, some applications where the gas stream being compressed is extremely dirty and internal washing cannot be used it may be necessary to shut down in order to clean the compressor internals to restore full flow capacity.

The internal seals that prevent leakage around the impellers are normally of the labyrinth type. They consist of a series of circumferential knife points that are positioned closely to the rotating impeller. In order for the compressor to maintain the design performance these knife points must not be damaged by rubbing, erosion, corrosion, or plugged-up with foreign matter. These seals are normal wearing parts and spares should be maintained in anticipation that they will need to be replaced after an extended operating period.

Labyrinth seals are used in compressors with various design features that will extend their life and make

Fig. 8—High-speed balancing machine.

them less susceptible to damage. One popular version is to machine the knives in the rotating part and to position a stationary sleeve of a soft material around the points to obtain a seal. The clearances between the rotating and stationary parts of the seal can be reduced because it is intended that the rotating points cut grooves in the adjacent material under normal operation. This type seal allows the compressor to operate with higher efficiency due to reduced leakage and longer life because the thin knife points are made of steel and resist erosion. Also in this design considerable radial rotor motion can be tolerated without altering the effectiveness of the seal. The stationary part of these seals have been manufacured in babbitt-lined steel, aluminum, non-metallic compounds, and compressed steel fiber materials. Each has its own advantages when considering the particular operating environment of the compressor.

In many compressors labyrinths are also used for the shaft seals. Where the small leakage allowed by labyrinth shaft seals can not be tolerated, either the oil film or the mechanical contact shaft seals can be used.

Compressor shaft seals of the oil barrier type present potential maintenance and operational problems. A recent study of a large cross-section of compressor users showed that approximately ⅓ of all compressor failures and loss of production was due to malfunctioning oil film type seals! Regardless of the cause, whether it was design, operation, or maintenance it points out a particular component of the compressor that commands respect.

Oil film seals consist of basically two stationary bushings which surround the rotating shaft with a few thousands of an inch clearance. Seal oil is introduced between the bushings and leaks in both directions along the shaft. "O" rings in the seal housing prevent leakage around the outside of the bushings. The seal oil is maintained at some pressure higher than the gas pressure inside the compressor. The differential pressure across the inner bushing is usually only a few pounds per square inch to limit the amount of inward oil leakage. This leakage is collected in a leakage chamber that is separated from the gas stream by a labyrinth seal and is drained away through a drain trap. If the gas being compressed contaminates the seal leakage, the leakage is discarded. The outer bushing takes the total pressure drop from seal oil pressure to the atmospheric drain.

Oil film seals can cause many different types of operating and maintenance problems. The most common is excessive inward leakage due to increased clearances between the bushing and the shaft. When these clearances become large enough the leakage chamber becomes flooded and oil spills over through the labyrinth and enters the compressor. This malfunction can cause many operating problems ranging from the nuisance of having to continuously add oil to the reservoir to contaminating the main process gas. The bushing clearances can increase in time due to corrosion or erosion of the inner surfaces of the bushing, wiping due to radial shaft vibration, or dirt in the seal oil.

Another shaft seal, which is not as common in process compressors, is the mechanical contact type. This seal has the advantages of being able to maintain low inner leakage rates with higher oil-to-gas differential pressures and therefore makes the seal oil pressure control system simpler. The seal has a spring loaded carbon face that runs against a face on the seal collar. The complete shaft seal either consists of two mechanical contact seals in a back to back arrangement or a combination of the mechanical contact seal on the gas side and a bushing seal on the atmospheric side. These mechanical contact seals have been used for many years to seal against pressure up to 1200 psi in natural gas service. They presently are being tested for pressures as high as 2500 psi differential across a single sealing face. The mechanical contact seal has an added advantage when applied to high pressure applications in that the radial sealing faces have a minimum effect on the rotor dynamics, whereas oil film bushing seals can lose their free floating feature and can upset the stability of the rotor when operating at high speeds.

High speed compression machinery must be properly balanced, especially when it is designed to operate between the *1st* and *2nd* lateral critical speeds. The correct method is to first dynamically balance each impeller on an arbor, and to check the rotor balance as each impeller is installed on the shaft. The runout of the shaft must be watched closely as each of the impellers are shrunk on the shaft. Abnormal change in shaft runout indicates that the impeller is not square on the shaft and must be adjusted before checking the balance. When the balance check indicates that an unbalance exists, the correction is made to the last impeller that was mounted. This procedure when strictly followed can produce a well balanced rotor even when the balancing is done in a low-speed balance machine. Some users routinely check the balance of complete rotors and make corrections on the first and last impellers. This practice is wrong for flexible rotors that have more than three impellers. The only sure way to balance a completely assembled high speed rotor is to use a high speed balance machine. The rotor unbalance must be checked throughout the operating speed range and any corrections that are made to the rotor must be made at the plane of unbalance. Most balance shops in this country have balancing equipment that operates below 1500 rpm. The photograph in Figure 8 shows a balance machine that will balance a 1000 lb. rotor at speeds up to 7000 rpm. The machine will handle most high speed flexible compressor rotors.

Experience has shown that high speed compressor rotors can be successfully balanced in a low speed balancing machine when the progressive impeller by impeller method is followed. The compressor is normally subjected to a mechanical test in the shop before shipping. During these tests the vibration amplitudes and frequencies are measured. When the compressor vibrates excessively the rotor is rebalanced in the high speed balancing machine, however this is normally not required. When a spare rotor is purchased or repaired after the compressor has been installed the rotor should be given a high speed balance check in order to prevent delays later in the users plant.

The alignment between adjacent shaft ends of all rotating equipment in the compressor train is very important to the operation and maintenance of the equipment. All compressors, gears, motors, turbines, etc. have some tolerance for misalignment. However, except for rotor balance, misalignment is the most frequent cause of running problems. Excessive misalignment can force the rotating equipument to vibrate and shorten the life of the bearings and gear type couplings. When misaligned shafts are rotated, the gear teeth in the coupling must slide back and forth on each other. This causes bending moments and forces to be imposed upon the shaft ends. The shaft end will fail if the conditions are severe enough or the coupling hub will come loose on the shaft.

Most couplings are mounted on a tapered shaft end. It is common that a key and keyway be provided to transmit the torque through the connection. The coupling hub is pushed upon the shaft taper by a nut which also prevents it from coming off the taper during operation. This type of coupling hub mounting is satisfactory for low speed-low horsepower applications but can be a limiting factor for modern high speed machinery. A much better method is to shrink the coupling hub upon the shaft taper using hydraulic pressure and jacking tools, as shown in Figure 9. This type of mounting allows for at least twice as much torque to be transmitted through the same size shafts with the same stress levels. This is possible since the key is no longer needed and the keyway stress concentration is eliminated. The misalignment forces from the coupling teeth will not loosen the coupling hub on the shaft because it is fitted on the shaft

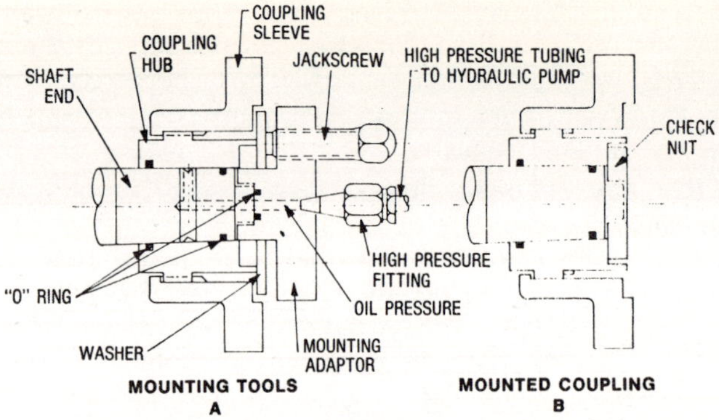

Fig. 9—Hydraulically-mounted coupling and tools.

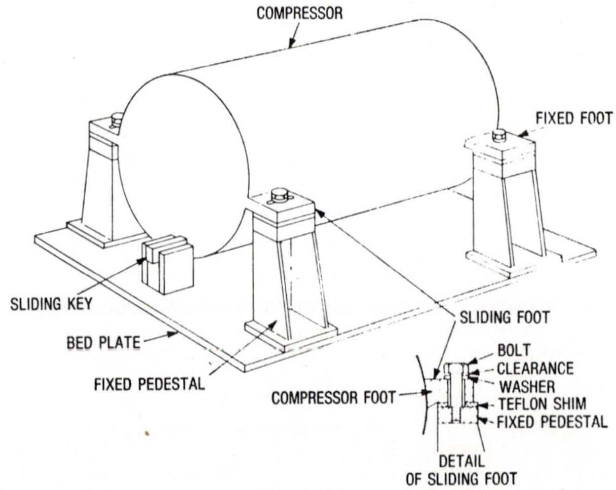

Fig. 10—Compressor case with sliding mounts.

taper with considerable interference. To mount a coupling hydraulically, the shaft end is drilled to provide an oil passage for pumping high pressure oil between the shaft taper and the coupling hub bore. A hydraulic hand pump is connected to the shaft end through an adapter. The hub is stretched with hydraulic pressure and the jack screws (Figure 9A), are used to push the coupling hub up on the taper a specified distance. The jack screws hold the hub in the correct position while the oil pressure is released. After the oil has drained the mounting plate is removed, a nut is mounted on the shaft end (Figure 9B) to protect the threads. To remove the coupling hub the procedure is reversed and the hub is "popped-off" against the tools with hydraulic pressure. This type of coupling mounting provides for an easily mounted hub that will transmit as much torque as the shaft material will allow.

Good shaft alignment must be maintained under dynamic conditions if trouble-free operation is to be achieved. To accomplish this the thermal growth of the individual elements must be taken into account when aligning compressor equipment in the cold conditions. For instance, the diameter and length of a compressor case will increase due to the heat of compression.

Normally the casing is supported at the horizontal centerline to allow the case to grow in diameter without changing the position of the shaft. The mounting feet on one end of the case are bolted and doweled to the foundation, as shown in Figure 10. The opposite end of

Fig. 11—Compressor case with centerline support.

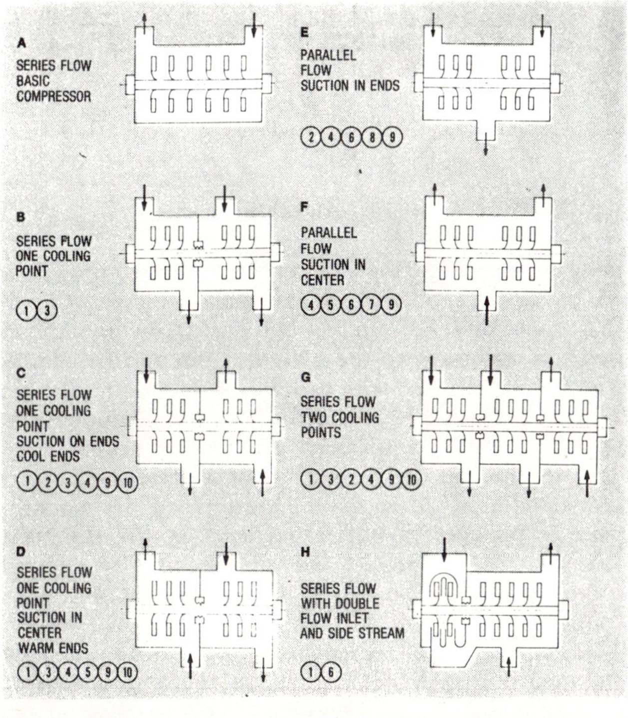

Fig. 12—Eight-compressor case and nozzle configurations.

the case must be allowed to move axially as the casing expands. This has been accomplished in the traditional compressor mount with holddown shoulder bolts that allow the case to slide axially upon lubricated shims but limit the vertical movement of the case. With this arrangement a vertical key and keyway are located between the case and foundation on the vertical centerline to prevent transverse movement of the case. This mounting requires a special foundation to support the vertical key and also regular lubrication of the shims.

The compressor case mounting shown in Figure 11 allows for thermal growth of the case in all directions, provides centerline support for the case, has four mounting pads that are all bolted and doweled to the foundation, and requires no maintenance. The ruggedness of this support will allow for considerable external forces to be exerted upon the casing without changing the shaft alignment.

Internal configuration. The process market places very demanding and ever changing requirements upon the selection and arrangement of the compressor internals and external casing nozzle configuration. Centrifugal compressor selections should be made with the ease of operation and maintenance aspects in mind as well as the compressing requirements. Figure 12 shows the most typical arrangements of the compressor internals and outer casing in the form of simplified diagrams.

The operation and maintenance advantages are listed for each as follows:

1. Single compressor body instead of two or more results in a simpler system.

2. Hot discharge at center of case to reduce lubrication and oil seal problems.

3. Reduced power required to compress gas results in a smaller driver.

4. Back to back impellers reduce natural rotor thrust and allows for more internal seal wear before overloading thrust bearing. Increase time between overhauls.

5. Cold inlet at center of case to reduce lubrication and oil seal problems.

6. Smaller compressor and higher speed to do the same compression job. Reduced foundation and maintenance problems.

7. Single inlet to better suit external piping arrangement.

8. Single discharge to better suit external piping arrangement.

9. No external balance piston leakage pipe. Compressor can tolerate increased balance seal wear without upsetting thrust balancing system and overloading thrust bearing.

10. Hot or cold sections of case are adjacent to reduces thermal gradients and distortion of the case. Makes alignment easier to achieve.

ACKNOWLEDGMENT

Originally presented at the 1st Texas A&M University Turbomachinery Symposium, October 1972, College Station, Texas.

LITERATURE CITED

[1] API 617—Centrifugal Compressors for General Refinery Service, Second Edition, 1963, American Petroleum Institute, Washington, D.C.

Unique Compressor Problems

Case histories on foundation and impeller resonance, thrust bearing loading, and two-phase flow describe centrifugal compressor problems experienced by today's HPI engineer.

Charles Bultzo

Humble Oil & Refining Co., Baytown, Texas

UNIQUE CENTRIFUGAL problems and the solutions developed are described in the following case histories. They are typical of those problems being experienced in industry today. However there is always something unique about each centrifugal compressor problem that requires a unique or different solution.

The Problem of Resonance. In one recent installation, a vibration problem was encountered on a centrifugal compressor and a two-motor drive train mounted on an elevated concrete structure. Vibration measurements on the supporting columns revealed high lateral sensitivity to the design running speed of 3,596 rpm. This was confirmed in part during initial test work, when the supporting columns were struck by a heavy timber. Reed tachometer measurements indicated a natural frequency of 3,550 rpm. Also, recordings of vibration levels on the foundation during deceleration consistently showed an increase to occur at approximately 3,500 rpm. A typical plot of a rundown recording is shown in Fig. 1. As a point of further confirmation, measurement taken at one-foot intervals of column height revealed that the point of maximum deflection in the columns occurred at close to the expected 5/9 L point for resonant columns. Since the foundation sensitivity manifested itself in the form of shaft deflection at the coupling between the two motors, a short-term repair was made by installing timbers as shown in Fig. 2. This reduced vibration levels on the motors and compressors by 50 percent or from five mils to two mils. The lateral deflection of the columns was reduced also. During a subsequent downtime, additional mass was added to the foundation in the form of larger cross section in the columns and deeper transverse beams. Vibration levels continued to be two mils or less on all shafts.

Another resonance problem occurred on the same installation. During inspection of the spare rotor prior to its

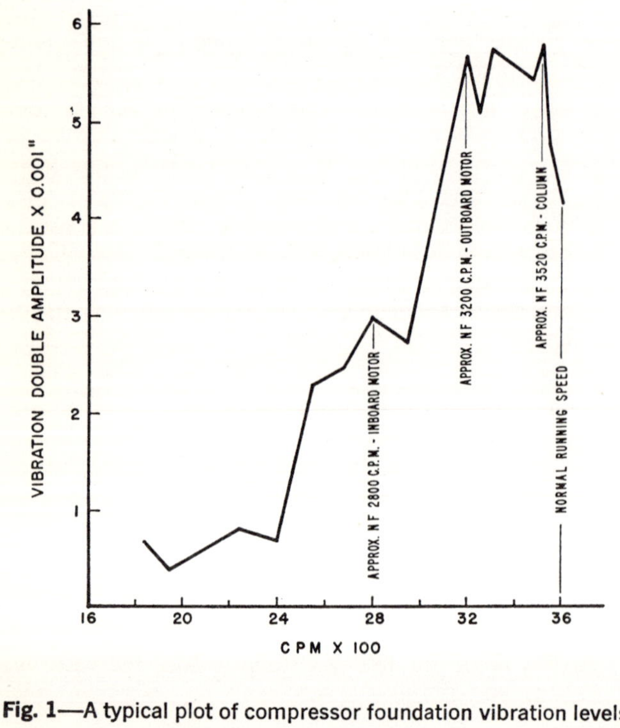

Fig. 1—A typical plot of compressor foundation vibration levels.

Fig. 2—Temporary repair of foundation using timbers.

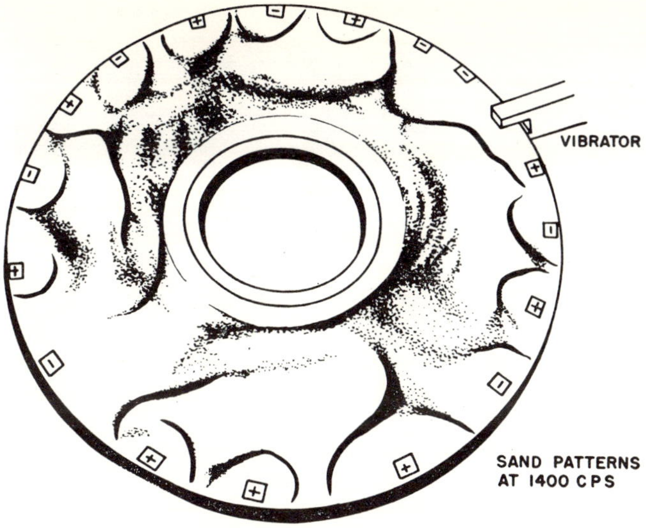

Fig. 3—Sand patterns on first stage impeller.

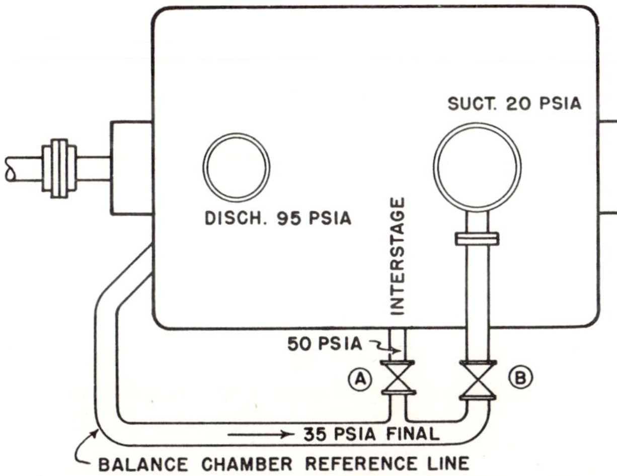

Fig. 4—Balance piston chamber piping arrangement.

installation, a test was made to determine the natural frequencies of the impellers. This was done to better understand the over-all frequency response range of a prototype impeller. One of the frequencies explored was the product of the number of variable inlet guide vanes and rps since this would be the environment in which the first-stage impeller would run. As shown in Fig. 3, the sand pattern shows a concentration to occur at each vane. The manufacturer agreed that this sensitivity could cause fatigue-type cracks on the disc side of this impeller. The problem was resolved by replacing the variable inlet guide vanes with a butterfly valve, thus eliminating the source of the problem. This was confirmed by an inspection of this impeller after eight months of operation which failed to show any distress in this area.

Thrust Bearing Loadings. One of the continuing problems in the application of centrifugal compressors is proper thrust bearing loading. This force is the result of differential pressure acting on the compressor impeller surfaces and is usually carried by a thrust bearing. Our present specification limits the thrust load on new compressors to 150 psi. Because of more general use of strain gages, the application engineer now has a tool which can be used to determine thrust loads when the equipment is new and to check any deterioration of internals after extended service. Also, the use of shaft position indicators and wear buttons has almost eliminated radical thrust bearing failures when they have been applied. Several case histories of thrust bearing problems follow.

Case 1. A centrifugal compressor in alkylation plant service had a history of frequent thrust bearing failures. After strain gages were installed on the thrust bearing support ring, the normal thrust load was found to be approximately 420 psi. However, during surge, the load would approach 600 psi. The rotor had no restraint in the form of seals that could help absorb the axial loading caused by the mass of the rotor moving during surge. Because of internal configuration, it was not possible to increase the size of the thrust bearing without machining. This would have required an extended downtime. Fortunately, the vendor had referenced the balance piston chamber pressure back to an interstage point rather than to suction. The configuration of this line was changed such that it could be referenced to the original interstage point or back to the main suction (See Fig. 4). After being returned to service, the valves were manipulated such that the original line (Valve A) to interstage was finally closed and the desired balance piston chamber pressure was achieved by partially closing the valve to the main suction (Valve B). After the precalculated pressure had been reached, strain gage readings confirmed a thrust load of approximately 200 psi. No thrust bearing failures have occurred since.

Case 2. In a more recent installation in propane refrigeration service, the compressor was equipped with strain gages in the thrust bearing support ring as well as a proximity-type shaft position indicator. "Zero" readings on the strain gages were taken prior to startup. After startup, the shaft position indicator showed the shaft to be running against the inactive shoes. Also, the strain gages did not show any change in output. A calculation of the expected thrust with the observed pressures on the suction, discharge, and three side loads indicated the thrust load to be approximately 150 psi and in the inactive direction. During a subsequent downtime, a test valve was installed in the balance piston chamber reference line and sealed open (See Fig. 5). After the unit was returned to service and normal flows had pressures established, the test valve was slowly closed. A table with the resulting data follows:

	1	2	3	4	5
Upstream Pressure, psig	36	44.5	50.5	52.5	44.5
Downstream Pressure, psig	35	34.5	34.5	34.5	34.5
Oil Temperature In, °F	122	122	122	120	119
Oil Temperature Out, °F	167	163	156	156	153
Shaft Position	−0.013	−0.0125	−0.008	−0.005	−0.003

At this point, the test was stopped since the valve was

Fig. 5—Test valve installed in balance piston chamber reference line.

almost closed. The data does show that as the valve was being closed, not only did the shaft move toward the active shoe but unloading of the inactive shoe was taking place as indicated by the significant change in the return oil temperature. The vendor is presently considering a redesign of the balance piston.

Two-Phase Flow. Many application engineers have considered the problems relating to two-phase flow in a centrifugal compressor. In some cases, liquid is injected for cleaning or antifouling purpose or to limit the discharge temperature. Tests run with liquid injection showed that substantial rates could be tolerated. In one case, it was interesting to note that the calculated polytropic efficiency changed one percent for each 2° F change in discharge temperature caused by liquid injection. However, in most cases, the presence of liquid is the result of inadequate knockout facilities in suction piping. An illustration of this is given below.

The piping to two centrifugal compressors is shown in Fig. 6. A question of capacity led to a test of both compressors. Although both had identical suction pressures/temperatures and a common discharge pressure, Unit B had a discharge temperature 9° F cooler than Unit A. Further, although the manufacturer had predicted a polytropic efficiency of 72 percent, the test indicated Unit A to be 74 percent efficient while data on Unit B indicated it to have an efficiency of 78 percent. Further examination of the suction facilities revealed that with the increased gas rates being experienced, the gas velocity in the suction knockout drum was now substantially above the required level for separation. Further, it was conjectured that because of this arrangement, the bulk of the liquid present was being centrificated into Unit B. Since the test, an inspection has revealed an accelerated loss of metal on the first-stage impeller caused by the passage of liquid. Plans have been made to increase the size of the knockout facility during the next downtime.

The high-pressure or second-stage portion of this in-

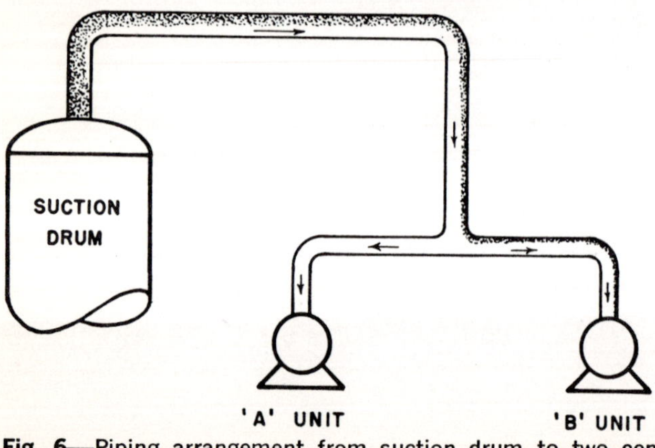

Fig. 6—Piping arrangement from suction drum to two centrifugal compressors.

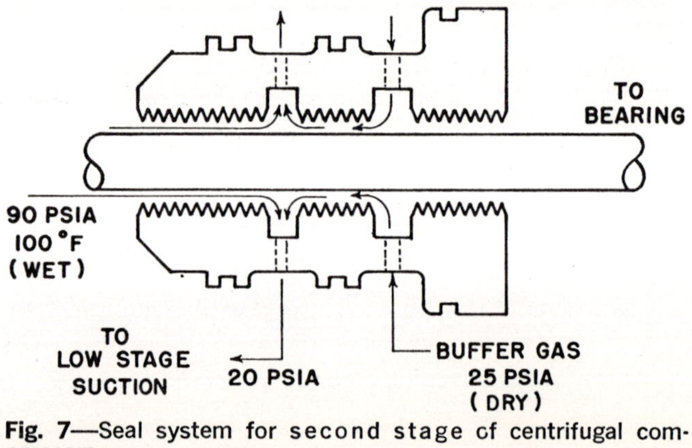

Fig. 7—Seal system for second stage of centrifugal compressors.

Fig. 8—The suction seals were badly cut.

stallation (two additional compressor casings) had a different type of problem that was attributed to two-phase flow. The sealing system used is shown in Fig. 7. The high-stage suction pressure is 90 psia. The sour/wet gas pressure is broken down to approximately 20 psia and bled to the low-stage suction (the other casing in the train). Sweet or buffer gas is injected to prevent the sour/wet gas from contaminating the lube system. After some 18 months of operation, a lowering of the lube oil viscosity in the system was observed. An analysis of the system revealed that little or no buffer gas was being injected because of an increase in sour gas reference pressure. After a proper differential had been established, the viscosity problem was improved considerably. Shortly after, both compressors were inspected. It was found that the discharge end seals were in excellent condition (gas conditions at discharge of 195 psia and 210° F). However, the suction end seals on both compressors were badly cut as shown in Fig. 8. The condition of the labyrinth was attributed to erosion caused by liquid present in the sour/wet gas. Further, a flash calculation of the gas being compressed confirmed that the two-phase condition existed at not only the suction (six percent liquid) but also at interstage in the labyrinth (two percent liquid). It was concluded that the seals had eroded to a point where some bypassing of the seal gas was occurring and the liquid was entering the bearing chamber. Also, the buffer gas temperature (80° F) was not high enough to reflash the liquid whose boiling point was calculated to be 90° F. Since that time, a buffer gas heater has been installed in an attempt to keep the combined interstage temperature above the boiling point of the liquid in the sour gas.

Originally presented to the ASME Petroleum Mechanical Engineering Conference, Philadelphia, Sept. 19, 1967.

A Closer Look at Turbomachinery Alignment

Successful shaft alignment requires consideration of all factors that influence movement of the machinery. Here's what to look for before and during the actual alignment.

Jack N. Essinger, Shell Chemical Co., Houston

Good turbomachinery alignment consists of three interrelated parts. Each part is an equally important step in realizing the goal of an equipment train capable of operating safely for long periods without the many problems resulting from severe misalignment.

1. The system survey. Many steps vital to good turbomachinery alignment are taken well ahead of the actual "cold alignment." Unless the person responsible for alignment has first-hand knowledge that pre-alignment preparations have been properly carried out and completed, verification is essential. Some of the items which warrant specific attention are:

1. Piping. A visual piping inspection by those responsible for alignment is vital. This inspection will assure that the piping is installed in apparent agreement with design criteria, is complete and in its functional state. Obvious items to look for are proper placement and adjustment of guides, anchors and supports; proper adjustment of tie-bolts on expansion joints; correct positioning of spring hangers; complete make-up of flanges with gaskets in place and bolts tightened; absence of slip-blinds which may have been installed for hydrotesting of pipe lines, and proper orientation of check valves. In short, it must be verified that the system is in order so post-alignment piping modifications will not nullify the alignment effort.

2. Grouting should be checked to be sure it is complete and apparently well done.

3. Foundation bolts should be checked for tightness.

4. Check all shimpacks. Shims are the vital link between the machine and the foundation and are essential to maintaining alignment over long periods. It is my experience that the only reasonable assurance of proper shimpacks is to physically remove and inspect the shims at every machine support just prior to final alignment. Obvious problems with shims include rust, improperly cut shims, folds and wrinkles, burrs, hammer marks and dirt. I once found a turbine installed on shimpacks wrapped in masking tape. It is good practice to use as few shims as possible and replace many thin shims with fewer shims of greater thickness. Stainless steel shims will pay for themselves many times over by minimizing alignment problems associated with shim deterioration. It is equally important to see that surfaces of equipment supports and soleplate/baseplate are clean and in good condition.

5. Check for misalignment of machine supports relative to the soleplate. A relatively simple test for problems in this area can be made when shimpacks are checked. Mount a dial indicator on the machine support with the indicator stem resting on the soleplate. Watch the indicator as the hold-down bolts are loosened. If movement of the indicator is more than 0.001 to 0.002 inches, it is an indication of a problem that must be defined and eliminated. Remove the shimpack and check with feeler gages to be certain the machine support is parallel with the soleplate. If not, re-grout, re-machine the support or prepare tapered shims.

Be suspicious if one support moves less than others. If three of the supports move 0.001 to 0.002 inches, and the fourth shows no movement, it may indicate that the support is carrying more than its share of the load.

6. Check the casing for distortion. Despite the massive appearance of turbomachinery casings, they are flexible and easily distorted. This relatively simple test for gross distortion can be made when shimpacks are checked:

a. With three supports tightened down, remove the shimpack from the fourth support.

b. Determine the total thickness of the shimpack and record the dimension.

c. Let this corner of the machine down (the machine is now supported only at three points). Using feeler gages, determine the distance from the soleplate to the machine support. Record the dimension.

d. Subtract the feeler gage dimension from the shimpack thickness. This is the total deflection of the machine casing with no support at the corner being checked.

e. Repeat the procedure at each of the four supports and compare the deflection of each. Gross differences in deflections at any of the four supports is an indication of probable casing distortion.

7. Check for piping strain. Piping strain is seldom detectable by visual observation. However, gross problems can be detected by a rather simple test. Following the check for casing distortion, place dial indicators on the machine to monitor both vertical and horizontal movement of the casing or shaft. Now loosen all the hold-down bolts. If the machine moves more than the average observed when checking individual supports, it is obviously the result of an external force—probably the piping.

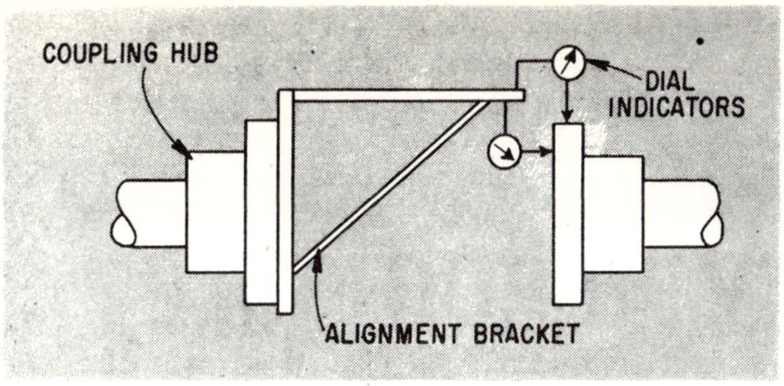

Fig. 1—Typical arrangement for "Face-OD" readings to determine cold alignment of shafts.

8. See that bearings are properly installed in the machines, that they are lubricated and that the bearing covers are properly tightened. These are rudimentary precautions made on the basis of experience through attempted alignment of machines with the bearings removed.

9. For turbomachinery trains with high-speed gearing, extraordinary precautions should be taken to set and align the gear itself. High-speed gearing is normally the most precisely made equipment in the train; it is also the most sensitive and most vulnerable to catastrophic failure as a result of poor installation. Unfortunately, installation instructions for most gears are grossly inadequate. One fundamental precaution is offered: if a gear is involved, be very, very careful.

Planning. Since turbomachinery alignment normally involves the participation of many people in various functions, it is imperative that an over-all plan be developed to assure that the goals of the program are known and understood and that the method of attack is consistent with the desired results. Moreover, such planning is valuable from the standpoint of minimizing the time involved (hence, the cost) to achieve satisfactory alignment. Plans should include:

1. Determine the desired placement of the shafts (cold settings) considering anticipated thermal growth of the various components. Define movement of shafts within bearing clearances (rising-pinion gears, for example), hydraulic loading and any other factors expected to produce relative movement of shaft centerlines when the machines are operated.

2. The sequence of alignment must be determined for multi-unit trains. For two-component trains, determine which of the two machines is to be moved.

3. Select a specific method for determining relative shaft positions.

4. Decide on tolerances for theoretical cold alignment settings. Allowable deviations which are unrealistically strict cost time and money, and may not be attainable. Tolerances which are too loose are an invitation to trouble.

5. Make a list of tools and instruments required for the alignment and make certain they are available and in good working condition. These include such items as alignment brackets, dial indicators, tools required for hot alignment checks, pre-cut shims of various thicknesses, tools for lifting the machines during shim changes, wrenches of the proper size for hold-down bolts, coupling tools and a device for rotating the shafts.

6. Provisions should be made for permanent recording of alignment data. Many good sets of alignment readings, written upon the side of a compressor, have been covered by the painter's gun.

2. Cold alignment. The term "cold alignment" refers to the position of a turbomachine's shaft centerline relative to the shaft centerline of a connected machine, with both machines in a non-operating or "cold" condition. Offset and angularity are both implied by the term. Cold alignment is important because it is normally the only check made to directly determine the relative position of the two shafts. Results of the check form the basis for determining shaft alignment during operation.

The only cold alignment techniques discussed here are those using dial indicators. Procedures are old and well established, but do have problems and pitfalls.

Fig. 1 illustrates the most widely used of the traditional alignment methods, commonly referred to as the "Face—OD" method. A bracket is attached to one shaft and extends near the coupling hub on the adjacent shaft. Dial indicators are attached to the bracket as shown, with the stem of one indicator resting on the face of the coupling hub. The stem of the other indicator is resting on the OD of the same hub. Offset of the shafts is determined by the OD readings. Angularity is determined by "face" readings.

The method has and continues to serve well in the vast majority of industrial alignment problems. Indeed, it is the specific method outlined in virtually every maintenance manual for industrial rotating equipment.

For high-speed, high-hp turbomachinery, where requirements for precise alignment are more stringent, shortcomings of the method become increasingly important. Consider the following points:

- The indicator readings from this method reflect not only misalignment of the shafts of the two machines, but inaccuracies in geometry of the coupling hub upon which readings are made. Typically, the measurements should have an accuracy approaching 0.001 inches. It is not uncommon to find runout of the face or the OD of a coupling hub in excess of this amount, even on good quality, high-speed couplings. This error can be eliminated, of course, by determining the geometry of the coupling hub and making appropriate allowances or by turning both shafts simultaneously such that indicator readings are always taken at the same place on the hub, but these precautions are difficult on large equipment and are seldom taken.

- When making face measurements it is absolutely

necessary that the axial floats in both shafts be accounted for. This is of little concern with small equipment where shafts are fixed axially by ball bearings. It becomes a problem, however, on machines equipped with hydrodynamic thrust bearings (or no thrust bearings at all). The normal procedure is to separate the shafts axially each time a reading is taken. The axial position of the shafts as they come to rest depends on how hard the craftsman pushes. Thus, consistent readings are difficult to obtain. The practice also leads to the use of makeshift tools to pry the shafts apart, and a coupling seldom comes through the alignment procedure unscathed.

- The face diameter upon which readings are taken to determine angularity of the two machines is relatively small, especially in high-speed equipment—sometimes no larger than 3 inches. For example, a reading of 0.002 inches (normally termed "acceptable") on the face of a 4-inch diameter coupling hub represents an angularity of 0.006 inches per foot or an error of 0.030 inches at the opposite end of a machine 5 feet long. Add to this a 0.001 inch error in the face of the coupling and the outboard end of the machine is 0.045 inches from the desired position. In most instances such an error is not catastrophic, but it is an error that should not exist.

- The tool normally used to hold the dial indicators is either a "universal" bracket, or makeshift device contrived on the spur of the moment. Brackets specifically designed for the machines being aligned are not normally considered necessary and are, therefore, not used. The assumption is usually made that the bracket being used is stiff; i.e., it does not deflect under its own weight. Normally there is no attempt to verify the assumption, which is a very poor one, especially for couplings with relatively long spacers. I have found many brackets which deflect 0.005 to 0.010 inches over a 12-inch span. In an extreme case (a machine in trouble), it was found that the indicator bracket which had been used deflected about 0.040 inches.

- The method requires that the coupling spacer be removed to make readings. For routine alignment checks this means added effort and the ever-present risk of coupling damage during disassembly and re-assembly. For new installations, it normally means that the coupling is left open until after cold alignment is complete. The interval is usually several weeks from the time equipment is first set upon the foundation until the alignment is completed, during which time the coupling is subjected to corrosion, dirt and physical abuse. Some couplings do not survive this initial treatment.

Reverse indicator method. In contrast to the "Face—OD" method, indicator readings can be taken by the reverse indicator method on the OD of the coupling hubs (or the shaft) only. Two brackets are used simultaneously, which is normally preferred.

Equally acceptable but less convenient is a method that uses a single bracket switched back and fourth for each set of readings.[1] Analysis of information gathered by this method yields sufficient data for determination of relative shaft positions. Advantages of this method over "Face—OD" are:

1. By proper design of the alignment brackets, the required removal of the coupling spacer is eliminated in most cases (shafts must be sufficiently in alignment to prevent binding of the coupling, of course). This feature provides three distinct advantages:

a. Wear and tear on the coupling is reduced.

b. Since both shafts turn as a unit with the spacer in-

Alignment hints

- Assign a responsible, technically astute individual to follow details of the alignment. It is too important to receive second-class attention.
- Spend adequate time in preparation. It is money well spent.
- Make sure the craftsmen understand both the method and the intent of the alignment procedures.
- Check the tools. A sticky indicator has no place among the alignment tools.
- Assure yourself that the indicator brackets are sturdy. The best test is a healthy shake; if the indicator readings do not return to the original position, the bracket is inadequate. The use of magnetic indicator brackets is **never** permitted for shaft alignment.
- Find a good way to turn the shafts. If you are lucky, the person who specified the machine had the manufacturer put wrench flats on the shaft extension at the outboard end of the machine. If not, find another means, such as a strap wrench or a clamp-on fixture made specifically for the purpose.
- When placing the indicator, be sure it starts out at mid-range, and be sure it rests squarely on the shaft. An indicator which sets askew gives bad readings. Likewise, be sure that the range of the indicator is not exceeded when readings are taken.
- Before making any moves (and before final acceptance), get at least two sets of identical readings. If the readings cannot be repeated, they are not acceptable.
- Check the data before making a shim change or horizontal move of the machine. It takes very little time to check the data—it takes a long time to move a large machine.
- Make one move at a time, then check the work. Don't worry about horizontal movements until the elevation is proper.
- Be alert to peculiarities of the specific machines being aligned. Tilting-pad bearings, for example, sometimes give trouble by permitting the shaft to rock back and forth on the lower pad, making repeatable measurements difficult. It is sometimes necessary to put temporary shims in such bearings to assure that the shaft is stabilized for alignment purposes.
- Stop the shaft at precise 90-degree increments for measurements. Turn the shafts in one direction only. If you pass the 90-degree mark, go around another complete turn. Extremely handy for the purpose of determining the quarter turn is a tool with four spirit levels mounted at 90-degree increments.
- Be wary of planned changes that do not turn out right. Alignment changes are calculable and predictable; if they consistently go wrong, look for something amiss in the system.

stalled, errors caused by coupling hub runout are entirely eliminated.

c. By spanning the entire coupling, angular misalignment is greatly magnified and more precisely diagnosed. For example, a span of 16 inches gives angular misalignment readings eight times as great as if measured on the face of a 4-inch-diameter hub.

2. Since face readings have been eliminated, there is no concern about axial float of shafts.

3. This method is not successful if makeshift tools are used—special tools must be constructed. Alignment results are simply better when proper tools are used. I wonder why alignment tools for high-speed couplings are not available in standard designs from the coupling manufacturer?

The single problem with the "Face—OD" method that is not solved by the "reverse indicator" method is deflection in the alignment bracket. This is a matter of real concern, and one which must not be ignored if proper cold alignment is to be achieved. The problem is readily handled, however, by determining the deflection in the alignment fixture and making appropriate corrections in alignment data. Fig. 2 indicates one method for determining the error. As shown, barstock of appropriate size is chucked in a lathe, and the end is turned to the proper diameter to accept the alignment bracket.

Without removing the barstock from the lathe, the alignment bracket is attached. With the bracket on top of the barstock, the indicator is set to zero. The entire assembly is then rotated 180 degrees, so the indicator now reads directly on the bottom of the barstock. The indicator reading thus obtained is reflective of the deflection in the alignment bracket.

Once determined, deflection should be stamped prominently upon the bracket, as should the designation of the equipment for which the bracket was specifically made. The alignment bracket becomes a special tool for that machine, and should be cared for as such.

3. The hot alignment check. It is now generally accepted that the traditional "hot check" for turbomachinery alignment is of little value. Not only is it costly and time consuming to bring a machine up to temperature, stop it, break couplings and attempt to determine alignment before it cools off, but results are highly questionable. In some instances a hot check is dangerous because it creates an unwarranted sense of security.

It is not possible to make the check quickly enough to accurately determine thermal growth of the equipment. For machines operating at temperatures well away from ambient, it is not uncommon to see a thermal change of 0.001 inch per minute when the machine is first shut down. At this rate, the effectiveness of the hot check is certainly lost before the alignment data can be obtained. Misalignment resulting from hydraulic forces and torque reactions, which can be significant, are never revealed by the traditional methods because the forces disappear when the machine is stopped.

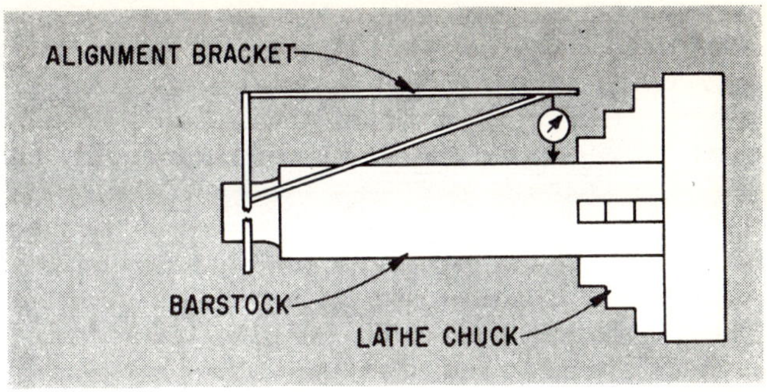

Fig. 2—Setup for determining deflection in alignment brackets.

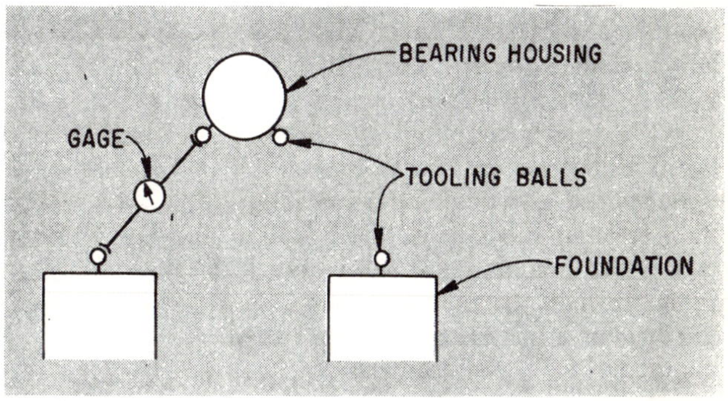

Fig. 3—Basic elements of mechanical device for checking hot alignment.

It is generally accepted that a superior alternative is to use the cold position of the shafts as a benchmark and deduce the hot alignment by monitoring the movement of the machine casings, or shafts, from the cold position to the hot position. Several methods have been used for this monitoring, with a variety of techniques applied to the actual determination of casing or shaft movements. The most widely known methods use optical or electronic techniques.

Details of these techniques have been well publicized.[2] However, a purely mechanical method for measuring hot alignment is also being used successfully. This method uses permanently mounted tooling balls for measurement location references. A spring-loaded mechanical gaging device measures relative movement between the reference tooling balls in increments of 0.001 inches. Fig. 3 shows the basic elements of this method as applied to a typical turbomachine. Again, the cold position is used as a benchmark for alignment corrections.

ACKNOWLEDGMENT

Abstracted from the paper "Alignment of Turbomachinery," originally presented to the Second Symposium on Compressor Train Reliability, Manufacturing Chemists Association, April 4, 1972.

LITERATURE CITED

[1] Jackson, C., "How to align barrel-type centrifugal compressors," *Hydrocarbon Processing*, 50 No. 9 p. 189 September 1971.
[2] Jackson, C., "Successful shaft hot alignment," *Hydrocarbon Processing*, 48 No. 1 p. 100 January 1969.

All the Facts About Machinery Alignment

Good alignment begins before a machine is installed and continues until the unit is operating reliably. Here are some frequently overlooked aspects of proper alignment and a discussion of some new tooling that can simplify the work and improve the results.

Malcolm G. Murray, Jr., Murray & Garig Tool Works, Baytown, Texas

ALIGNMENT is a broad term, covering all sorts of things, from human posture to tunnel bores. For this article, however, we will concentrate on alignment of horizontal, flexibly coupled machine shafts. Such shafts, when in perfect alignment, will have their extended centerlines coincide along the same straight line (or curve, in the case of long, slender shafts), in space.

Perfect alignment seldom occurs, but good alignment can and should be attained for any machine needing continuous, reliable operation. Poor alignment can cause excessive vibration and premature failure of bearings, seals, couplings, rotating elements, stationary machine structure and connected piping and ductwork.

Alignment phases. The alignment task can be divided into three main phases—measurement, calculation and movement. Measurement and calculation can be further subdivided into basic, or "cold" and thermal growth, or "hot" categories. Before getting into the alignment procedures, though, a few things need to be done first.

Pre-alignment requirements. The most important requirement for alignment, or any other job, is to have someone who knows what he is doing and cares enough to do it right. Years ago, when machines operated at slow speeds and were extensively spared, adequate alignment could be done by a relatively unskilled man, using a straightedge and hammer. Now, this is seldom the case. Higher speeds demand precise alignment, including consideration of relative thermal growth between the coupled machine elements. Successful alignment now depends not only on the field machinist, but on the engineer who writes the purchase specifications for the machine. Unfortunately, these two men often do not become as well acquainted as they should. The result may be a machine needlessly difficult to align. The solution to this problem is easy and obvious, but not very commonly used. That is, conduct a good training course in machinery alignment, and have the machinery purchasing engineers attend along with field machinists. Encourage plenty of discussion, and do some actual "hand-dirtying" field alignment. If necessary, a consultant can be engaged to conduct the course, until adequate in-house talent has been developed. References 6, 11 and 23 are useful supporting material for such training. With the aid of such a course, the purchasing engineer can do a lot to minimize field alignment problems.

Let us suppose we are past the foregoing hurdle and have competent and conscientious machinists and engineers. What now? There are still a number of things that need to be done before attempting field alignment. Some of these items are as follows:

- *Foundation* must be of adequate size and strength. A rule of thumb calls for concrete weight equal to three times machine weight for rotating machines, and five times for reciprocating machines.

- *Baseplate* must have adequate rigidity. Mounting pads should be level, flat, parallel and clean. Shims should be clean and of corrosion and crush resistant material. Machine hold-down bolts must be of adequate size, with clearance to permit alignment corrective movement. Clearance beneath machine of 2 inches for jacking is desirable. Jackscrews, if provided, should be rigid and corrosion resistant. Any "soft foot" condition should be eliminated. For further baseplate recommendations, see Reference 19.

- *Grout* should be of suitable material, in good condition, with no voids remaining beneath the baseplate. Tapping with a small hammer can detect hollow spots, which can then be filled by epoxy injection or other means.[20]

- *Piping* should be well fitted and supported, and sufficiently flexible, so that no more than 0.002-inch movement occurs at the flexible coupling when the last pipe flanges are tightened.

- *Flexible couplings* should be installed with an interference fit. Recommendations vary from 0.0003-inch overall, to 0.0005-inch per inch of shaft diameter. Couplings should **not** be installed loose, or depend on a setscrew to hold them on the shaft. Coupling cleanliness, and lubrication for some types, are important. Sending a repaired machine to the field with its coupling-half unprotected invites lubricant contamination, rusting, dirt accumulation and premature failure.

- *Overhung pump* radial flange bolts should be tightened equally to recommended torque, and then left alone —not used to make coupling alignment movements.

For an excellent discussion of pre-alignment requirements, see Reference 7.

Tools. Good tools are a key factor in doing the alignment quickly, and doing it well. Many plants can benefit from a custom-designed alignment tool kit, as shown in Fig. 1 and described further in Reference 21. If space were available, this would be a good place to provide a comprehensive list of recommended alignment tools. However, I will omit such a list, and direct anyone interested to Reference 23, where such can be found. Many tools on this list are standard items that would be carried by most field machinists. Later, however, as we get into certain alignment operations, mention will be made of several useful tools that are not yet standard in most machinist tool kits.

Axial gap and end float. Check the requirements of the machine elements being coupled, and their coupling, and adjust axially as required to stay within specified gap and float limitations. Failure to do this may cause premature wear of coupling and bearings, due to excessive thrust. References 17, 23, 28 and 32 give detailed procedures for handling axial adjustment.

Alignment tolerances. Tolerances must be established before alignment, in order to know when to stop. Various tolerance bases exist. One authority[12] recommends ¼ to ½-mil maximum centerline offset per inch of coupling length, for hot running misalignment. A common tolerance in terms of face-and-rim measurements, is 0.003-inch allowable face gap difference and centerline offset. This ignores the resulting accuracy variation due to face diameter differences, but works adequately for many machines. The real criterion is the running vibration. If excessive, particularly at twice running frequency and axially, further alignment improvement is probably required.

BASIC MEASUREMENT

For precise cold measurement, dial indicators are normally employed. Other tools, such as dial calipers and feeler gages, may be advantageous. Machine type and configuration will influence the method or "setup" for taking the measurements. Although the "reverse-indicator" setups have become popular in recent years, I do not advocate their use exclusively—there are situations where a "face-and-rim" setup is more logical. There are several common dial indicator setups. Most of these require that one or both shafts be turned to take the measurements. If the shaft cannot be turned by hand, a wrench may be used. To avoid scarring the shaft, grip it for turning with a full chain type wrench, preceded by a wrap of compressed asbestos sheet packing. By full chain, I mean the design which grips with roller chain only, and does not use serrations or teeth. These same wrenches, modified slightly, can be used to prevent the shaft from turning. This is sometimes desirable when attaching clamp-on jigs. Another shaft-stopping technique, for the smaller machines, is to use a double indicator stand magnet arrangement, with V-surface contacting the shaft.

Data recording. Taking measurements is next to useless, if data recording is haphazard. "Back-of-envelope" figures are likely to be misinterpreted and lead to error. A good way to avoid such problems is to use a prepared alignment data form. This will promote systematic data acquisition, correct interpretation of the recorded figures, and proper calculation of required corrective movements. Fig. 2 shows an example of such a form, corresponding to the setup in Fig. 3.

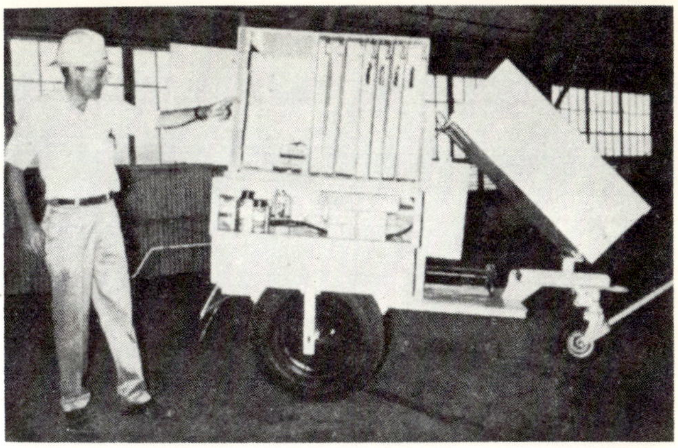

Fig. 1—Alignment tool kit used at a large overseas refinery. Inner cabinet lowers to give a stable towing configuration.

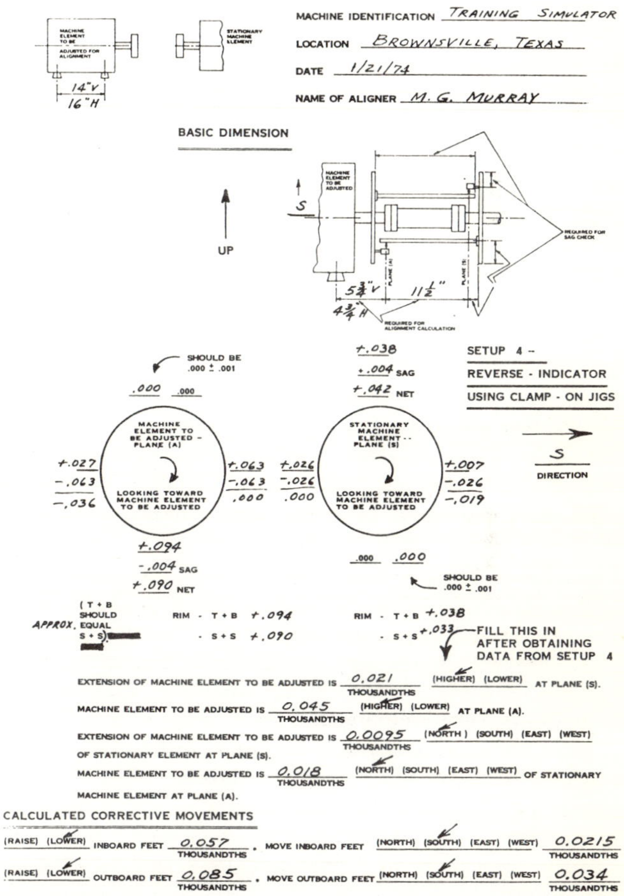

Fig. 2—Example of alignment data form.

Fig. 3—Clamp-on alignment measurement jigs, with reverse indicator setup, shown mounted on alignment training simulator. This single set of jigs will handle diameters from ⅞ inch to 8½ inch and spans up to 26 inches.

Fig. 4—Experimental 61-inch span "face-and-rim" setup for long span alignment measurement using clamp-on jigs. Pendant counterweights counteract deflection of clamp plates. To avoid effect of indicator weight, measurements are made with hand-held dial calipers for "face" measurements and plain ID calipers plus OD micrometer for "rim" measurements. Variation for full rotation will stay within 0.002 inch.

Sag. As you can see, we have a place in the form for recording sag. Sag does not affect horizontal measurements. It may or may not be significant for vertical measurements, depending on such things as spacer length, bracket or jig stiffness, extension arm weight, etc. If in doubt, sag should be checked. For this purpose, we can use a shop arbor, or merely a horizontal pipe on sawhorses. We install a setup similar to that which is to be used in the field, and rotate it from top to bottom through 180° and note the sag. Generally, sag should be checked when using clamp-on jigs with reverse indicator method for spans greater than 6 inches. Also for rim measurements when using face-and-rim method for spans more than 6 inches, or when bracket stiffness is in doubt. Face measurements are not affected greatly by sag, and generally need not be checked for its effect. An exception to this will be noted shortly, in the discussion on measurement across very long spans.

All of the foregoing refers of course to bracket sag. In long machines, we will also have shaft sag. This is mentioned only in passing, since there is no need to do anything about it at this time. "Point-by-point" alignment with dial indicators will automatically take care of shaft sag. For initial leveling of large turbogenerators, etc., especially if using precision optical equipment, shaft sag must be considered. Manufacturers of such machines know this, and provide their erectors with suitable data for sag compensation.

Bracket sag, if measured and accounted for, offers no great problem. Nevertheless, it seems to bother a lot of people—as evidenced by the repeated raising of this topic for discussion at the 1973 Texas A&M Turbomachinery Symposium. A valid point was made, concerning the difficulty of calibrating sag on certain complex long-span brackets. One alignment discussion leader described his experiments with various bracket materials, including glass, in unsuccessful efforts to eliminate sag completely. This brought forth a suggestion for using hollow tubing pressurized with hydrogen or helium, for stiffening plus lift!

The span limitations, due to sag and hysteresis, for compact clamp-on jigs such as shown in Fig. 3, were also noted. One engineer mentioned a machine with a 40-inch coupling gap, for which his company had been unable to devise a completely satisfactory bracket or jig. As a result, I have experimented and arrived at a simple self-compensating setup which seems to overcome the long-span sag problem. Fig. 4 shows this setup. Surprisingly enough, the setup uses the face-and-rim measurement technique, which most engineers associate with close-coupled situations, if indeed they have not abandoned it altogether in favor of reverse-indicator measurement.

CALCULATING THE CORRECTIVE MOVEMENTS

Having taken, recorded and interpreted the basic measurements, and corrected for sag, our next step is to calculate the required corrective movements. Various mathematical and graphical techniques are available for doing this. References 1, 2, 6, 11, 12, 13, 15, 16, 22, 23, 27 and 31 describe several such methods. These are making headway in replacing the trial-and-error and intuitive approaches still used by many field machinists.

A recent development in simplifying the calculation of alignment corrective movements is the plotting board, shown in Fig. 5. This is a patented portable graphical calculator, which can be used quite easily to obtain quick and accurate answers, using measurements taken by either the reverse-indicator or face-and-rim technique. With a grid allowing use of any consistent English or metric measurements, a transparent overlay is marked, then slid

Fig. 5—Machinery alignment plotting board for graphical determination of required corrective movements.

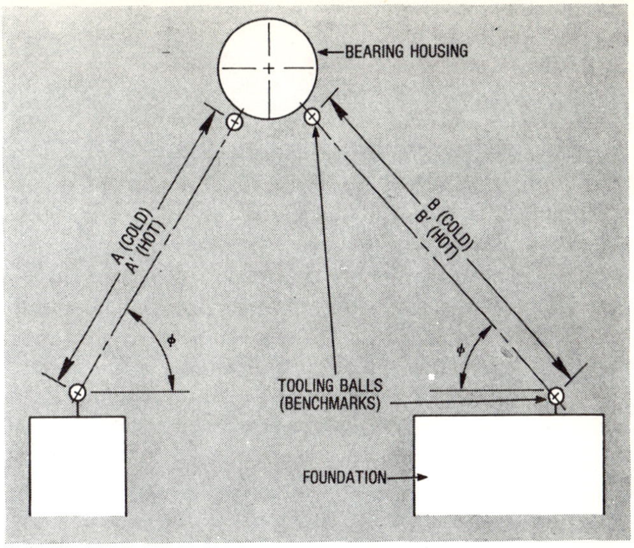

Fig. 6—Original Essinger triangulation setup for measuring thermal growth relative to fixed points on the machine foundation.

and/or pivoted to represent the misaligned condition. The required corrective movements are then read directly. As with the better mathematical methods, angular and parallel misalignment are corrected simultaneously, and pre-positioning one element low is not required. The plotting board is becoming increasingly popular among machinists and engineers in many industries.

Thermal growth. In some machines, thermal growth must be considered in doing the alignment. "Hot alignment" is a common term here, although most of the alignment work may actually be done with the machine at ambient conditions.

Before discussing thermal growth, mention should be made of situations where it may safely be ignored. These, despite the attention given hot alignment in recent years, are still a numerical majority, and include most pump-and-motor installations in process plants. Reference 10 has a good discussion on a related aspect of this—integral pump design to compensate fully or partially for thermal growth.

Methods of accounting for thermal growth are numerous. They vary from "rules of thumb" of dubious accuracy, to sophisticated techniques using precision optics,[3, 4, 13, 15, 18, 29, 30, 33] lasers,[9] and non-contact electronic proximity probes.[5, 13, 14, 15] Among the more questionable are formulas based on coefficient of thermal expansion of the machine material, which assume a uniform temperature rise from one end to the other. A better way, where feasible, is the use of quickly attached clamp-on jigs (see Fig. 3 and Reference 25), for measurement a few minutes after shutdown. A series of three such readings, a few minutes apart, can be extrapolated graphically back to zero time, for a fair approximation of running alignment measurements. Another approach, used and advocated by one consultant, and condemned as risky by several others, is to align the machine during operation, using vibration as the primary reference. Table 1 contains a brief discussion on using modified leveling wedges to achieve this result, safely.

Among the best methods, where cost and complexity can be justified, is Ray Dodd's proximity probe setup measuring in reverse-indicator manner between machine elements.[5] This allows continuous running monitoring where desired, and can be made largely immune from tampering or accidental bumping. It is also independent of relative foundation movement, which can introduce error with methods using foundation points as fixed references.

Another good approach, which has the advantage of simplicity, is Jack Essinger's Invar rod triangulation, shown in Fig. 6, and described in Reference 8. In its original configuration, this method had the possibility for error, mentioned earlier, of any system relying on fixed reference points along a foundation. Foundations, especially long, elevated ones, sometimes grow (or shrink) unequally along their length. A modified setup using the basic Essinger tooling, is shown in Fig. 7. This eliminates the foundation as a source of error, and gets back to relative growth between the machine elements, which is what we are really interested in. As a bonus, this modified setup also gives us axial growth, which can be useful information, particularly when using the new metal diaphragm couplings.

Extensions of the mathematical, graphical and plotting board methods described earlier can be used to incorporate the thermal growth data into the calculations for over-all corrective movements. Examples of this can be found in References 6, 11, 12, 13, 15, 22 and 23.

MOVEMENT

Before attempting movement, by any method, a check should be made to see if desired movement is **possible.** If the machine is to be lowered, and insufficient shims are installed to permit lowering the desired amount, it will be necessary to change the approach and shim up the other machine element. If insufficient clearance is available be-

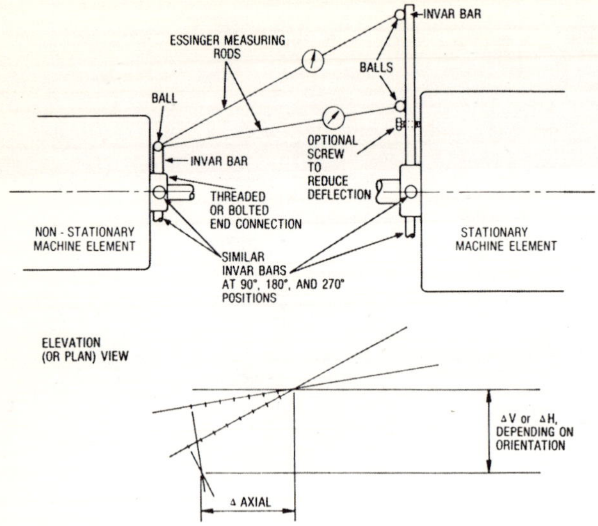

Fig. 7—Modified triangulation setup. Measurement is in planes parallel to the machine axis and between the two machine elements. Being independent of fixed reference points on the foundation, accuracy is not impaired by unequal foundation growth.

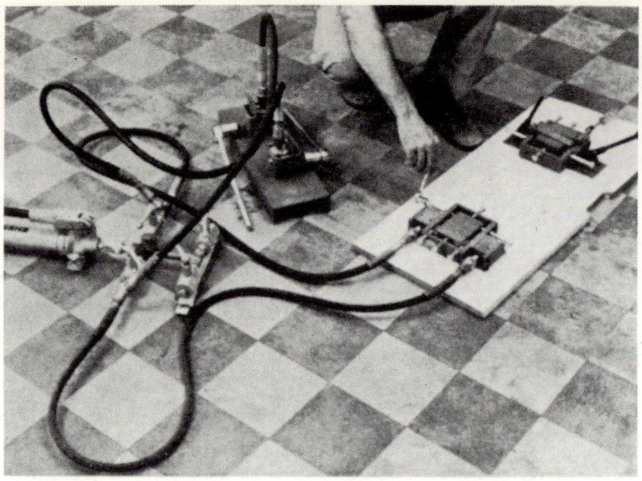

Fig. 8—Machinery alignment positioners. These portable devices slip under the center of a machine near each end, raise it for shimming, and move it axially and laterally for horizontal alignment. The heart of the device is a planar ball bearing assembly which eliminates most friction and allows easy and precise movement of heavy loads.

tween hold-down bolts and their holes, to permit desired horizontal movement, either the bolt shanks must be reduced, or the holes enlarged. Bolt shanks may be safely reduced to the thread minor diameter. Beyond this, they may be reduced further if on the side of the machine which will unload them due to machine torque. If the driven machine has a rapid deceleration characteristic, be cautious in applying this latter criterion to the driver bolts.

The movement phase of the job, surprisingly, has been largely neglected or taken for granted by those writing on the subject of alignment. Yet, what good are precise measurement and calculation, without some equally precise means of making the required corrective adjustments?

A look at the pumps in the typical process plant will show that most of their drivers lack alignment jackscrews. Battered feet, seen all too often, indicate that alignment via steel sledgehammer is still quite common.

A number of alignment movement methods exist. These are shown in Table 1 along with their costs, advantages and disadvantages.

Of particular interest is a relatively new device, the portable alignment positioner. This is available in several versions—both screw-lift and hydraulic (Fig. 8), and has a number of advantages over older movement methods. Further information on this tool can be found in Reference 26.

A useful accessory, for monitoring horizontal movement, is a foot reference attachment giving a smooth vertical surface for the indicator tip to ride on. Without such an attachment, foot slope and surface irregularities can cause errors.

LITERATURE CITED

[1] Alignment Procedure, Revised Edition. Buffalo, N.Y.: Joy Manufacturing Co., 1970.

[2] Baumann, Nelson P., and Tipping, William E., Jr. "Vibration Reduction Techniques for High-Speed Rotating Equipment—ASME Paper 65-WA/

TABLE 1

Vertical adjustments (to raise machine for shimming)

Feature→ Device ↓	Initial cost for typical machine	Durability	Ease of use	Will vertical movement upset previous horizontal adjustment?	Practical to install in field where welding is not permitted?	Can same device be used for vertical and horizontal adjustments?	Remarks
Permanent vertical jackscrews at each foot	$120	Good, if made of 316SS and/or kept lubricated. Otherwise, corrosion may occur.	Easy. No clearance problems. Easy access.	No, if installation is Good and user is careful, but problems could occur if similar machines are interchanged among several positions, due to recess mismatching.	Situation varies, but impractical more often than not.	No	May not be practical on some cast iron feet. Have seen it used mainly on fabricated steel motor feet. Even then, such installations are rare. Note that cost covers only one machine, since device is not portable.
Permanent leveling wedges at each foot	$120 (60-80 ton max.)	Good	Appears easy, if kept lubricated, and if vertical measuring reference points are available.	Maybe, but should be minimal with the better units.	Yes, if the 2-3-inch extra height poses no problems.	Not with present designs, but modification appears possible to permit this in some cases.	Cost covers only one machine, since once installed, device ceases to be portable. The better units have tilting pads, and guides to achieve straight lift without horizontal drift. With provision of vertical measuring reference points, shims can be omitted. This device seems to have the potential for becoming a safe means of making vertical and horizontal alignment adjustments *while the machine is running*. Several modifications would be required first, however.

TABLE 1 *continued*

Device ↓ / Feature→	Initial cost for typical machine	Durability	Ease of use	Will vertical movement upset previous horizontal adjustment?	Practical to install in field where welding is not permitted?	Can same device be used for vertical and horizontal adjustments?	Remarks
Hydraulic jack	$150 (5 ton lift)	Good	Easy. 1¾-in. minimum clearance for 5-ton lift. 2⅛-in. for 20-ton lift.	Usually	Does not apply, since device is portable.	No, unless backup support is provided, which is not usually done.	
Steel wedges and hammer	$10	Good	Fair. Limited to small machines.	Yes	Does not apply, since device is portable.	No, unless backup support is provided, which is not usually done.	
Pry bar	$5	Good	Fair. Limited to small machines. Several ft. horiz. clearance required.	Yes	Does not apply, since device is portable.	No, unless backup support is provided, which is not usually done.	
Crane or hoist	Highly variable	Good	Awkward	Yes	Does not apply, since device is portable.	No, but under some conditions it can *assist* with horizontal adjustments, by reducing effective machine weight. See horizontal adjustment methods tabulation.	
Machinery alignment positioners	$650 (6 ton max.) $2550 (20 ton max.) $4000 (80 ton max.)	Good	Easy. 2-inch minimum clearance up to 20 tons. 3-inch minimum clearance from 20 to 80 tons. 6 ton rating uses screw lift. Larger ratings use hydraulic lift.	No, if operated properly.	Does not apply, since device is portable.	Yes, and at the same time.	Prices and load ratings are for one pair of positioners — suitable for symmetrical machine elements such as electric motors. For non-symmetrical elements, three or more positioners may be required. These devices are portable, and one set will handle the typical process plant, even with several thousand installed machines.

Horizontal adjustments (lateral and axial)

Device ↓ / Feature→	Initial cost for typical machine	Durability	Ease of use	Practical to install/use in field where welding is not permitted?	Possibility of damaging machine being aligned	Can same device be used for vertical and horizontal adjustments?	Remarks
Permanent horizontal jackscrews at each foot	$100—lateral only $160—lateral plus axial Prices based on factory installation, using heavy lugs and 316SS nuts.	Good, if made of 316SS and/or kept lubricated, otherwise, corrosion may be a problem.	Easy and precise if well designed for adequate rigidity. Sometimes this rigidity is lacking, and stick-slip effect gives poor precision.	Could be installed by drilling and tapping, for about 50% of machines. Estimated cost $150 for typical machine, for lateral screws only. Time-consuming to install.	Nil	No	All cost figures are for a single machine only, since device is not portable.
Portable clamp-on jackscrews at each foot	$130	Good	Easy, where clamp rigidity can be adequately maintained. On about 50% of machines, this is a problem.	Yes, on about 50% of machines. Generally works better on small machines than on large ones.	Nil	No	
Steel sledge hammer or ram	$10	Good	Fair. Depends on access for swinging and suitable surface for hitting. Precision not always adequate.	Estimated usable for about 65% of machines (which is not meant to imply that I *recommend* it).	Can crack cast iron feet, or bend steel feet. Impact could conceivably damage other machine parts.	No	Not a recommended method, but a common one nevertheless, due to ready availability of steel sledge hammers.
Soft head sledge hammer or ram	$10 to $50	Poor to good, depending on type.	Fair to good. Depends on access for swinging and surface for hitting.	Estimated usable for about 65% of machines.	Damage to feet is unlikely. Other impact damage more difficult to assess.	No	Two types available. Lead head type costs least, but deforms readily and requires frequent recasting. Plastic-tip, hollow-head, cast-iron shot loaded "dead-blow" type costs more, but lasts longer. Both types work about equally well.
Comealong	$50	Fair	Poor. Requires multidirectional large clearance and anchor points. Stick-slip effect likely, with poor precision.	Estimated usable for about 50% of machines, but for rough adjustments only.	Unlikely	Yes, with suitable anchors (est. 25% of machines) but vertical lift will upset previous horizontal adjustment.	
Machinery alignment positioners	$650 (6 ton max.) $2550 (20 ton max.) $4000 (80 ton max.)	Good	Easy. 2-in. minimum clearance up to 20 tons, 3-in. minimum clearance from 20 to 80 tons. 6 ton rating uses screw lift. Larger ratings use hydraulic lift.	Estimated usable for 80% of machines 100 hp and larger, and for 50% of machines below 100 hp.	Nil	Yes, and at the same time. Vertical movement can be made without upsetting previous horizontal adjustment.	Prices and load ratings are for one pair of positioners — suitable for symmetrical machine elements such as electric motors. For non-symmetrical elements, three or more positioners may be required. These devices are portable, and one set will handle the typical process plant. It can thus become economically attractive by saving the cost of permanent jackscrews on many machines. Comes complete with foot reference attachments for accurate monitoring of horizontal movements.
Crane or hoist, plus tension dynamometer, plus portable clamp-on jackscrews, soft-head sledgehammer, or alignment positioners	Highly variable	Good	Fair. Requires more manpower than other methods.	Yes	Unlikely	Yes, in a manner of speaking.	This is a "last resort" method, for very heavy machines. By applying a measured lift, the effective weight of the machine element can be reduced, without actually lifting it off its supports. When the effective weight is low enough, one of the methods mentioned can be used to accomplish the alignment adjustments.

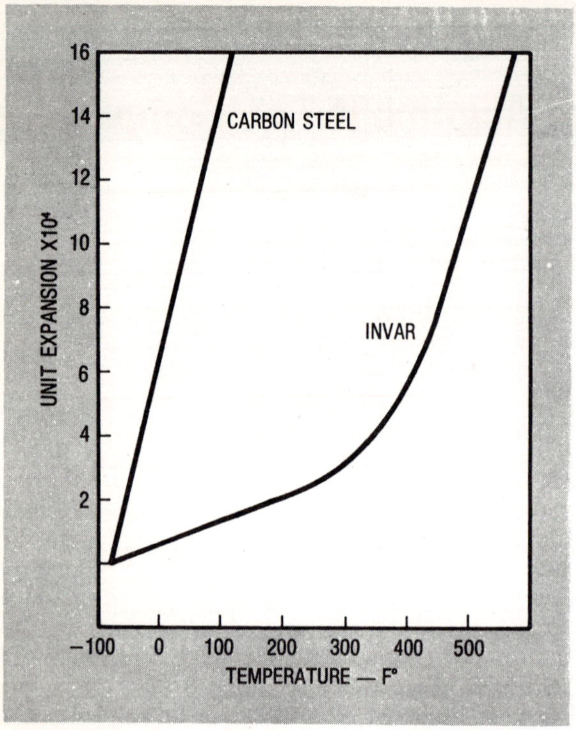

Thermal expansion characteristics of Invar compared to carbon steel.

It is appropriate, at this point, to mention some facts about the thermal expansion characteristics of Invar. This material is used in the Dodd, Essinger and modified Essinger systems because of its very low rate of thermal expansion. However, it is a common misconception that Invar remains dimensionally stable at all temperatures and therefore solves all thermal expansion problems. This is true up to about 200-300° F but above this temperature, the rate of thermal expansion increases sharply. The curve above, reproduced by permission of Carpenter Technology Corp., shows the thermal expansion versus temperature of Invar compared to plain carbon steel. This material is excellent for hot alignment systems, or other tooling requiring low expansion material, but the designer must be aware of its applicable temperature range.

PWR-3. New York: The American Society of Mechanical Engineers. 1965.

3 Blubaugh, R. L., and Watts, H. J. "Aligning Rotating Equipment." *Chemical Engineering Progress.* Vol. 65, No. 4 (April 1969), pp. 44-46.

4 Campbell, A. J. "Optical Alignment of Turbomachinery." *Proceedings, 2nd Turbomachinery Symposium.* Gas Turbine Laboratories, Texas A&M University, 1973.

5 Dodd, V. R. "Shaft-Alignment Monitoring Cuts Costs." *Oil and Gas Journal,* Vol. 70, No. 39 (Sept. 25, 1972), pp. 91-96.

6 Dreymala, James. Factors Affecting and Procedures of Shaft Alignment. Houston: Dreyco Mechanical Services, 1970.

7 Essinger, J. N. "A Closer Look at Turbomachinery Alignment." *Hydrocarbon Processing.* Vol. 52, No. 9 (September 1973), pp. 185-188.

8 Essinger, J. N. "Hot Alignment Too Complicated?" *Hydrocarbon Processing.* Vol. 53, No. 1 (January 1974), pp. 99-101.

9 Hamar, M. R. "Laser Alignment in Industry—ASTME Paper MR68-408." Dearborn, Mich.: The American Society of Tool and Manufacturing Engineers, 1968.

10 Hancock, W. P. "How to Control Pump Vibration." *Hydrocarbon Processing.* Vol. 53, No. 3 (March 1974), pp. 107-113.

11 Hughes, T. Training Filmstrip—Reverse Indicator Method of Shaft Alignment—Part I & II. Houston: Hughes & Associates, 1974.

12 Jackson, C. "Start Cold for Good Alignment of Rotating Equipment." *Oil and Gas Journal,* March 11, 1974, pp. 124-130.

13 Jackson, C. "Cold and Hot Alignment Techniques of Turbomachinery." *Proceedings, 2nd Turbomachinery Symposium.* Gas Turbine Laboratories, Texas A&M University, 1973.

14 Jackson, C. "Successful Shaft Hot Alignment." *Hydrocarbon Processing.* Vol. 48, No. 1 (January 1969), pp. 100-104.

15 Jackson, C. "How to Align Barrel-Type Centrifugal Compressors." *Hydrocarbon Processing.* Vol. 50, No. 9 (September 1971) (corrected reprint).

16 King, W. F., and Petermann, J. E. "Align Shafts, Not Couplings!" *Allis-Chalmers Electrical Review.* Second Quarter 1951, pp. 26-29.

17 Mills, E. B., and Petermann, J. E. "Solve Axial Thrust Problems with Limited End Float Couplings." *Allis-Chalmers Electrical Review.* Third Quarter 1957, pp. 23-26.

18 Mitchell, J. S. "Optical Alignment—An Onstream Method to Determine the Operating Misalignment of Turbomachinery Couplings." Paper presented at Second Symposium on Compressor Train Reliability Improvement, Manufacturing Chemists Association, Houston, April 4, 1972.

19 Murray, M. G. "Better Pump Baseplates." *Hydrocarbon Processing.* Vol. 52, No. 9 (September 1973), pp. 157-160.

29 Murray, M. G. "Better Pump Grouting." *Hydrocarbon Processing.* Vol. 53, No. 2 (February 1974), pp. 101-104.

21 Murray, M. G. "Improve Your Machinery Maintenance Tool Kits." *Hydrocarbon Processing.* Vol. 53, No. 1 (January 1974), pp. 111-114.

22 Instructions—Machinery Alignment Plotting Board." Murray & Garig Tool Works, Baytown, Texas, 1974.

23 *Manual: Alignment of Horizontal, Flexibly-Coupled Machine Shafts.* Murray & Garig Tool Works, Baytown, Texas. In preparation.

24 *Bulletin: Machinery Alignment Plotting Board.* Murray & Garig Tool Works, Baytown, Texas, 1973.

25 *Bulletin: Machinery Alignment Measurement Jigs.* Murray & Garig Tool Works, Baytown, Texas, 1973.

26 *Bulletin: Machinery Alignment Positioners.* Murray & Garig Tool Works, Baytown, Texas, 1973.

27 Nelson, Carl A. "Orderly Steps Simplify Coupling Alignment." *Plant Engineering,* June 1967, pp. 176-178.

28 *NEMA Standards, Application Data, A-C, Section MG 1-14.38, Coupling End-Play and Rotor Float for Horizontal Alternating-Current Motors.* National Electrical Manufacturers' Association, September 1970, Part 14, page 10.

29 "Optical Alignment—A Maintenance Service to Reduce Your Machinery Downtime." Houston: Dresser Industries, Inc., Machinery Group, 1969.

39 *Optical Alignment Manual, No. 71 1000.* Hoboken, N.J.: The Keuffel and Esser Co., 1969.

31 "Service Memo SD-5-69; Reverse Reading Coupling Alignment." Houston: Dresser Industries, Inc., Machinery Group, 1969.

32 Shaw, G. V. "Employing Limited End-Play Couplings Saves Motor Bearings —Sometimes." *Power,* November 1955, p. 120-121.

33 Yarbrough, C. T., "Shaft Alignment Analysis Prevents Shaft and Bearing Failures." *Westinghouse Engineer,* May 1966, pp. 78-81.

Hot Alignment Too Complicated?

Existing measurement techniques for hot alignment of rotating machinery are complex and difficult to use. Here is a simplified way to obtain the necessary data.

Jack N. Essinger, Shell Chemical Co., Houston

INFORMATION FOR THE hot alignment check is normally obtained by measuring vertical and horizontal movements of shafts or bearing housings of turbomachinery components relative to fixed references. When optical alignment techniques are employed, the references are vertical and horizontal planes of sight established by the optical instruments. For electronic or mechanical techniques, the fixed reference is normally the machine foundation. For the typical installation, however, the foundation is not in the proximity of the bearing housing, and it is necessary to install pedestals extending from the foundation to a point near the bearing housing, these pedestals becoming the reference from which measurements are taken. To assure the accuracy of the alignment data, the pedestals must be substantial, and they must be maintained at constant temperature to minimize measurement inaccuracies resulting from thermal growth of the pedestals themselves. The pedestals are normally water cooled.

The alternate method presented here also uses the machine foundation as a reference, but eliminates the requirement for alignment pedestals by using permanent reference points which are affixed directly to the foundation and to the bearing housings (or the machine case), as shown in Fig. 1. All four reference points lie in a plane perpendicular to the centerline of the machine shaft. Similar reference points are mounted at each bearing housing in the train.

Following cold alignment of the compressor train, reference dimensions A and B, and angles Θ and ϕ are determined at each bearing housing, and are recorded. When the machine is brought on line, dimensions A′ and B′ are measured at each position in the hot running condition. The data thus obtained is adequate for determining the vertical and horizontal movement of each bearing housing in the machine train relative to the foundation. See Fig. 2. Using common grid paper (4 x 4 grid is usually a convenient size), lay out reference vectors A and B at angles Θ and ϕ, having these vectors cross at one of the grid intersections. The intersection of these vectors represents the centerline of the machine shaft in the cold position. Now refer to the cold and hot measurements previously made (A, A′, B, and B′), and determine the movement of the bearing housing along vectors A and B by taking the differences between cold and hot measurements (ΔA and ΔB) for each location. Lay out these movements along vectors A and B using any convenient scale, say ¼ inch equals 0.001 inch, to establish points a and b. Now draw lines through a and b perpendicular to vectors A and B. These lines represent arcs of radiuses A′ and B′ drawn from the foundation benchmarks. The intersection of these lines defines the position of the machine shaft centerline in the hot position relative to the cold position. To determine the movement in vertical and horizontal directions, it is necessary only to scale off the dimensions referred to as ΔH and ΔV, using the same scale as used in plotting ΔA and ΔB. A similar plot for the data secured at

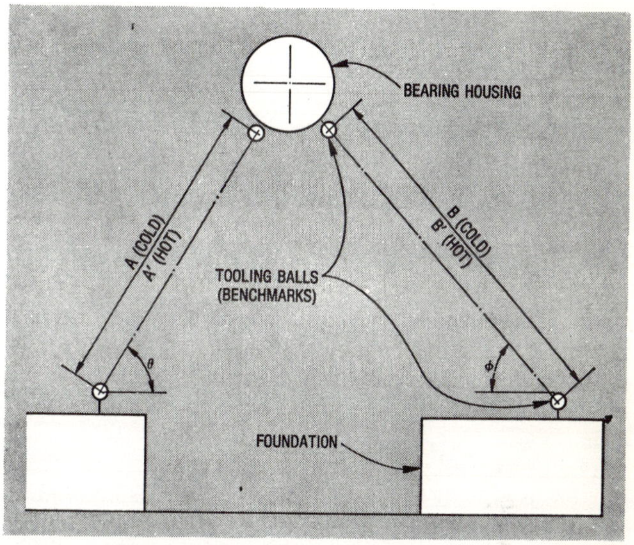

Fig. 1—Typical placement of benchmarks on foundation and bearing housing.

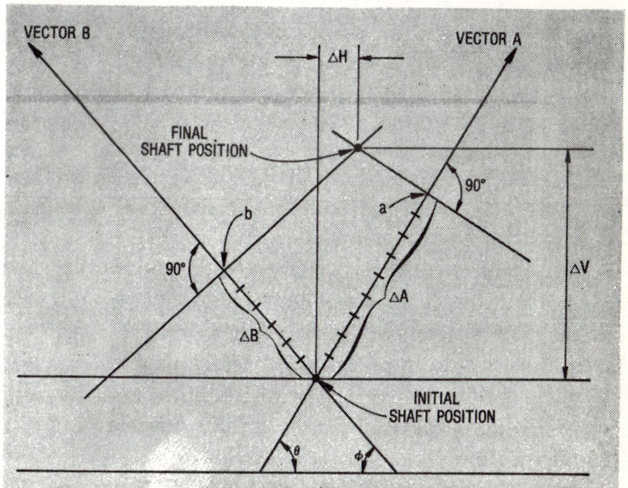

Fig. 2—Graphical determination of shaft in hot position relative to cold position.

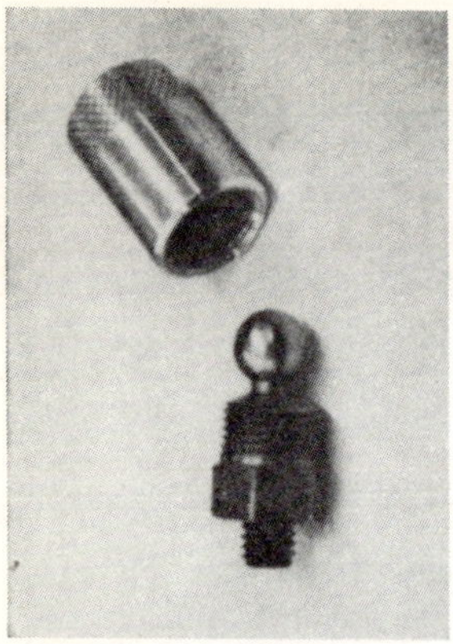

Fig. 3—Stainless steel ball with protective cover.

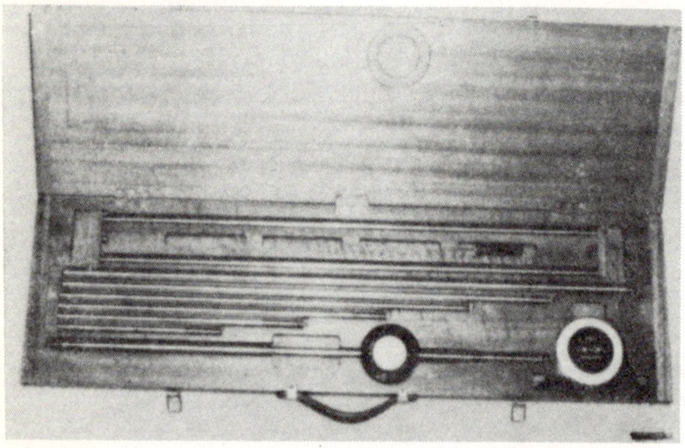

Fig. 4—Precision tool set for hot alignment checks.

Fig. 5—Calibration of the alignment gage with the Invar standard.

each bearing housing affords sufficient information for plotting the hot alignment of the entire turbomachinery train.

The benchmarks for this technique are ½ inch diameter precision balls. Because these benchmarks become an integral part of the installation, and because the accuracy of alignment records, over the long term, are dependent upon these references, it is recommended that they be made of stainless steel to prevent corrosion and that they be mounted substantially to avoid inadvertant movement. Fig. 3 shows a benchmark and protective cover specifically designed for turbomachinery work.

Fig. 4 shows a set of precision tools specifically made for the purpose of turbomachinery alignment by this technique. The alignment gage employs a long range dial indicator set into a spring-loaded telescoping column with ½ inch diameter spherical seats at either end. By using six extensions, the tool covers a measurement range of 25-60 inches, in increments of 0.001 inch. This range is adequate for nearly all turbomachinery in use today. The standard provided for calibration of the basic tool is constructed of Invar, as are all of the extensions, thereby eliminating the concern for thermal growth of the alignment gauge. The inclinometer is used for determining angles at which the measurements are made.

A stepwise procedure for alignment by this method is as follows:

1. Affix benchmarks to the machine and to the foundation.

2. Align the turbomachinery train in the normal manner, recording coupling alignment data.[1,2]

3. Using the Invar standard, calibrate the alignment gauge as shown in Fig. 5.

4. Using the appropriate gauge extension, determine and record cold measurements at each location, as shown in Fig. 6.

5. Determine the angles at which measurements are being made by using the inclinometer as shown in Fig. 7. Record the data.

6. When the train is brought on line, obtain and record hot alignment data by repeating Steps 3 and 4 at each reference station.

7. Plot the data as indicated in Fig. 2 to determine vertical and horizontal movement of each bearing housing.

8. Using combined data from the individual locations, plot the hot alignment of the entire equipment train.

9. Make appropriate alignment changes as dictated by the above data.

Fig. 8 shows typical placement of benchmarks at the various bearing housing locations in a turbine-compressor train.

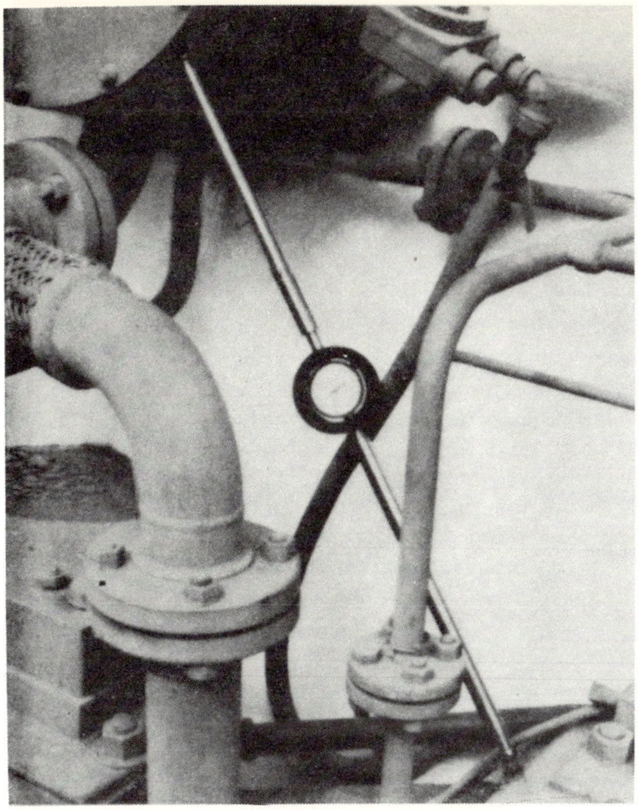

Fig. 6—Reference measurement between bearing housing and machine foundation.

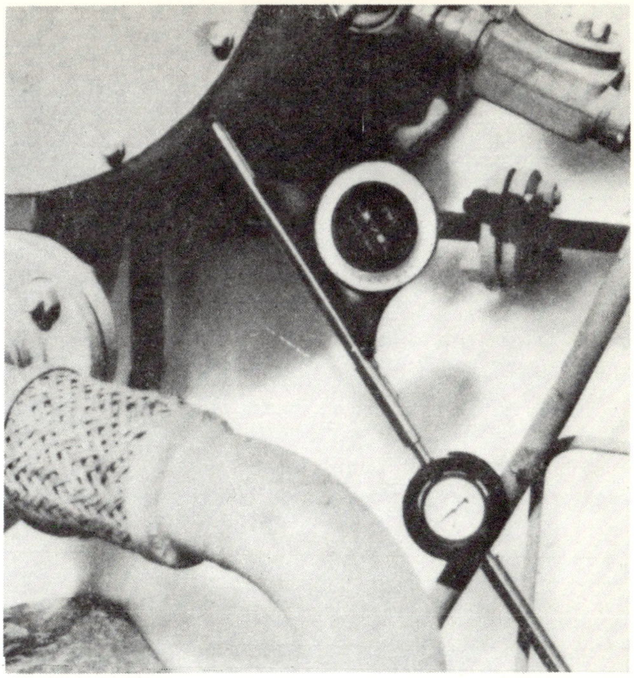

Fig. 7—Inclinometer for determining angle at which measurements are taken.

This alignment technique has been used successfully on several turbomachinery trains over the past three years. Some of the apparent advantages of the method are:

- The technique is simple, and no specific technical skills are required.
- Vertical and horizontal movements of the machine are obtained with equal ease.
- Once the benchmarks are established, setup time for hot checks, or re-checks, is minimal.
- The benchmarks are permanent, and are readily maintained for long-term monitoring of alignment.
- The method is highly resistant to accidental loss of reference; i.e., to being "bumped" during use.
- It is readily adaptable to cramped quarters.
- Datataking is a one-man operation.
- Initial investment in hardware is modest. Re-use is high.
- The equipment is highly portable.
- The hardware is simple, reliable, and requires minimal calibration.

Fig. 8—Typical benchmark placement on a turbocompressor train.

LITERATURE CITED

[1] Jackson C., "How to align barrel-type centrifugal compressors," Hydrocarbon Processing, 50, No. 9, p. 189, September 1971.
[2] Essinger, J., "A closer look at turbomachinery alignment," Hydrocarbon Processing, 52, No. 9, p. 185, September 1973.

How to Prevent Turbomachinery Thrust Failures

Thrust failures in high speed turbomachines can cause a considerable loss of plant production. These failures can easily be prevented if thrust bearings are installed correctly and rotor axial movement is monitored. Here are the important facts.

Charles Jackson, Monsanto Polymers and Petrochemicals Co., Texas City, Texas

MACHINES such as turbines and centrifugal compressors can be severely damaged if the thrust bearing fails unexpectedly. Failure can easily cost over $1/2 million including lost production. Such a failure can be prevented if careful consideration is given to the machinery and its protective instrumentation.

Most machines of this type have a thrust bearing of the hydrodynamic (slider bearing) type which develops an oil wedge lifting the thrust collar or runner from a stationary bearing, which could be of the straight flange, tapered land or tilting pad type. Actually, a thrust bearing is really an assembly which has two bearings. One bearing takes the normal or expected thrust loads imposed by the rotor. The other bearing would therefore be located, in design, to take abnormal or unexpected thrust loads. Many vendors will refer to one as the active (normal) thrust bearing and the inactive (counter or abnormal) thrust bearing.

Because of thermal expansion of the bearing elements, the two bearings could grow into one another causing severe loading and immediate failure. For this reason, the contact on each bearing is separated a certain distance based on the operating environment temperature. This separation distance is often referred to as the float zone between active and inactive thrust shoes (pads, plates, surfaces). For steam turbines, this distance might be 0.009 inches up to 0.014 inches. Thrust bearings for steam turbines will normally be at the steam inlet end since the rotor will expand from the thrust bearing. The blading (rotor) should move away from the nozzles and stationary blading as this is a critical and sensitive dimension.

On the other hand, centrifugal compressors are more lenient, with cold rotor floats of 0.015 inches to 0.022 inches typically. Total rotor float of a centrifugal compressor (with thrust bearings removed) moving axially in the case from one staging to the next could be in the order of ¼ inch (0.25 inches).

There is nothing unusual when a machine, in operation, floats randomly between the active and inactive bearings. I would consider this quite healthy exposing basically no thrust bearing loads on the bearings, provided an axial vibration is not set up which could cause fatigue or efficiency swings.

Thrust loads can be measured in many ways: Bearing metal temperatures, oil exit temperatures or heat differences, load cells behind the pads or bearings, or by deflection of the pads. Temperature and load measurement are recommended for actual engineering data or load distribution for attitude and alignment studies. Rotor deflection in an axial direction is recommended for machine protection monitoring with automatic shutdown to prevent severe failure damage.

Thrust failure protection is the topic of this discussion, and a systematic procedure is outlined with case histories of "saves" from the worst type of machine failure.

Different types of sensors can be used on a continuous basis to measure thrust movement or deflections. Earlier developments used both air and hydraulic nozzles which varied in back pressure as the distance between the nozzle and a part on the rotor (often the thrust collar or trim balance ring) changed. Calibrations could be obtained with sensitivities of 1 psi per mil (0.001 inch). However, sensing ranges were often limited to 25 to 30 mils (0.030 inches). The better sensor is the eddy current or inductive probe which can be installed in rather small tapped holes, ¼ to ⅜ inch, and can sense movements in excess of 80 mils with sensitivities of 100 to 200 mv/mil. (See calibration curve at 100 mv/mil—Fig. 2.)

At this point, one must develop a proper perspective of the problem and the solution. Several things are important to consider:

- Thrust bearings are known to fail in about 30 seconds; yet, immediately prior to failure the bearing on examination may have zero defects until the supporting oil film breaks down.

- Cost of the thrust bearings may be $700 or higher for pads only and may be $5,000 for a total assembly. A machine wreck by thrust bearing failure will intermarry the rotating parts to the nonrotating parts with a repair cost of $1/2 million or more. Therefore, the thrust bearing (failed) should be sacrificial to the big machine.

- Trying to save the thrust bearing will only cause severe confusion. First, if shutdown occurs before babbitt failure, inspection will indicate the thrust bearing was okay, but the sensor and related instrumentation indicated the onset of a failure! Unnecessary shutdowns will lead to a disarming of the shutdown system rendering the system incapable of proper protection. Second, the thrust bearing, being constructed of pads, for example, with self-leveling articulating links, and possibly shim packs of ⅛ to ¼ inch, can deflect 4 to 8 mils under load increases well within the design of a thrust bearing. Large deflections can be reduced and downtime minimized by machining single element shim plates to replace large packs of thin shim.

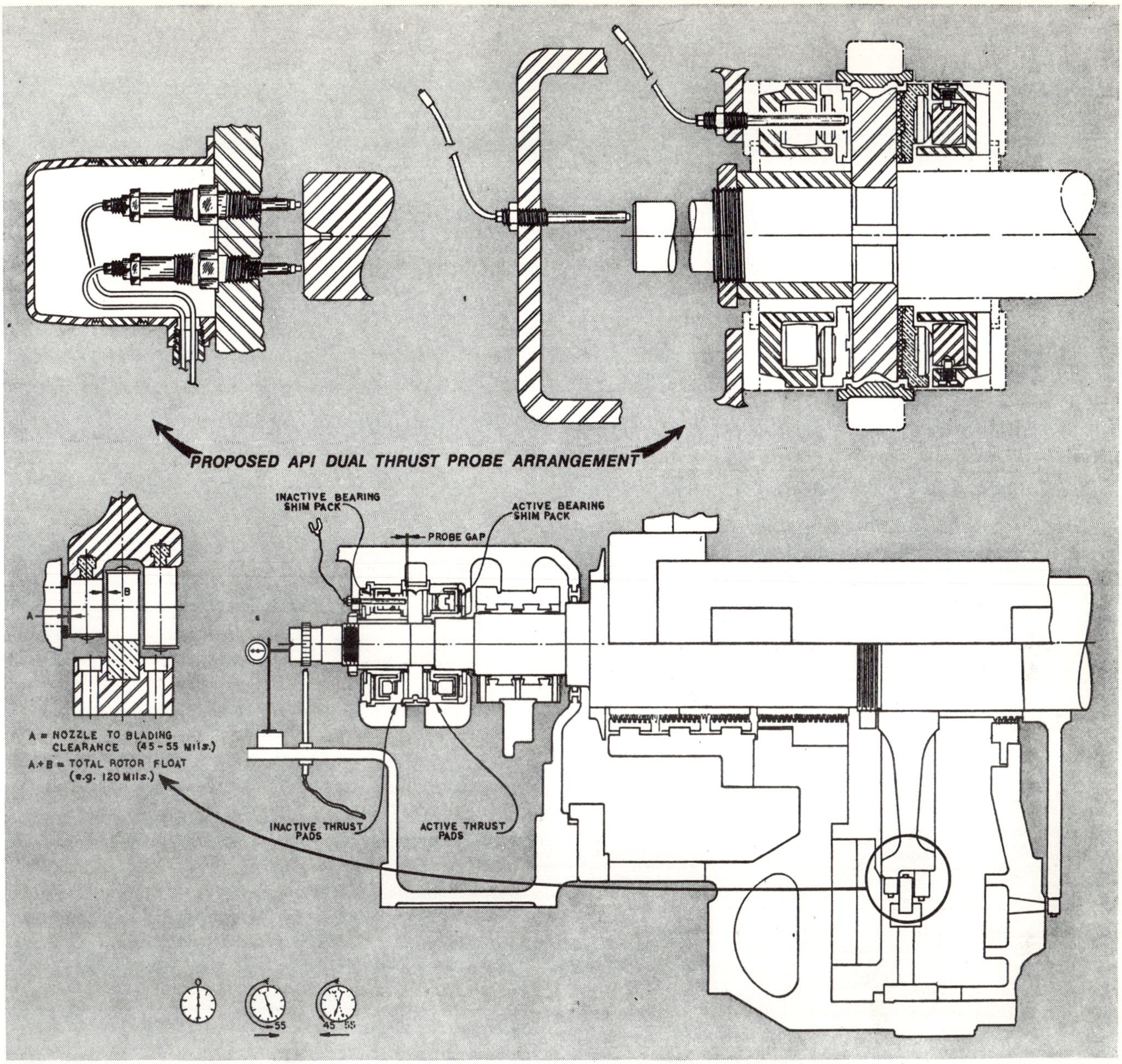

Fig. 1—Dual thrust probe arrangement and thrust bearings relative to blading in a steam turbine.

• Machine builders are naturally going to set short limits of rotor movement in their instructions for thrust shutdown instrumentation to protect their machine designs from possible criticisms. The builder doesn't care how many times you may shut down for inspections on movements of 5-6 mils (his suggested limits, but within expected deflections for rated loads).

• Limits for shutdown should definitely indicate that babbitt is being removed yet the shutdown response must be sufficient to bring the machine to rest within the babbitt thickness and definitely within the rotor travel clearance before contact occurs.

• A sensor system should be correlated with normal bump checks of rotors and dial indicator measurements. This is a very good way to improve everyone's confidence in the machinery protection instrument system.

THRUST BEARING FITTING PROCEDURES AND SENSOR POSITIONING

Steam turbine example:

Specified thrust float by manufacturer = 0.009 inches – 0.012 inches.

Specified shroud (rotor blading) standoff to nozzle ring (closest contact) = 0.045 inches – 0.055 inches.

Specified total rotor float, minimum = 0.120 inches.

Procedure (see Fig. 1):

• With thrust bearing removed, move rotor until shrouds contact nozzle ring ("A" clearance = 0). Indicate with dial indicator (compressed more than 200 mils) and set indicator to read zero. Move rotor towards exhaust until first stationary blading stops the rotor on axial travel

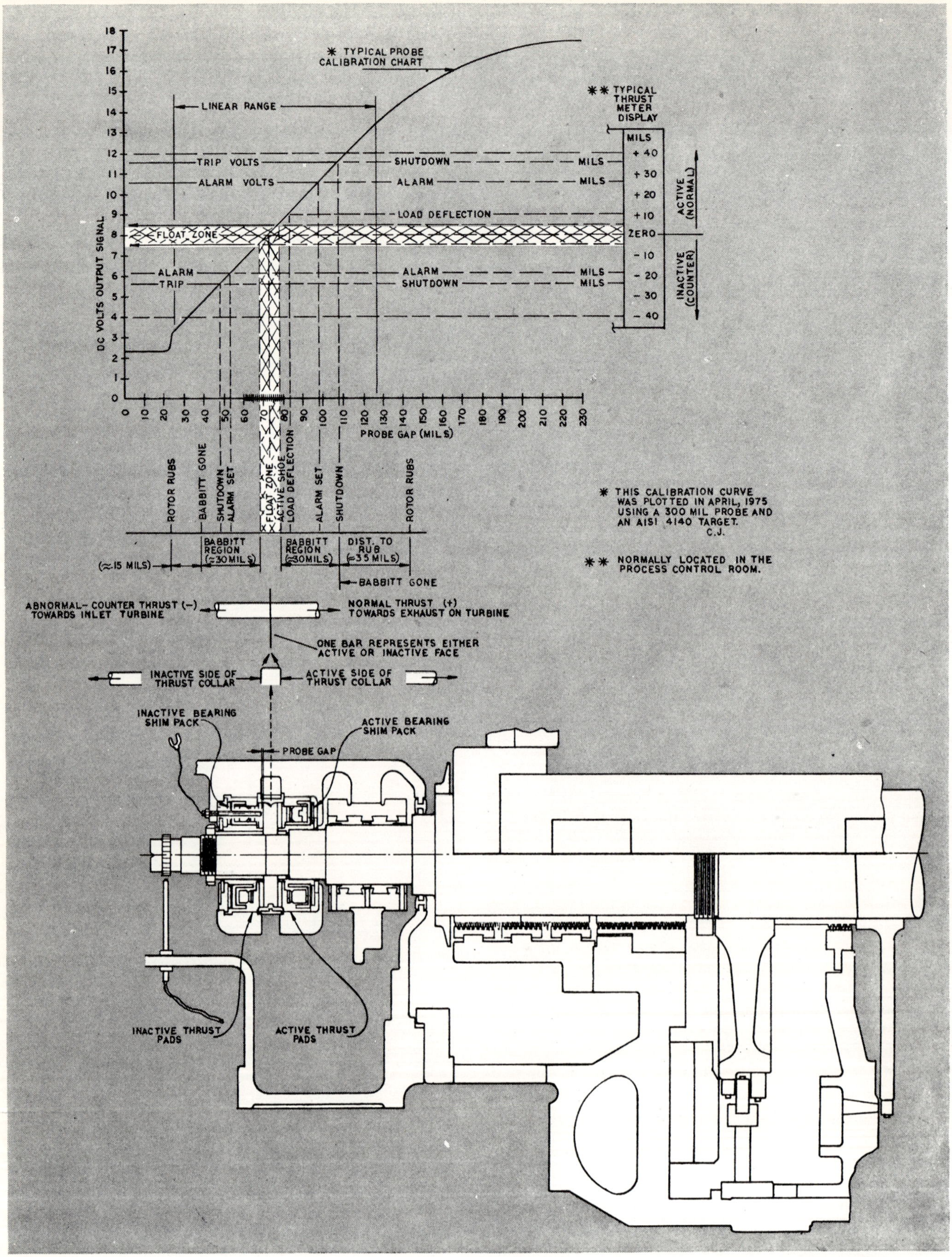

Fig. 2—Thrust bearing float and probe gap projected through calibration curve to rotor position readout instrumentation.

contact. Record this travel on maintenance setup sheet after assuring that greater than 120 mils was obtained (example specification from above).

- Return rotor to the zero reading (against the nozzle block), install only the active thrust bearings with shims. Move rotor towards exhaust until contact is made with the thrust bearings (lower half only is used at this point). If indicator reads 0.055 inches good; if greater, increase active bearing shim thickness until 55 mils is obtained. Place equal amount of shims in upper half active thrust bearing. If less than 55 mils, reduce shims until 55 mils travel is obtained.

- Place inactive thrust bearings with shims into place. With the rotor against the active thrust bearing, move the rotor towards the nozzle block shimming for a 46 to 43 reading. This yields a float zone of 9 to 12 mils.

This procedure could have been reversed, i.e., setting the inactive bearing shims for 45 mils standoff and then the active shims for 9 to 12 mils float. With a large float tolerance, this would be preferred.

Confirmation of this total float by a bump check with the top half bearings in place is necessary. Lack of confirmation, e.g., less travel, can be a real problem in troubleshooting machine fits and will not be covered here. This would fall within a maintenance paper concentrated at setting thrust bearings.

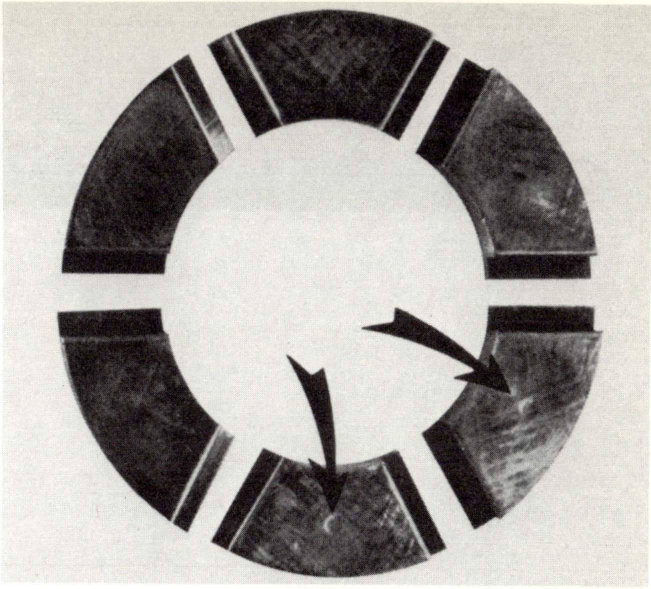

Fig. 3—Failed thrust bearing due to improper venting of buttons supporting thrust bearing pad. This is typical of a failure to properly bond temperature sensor to bearing babbitt (7,600 hp @ 6,000 rpm).

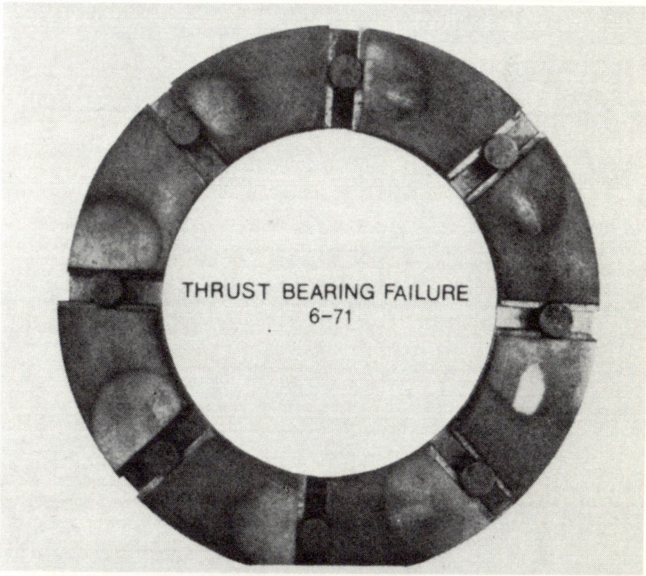

Fig. 4—Failed thrust bearing on compressor due to overload. Trip-out occurred manually at 17 mils axial deflection. Eight to 10 mils of babbitt removed by failure (5,000 hp @ 11,000 rpm).

Sensor placement. A calibration curve (Fig. 2) is plotted for the eddy current sensor and its related readout device. This calibration curve will be for the metal of the shaft to be sensed and the plot will be volts (ordinate) versus gap (abscissa). Gap implies the space between the probe (eddy current sensor face) and the metal used in calibration.

The calibration curve is at a sensitivity of 100 mv/mil (Fig. 2) with over 80 mils of linear range, and the voltage at the center of the linear range is 8 volts. At +40 mils from the center of linear range, the output voltage is 12 volts. At −40 mils from the center, the voltage is 4 volts. (Note: At 5 mils on either side of the center, the voltage is 8.5 volts and 7.5 volts, respectively.)

For further convenience, assume the bump checks above yielded 45 mils to 55 mils or exactly 10 mils float.

Because the center of the calibration range of the sensor system from the calibration curve occurred at 73 mils gap with an output voltage of 8 volts, this point will be used for the center of float position. Two methods can be used to secure this position:

1. Set the rotor in the center of the float zone by dial indicator and adjust the sensor to thrust collar (or shaft end) gap until 8 volts output is obtained at the thrust monitor output jack. The zero adjust would manually be moved until the readout meter indicates zero (center of scale). Bump the rotor to confirm both indicator and voltage agreement, i.e., −5 mils = 7.5 v, zero = 8 v, +5 mils = 8.5 v.

2. Move the rotor against the active thrust shoes and adjust the sensor gap until a voltage that corresponds to 5 mils active direction is obtained at the thrust monitor output jack, e.g., 8.5 volts. The zero adjustment calibration would manually be moved until the readout meter indicates 5 mils (active). A bump check should indicate 10 mils of travel on the indicator and 10 mils (+5 mils to −5 mils) on the readout meter of the sensor system.

Either method will work; however, the author prefers the last method with present instrumentation. It is easier to hold the rotor against one end of the travel. Further, should you wish to set up all machines against the active bearing and have the same meter reading regardless of float zone; then Method 2 (above) becomes a standard practice.

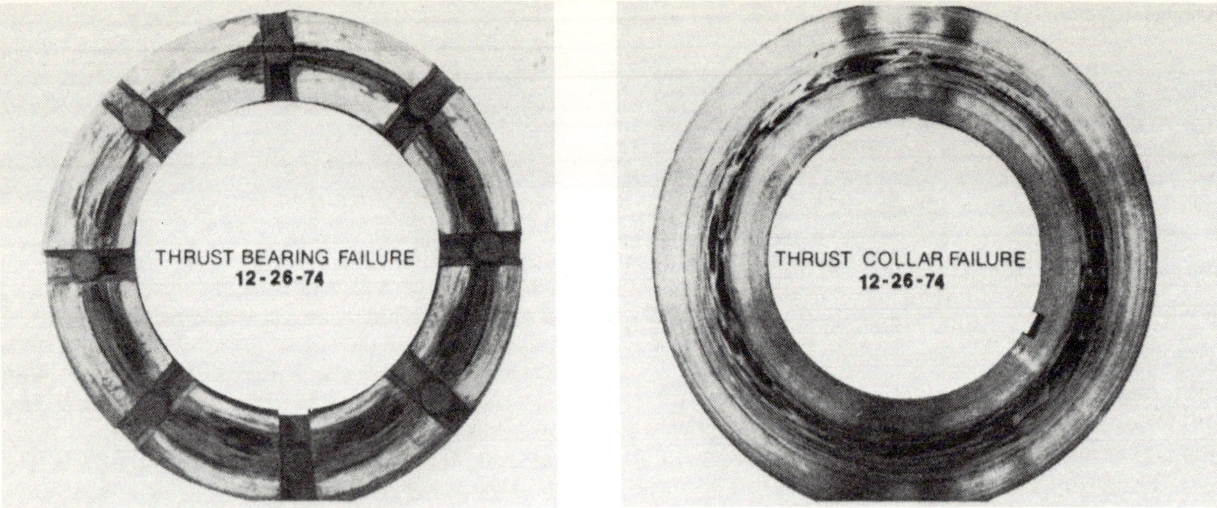

Fig. 5—Failed thrust bearing due to overload but primarily from cavitation caused by thrust reversals of compressor operating in a surge condition for over 20 minutes. Automatic trip-out occurred at 28 mils axial deflection. Sixteen mils of babbit removed by failure.

Fig. 6—Failed steam turbine thrust bearing due to overload from selective fourth and fifth stage salt deposition on blading. Salting caused by impure boiler feed water (15,000 hp @ 4,500 rpm).

Most turbines will require that the sensor be placed on the steam inlet shaft-end or at least outside (outboard) the thrust collar. Further, the thrust will normally be toward the exhaust so that on turbines an increasing probe gap will indicate a movement in the active load direction. Conversely, a decreasing gap will indicate an inactive (counter) movement. One arrangement of shaft-end (near thrust bearing) dual probes used for thrust movement sensing is shown in Fig. 1 as taken from a current task force effort of the API Subcommittee of Refining Equipment. A current proposal to API has been presented for adoption as a complete specification for vibration and thrust movement monitors.

Because a compressor's normal thrust can easily be in either a gap decreasing or increasing direction, it seems normal to use the turbine direction, calibration, polarity, etc., as standard and call out the compressor system by specific design.

It is worth nothing that new instrument designs may soon require limited adjustment of the zero meter indicator in favor of self-centering zero meter convention. This will commit Method 1 above to be the preferred method and give equal travel range to active and inactive movements of the rotor. This presents no problem to the user. However, to meet the requirements of all current types of turbomachinery, it shall require a minimum of 80 mils linear range by the sensor system builder.

Centrifugal compressor. Centrifugal compressor thrust bearing shims are set up similarly. However, the compressor rotor may have 250 mils of total float and will be set up in the center of float with a spacer or shims to give 15-20 mils (typically) of float. Efficiency of the compressor and possible aerodynamic initiated instability would be adversely affected by sizeable excursions from this position.

The operating, maintenance and diagnostic people have found that it helps to place two strips of black tape on the

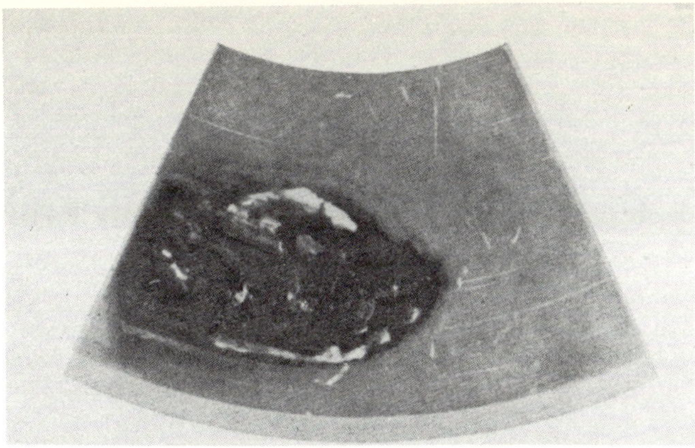

Fig. 7—Failed expander thrust bearing (one pad only) after 4,000 hours operation. Failure caused by conversion to synthetic lubricating oil with higher heat generated because of oil's different specific heat.

thrust meters to indicate the limits of the free float positions when a machine has been calibrated and checked.

THRUST ALARM AND SHUTDOWN LIMITS

Based on more than 20 years of experience with thrust bearing failures (over 50 by count) and use of various thrust indicators from "squallers" (electric touch method) to hydraulic back pressure indicators for gap, to pneumatic nozzles[1] with back pressures versus gap, to present eddy current sensor systems; our limits have been well defined, and countless *saves* have been accumulated.

Thrust *alarms* have been set up as 15 mils from the normal commissioning rotor position at design conditions (generally with good practice this will be within 5 mils of the active bump position). However, one can exert little force on a thrust bearing during the process of manual prying to bump the rotor.

Thrust *shutdowns* have been set up as 5 mils past the alarm point or 20 mils from normal load deflection points. Given the proper instrument system, this shutdown point can be extended 5 mils for a 25-mil set point value. At present, this has not been done with existing instrumentation systems. New instrumentation systems allow sufficient range.

Examples of thrust failures are shown in the photos (Figs. 3-7). These bearings while partially failed did not cause rotor damage, with the exception of Fig. 6. This failure did rub all trailing edges of each of 10 blading shrouds with repair allowed within one week. This failure occurred over approximately 2½ minutes (a true exception) and though the horsepower was 15,000, the speed was only 4,500 rpm.

Total thrust bearing failures generally show nothing but an awesome puddle or splotch of melted babbitt. The other extreme, a thrust bearing not taken to complete failure, often shows (Fig. 4) either some darkened areas at the minimum clearance point, spalling, flakings, etc., or the bearing looks to be in original as-received conditions.

CONCLUSION

It is hoped that this review of normal thrust setting procedures, thrust or rotor movement monitoring, plus some of the practices and beliefs in thrust failure prevention, will be helpful to those establishing a failure prevention system in this area.

LITERATURE CITED

[1] Jackson, C., "An experience with thrust bearing failures," *Hydrocarbon Processing*, January 1970, pp. 107-110.

Why Properly Rated Gears Still Fail

High speed gears receive a lot of attention from users and designers, but many units still fail in the field. Here's why they fail and what to do about it.

Heinz P. Bloch, Exxon Research and Engineering Co., Florham Park, N.J.

PARALLEL double helical gears frequently are used as speed increasers or speed reducers in high speed machinery. The design of the gear teeth for desired strength and durability is generally given close attention by both purchaser and manufacturer and service factors are applied per AGMA 421.06 to take care of expected overload conditions.

However, gear tooth failures still occur. One basic but frequently overlooked cause of gear failures is tooth overload due to axial thrust transmission to the mesh. The most common source of externally imposed axial thrust arises from gear couplings used to connect the gear box to the driver and driven equipment. Gear unit thrust bearings are usually rated to absorb the maximum anticipated axial thrust. However, due to the transient nature of coupling transmitted axial thrust, it is rarely included in the tooth strength and durability rating.

This article shows how to assess the overload effects of axial thrust transmission on double helical gears as an aid to failure investigation. It also outlines a number of options to remove coupling transmitted excessive axial thrust from contributing to a gearing problem.

Thrust is transmitted through gear tooth couplings. Friction between the teeth of gear-type couplings is responsible for the transmission of axial thrust between two coupled machines. This friction resists the normal relative movement between coupled shafts that occurs due to thermal expansion or hydraulic forces and results in transmission of axial forces from one rotor to another through the coupling. The maximum axial thrust transmitted is $F = 2T\mu/Dp \cos \Theta$, where T is the torque at the coupling (in pound-inches), μ is the coefficient of friction between the coupling teeth, Dp is the coupling tooth pitch diameter, and Θ the coupling tooth pressure angle. The coefficient of friction, μ is assumed to be 0.15 (API 613), although some petrochemical industries report values as high as 0.30. Higher μ-values are simply a reflection of conditions frequently encountered in petrochemical plants with less than optimum maintenance coverage.

The coupling friction forces are generally transient in nature. However, relatively high friction factors have actually been observed to persist for long periods of time on poorly lubricated gear couplings. The actual value depends on such variables as tooth interference due to thermal and sometimes centrifugal growth, viscosity and adhesion of lubricant, or lubricant film interruption because of either severe coupling misalignment or long-term operation with virtually perfect alignment.

Location of thrust bearing must be considered. The location of the gear box thrust bearing will determine whether thrust transmission will occur for a given machinery train arrangement. For the train configuration shown in Fig. 1, the thrust bearing is generally part of the high speed pinion shaft assembly. Using a coupling friction coefficient of 0.3, the pinion thrust bearing could be exposed to a maximum axial thrust of $F_1 = \dfrac{(2)\ (0.3)\ (T_{HSS})}{D_{PHS} \cos \Theta}$ (where T_{HSS} = torque acting on high speed shaft, and D_{PHS} = tooth pitch diameter of high speed coupling), plus M_A, the magnetic centering force of the motor.

However, the maximum possible axial force imposed on the gear *mesh* of Fig. 1 is not related to F_1, but rather,

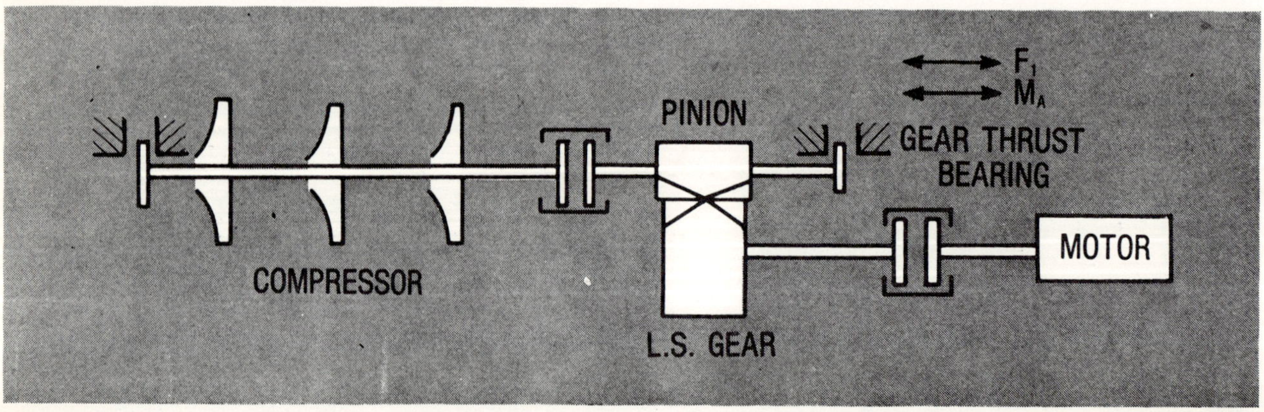

Fig. 1—Forces acting on thrust bearing mounted on pinion shaft.

as will be seen later, to the tangential driving load F_T and the magnetic centering force M_A of the motor.

If the thrust bearing is mounted on the low speed gear shaft as shown in Fig. 2, it could be exposed to the same maximum thrust as the bearing shown in Fig. 1. However, the thrust force is now transmitted across the gear mesh. The maximum possible axial force imposed on the mesh is now not only a function of F_T but of F_1 as well.

The analysis can, of course, be extended to drivers other than motors. Unlike motors, which are generally furnished with axially free-floating rotors, these other drivers will probably incorporate thrust bearings.

Axial thrust transmitted through gear mesh can overload one helix. How the axial thrust transmitted through the gear mesh may possibly overload one helix of a double helical gear can best be shown by sample calculations.

Let us assume we are dealing with a drive arrangement as shown in Fig. 1. The compressor absorbs 5,000 hp at 6,400 rpm and the step-up gear has been purchased with a 1.5 service factor. From $T = \frac{63{,}000 \text{ hp}}{\text{rpm}}$, $T = \frac{(63{,}000)\ (5{,}000)}{6{,}400}$, or 49,200 lb.-in. Using a pinion pitch diameter of 12 inches, the tangential driving load would be $F_T = \frac{(2)\ (49{,}200)}{12}$, or 8,200 pounds. This tangential driving load can be resolved into axial thrust components by multiplication with the tangent of the helix angle. For $\Psi = 30°$, the axial thrust per helix will equal $\frac{(8{,}200)}{2}$ tan 30°, or 2,265 pounds. In addition, it is readily evident that the gear mesh can be exposed to an external force, F_A. The magnitude of F_A is the lesser of the two values, F_2 or M_A, with M_A typically around 1,200 pounds for a 5,000-hp induction motor turning at 1,780 rpm, and $F = \frac{(2)\ (0.3)\ (T_{LSS})}{D_{PLS} \cos \Theta}$

Since the motor torque T_{LSS} is 177,000 lb.-in. and low speed coupling pitch diameters D_{PLS} are probably in the 10-12 inch range, F_2 will no doubt be significantly larger than M_A. Accordingly, the magnetic centering force of the motor constitutes the external force imposed on the gear mesh, i.e., it may create an axial push or pull acting through coupling teeth that refuse to slide.

A free body diagram can now be constructed to investigate if the various forces can overload one helix of a typical double helical gear.

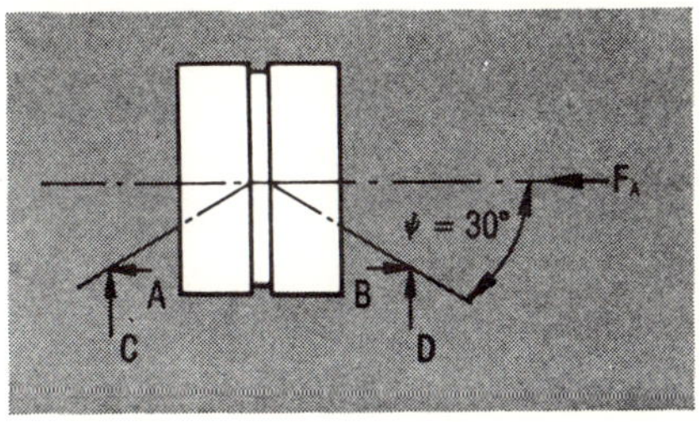

$$A\text{-}B + F_A = 0$$
$$A\text{-}B + 1200 = 0$$
$$A = C \tan \Psi,\ C = A/\tan \Psi = 1.72A$$
$$B = D \tan \Psi,\ D = B/\tan \Psi = 1.72B$$
$$C + D = F_T = 8{,}200 \text{ lbs.}$$
$$A + B = F_T \tan \Psi = 4{,}730 \text{ lbs.}$$
$$2A = 3{,}530 \text{ lbs.},\ A = 1{,}765 \text{ lbs.}$$
$$B = 2{,}965 \text{ lbs.},\ C = 3{,}060 \text{ lbs.}$$
$$D = 5{,}140 \text{ lbs.}$$

Rated tangential load per helix: $\frac{8{,}200 \times 1.5}{2} = 6150$

Tooth loading with max. possible thrust: $= \frac{(100)\ (5140)}{6150}$

$\approx$ 84% of rated tangential load.

This analysis shows that even with the maximum anticipated external force, the gear teeth are safe for the particular train configuration shown.

For the drive arrangement shown in Fig. 2, the external force F_A acting on the gear mesh can be as much as $F_1 = \frac{(2)\ (0.3)\ (T_{HSS})}{D_{PHS} \cos \Theta}$. This condition can exist because the low speed shaft thrust bearing constrains the axial move-

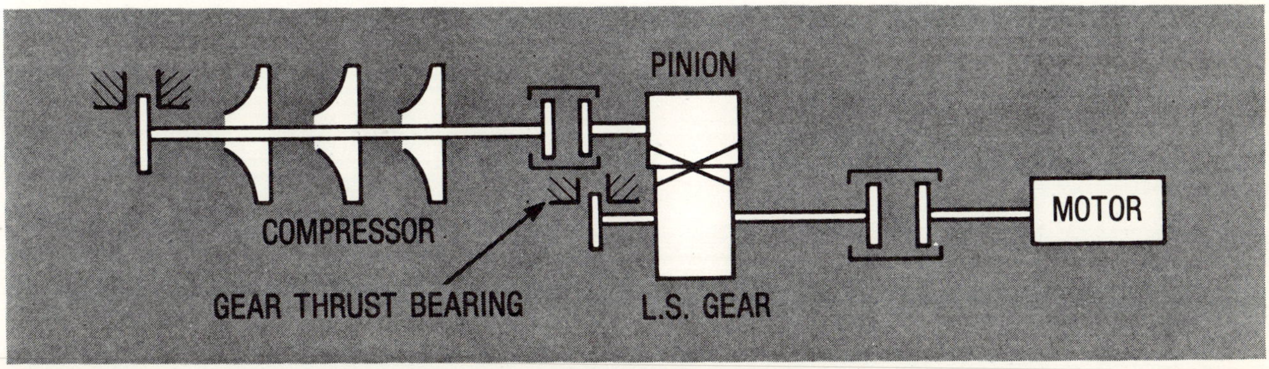

Fig. 2—Forces acting on thrust bearing mounted on low-speed gear shaft.

ment of the input gear. Thus, $F_1 = F_A = \frac{(2)\ (0.3)\ (49200)}{(7)\ (0.967)*}$ $= 4{,}380$ pounds. We now have the following force equations:

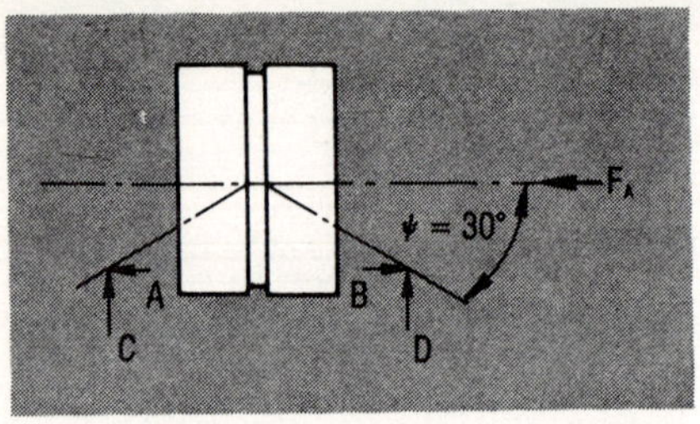

$$A\text{-}B + F_A = 0$$
$$A\text{-}B + 4{,}380 = 0$$

$$A = C \tan \Psi,\ C = A/\tan \Psi = 1.72A$$
$$B = D \tan \Psi,\ D = B/\tan \Psi = 1.72B$$
$$C + D = F_T = 8{,}200 \text{ lbs.}$$
$$A + B = F_T \tan \Psi = 4{,}730 \text{ lbs.}$$
$$2A = 3{,}530 \text{ lbs.},\ A = 175 \text{ lbs.}$$
$$B = 2{,}965 \text{ lbs.},\ C = 300 \text{ lbs.}$$
$$D = 7{,}900 \text{ lbs.}$$

Rated tangential load per helix: $\frac{8{,}200 \times 1.5}{2} = 6{,}150$

Tooth loading with max. possible thrust: $= \frac{(100)\ (7900)}{6150}$

$\approx 128\%$ of rated tangential load.

This analysis demonstrates that a gear-type coupling with an anticipated maximum coefficient of friction of 0.3 may allow the transmission of excessive axial thrust into the gear mesh. In other words, if a gear required all of its designed-in service factor initially, the additional thrust transmitted through gear-type couplings may further reduce the service factor to a serious degree. While not usually causing immediate tooth failures, the increased loading may result in gear failure after a relatively short time of operation should the axial thrust be continuously present.

If the failure analysis concludes a gear mesh failure due to axial thrust, one has a number of options, some of which are included here.

- One can examine the coupling and coupling lubrication system to determine the cause of excessive frictional forces.

- One can purposely install a thrust bearing with lower load carrying capacity. This, in effect, makes the thrust bearing the weak link, or sacrificial component. Using adequate safeguards, such as temperature thermocouples embedded in the thrust bearing material and connected to a shutdown system, potential failures thus confined to the thrust bearing replacements are certainly less costly and less time consuming than gear tooth failures.

- Replace couplings with large tooth pitch diameter gear couplings.

- One can replace the coupling with an axially soft diaphragm coupling which transmits little or no axial thrust. Diaphragm couplings accommodate misalignment while transmitting power by material flexure. However, the purchaser should recognize that high axial movement cannot be accommodated by diaphragm couplings. This requires that the initial system axial alignment take into account the thermal expansion of both coupled shafts.

- One can purchase replacement gear elements with a larger service factor than listed in AGMA 421.06. The above calculations can actually be performed to define the service factor necessary to prevent gear mesh failures resulting from axial thrust transmission.

* Assuming a coupling tooth pitch diameter of 7 inches and a pressure angle of 14.5°. Note that some gear couplings have pressure angles as high as 40° which may cause a very significant increase in F_A.

Extend Gear Coupling Life—Part 1

Extended service of gear couplings can be achieved by understanding the principles of wear, how lubricants function and proper maintenance.

Michael M. Calistrat, Koppers Co., Inc.
Baltimore, Md.

PROPER LUBRICANT selection greatly increases the life of gear couplings. Once properly installed periodic lubrication is the only maintenance required. In order to select proper lubricants the principles of gear wear and lubricant properties must be understood.

Since maintenance is a disadvantage of gear couplings compared with "dry" couplings, why are they so popular? The principal reasons are: Gear couplings can transmit more power per pound, and per inch of diameter than any other type of coupling: Gear couplings are forgiving; even when improperly installed, or mishandled, they continue to perform better than most other types of couplings; gear couplings are reliable and safe. Their main cause of failure is teeth wear, which is gradual rather than abrupt. Even when the teeth wear out completely the only effect is loss of power without the coupling flying apart.

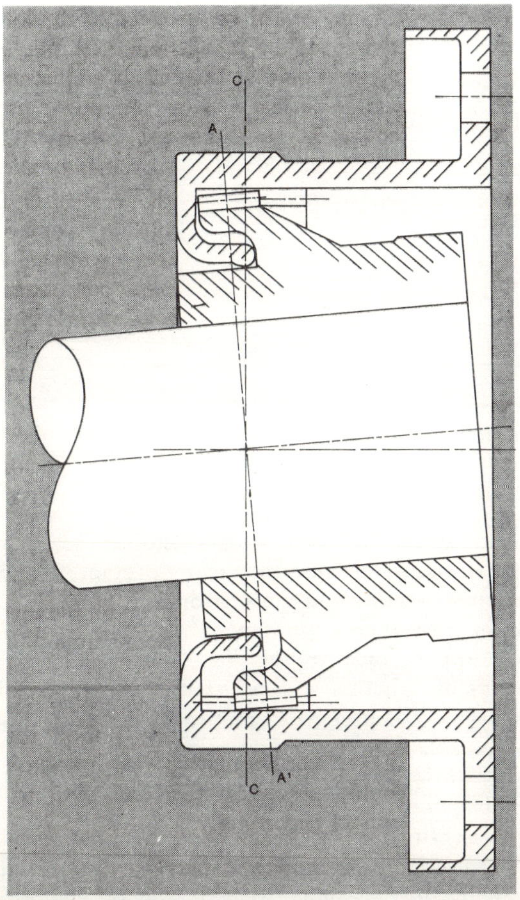

Fig. 1—Misaligned gear coupling.

Couplings in general can be divided into low speed, or standard couplings, and high speed, or high performance couplings. Maintenance requirements of the two types are very different. This article will describe maintenance of the standard type.

Standard couplings can be defined as the ones "off the shelf," not made to order. They generally operate at "motor speeds" and are seldom dynamically balanced. The gear coupling was invented by Gustave Fast of Baltimore, Md. in 1920. His basic design is still used today. Many of the original dimensions for sleeve flanges and bolt sizes became standards of the industry. As a consequence, couplings made today by any manufacturer can be bolted, by half-coupling, with other manufacturer's couplings. Standard couplings are sized for the nominal shaft diameter in ½-inch increments. A size 1½ (or 15 in some cases) is designed to accept a 1.5 in. shaft, however most hubs can be overbored as much as ½-in. The size of a coupling also represents, with few exceptions, the pitch radius of the gear mesh.

Why do gear couplings require lubrication? Although standard couplings are made in a large variety of constructions, they can be divided into two categories depending on the method used to accommodate misalignment: through flexing or through sliding. Gear couplings fall into the second category. Fig. 1 shows a section of a gear coupling in which the misalignment is intentionally exaggerated. Let us follow the path of the hub tooth "A" as the coupling makes one revolution. Without misalignment the tooth will stay in the plane C-C. Because of the misalignment the tooth is to the left of the plane C-C in the upper position, but after one half revolution it moves to the right of plane C-C in the position marked A^1. After the coupling completes the revolution the tooth assumes again the position A. Although the hub tooth remains constantly engaged with the same sleeve tooth, it moves back and forth over the surface of the mating tooth. The frequency of this oscillatory motion is the same as the shaft's speed. The amplitude of the motion (i.e. the total travel of the hub tooth from one extreme to the other) is very small and is a function of the pitch diameter and the angle of misalignment:

$$a = \text{P.D. } (\text{Tan } \alpha) \quad (1)$$

The velocity of the oscillatory motion is a function of the amplitude and of the shaft rpm.

$$V_s = \frac{\pi N}{60}(a) = \frac{\pi (N)\ P.D.}{60}\ (\mathrm{Tan}\ \alpha) \qquad (2)$$

To prevent the wear that this rubbing of one tooth against the other generates, gear couplings must be lubricated.

How does the lubricant work? The forces between the mating teeth as well as the oscillatory motion tend to wipe off the lubricant from the contact surfaces. However, when one tooth slides in one direction the lubricant has a chance to wet the area left uncovered behind the tooth. To do that the lubricant has to work against two elements: its own viscosity, and the very short time available during one cycle. The teeth would operate dry if it wasn't for the high centrifugal forces created by the couplings' rotational speed. This forces the lubricant between the teeth insuring good lubrication.

It is customary to measure the effect of centrifugal forces with a factor *G* which indicates the apparent increase in an object's weight when the object rotates around an axis.

$$G = 28.4\ (10^6)\ \ R\,N^2 \qquad (3)$$

inside a coupling we can write

$$G = 14.2\ (10^6)\ \ (P.D.)\ N^2 \qquad (4)$$

The graph from Fig. 2 helps in finding the *G* level in couplings. For example, a Size 4 coupling, with an 8-in. pitch diameter, operating at 3,500 rpm has a *G* of 1,390 or, as it is common to say, has 1,390 *G*'s. This large apparent increase in weight forces the lubricant to flow rapidly and replenish the voids left behind by the moving teeth. The influence of centrifugal forces on the quality of lubrication was confirmed in lab tests. As Fig. 3 shows, the wear rate of a coupling decreases rapidly as the speed increases up to a point where no further benefit is obtained from an increase in the *G* level.

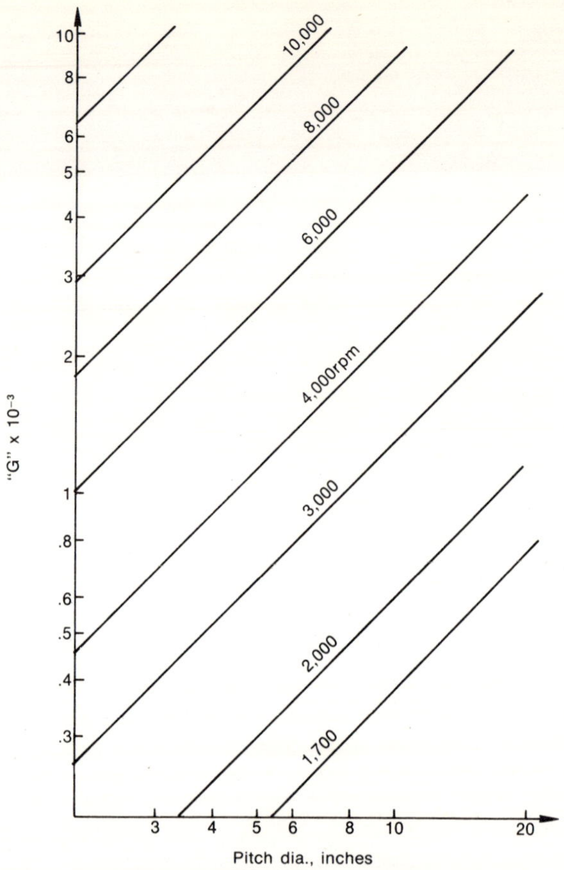

Fig. 2—Centrifugal forces in couplings.

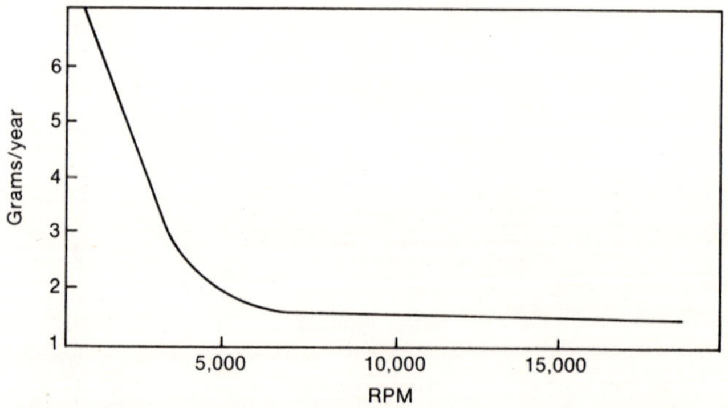

Fig. 3—Gear coupling wear rates as a function of speed.

Oil versus grease—which is better? To compare oils with greases we have to look at two of their features: lubrication properties and convenience. One must first understand that greases are nothing but oils to which a thickening agent was added. The vast majority of greases (about 97 percent of total U.S. production) use a metallic soap as the thickening agent. Generally, these soaps have no lubricating properties, hence a grease is no better than the oil used in blending it. The biggest advantage of greases is that they are easier to seal, thus they stay where the lubricant is needed.

Figs. 4 and 5 illustrate two of the most popular type of gear couplings; the Fast type, with an all metal seal, and the elastomer seal type. When the couplings rotate, centrifugal forces cause the lubricant to form an annulus inside the sleeve. The inside diameter of the lubricant annulus is always the seal diameter because none of the coupling seals can retain more lubricant when under high *G*'s. The seal of the Fast type is one of the few so designed that, under static condition it will retain enough oil to form an adequate annulus when the coupling is rotating.

Oils cannot be used with elastomer seal couplings (Fig. 5) mainly because they cannot retain enough lubricant under static conditions. Either oil or greases can be used in Fast type couplings. The question of which is better applies only to the Fast couplings. The answer depends on whether a good grease is available. If it is, then a grease is preferable simply because more lubricant can be packed in the coupling, even though a small part of the lubricant is a thickener. If a good grease is not available, then any high viscosity oil is preferable. Why high viscosity? Studies performed in the lab with standard couplings indicated that the most influential parameter for coupling wear is the oil viscosity. The graph (Fig. 6) shows that the higher the oil viscosity,

Fig. 4—All metal seal coupling.

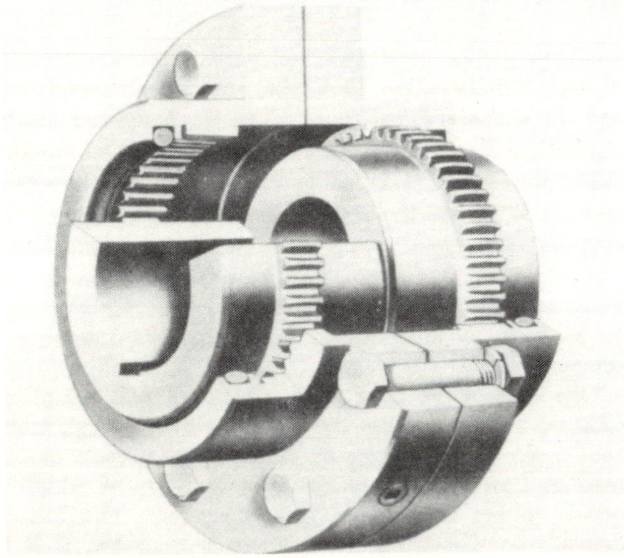

Fig. 5—Elastomer seal coupling.

the smaller the wear-rate. When greases are considered the consistency of the grease should not be confused with the viscosity of the base oil. The same grease consistency can be obtained with a wide variety of oils of different viscosities.

Greases and centrifugal forces. Let us first see what happens when the centrifugal force is very small, which can be the case either at low speeds or when small couplings are used. Because the grease is subjected to a small centrifugal force it needs more time to flow into the voids between the teeth. Fortunately, at lower speeds more time is available. While the time available increases in the same ratio that the speed decreases, the centrifugal forces decrease with the square of the speed, as shown in Equation 4. Thus, the forces that make the grease flow decrease more rapidly than the time available increases. At some speeds the coupling could operate without sufficient lubrication, even when it is full of grease. This condition becomes particularly critical when stiff greases (NGLT No. 2 or No. 3) are used, because they flow less readily than a No. 1 or No. 0 grease. It was determined experimentally that a No. 2 (or stiffer) grease should not be used when the coupling operates below 10 *G*'s, i.e. when the speed is below:

$$N_c = 850 \sqrt{(P.D.)} \quad \text{(Remember: P.D. is approximately twice the coupling size)} \qquad (5)$$

Now let us see what happens under normal operating conditions. It is known that greases have a tendency to partially lose their oils, a phenomenon known as "bleeding." There are three primary forces responsible for oil separation from greases[3]: syneresis, gravity and external pressures. In couplings the apparent gravity generated by centrifugal forces is so high that the other forces responsible for bleeding become insignificant.

Centrifugal forces have a larger effect on thickeners from the grease than on the oils because thickeners have a higher specific gravity. Hence, the thickeners are forced away from the axis of rotation, and will eventually coat the inside surface of the coupling sleeves. In observing Figs. 4 and 5 one can see that the gear mesh (the hub and sleeve teeth) is located on the sleeve's inside diameter.

Hence, the soap separated from the grease will surround the coupling's teeth and *WILL PREVENT THE OIL FROM LUBRICATING THE TEETH SURFACES*. In other words, although the coupling is full of grease it can be starved of lubricant. Understanding this phenomenon is very important for proper coupling maintenance.

For many years it was assumed that the amount of oil separation was a function of the *G* level only. Consequently, some greases that were not acceptable for high speed operations were considered acceptable for standard couplings. Recent studies[4, 5] have shown that rather than being a function of the *G* level, the maximum oil separation from a grease is mainly a function of time. This means that a grease will bleed a given percentage of oil at any *G* level, if it is subjected to these centrifugal forces for a long enough period of time. We use a standard test method[4] for evaluating greases: a sample is subjected to a high centrifugal force (36,000 *G*'s) in a laboratory centrifuge, and the oil separation is closely monitored. The test is interrupted when the separation stabilizes. The greases are then given a (K36) rating which is a fraction. The numerator represents the maximum oil separation in volumetric percentage, and the denominator represents the time required to attain that separation. For example, if a grease has K36 = 70/35 it has a maximum oil separation of 70 percent and it needs 35 hours of operation at 36,000 *G*'s to attain this separation. When the grease is subjected to a smaller centrifugal force, the time required for the maximum separation can be calculated with the experimental formula:

$$t = 2{,}400{,}000\ T/G^{1.4} \qquad (6)$$

where *T* is the number of hours used to determine the K36 factor.

Which grease is best? Coupling manufacturers rely both on lab tests and field experience when making grease recommendations. Depending on the application some greases are better than others. Just because a particular grease is not on a list of recommended lubricants does not necessarily mean that it is not suitable.

There are four basic parameters that can be used for evaluating greases for gear couplings:

A grease should have more than 92 percent oil. It was shown that most greases will eventually separate into oil and soap. The more soap a grease has, the higher the chances of the teeth being starved of lubricant. The 92 percent level is a desirable number rather than an excluding one, but it is not unrealistic; there are many greases that have more than 92 percent oil.

A grease should have a base oil with a high viscosity. It was shown (Fig. 6) that the higher the viscosity of the oil, the lower the wear-rate of the coupling. Oil viscosity should not be confused with the grease's consistency. To blend a given grease, let's say a NLGI No. 1, oils with viscosities as low as 200 or as high as 4,000 SSU at 100° F can be used. If the oil is lighter more thickener is used in order to attain the desired grease consistency. An oil viscosity higher than 900 SSU at 100° F is desirable.

A grease should resist centrifugal separation. Very few companies are equipped to determine this characteristic, however K36 values are available for many greases. For a good choice a grease should not only bleed as little oil as possible, but should also bleed the oil slowly.

A gear coupling grease should not have a consistency higher than NLGI No. 1. It was shown that a more consistent grease could cause problems at low *G*'s. In order to avoid misapplications one should not use No. 2 or No. 3 greases.

Some good quality greases incorporate special additives or fillers. Experience has shown that EP additives are beneficial only in highly loaded couplings, and in all couplings during the breaking-in period. Under normal conditions couplings do not operate under high contact pressure. Lead, molybdenum disulfide and other high density compounds will separate immediately inside the coupling, and as such are beneficial only in very low speed, high load applications.

Coupling's breaking-in. Usually the gear mesh of a coupling has 60 teeth at the hub and as many at the sleeve. The number can vary slightly depending on the coupling's size. Because of the many teeth, and depending on the precision of manufacture, more or less of the matching teeth will be in contact when the coupling first starts transmitting the torque, even if the coupling is perfectly aligned. The teeth that do transmit the torque are subjected to larger than normal forces, so they wear relatively fast. However, as these teeth wear, the other teeth that originally were not in contact start sharing the load. Eventually all the teeth are in contact and the coupling is broken-in. The duration of the breaking-in depends on the load, the quality of manufacturing and the materials used. Experience and laboratory tests show that a standard

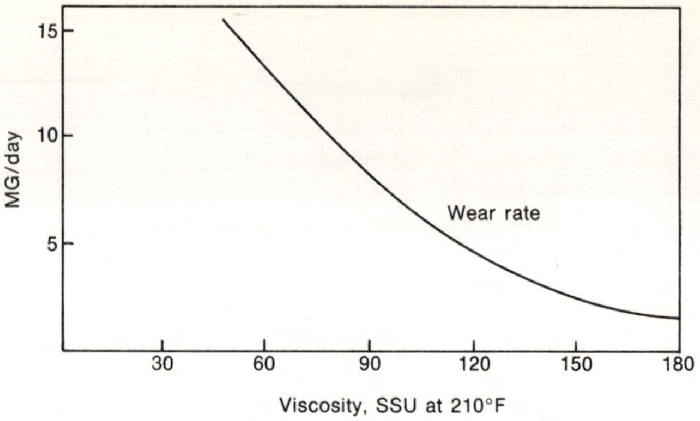

Fig. 6—Effect of lubricant viscosity on coupling wear.

gear coupling is completely broken-in in three to five days.

Breaking-in of a coupling creates a special problem for the lubricant: high temperatures. Rapid wear of the teeth that are originally in contact generates heat at a larger rate than the coupling can normally dissipate. Temperature at the surface of the sleeve can reach as much as 120° F above ambient. If the ambient temperature is high the grease in the coupling could operate close to, or even above its dropping point. For this reason greases with a low dropping point should be avoided. It is also possible that during breaking-in the flange bolts will lose some of their pretorque. If the equipment can be stopped approximately a week after start-up the coupling should be inspected and the bolts retightened. Unless there is evidence of grease loss between the flanges there is no need for relubrication at that time.

Maintenance procedure. In the order of increasing complexity, maintenance can consist of: relubrication, relubrication with realignment and coupling replacement.

Even the best coupling greases cannot perform too long; besides the inevitable soap separation, oils oxidize and lose some of their lubricating properties. The possible loss of lubricant should also be considered. Many users do not install the paper gasket or O-ring between the flanges, which is a mistake. Manufacturers recommend relubrication at either six months or every year. Actually, depending on the grease used and the *G* level in the coupling, relubrication periods could be longer, or shorter. Independent of how often the coupling is lubricated, it should first be opened and as much of the old grease as possible should be removed. Even if cleaning is sometimes difficult it should be done because it prolongs coupling life and prevents unscheduled equipment breakdown.

Here are a few rules to be observed when opening a coupling: Remove and save the bolts and nuts. In most cases the bolts have special body tolerances and should not be replaced with "similar" bolts. Lock-nuts should be replaced after six installations, or when they are loose on the bolts. Do not separate the sleeves by forcing a chisel or screwdriver between the flanges. By damaging the flange surfaces a good seal can no longer be made and one risks losing the lubricant when the coupling is restarted. Remove as much of the old grease as possible.

When reassembling the coupling use a new gasket if the old one is damaged. Proper bolt torqueing is very important not only to form a good seal at the flanges, but

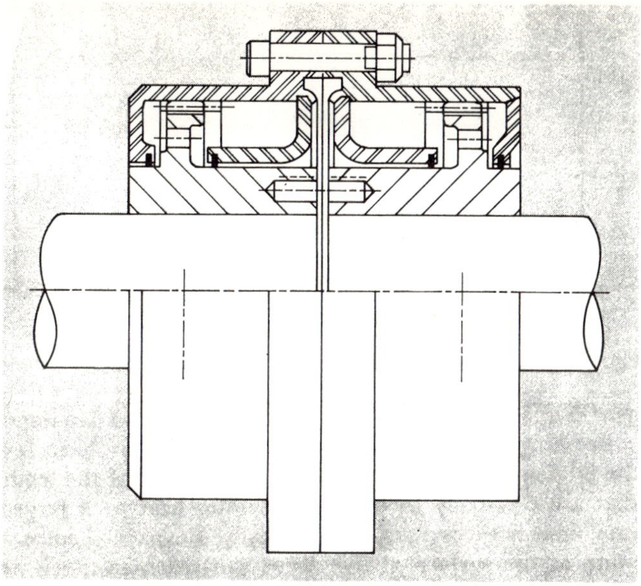

Fig. 7 — Permanently sealed gear coupling.

also for proper torque transmission. One should keep in mind that in a properly installed coupling the torque is transmitted from one flange to the other mainly through friction; only a small part of the torque is transmitted through the bolts. If the bolts are not properly tightened then they are subjected to shear forces that can damage the bolts and elongate the flange holes. Bolt torqueing values are available from coupling and fastener manufacturers and depend on coupling size, type and manufacturer.

The lubricant capacity can vary anywhere from a few ounces to over 25 lbs. In order not to waste lubricant the manufacturer's recommendations should be followed. If this information is not readily available it is safe to remove two opposite lube plugs, rotate the coupling until the lube holes are about 45 degrees to the vertical, and pump lubricant through the upper hole until it comes out through the bottom one. In no case should grease be pumped into a coupling if a second lube plug was not removed. The pressure developed by a grease gun is very high and a closed coupling can be damaged. If the coupling is new, or if an old coupling was completely cleaned of the old grease, it is very important to hand pack the teeth with some lubricant to insure proper coupling performance at start-up. There is no danger in overfilling a coupling with lubricant!

If a long spacer is used between the sleeves, and if the spacer is not provided with plates at both ends, a full disk gasket and back-up metal plate should be used between each sleeve and the spacer. The gasket should be installed between the sleeve and back-up plate rather than the other way. Without a back-up plate the disk gasket will most likely be ruptured by the pressure developed by the centrifugal forces. When spacers are used it is important that each half-coupling be lubricated individually.

At least at the first maintenance stop the original alignment should be checked to verify if either machine moved on its base. Many good articles, and good instructions on alignment are available.

If the coupling must be removed, but will be reused, the hub should not be pulled by the sleeve as permanent damage at the sleeve could occur. If the hub does not have puller holes a plate that does not touch the sleeve should be used behind the hub.

Many questions have been asked about the use of heat for hub removal. Heat is beneficial only if the hub can be heated faster than the shaft, which is seldom the case. A welding torch will most likely damage both the coupling and the shaft. To avoid difficulties in hub removal the best procedure is to use an anti-seize compound on the shaft when the hub is first installed. If an anti-seize compound is not available, a grease with a zinc-oxide filler can be used successfully.

A maintenance free gear coupling. In some applications the equipment to which a coupling is installed cannot be stopped when relubrication is scheduled. While this causes no problems in lubricating the equipment's bearings, it prevents coupling lubrication. It is then possible that the coupling is not serviced until the next scheduled maintenance, or the next one, or never.

A coupling[6] which incorporates the advantages of a gear coupling, but does not require periodic maintenance is shown in Fig. 7.

Each half-coupling is double sealed; the lubricant will not be lost or contaminated when the two flanges are separated.

The coupling has a larger lubricant capacity than the Fast's and does not require relubrication for a period generally longer than the one of the connected machinery. A few prototypes have been in continuous and successful operation in selected petrochemical plants for about three years.

NOMENCLATURE

a = amplitude of axial motion, inches
$P.D.$ = pitch diameter of the gear mesh, inches
α = angle of misalignment, degrees
V_s = axial sliding velocity, inches/second
N = rotational speed, rpm
R = radius of rotation around an axis, inches
N_c = critical speed for some greases, rpm
t = time required for centrifugal separation, hours

LITERATURE CITED

1 Wright, John, "Which shaft coupling is best . . . lubricated or non-lubricated?" *Hydrocarbon Processing*, April 1975.
2 Calistrat, Michael, "What causes wear in gear-type couplings?" *Hydrocarbon Processing*, January 1975.
3 Calhoun, S. F., "Fundamental Aspects of Grease Bleeding." NLGI Spokesman, January 1966.
4 Calistrat, Michael, "Grease Separation Under Centrifugal Forces." ASME Paper No. 75-PTG-3, ASME-ASLE Lubrication Conference, October 1975.
5 Clapp, A. M., "Fundamentals of Lubrication Relating to Operation and Maintenance of Turbomachinery," First Turbomachinery Symposium, Texas A & M University, October 1972.
6 U.S. Patent No. 3,953,986, May 1976.

Extend Gear Coupling Life—Part 2

Michael M. Calistrat, Manager, Power Transmission Development Section, Koppers Company, Inc.

PROPER LUBRICATION and alignment are the two factors which determine the service life of a high speed coupling. While standard couplings have a regular maintenance schedule, high performance couplings must operate maintenance-free for two or three years, sometimes even longer.

The main cause of failure of standard couplings is tooth wear. High speed couplings, however, have a useful life limited by the amount of vibration imposed on connected machines. As a rule, the teeth of modern, high performance couplings are hardened and the wear-rate is very small. It is wear at the tip of the teeth that causes pilot clearances to increase allowing the floating member of the coupling to shift laterally and cause unbalance resulting in increased vibrations.

METHODS OF LUBRICATION

While most standard gear couplings are grease lubricated, the majority of high speed couplings are lubricated by a continuous flow of oil (Fig. 1). There are, however, many high speed couplings which are packed with grease. The main reason a continuous flow of oil is preferred is because the oil, besides lubrication, provides means to remove heat from couplings. Contrary to what some people believe, couplings are not just hardware; they are machines which transmit mechanical power and as such have an efficiency. The efficiency of couplings is excellent: under normal conditions it is better than 99.9 percent. If we realize, however, that high speed couplings also transmit high power, the amount of power loss can be considerable. For instance, a coupling transmitting 20,000 hp, at 99.9 percent efficiency, will absorb 15 kw, which it will transform into heat. If this heat is not dissipated it will damage the coupling.

If a coupling is under a horse-shoe cover, air flow generated by the high rotation speed of the coupling is sufficient to dissipate the heat. Couplings which are within a sealed enclosure must be provided with additional cooling means[2].

Grease packed couplings. The qualities required for a coupling grease were discussed in a previous article[3], and are not different for high speed couplings. For high speed couplings the most important criterion for grease selection is resistance to centrifugal separation. Only a few greases have good resistance, and coupling manufacturers' recommendations should be followed. Sealing of the coupling, particularly between the flanges, must be perfect. High centrifugal forces generated by high speeds result in pressures that can exceed 100 p.s.i. Even a small leak could quickly deplete the coupling of lubricant.

Oil lubricated couplings. It was shown previously that for optimum lubrication, couplings should use a very viscous oil. In turbomachinery, however, it is customary to use only one lube circuit to lubricate the bearings of the machines and the couplings. Unfortunately, turbine oils have a low viscosity and do not provide good lubrication for couplings. To offset this marginal lubrication, high speed couplings are made of heat treated alloyed steels for strength and nitrided to reduce wear and the friction coefficient. Surface harness of nitrided teeth usually exceeds 50 R_c.

Oil is brought to the coupling by one or two nozzles, which direct a jet of oil to collector lips machined either in the hubs or in the spacer. The oil then flows through the coupling due to centrifugal forces and generally is discharged at the end opposite to the entrance side.

SLUDGE ACCUMULATION

The high centrifugal forces present in a coupling do not affect the oil, as long as it is a homogenous fluid; however, the oil not only has components which are added for the purpose of enhancing its properties, but also has unwanted impurities. Under centrifugal forces, some of the additives and most of the impurities are separated and retained within the coupling. The problem is as old as the gear coupling. In 1925, Gustave Fast, the inventor of the gear coupling, obtained U.S. Patent No. 1,616,034, which states: "In such circulating systems trouble has been experienced because of the collection of dirt within the coupling due to the fact that

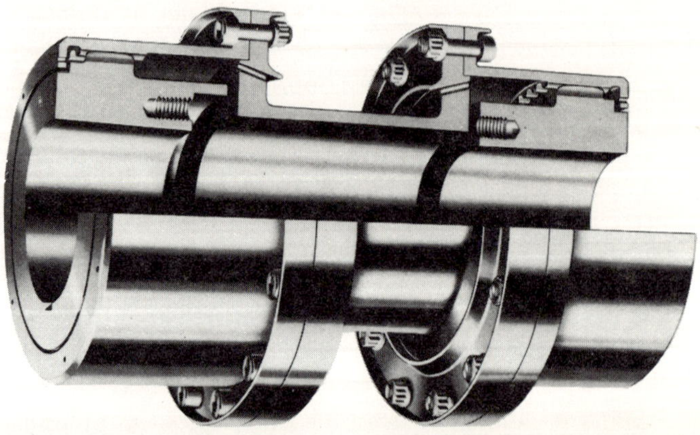

Fig. 1—Typical high speed coupling.

centrifugal force resulting from the revolution of the coupling with its retained oil gives the effect of a centrifugal separator . . ." Sludge accumulation is not of concern if it does not affect coupling performance. But when the quantity of sludge in a coupling becomes excessive, it will:

- Impair the axial movement of one shaft with respect to the other
- Corrode the teeth and accelerate their wear rate
- Reduce or completely stop the flow of oil; with high temperatures and high wear rates as a consequence.

Sludge can be of two types: a moist sludge with the appearance and consistency of a NLGI No. 3 grease, or a dry sludge which is hard and crumbles like clay. The moist sludge is caused by an accumulation of oil additives in the coupling, in particular by silicone anti-foaming compounds. Moist sludge is a lubricant and its only drawback is that it prevents oil from circulating through the coupling. Dry sludge is caused by an accumulation of impurities from the oil. The type of impurities varies as a function of plant location such as sand in the desert, or fertilizer when near a chemical plant.

Whether the sludge is moist or dry, the worst conditions are caused by water or other corrosive liquids that may enter the coupling. Besides causing rapid wear, and shortening the coupling's useful life, corrosion can create dangerous conditions because it reduces the fatigue resistance of the coupling sleeves, which in turn can result in catastrophic failures.

How to reduce the rate of sludge accumulation. The first thing to do is to make sure the oil is clean when it is delivered. Oil purity is very important because the coupling will eventually retain all the impurities. The following example might be extreme, but enlightening: A coupling requiring an oil flow of three gallons per minute will have a total oil circulation of 1,576,800 gallons per year! If we assume a nearly perfect oil purity of two parts per million, and if the coupling's centrifugal effect would separate all of these impurities, then in one year the coupling would accumulate three gallons of sludge!!

The lube circuit filter will retain all particles larger than the element mesh size. Obviously, with a smaller mesh size fewer particles will pass through the filter. A particle count was made from a sludge sample taken from a system using an eight micron filter, with the following result:

Particle Size	Percent Content
0 to 1.76 microns	47.5%
1.77 to 3.52	40 %
3.53 to 7.01	8 %
7.05 to 10.56	3 %
10.57 to 14.0	1.5%
	100.0%

Many installations solved the sludge problem by using a ½ micron filter. The main drawback of such a fine

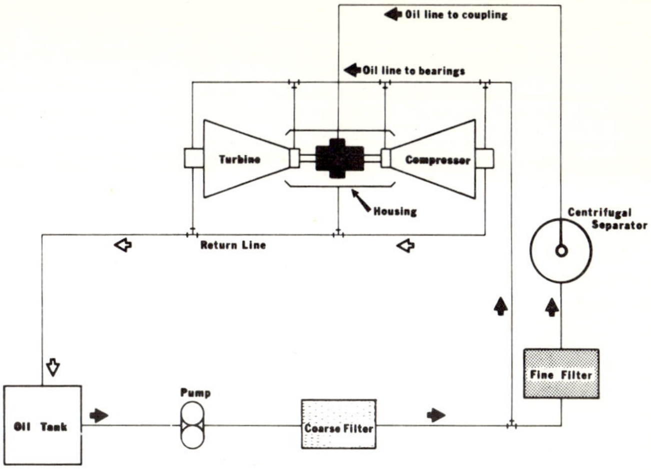

Fig. 2—Improved oil circuit.

filter is its size. To reduce the filter size and make element replacement easier, a separate lube circuit to the coupling is advantageous, as shown in Fig. 2. The filter should be of good quality, otherwise it can be a source of sludge. Wash-out fibers from the element will be retained within the coupling and plug the coupling's oil passages.

The piping system can contaminate the oil, particularly when water is present. In one known application sludging was stopped when the piping was replaced with stainless steel tubes. The lid of the oil tank in particular can generate dirt because water vapor can condensate on it during the cold hours of the night. The lid should be painted with a good quality paint and inspected periodically.

Air borne dust can contaminate the oil through vents in the oil tank and coupling guards. Fine filters should be provided at all vents. These filters cannot, however, prevent water vapor from entering the tank. In one known application sludging was eliminated by nitrogen purge at the oil tank and coupling housings.

Centrifugal separators installed in a by-pass circuit will help in removing most of the dirt and water from the oil. These separators should be operated only periodically because once the oil is cleaned and the sources of contamination eliminated, the oil will stay clean.

Excessive use of oil additives can cause more problems than they prevent; anti-foaming additives in particular have a tendency to separate within couplings. Anti-rust additives react with metal particles existing in the coupling and form soaps, which again are retained in the coupling and form sludge. In one ammonia plant sludge accumulation was completely eliminated after mineral oils with no additives were used.

Anti-sludge couplings. In parallel with users' efforts to minimize the causes of sludge accumulation, coupling manufacturers intensified their efforts to design a coupling that would not accumulate sludge. They started by drilling very small radial holes in the sleeve's tube so that water would be thrown out of the coupling, but

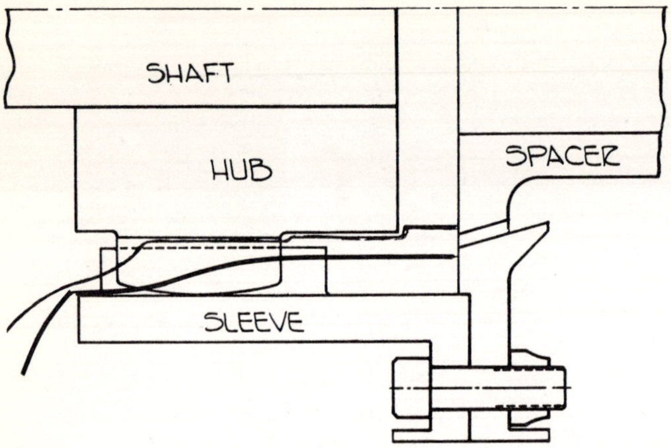

Fig. 3—Oil flow through dam-less couplings.

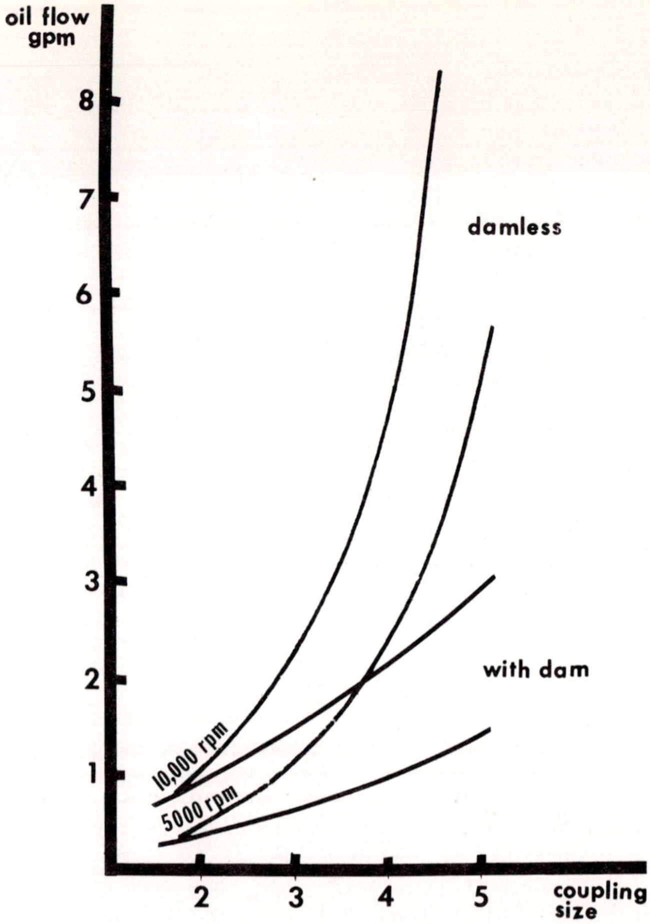

Fig. 4—Lubrication requirements (Koppers couplings).

the oil, with its higher viscosity, will have a very small flow. Upon inspection it was found that sludge plugged these small holes and made them ineffective. Hole size was gradually increased until it became so large that all oil supplied to the coupling was drained before it had a chance to form a lubricant annulus to flood the gear mesh. This fact completely changed the principles of coupling lubrication by means of a continuous oil flow. While in a dam-type coupling lubrication is insured by the oil annulus and the oil flow is determined by the cooling requirements, the dam-less coupling needs a large oil flow to be properly lubricated. Fig. 3 illustrates oil flow through a dam-less coupling. It can be seen that the only restriction to oil flow is the gear mesh. If flow is insufficient, oil will wet only the portion of the teeth under the lower line. The upper line indicates a correct flow. It became evident that recommendations for oil flow through couplings needed to be reconsidered. Fig. 4 shows comparative flows for various couplings, as recommended by Koppers. This graph clearly indicates that dam-less couplings, particularly the large ones, require a significantly larger oil flow than dam-type couplings.

The oil supply. While a dam-type coupling that receives less than the recommended oil flow will operate hotter, a dam-less coupling without sufficient oil flow can fail rapidly[1]. Establishing nozzle size should not be treated lightly and the flow to each coupling should be actually measured before start-up. A simple method is to slide a hose over the nozzle and use a bucket and stop-watch with the oil at the normal pressure and temperature.

What can sometimes happen is shown in Fig. 5 which illustrates nozzles used at one chemical plant. The drive train consisted of a gas turbine driving a high pressure compressor through a Size 5 coupling and a low pressure compressor through a Size 4 coupling. Recommended and actually measured oil flows are indicated. What happened is easy to guess; the mechanic could not obtain the required oil pressure so he crimped the oil tubes starting from the turbine end. Once he obtained the desired pressure he left the last oil tube uncrimped. As a consequence, very little oil was supplied to some of the couplings, and too much to the last one in the line.

The graph in Fig. 6 can be used to determine proper nozzle sizes.

Another important aspect of oil supply lines is orientation and rigidity of the tubes. It was found through lab tests that not all the oil delivered by the nozzle penetrates the coupling. If the oil jet is properly orientated (Fig. 7), about 98 percent of the oil enters the coupling. If the oil jet is orientated against coupling rotation less than 75 percent will enter the coupling.

The oil tube should be short and rigid as shown in Fig. 8. A poor design is shown in Fig. 9. Long thin tubes can vibrate and very little oil will enter the coupling.

Oil supplied to the couplings should be cooled, not only because a cool oil does a better job in removing heat, but because viscosity of the oil increases as temperature is lowered. As shown previously, gear couplings wear less when lubricated with highly viscous oil.

Maintenance procedure. Whenever the equipment is down, couplings should be opened, inspected and cleaned of any sludge. Teeth should be checked for unusual wear on the flanks and tips. If the coupling was in operation

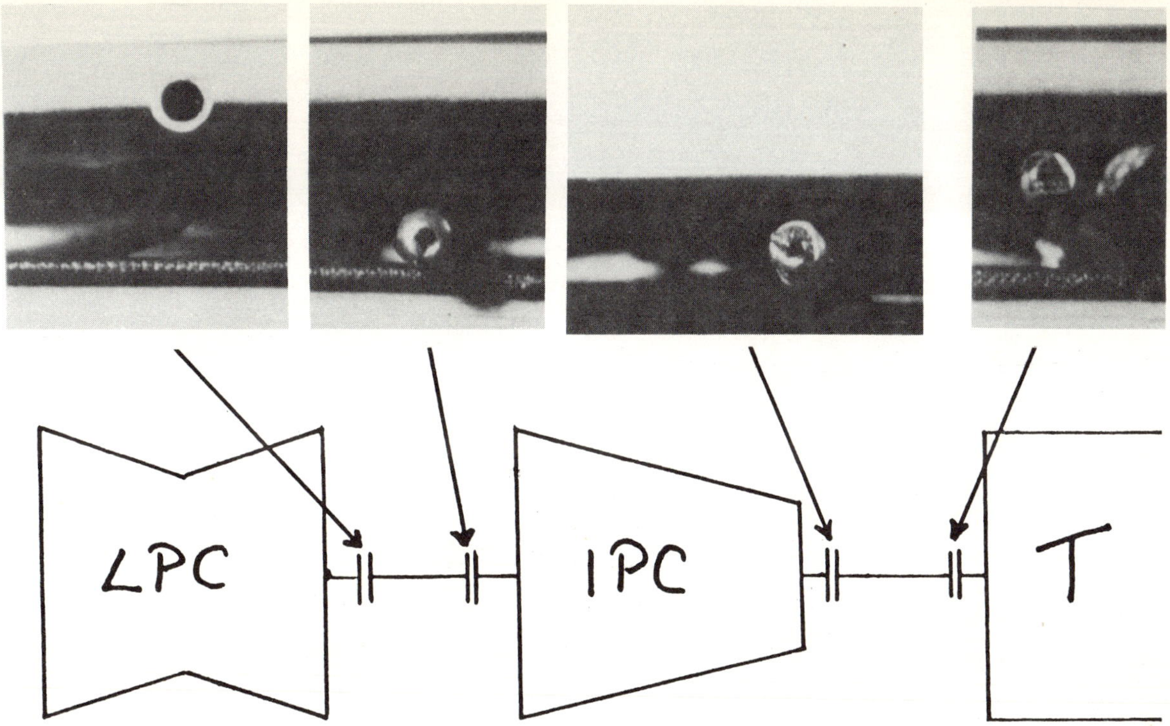

	NOZZLE OIL FLOW (USGPM)			
DESIGN	1.25	1.25	1.8	1.8
ACTUAL (AT 20 PSI)	0.96	0.15	0.27	0.18

Fig. 5—Oil nozzles in a compressor train.

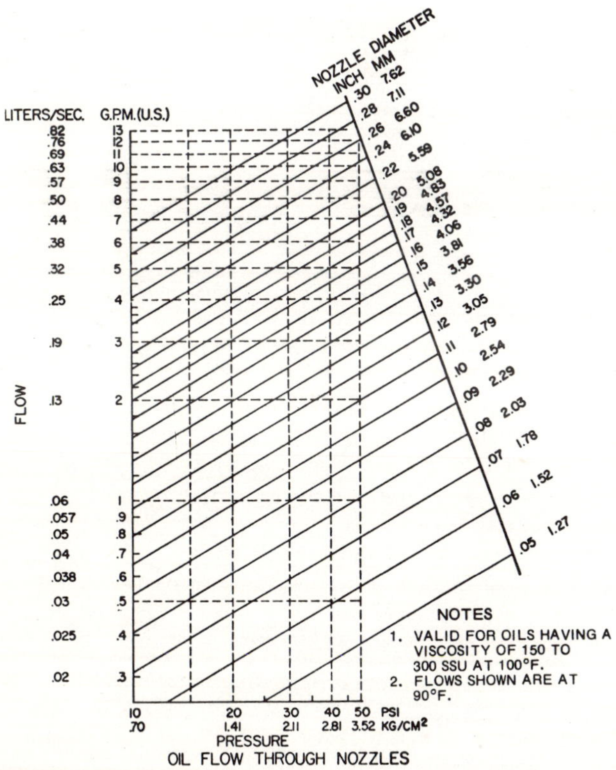

Fig. 6—Oil flow through nozzles.

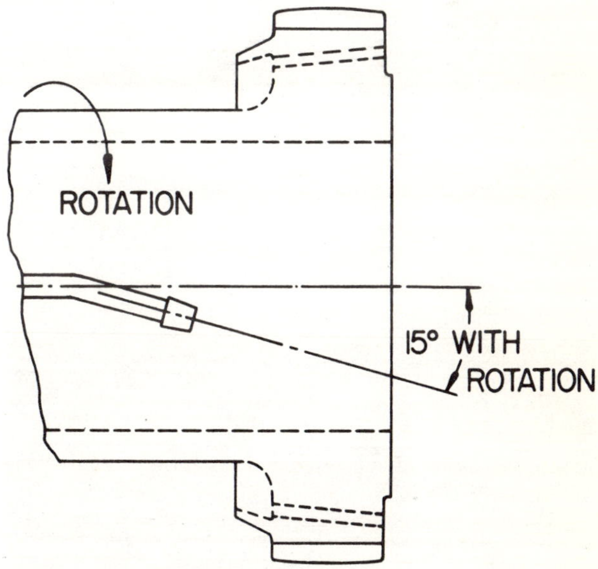

Fig. 7—Optimum nozzle orientation.

Fig. 8—Good piping system.

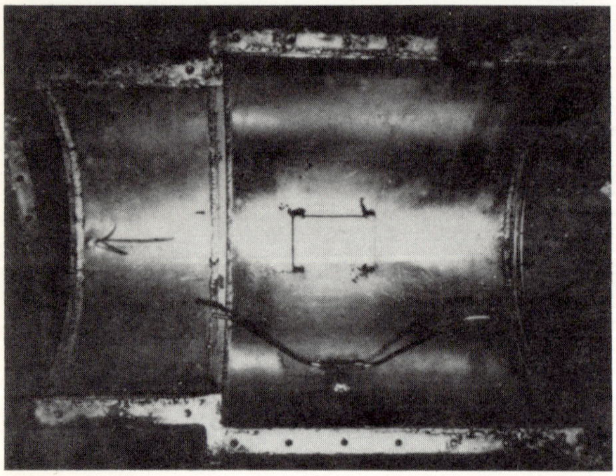

Fig. 9—Incorrect piping system.

a relatively short period of time, minor galling on some of the teeth is not abnormal. This galling, when superficial, is caused by the breaking-in process and is usually self-healing.

The importance of reassembling the coupling in the original condition cannot be overemphasized. High performance couplings are dynamically balanced and this balance can be disturbed if the match marks are not respected. Also, bolts from one flange are weight matched with their nuts to 0.1 grams. Mixing bolts, or even only the nuts can disturb dynamic balance of the coupling. Machine alignment should again be checked before start-up.

LITERATURE CITED

[1] Calistrat, Michael "What causes wear in gear-type couplings?" *Hydrocarbon Processing*, January 1975.
[2] Filepp, Les "Lubricant as a Coolant in High Speed Gear Couplings." *Journal of Lubrication Technology*, Transactions of ASME, January 1970.
[3] Calistrat, Michael "Extend Gear Coupling Life." *Hydrocarbon Processing*, January 1975.

ACKNOWLEDGMENTS

Figures 1, 6 and 7, Koppers Company, Power Transmission Division; Figures 5, 8 and 9, Imperial Chemical Industries Ltd.

Repair Techniques for Machinery Rotor and Case Damage

Only machinery in effective use can create profit. When failures occur, emergency repairs must frequently be made. Follow these guidelines to make effective repairs and get the machinery back in a profit-making condition.

Harry A. Erb, Consultant, Elliot Co., Greensburg, Pa.

TODAY'S MACHINERY can be characterized by rapid increase in range and complexity, as well as greater refinements and exactness in manufacture. These factors, together with advancing automation, demand a far greater versatility of the people charged with and responsible for the activities of installation, operation and maintenance.

In the past, an occasional malfunction of machinery resulting from any unusual condition (operating or otherwise) could perhaps be considered acceptable as an operator was probably available to quickly get the machine back into normal operation. In the largely automatic plant today, such malfunctioning of machinery is intolerable!

Even though many skilled operators perform these activities almost instinctively, aided and abetted by experience or repetition, both terms must be defined. "Experience" has been defined as learning by doing and/or observing by our senses as contrasted to learning by thinking. "Repetition" means repeating the same experience. For example, we all know many men who are credited with, say, 20 years' experience when in reality they have one month's experience repeated 240 times. These are the people who have had plenty of experiences but have not profited by them.

The point is that no one can learn by experience alone. We must think out the experience; otherwise, we only repeat. More simply, experience is often negative—finding out how not to do. Thinking is always positive—thinking out how to do. Positive experience after thinking out the problem and then taking action gives confidence to the individual; repeating the action gives skill.

Proper installation is the first and most important requisite for satisfactory operation of high speed rotating machinery—and this fact cannot be over-emphasized. The rugged and simple construction of this class of machinery permits its operation under adverse and unfavorable conditions, but it is of primary interest to the purchaser to provide the best conditions possible in order that maximum service at minimum cost can be obtained.

INITIAL PREPARATION AND PLANNING

Having assumed that the machinery is properly designed and carefully built, and assured of a proper and careful erection and startup, we will now outline the major repair techniques to take care of repairs to both the rotor and the stator. As indicated earlier, we want to consider repairs during a planned "turnaround"—not "random" repairs which are frequently done on an "emergency" basis and where techniques are sometimes used which are questionable and should only be used in emergencies. These will be covered later.

To plan for a "turnaround," one must be guided by the operating history of the given plant and, if it is the first "turnaround," by conditions found in other plants utilizing the same or similar process and machinery. This is how the time between subsequent "turnarounds" has been extended to three years or more in many instances.

By utilizing the operating history and inspections at previous "turnarounds" at this or similar installations, one can get a fair idea of what parts are most likely to be found deteriorated and, therefore, must be replaced and/or repaired, and what other work should be done to the unit while it is down. It should be pointed out that, with modern turbomachinery, items such as bearings, seals, etc., which are precision made, are seldom if ever repaired except in an emergency; such items are replaced with new parts.

This means that parts must be ordered in advance of the "turnaround" and other work must be planned for so that the whole operation will proceed smoothly—and without holdups which could have been foreseen. This usually means close collaboration with the manufacturer or his service shop representative so that handling facilities, service men, parts, cleaning facilities, inspection facilities, chrome plating and/or metallizing facilities, balancing facilities, etc., are available—and will be open for production at the proper time required.

DISASSEMBLY OF THE UNIT

A starting point in major repair operations is, of course, disassembly of the unit; and this gives us a rational starting point.

Hopefully, proper overhead cranes and laydown space have been provided. To this we must add ancillary equipment such as lifting rigs with at least three hand-operated chain hoists or equivalent so that the top half can be properly leveled. The clearances on turbomachinery are as small as possible so fairly precise leveling is a must. Secondly, the crane should be geared for very slow and steady hoisting or lowering; and this precludes the use of "long boom" mobile cranes although they are frequently used but are certainly not recommended.

As the top half of the unit is being raised, the gaps between top and bottom flange faces should be periodically checked to assure that the top half is indeed coming up square. Guide studs are generally recommended to assure a straight up lift after the top half is free from the dowels. Once the top half is free from the rotor, it can be lowered to nail-free cribbing and prepared to roll over.

To roll over such a unit, assuming that the length is greater than the width, it is preferable to use a crane with two hoists and roll over parallel to the axis of the crane using the main hoist and a single looped cable so attached to the casing that it cannot slip off as the top half rolls over center and starts downward. A similar looped cable should be attached to the other end and using the auxiliary hoist slowly lift the down end and, either by operation of the auxiliary hoist or by operation of the main hoist, approximately level and set the inverted top half on cribbing so that it can be conveniently worked on.

Depending on the internal conditions found, the diaphragm labyrinth or other packing should be removed and, if it shows no appreciable wear, cleaned up, inspected, and stored in marked packages in a floor area reserved for the parts.

Diaphragms should also be removed, cleaned up, inspected, and stored for reassembly. This same procedure applies to all other parts from both halves. Inspection should be visual, dimensional, and, if any cracks are "suspected" in the diaphragms, magnaflux-inspected as well.

After the casing seals have been removed, the rotor is ready to be lifted from the casing. A lifting rig is helpful here, but otherwise one end can be attached directly to the main hoist and the other end to the main hoist also but through a manual hoist to assure leveling as the unit is raised. Again the lift must also be square and a careful watch made as the rotor lifts off the bottom half bearings to assure that it is indeed coming up straight and square. The rotors in most cases are not so heavy as the other assembled stator parts and, consequently, can be guided by hand.

Stands should be available for lowering the rotor onto. The rotor should rest on the bearing journals which must be protected by soft packing or equivalent to avoid any marring of the journals.

If the rotor is reasonable clean, it is ready for visual inspection. The things to look for are journal scoring, packing rubs, wheel rubs, etc. If these appear to be very light, the rotor is ready for indication using a device as shown in Fig. 1, and using a ball axial locating device, i.e., a spring, say 5-10 pound spring scale, lightly compressed to hold the ball up against a ground flat surface such as a ¾-inch x ¾-inch lathe tool bit. The parts to be indicated are wheel wobble and wheel roundness at the packing surfaces and thrust collar run-out. The rotor shaft should be indicated at about the middle of the rotor and should run true about 0.002-inch to 0.004-inch T.I.R. Most manufacturers have established standards and have limits for these run-outs. These should be followed and, if any readings are questionable, the manufacturer should be consulted.

If the shaft does indeed have a permanent bow in excess of the manufacturer's limit or if there is evidence of wheel distress, the rotor must be disassembled. Similarly, if the journals or seal surfaces on the shaft are badly scored, disassembly in most cases is indicated.

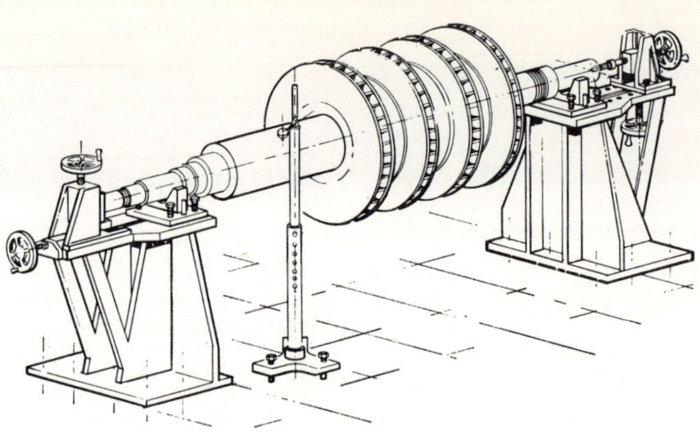

Fig. 1—Rotor testing fixture.

SHAFT STRAIGHTENING

Successful straightening of bent rotor shafts which are permanently warped has been practiced for the past 40 or more years, the success generally depending on the character of the stresses which caused the shaft to bend.

In general, if the stresses causing the bend are inherent in character, resulting from improper forging, rolling, heat treating, thermal stress relieving, and/or machining operations, then the straightening will generally be only temporary in character and will generally be unsuccessful.

If, however, a bent shaft results from stresses set up by a heavy rub in operation—by unequal surface stresses set up by heavy shrink fits on the shaft,—by stresses set up by misalignment,—or by stresses set up by improper handling, then the straightening will generally be permanent in character and may be attempted with good chances of success.

Before attempting to straighten a shaft, first determine, if possible, how the bend was produced. If the bend was produced by an inherent stress, relieved during the machining operation; during heat proofing or on the first application of heat during the initial startup; or by vibration during shipment—then straightening should only be attempted as an emergency measure, with the chances of success open to serious question.

The first thing to do, therefore, is to carefully indicate the shaft and "map" the bend or bends to determine exactly where they occur and their magnitude. In transmitting this information, care should be taken to identify the readings as "actual" or "indicator" values. With this information, a knowledge of the shaft material available, the method used for straightening can be selected.

For medium carbon steel shafts (carbon 0.30 to 0.50) three general methods of straightening the shaft are available. Shafts made of high alloy or stainless steel should not be straightened except on special instructions which can only be given for individual cases.

The peening method used for straightening carbon steel shafts consists of peening the concave side of the bend at the bend. This method is generally the most satisfactory where shafts of small diameters are concerned—say shaft diameters of 4 inches or less. It is also the preferred—in many cases, the only—method of straightening shafts which are bent at the point where the shaft section is abruptly changed such as at fillets, ends of keyways, etc. By using a round end tool ground to about the same radius as the fillet and a 2½-pound machinist's hammer, shafts which are bent in fillets can be straightened with hardly any marking on the shaft. Peening results in cold working of the metal, elongating the fibers surrounding the spot peened and setting up compression stresses which balance stresses in the opposite side of the shaft thereby straightening the shaft. The peening method is the preferred method of straightening shafts bent by heavy shrink stresses such as sometimes occur when shrinking turbine wheels on the shaft. Peening the shaft with a light (½ pound) peening hammer near the wheel will often stress-relieve the shrink stresses causing the bend without setting up balance stresses.

The heating method used for straightening carbon steel shafts consists of applying heat to the convex side of the bend. This method is generally the most satisfactory where shafts of large diameters are concerned—say shaft diameters of 4½ inches or more. It is also the preferred method of straightening shafts of very large diameters or of straightening shafts where the bend occurs in a constant diameter portion of the shaft (say between wheels); but is generally not applicable for shafts of small diameter or if the bend occurs at a region of rapidly changing shaft section. Because this method partially utilizes the compressive stresses set up by the weight of the rotor, its application is limited and care must be taken to properly support the shaft.

The shaft bend should be mapped and the shaft placed horizontally with the convex side of the bend placed on top. The shaft should be supported so that the convex side of the bend will have the maximum possible compression stress available from the weight of the rotor. For this reason, shafts having bends beyond the journals should be supported in lathe centers. Shafts with bends between the journals can usually be supported in the journals; however, if the bend is close to the journal, it is preferable to support the shaft in centers so as to get the maximum possible compression stress at the convex side of the bend. *In no event should the shaft be supported horizontally with the high spot on top and the support directly under the bend as this will put tension stresses at the point to be heated and heating will generally permanently increase the bend.* Shafts can be straightened by not utilizing the compressive stress due to the weight of the rotor but this method will be described later.

To straighten carbon steel shafts using the heating method, the shaft should be placed as outlined above and indicators placed on each side of the point to be heated. Heat should be quickly applied to a spot about 2 to 3 inches in diameter using a welding tip of an oxy-acetylene torch. Heat should be applied evenly and steadily carefully watching the indicators until the bend in the shaft has about tripled its previous value. This may only require perhaps 3 to 30 seconds, so that care must be taken to closely observe the indicators. The shaft should then be evenly cooled and indicated. If the bend has been reduced, repeat the procedure until the shaft has been straightened. If, however, no progress has been made, increase the heat bend as determined by the indicators in steps of about 0.010 inches to 0.020 inches or until the shaft spot heated approaches a cherry red. If results are not obtained on the third or fourth try, using heat, a different method must be substituted.

The action of heat applied to straighten shafts as described above is that the fibers surrounding the heated spot are placed in compression by the weight of the rotor; the compression due to the expansion; and the resistance of the other fibers in the shaft. As the metal is heated, its compressive strength decreases so that ultimately the metal in the heated spot is given a permanent compression set. This makes the fibers on this side shorter and by tension they counterbalance tension stresses on the opposite side of the shaft thereby straightening it.

The heating and cooling method of straightening shafts is especially applicable to large shafts, which cannot be supported so as to get appreciable compression stresses at the point of the bend. This method consists of applying extreme cold (say, using dry ice) on the convex side of a bend and then quickly heating the concave side of the bend. This method is especially applicable to straightening shaft ends beyond the journals or of large vertical shafts which are bent anywhere.

The action is that the shaft side having the long fibers is artificially contracted by the application of cold. Then this sets up a tension stress in the fibers on the opposite side which, when heated, lose their strength and are elongated at the point heated. This sets up compression stresses in the concave side which balance the compression stresses in the opposite side. Indicators should also be used for this method of shaft straightening, —first bending the shaft in the opposite direction from the initial bend, about twice the amount of the initial bend,—by using dry ice on the convex side,—and then quickly applying heat with an oxy-acetylene torch to a small spot on the concave side.

Various shafts of turbines and turbine-generator units have been successfully straightened using one method or another of straightening. These include several 5000-kw turbine-generator units, one 6000-kw unit, and many smaller units. Other manufacturers of turbines and other equipment have long used these straightening procedures. The same procedures have also been used by the U.S. Navy and other users to straighten shafts. With sufficient care, a shaft may be straightened to 0.0005 inches or less, (0.001 inch indicator reading). This is generally satisfactory.

RESTORING WORN, ERODED AND/OR CORRODED SHAFTS ON TURBOMACHINERY

Wear in machine parts is reduced by proper design, proper operating procedures, and proper preventive maintenance. In spite of this, certain parts on turbines and associated high speed rotating machinery shafts such as journal surfaces, packing ring surfaces, etc., may become worn, scored, eroded, and/or corroded to such a degree that refinishing of the surfaces is the only alternative to renewal of the shaft in order to restore the unit to a normal serviceable condition.

Within limits, these journal or packing ring surfaces of shafts can be refinished to a good surface by turning down and grinding to a good finish; but this immediately introduces the requirement of special odd-size bearings and/or shaft sealing rings, not generally carried in the machine builder's stock. Such special odd-size renewal parts are more costly than standard size parts—and in general cannot be furnished by the machine builder except on long term shipping basis. For these reasons the turning down of shaft journals or packing ring surfaces is not recommended except in an extreme emergency.

Fusion or braze welding cannot safely be resorted to to restore these worn or scored shaft surfaces to their original dimensions because of the possibility of shaft warpage. New shafts, even if separate from the other parts of the rotor, present problems in field disassembly and assembly, requiring specially skilled personnel and equipment. In addition, in many of these units the shaft is integral with the rotor, hence obtaining a replacement shaft is extremely costly and may be prohibitive from an outage time standpoint.

The progress in the art of metal spraying within the past 40 years has made it possible to safely and quickly restore these worn or scored shaft surfaces to their original dimensions, and this alone is prehaps the greatest single factor in reducing maintenance and repair costs of this class of equipment. Moreover, since most of these shafts are of ordinary carbon or alloy steel and are subject to corrosion and/or abrasion at the journal and packing ring surfaces, metal spraying of these surfaces with an abrasion- and corrosion-resisting metal may result in a restored shaft which is superior to the original shaft. In fact, many rotors are now built with the packing surfaces metal-sprayed with a hard stainless steel to eliminate corrosion and reduce abrasion at these surfaces.

To better understand the useful applications and limitations of sprayed metal, it is necessary to briefly examine its structure. Sprayed metal consists of tiny laminar particles joined together and to the parent metal by interlocking and oxide cemetation. The surface of the parent metal must, therefore, be prepared in such a way that it is covered with tiny anchorages into which the laminar particles can interlock themselves. The surface must also be free of rust, dirt, oil, grease, etc.

Sprayed metal is inherently more porous than the usual forms of the same metal and, therefore, minimum thickness coatings are required to make them impervious to moisture and corrosive agents which might attack the parent metal. The minimum thickness for packing surfaces on rotor shafts is approximately 0.030 inch.

The porosity of sprayed metal reduces its resistance to indentation but its abrasion-resistance is generally greater than the same metal in the usual forms. This makes it particularly applicable for packing and journal surfaces. Similarly, it is not suitable for applications under concentrated or very high compressive stress intensity as would occur on cams or cam rollers or on turbine shafts under the turbine wheels. Also, in general, metallizing is not recommended for applications under more moderate compressive stress—but which are required to transmit torque. Therefore, it is not to be used to build up shafts or couplings to restore coupling fits or under gears or pinions to restore those fits.

Under special circumstances, metallizing may be used to restore shafts under light anti-friction bearing fits; but it is usually simpler and generally more satisfactory to use chrome plating for this purpose as the amount to be built up is usually only of the order of a few thousandths on the radius. Chrome plating is, therefore, the recommended method to be used for building up shafts, couplings, turbine wheels, or blower impellers to restore fits; **provided** that the plated surfaces are accurately ground to size; and **further provided** that the final ground size radial thickness of the chrome plating does not exceed 0.007 to 0.010 inch. Chrome plating for radial thickness in excess of these may require multiple chrome plating operations at, say, 0.015 inch steps with intermediate grinding operations. This extends the time of chrome plating and should be investigated before any chrome plating is attempted.

Chrome plating may also be used to restore journal or packing surfaces to original size provided that the wear or abrasion or erosion is limited so that the chrome plate does not exceed the limit of 0.007 to 0.010 inch radial depth.

The reason for limiting the radial thickness of chrome plating is because if heavy plating is attempted the end junction of the plating will usually result in a "V" notch. Although this is of no serious engineering consequence, many customers and insurance inspectors will raise objections based on the appearance of this "V" notch.

A way of minimizing this "V" notch effect is to join the bottom of the underground surface to the shaft surface with a double fillet. This, however, extends the length of the chrome plated area; and this must be checked if there are many proximity vibration pickups in the area of chrome plate. We recently had a case where the shaft was chrome plated under the pickup and showed a runout of 0.0025 inch. We were unable to explain this until the chrome plate was discovered.

Therefore, the principal application of metallizing on turbines and associated rotating machinery is for building up and restoring to original size, packing and journal surfaces which have been worn or abraded beyond the limits for successful application of chrome plate. This can be done but the rotor must be moved to a shop equipped with metal spraying, turning, and grinding equipment. Assuming that the shop has adequate equipment and a skilled spray gun operator, the two factors more responsible for a successful metallizing job are proper preparation of the surface and keeping the surface clean after it is prepared. In humid atmospheres, the metal spraying should be done immediately after the surface is prepared. In certain instances metallizing troubles have developed by permitting the prepared surface to stand for a matter of several hours before metallizing. Even finger marks on a prepared surface have resulted in faulty jobs. Therefore, the requirement for cleanliness cannot be emphasized too strongly. Here again, be careful to investigate whether or not there are proximity vibration pickups in the area of the metallizing. We have no direct experience on how much this would affect the readings, but, if possible, move the vibration pickups so that they are at least ⅛ inch axially from the metallized or chrome plated area.

The above general notes are intended to cover only the application where metallizing is recommended and to give a brief description of the technique. The work should be done only in shops which have considerable experience in metallizing.

Where an insurance carrier is involved, it is advisable to consult the insurance carrier before any work of this nature is attempted. In general, insurance carriers will approve metallizing; but specific cases should always be brought to their attention before the work has been done.

OTHER ROTOR REPAIRS

Other rotor repairs such as cracks in welded centrifugal impellers, blade cracks in turbine or axial flow impellers, erosion and/or corrosion wear in general require new impellers, new blades, etc. Unless spare rotors or spare wheels are available, this generally requires repairs at the factory or at an authorized service shop in the local area.

In some cases, minor cracks in welded impellers can be repair welded with appropriate heat treatment subsequent to welding. Here again this requires special knowledge and facilities usually only available at the factory or at an authorized service shop.

If the "turnaround" sequence has been established by the results of previous inspections, it is usually possible to have the long-term items available at the local service shop so that rotor repairs can be completed without undue delay. In the event that spare rotors are provided, the rotor to be repaired can be done in normal sequence to be ready for the next scheduled turnaround in case it is needed.

STATOR REPAIRS

Stator repairs are usually associated with erosion and/or corrosion or from failures resulting from improper installation, operation and repair, or faulty handling.

For stator repairs, it is difficult to generalize as conditions vary over a wide range. We will, therefore, only include what might be called emergency repairs to stator parts such as fractured diaphragms and/or casings. These have on occasions been repaired by Metalock, but this requires a Metalock expert. In some cases where steel parts are involved, welding might be employed. Actually for fractures, specific repair techniques can only be specified after a study of the actual failure.

BALANCING ROTORS

We will not go into the details of balancing rotors except to state that all elements of a rotor should first be balanced individually—and then fully assembled; and the assembled rotor balanced in three planes.

The residual dynamic imbalance should be corrected at the ends of the rotor, and the remaining residual static imbalance should be corrected at about the middle of the rotor.

Most manufacturers today give allowable limits for residual imbalance. Here is a simple expression developed from $CF = Mr\omega^2$ in case it is not specified.

Residual unbalance in inch ounces for 20 percent gravity is

$$\text{Inch ounces} = \frac{0.1125}{\left(\frac{N}{1000}\right)^2} \times W$$

where N is maximum normal rpm of rotor and W is rotor weight in pounds.

Incremental balancing of a rotor after all the parts have been individually balanced is not recommended because of the well-known fact that the true center of the rotor moves about in a random fashion as heavy shrunk wheels are assembled on the shaft. If it were not for the random movement of the true center, an assembled rotor would not require any balance correction. Experience shows this is not true, so with incremental balancing of wheels as they are assembled—the first one may move toward, say, 12:00 so that metal must be taken off at 12:00. When the next wheel is shrunk on, the whole shaft may move toward 6:00; so now the balancing done on the preceding wheel must be counter-balanced on the second wheel, remembering that after a wheel is balanced no further balancing is permitted on it. This can only result in couples which certainly make this method open to question.

CONCLUSION

The basic repair techniques apply largely to disassembly and assembly of turbomachinery units and to disassembly and assembly of turbomachinery rotors. The assembly of seals, thrust bearings, etc., must follow the manufacturer's instructions. You will note that we have to a degree downgraded experience. This is because an experienced man is more likely to not follow laid down techniques than a completely inexperienced man who has no way to go except to implicitly follow written techniques.

It is not the purpose of these notes to train maintenance men to repair turbomachinery although these notes were originally made up for this purpose. Our purpose is to outline these repairs to acquaint operating people of what can be done if repairs become necessary for any reason and how these repairs can be properly handled. The most important point, in my judgment, is if excessive maintenance costs are encountered—don't start out by replacing these parts until the reasons for the failure of these parts has been determined.

The basic philosophy is "correct the cause if at all possible—do not attempt to minimize the effect."

ACKNOWLEDGMENT

Abstracted from a paper originally presented at the Second Turbomachinery Symposium, Texas A&M University, College Station, Texas, October 1973.

How to Control Compressor Noise

Understanding possible sources of compressor noise is the first step in dealing with the problem. And, consideration must be given to the economics of available solutions. Here is a discussion of the pertinent facts.

G. M. Diehl, Ingersoll-Rand Research, Inc., Phillipsburg, N.J.

LARGE NUMBERS of compressors, of many different types, are used in all major industries and on all large construction sites. It is almost impossible to imagine steel production, refineries, chemical plants, generating stations and large process industries without compressors.

Many types of compressors have been identified as major noise sources. This is not because they produce higher sound levels than other classes of machinery, but because their levels are fairly high and so many are used in industry.

The Environmental Protection Agency has designated portable air compressors the second noisiest class of construction equipment, right after medium and heavy trucks, for the same reason, and has written "Noise Emission Standards for Construction Equipment—Portable Air Compressors:"

Compressor sound control requires an understanding of the noise generating process, and various techniques have been found to be effective in reducing the noise of centrifugal, axial, axi and reciprocating compressors.

COMPRESSOR NOISE SOURCES

Noise radiated from a compressor is complex and consists of components from many sources. In order to reduce total noise, the various contributions must be identified and evaluated, and the largest ones worked on first. The ideal approach is to prevent generation of noise by better designs, but this is not always the most economical solution. Noise reduction after a machine has been built is often the most practical procedure.

Turbulence. Consider turbulence for example. Turbulence is the most important source of noise in centrifugal compressors. It is really a combination of two effects, (1) vortex shedding and (2) up-stream turbulence. The boundary layer over each blade is turbulent by the time it reaches the trailing edge. Turbulent layers on the top and bottom surfaces produce a fluctuation in lift, and this fluctuation has a broad frequency spectrum. Application of a fluctuating force to a gas generates sound at the same frequency. Therefore, broad band noise is radiated. If flow is turbulent when it enters a blade row, turbulence is increased and noise is greater.

Turbulence noise is radiated through the compressor casing, and it can be controlled by an acoustic enclosure—after the compressor has been installed. It is almost impossible to eliminate turbulence by design.

Piping noise, produced by the same source can be reduced effectively by lagging the pipes with 2 to 3 inches of Fiberglas. Ultracoustic, Rockwool or similar material with a density of about 4 pounds per cubic foot, and covering this with a jacket weighing about 1 pound per square foot. The jacket can be #24-gauge steel, or the equivalent weight of aluminum, lead or leaded-vinyl.

To be most effective, enclosures and pipe lagging must be tight. Leaks in an enclosure greatly reduce its effectiveness. It is unfortunate that a small leak in a high quality enclosure is more damaging than the same leak in a poor enclosure. For example, an opening with an area of 1 percent of a wall whose transmission loss is 50 dB reduces overall transmission loss to only 20 dB. A 1 percent leak in a 25dB wall results in a final transmission loss of 19dB. That is, the effectiveness of the 50dB enclosure is reduced by 30dB while effectiveness of the 25dB enclosure is reduced by 6dB. This shows that if you plan to have leaks in the enclosure there is no point in paying for a high quality one.

Installation of inlet and discharge silencers is another example of effective noise reduction after a compressor has been built. Silencers reduce the noise entering inlet and discharge pipes and makes piping noise reduction easier. In some instances, pipe lagging is not necessary, depending, of course, on the final noise level required. When silencers are used they should be as close as possible to compressor inlet and discharge flanges. *It should be noted that it is almost impossible to make a significant reduction in the noise from centrifugal compressor installations unless the piping is treated.*

Interaction of rotating and stationary vanes. On the other hand, centrifugal compressors have other major noise sources which can be reduced by design. An example of this is noise produced by interaction of rotating impeller blades with stationary vanes.

Every time a blade passes a given point, the air or fluid at that point receives an impulse. Therefore, that point will receive impulses at a frequency equal to the number of impeller blades times the number of revolutions per second. In axial flow compressors the magnitude of this blade-passing frequency component is one of the largest in its generated sound spectrum. It is present also in centrifugal compressors with diffusers, but in most cases it is not as important as the blade-rate frequency. This is calculated as follows:

$$f = \frac{N_r \times N_s}{K} \times \frac{\text{revolutions}}{\text{second}}$$

where

f = Frequency in Hertz (1 Hertz = 1 cycle/sec.)
N_r = Number of rotating (impeller) blades
N_s = Number of stationary (diffuser) vanes
K = Highest common factor of N_r and N_s

For example, let $N_r = 6$, $N_s = 8$ and the speed equal to 6,000 rpm. Then $f = \frac{6 \times 8}{2} \times 100 = 2{,}400$ Hz. That is, there are 24 times in each revolution when impeller blades line up with diffuser vanes, and each time this happens, 2 impeller blades match 2 diffuser vanes. Therefore, the frequency will be 24 times the number of revolutions per seconds, and the pulses will be of double strength.

When there are 6 impeller blades and 9 diffuser vanes, there are 18 times when impeller blades are in line with diffuser vanes. Each pulse is 3 times as strong as it would be with a single coincidence, because 3 rotating blades match 3 stationary vanes. The frequency is 18 times the number of revolutions per second.

Obviously, combinations like 12 and 12 are not recommended because of the many points of coincidences and the strength of the pulses.

It can be seen that it is better to use unequal numbers of rotating and stationary vanes. Prime numbers are the best of all because they have no common factor. This produces high frequency pulses which are easier to control than low frequency ones. Furthermore, when prime numbers are used, each pulse is only of single strength.

Impeller-diffuser distance. Increasing radial distance between impeller blades and diffusers reduces noise. It is particularly effective in reducing blade-passing frequency and blade-rate components. Unfortunately, this procedure also decreases performance, but for close initial spacing, the decrease in noise is greater than the decrease in performance. That is, noise increases rapidly as the spacing becomes smaller and smaller.

Effect of horsepower. Centrifugal compressor noise is affected by many operating parameters. There is a direct relation between horsepower and noise, but the relation is not the same for all types. On one particular class, the overall noise can be predicted quite accurately by the following:

$$\text{Increase in dB} = 17 \log \text{horsepower ratio}$$

In most instances, doubling the horsepower results in an increase of about 4 to 5 dB in overall noise.

Effect of speed. Rotational speed has a definite effect on noise. For any particular design, sound level will increase anywhere from 20 to 50 times the logarithm of the speed ratio. At lower speeds, centrifugal compressor noise will increase about 20 log speed ratio. At high speeds, the increase will more nearly equal 50 log speed ratio. Increase in sound with speed applies to overall noise and to the component of highest level—usually blade-passing frequency or blade-rate frequency. The increase at other frequencies is not as great, and may be of the order of 10 to 15 times the log of the speed ratio.

These same relations apply to impeller tip speeds. But, in general, less noise will be produced with large diameter, slow speed units than with small diameter, high speed machines, even though the impeller tip speeds are the same in both cases. There are several reasons for this:

- Not all turbulence is produced by the impeller. Even though tip speeds are the same, the slow speed machine will have larger areas in internal passages with lower velocities and less restriction. Extra care in producing fine interior finish in centrifugal compressor casings to reduce noise is not justified. There is no detectable difference when passages are hand finished.
- Mechanical forces due to unbalance are proportional to the square of the speed. Low speed will produce less structural vibration.

Number of stages. Noise generated by centrifugal compressors can be reduced by decreasing the work per stage —that is, by increasing the number of stages.

Effect of gas molecular weight. Molecular weight of gas in a compressor system has a pronounced effect on generated noise. Very little test data are available on this, and it is difficult to predict mathematically. It is certain though that more noise is produced with high molecular weight than with low molecular weight gas.

Head-capacity operating point. Mass flow and discharge pressure both have a profound effect on noise produced by a compressor. As mass flow is reduced noise decreases until a point near surge is reached. Beyond this point noise increases rapidly.

Inlet and discharge. The inlet to a centrifugal compressor plays an extremely important part in the generation of noise and pressure pulsations.

It is sometimes thought that if the inlet and discharge areas are increased, gas velocities will be reduced and, therefore, less noise will be generated. On the contrary, if the inlet opening is increased, it may actually create more noise rather than reduce it. In order to obtain proper entrance, the gas should be as near as possible to the impeller center line. This, of course, indicates a small inlet. When gas enters near the center line on its way to the impeller vane there will be relatively low shock and turbulence. If the suction opening is enlarged, gas is admitted farther up on the impeller vane where linear speed is higher. A sudden change from low velocity in the large inlet to high velocity part way up on the vane causes shock, turbulence and increased noise.

Reducing flow velocity in discharge piping is an effective way to reduce piping noise. This should not be accomplished by enlarging the opening at the cutwater, but by using a properly designed pipe increaser between the correct discharge opening and the piping.

AXI COMPRESSORS

Axi compressors are helical rotor, positive displacement compressors. Gas enters the intake ports and flows into a pocket formed between the rotors and the wall of the casing.

The pocket rotates away from the intake ports, and as the lobes and grooves roll into each other the pockets shorten and compress the gas. It moves axially and is carried toward the discharge end.

The major noise source in compressors of this type is the discharge. Next in importance is the inlet, and then noise radiated from the compressor casing. Sound pressure level measured in the discharge of larger, higher horsepower units may be in the range of 140 to 145 dB. Without silencers or an enclosure, sound level three feet from the compressor may be about 132 dBA, and it consists of a fundamental frequency equal to the number of lobes times revolutions per second, plus higher order harmonics of this frequency.

The amount of necessary sound control depends on final acoustic design criterion. For maximum noise reduction, both discharge and inlet silencers should be installed, as close to the compressor as possible. Inlet and discharge piping and inlet and discharge silencers should be lagged, and an acoustic enclosure should be installed on the compressor itself to reduce casing radiation.

RECIPROCATING COMPRESSORS

In general, the sound level of reciprocating compressors is not very high. It usually consists of multiples of piston movement and is generated by both aerodynamic and mechanical forces.

Inlet and discharge noises are major components, but pulsating flow can be reduced effectively by a chamber-type silencer or snubber.

Inertia forces are major exciting factors causing vibration and noise in reciprocating compressors. These forces are due to the motion of pistons and related parts and imbalance of the connecting rod and crank mechanism. Forces produced by unbalanced masses also appear in rotating parts of the machine, as both static and dynamic unbalance.

Impacts in the crank-connecting rod system and knocking of pistons against cylinder-liners during cross-over, are major sources of noise in reciprocating compressors. During each revolution of the crankshaft, the piston shifts several times from one side to the other moving in the plane of connecting rod motion. The gap between piston and cylinder liner permits the piston to move with a certain velocity in the transverse direction, impacting against the cylinder wall. These knocks produce an intense vibration of the cylinder walls at their resonant frequency.

This discussion is confined to some of the major noise sources in compressor systems and does not include those attributed to motors, turbines, gears or valves. It does not include foundation noise either, even though improper mounting on an inadequate foundation is often the cause of very high vibration and noise.

ACKNOWLEDGMENT

Originally presented at the 79th National Meeting, American Institute of Chemical Engineers, Houston, March 16-20, 1975.

Reclaim Compressor Seal Oil

Sour seal oil from compressors can be reclaimed by sparging with air or nitrogen. Mass transfer data are given for design.

Heinz P. Bloch, Exxon Research and Engineering Co., Florham Park, N.J.

SERIOUS corrosion attack on the babbitt-lined seal rings of a compressor can be caused by contaminated seal oil if the process gas is sour—contains hydrogen sulfide. The compressor seal oil, after passing through the inner seal sleeve, comes into contact with the process gas and absorbs H_2S in the general region of the sour oil drain cavity.

Three measures are available to circumvent corrosion attack on compressor seal elements: inject buffer gas, discard the sour seal oil or purify the sour seal oil. The first measure requires a low-cost, reliable supply of buffer gas injection. However, available buffer gas supplies often prove costly or unreliable. Discarding of sour seal oil, which is normally practiced at an average rate of 20 gallons per compressor stream day and quantities reaching as high as several hundred gallons per day under failure conditions, will also present serious ecological and cost problems. If discharged to the sewer, it adds to the cost of operating the effluent cleanup equipment. Sour seal oil purification thus stands out as the most desirable measure.

Sparging with air or nitrogen into a sour seal oil collection vessel is an effective, low cost means of purifying

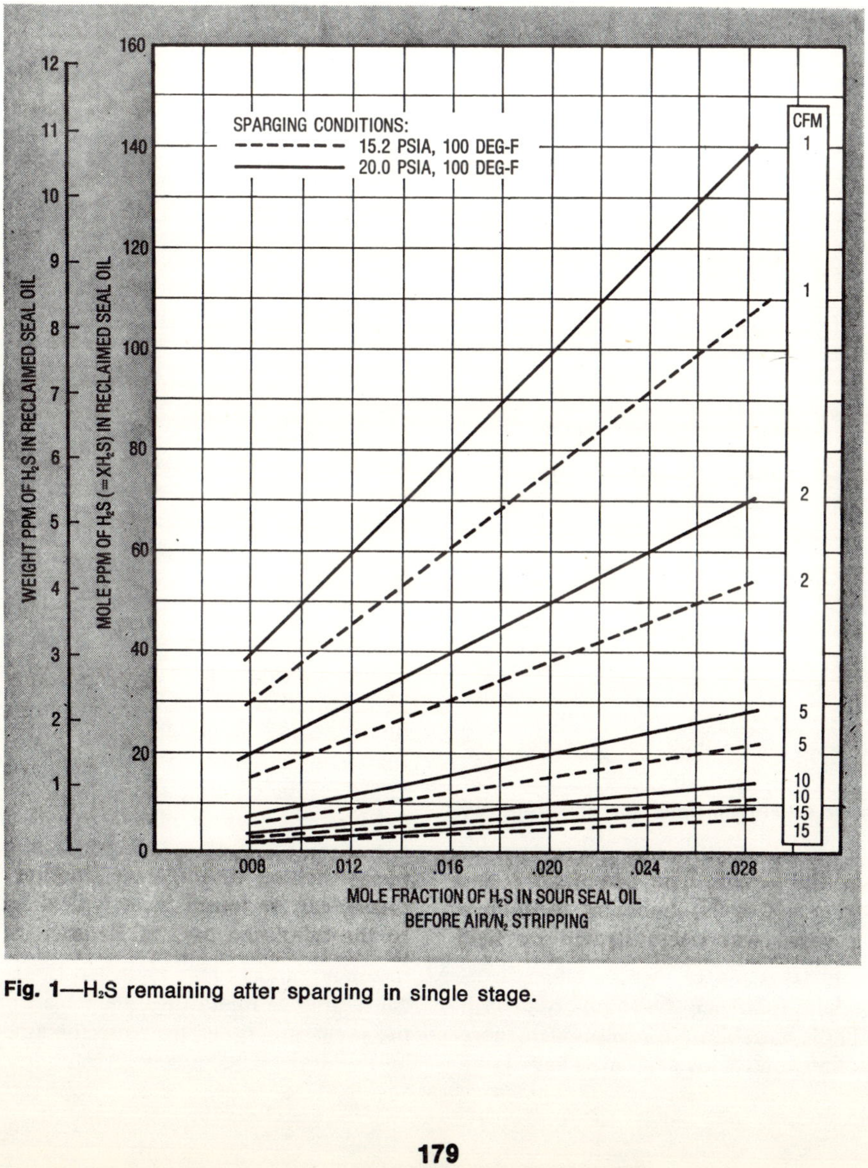

Fig. 1—H_2S remaining after sparging in single stage.

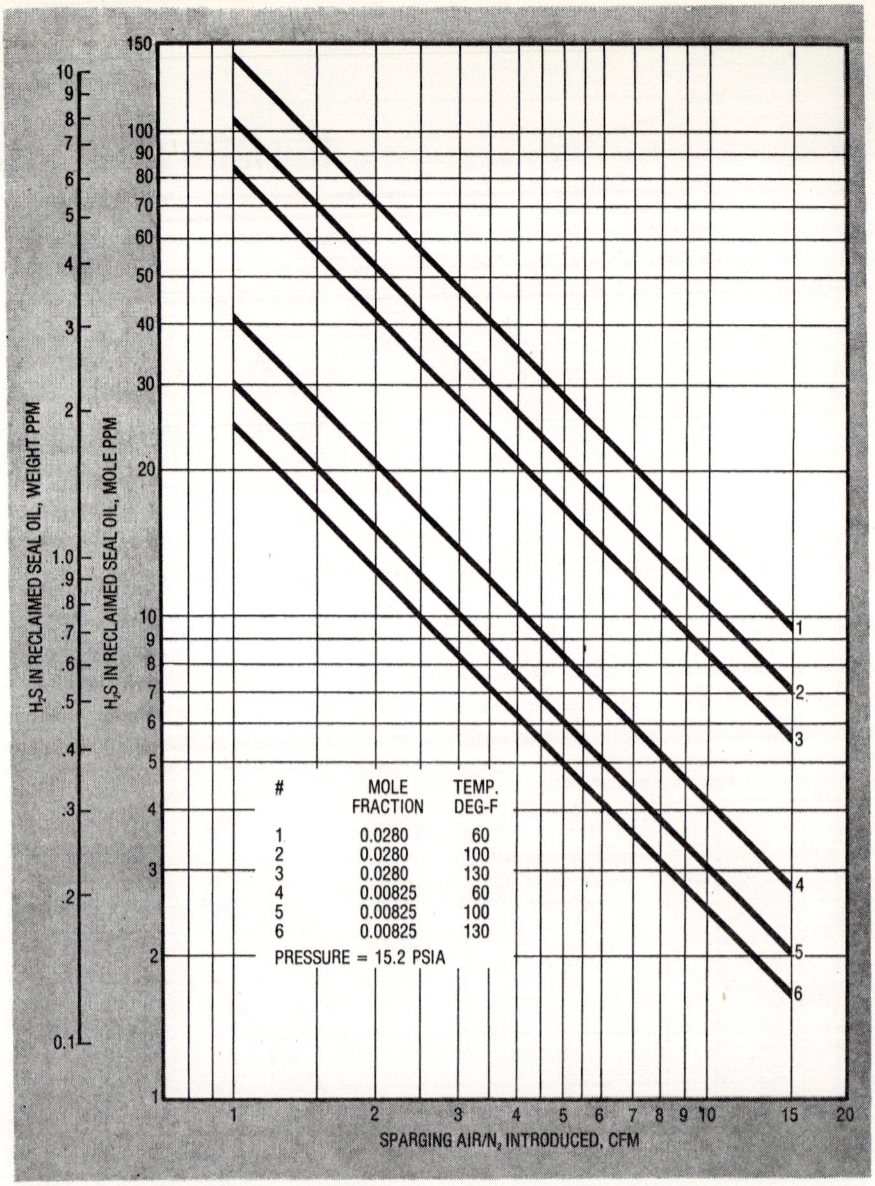

Fig. 2—To show temperature effects.

sour seal oil. Since the solubility of H_2S in lube oil and water is low, air/N_2 sparging can be employed to strip out the H_2S. As air or N_2 is continuously injected into the liquid and removed from the vapor space, the partial pressure of H_2S is lowered, which in turn reduces the H_2S concentration in the liquid. The practicality and effectiveness of sparging air or N_2 bubbling through a batch-type collecting vessel was demonstrated by field testing and subsequent sample analysis. This work further led to the development of correlations for continuously operating purification systems employing the sparging principle.

Seal oil contamination. The total amount of H_2S dissolved in the seal oil is influenced by the surface area over which gas-liquid contact takes place. Two variables, temperature and pressure, also influence the solubility of H_2S gas in the seal fluid. These variables are investigated more closely to see if contacting conditions can cause very large amounts of H_2S to dissolve in pure seal oil and also pure water. The solubility in water is of interest because condensed water vapors are present in many lube or seal oil reservoirs.

The K-value, or mole fraction of H_2S in the vapor phase divided by the mole fraction of H_2S in the liquid phase, can be found from typical solubility charts similar to the tabulated data of Reamer, Sage and Lacey.[1] The chart can be entered at a contacting pressure and a vertical line drawn to a given (or assumed) contacting temperature. The approximate K-value can then be read off the upper horizontal scale.

If the contacting pressure is maintained constant, an increase in contacting temperature will cause a decrease in the amount of H_2S absorbed by the seal oil. At constant contacting temperature, a decrease in pressure will have the same effect, i.e., decrease the amount of H_2S absorbed by the seal oil. Changes in solubility are roughly propor-

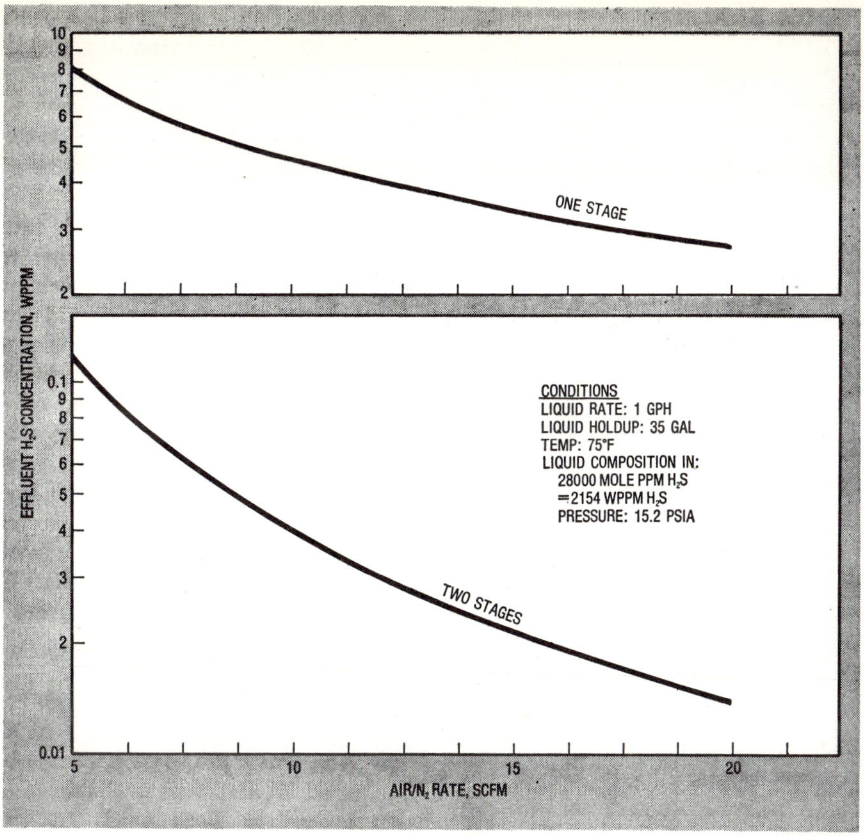

Fig. 3—Two stages much better than single stage.

tional to changes in temperature or pressure. Some high pressure compression services operate in pressure regions where the mole fraction of H_2S in the liquid phase is equal to or greater than the mole fraction in the vapor phase. The seal oil could thus become severely contaminated.

The solubility of H_2S in water was determined from Henry's Law Constants for H_2S in water. Results were quite similar to those obtained for H_2S dissolved in oil. As in the case of H_2S solubility in seal oil, changes in the solubility of H_2S in water were found to be roughly proportional to changes in temperature or pressure.

Conditions for sparging. No data are available regarding maximum safe allowable H_2S concentrations in lube oils. However, an arbitrary limit of 0.5 weight parts per million (wppm) is considered practical and achievable. There is some controversy as to whether or not this value is below the threshold of perceptible odor; but, in any event, it is highly unlikely to seriously endanger compressor component life. Assuming a seal oil molecular weight of 440, 0.5 wppm of H_2S equals 6.5 molal parts per million (mppm) in the seal oil. Compared with literature data,[1] we note considerably higher equilibrium K-values to exist at near-ambient pressures than at the higher contacting pressures known to prevail in most compressor seal regions.

At 15 psia and approximately 105°F, the molal concentration of H_2S in the vapor phase would have to be $y \leqq$ (K) (6.5) (10^{-6}), or 195 mppm. At ambient pressure, near ambient temperature contacting conditions anticipated in the sour seal oil collecting vessel, partial pressures not exceeding (195×10^{-6}) (14.7), or 0.00287 psia correspond to the 0.5 wppm which we arbitrarily consider "maximum permissible." Using a similar calculation procedure to determine levels in water, a partial pressure of 0.0037 psia would correspond to the 0.5 wppm limit in water with disengagement conditions again maintained at near ambient.

Our goal is to achieve low partial pressures of H_2S in the sour seal oil collecting vessel since this would result in the desired low concentration of H_2S dissolved in the liquid. A series of equilibrium flash calculations allowed us to determine how much H_2S would remain in the reclaimed sour seal oil if sparging air or N_2 were introduced into single stage and two-stage stripping drums. The calculations were performed for a series of assumed air/N_2 flows, pressures and initial H_2S concentrations. They showed that the H_2S content of seal oil could be reduced below 0.5 wppm, as desired, by sparging air or N_2 through the sour oil at a sufficiently high temperature. The necessary temperature (60-130°F) and air or N_2 (1-15 scfm) depend on the initial concentration of H_2S.

Effectiveness of stripping can be represented graphically as in Fig. 1 which shows the residual H_2S concentration in reclaimed seal oil as a function of the original concentration of H_2S in the oil. The data were plotted

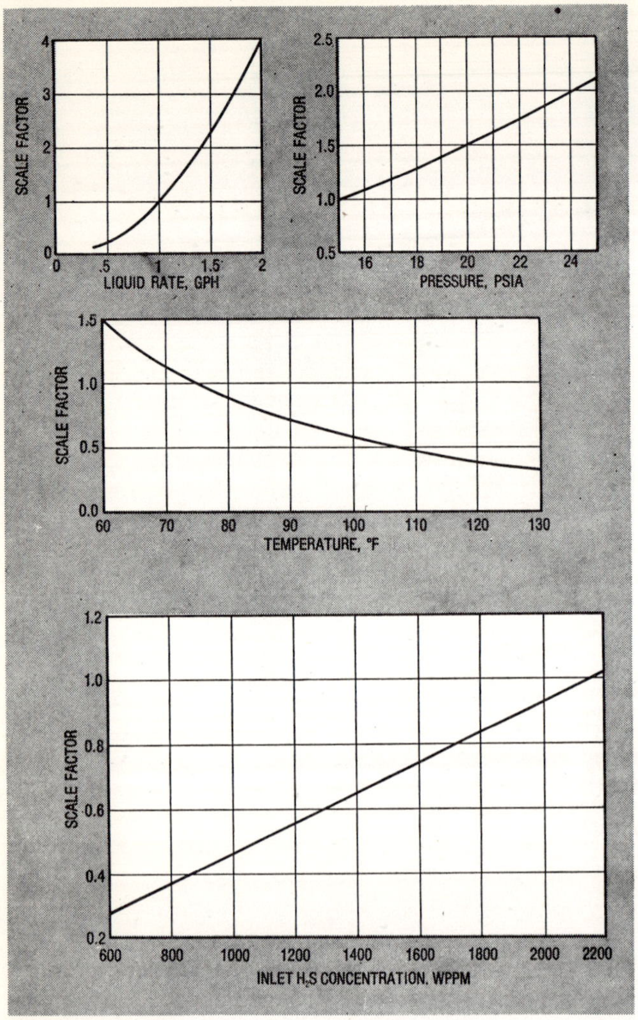

Fig. 4—Scale factors for two stage conditions.

for a single equilibrium stage and the graph illustrates how sparging at different pressure levels or at different air/N_2 injection rates affects the degree of oil purification achievable.

The significance of operating in different temperature environments can be estimated from Fig. 2. H_2S-disengagement is more effective at higher temperatures. However, air or N_2-sparging at temperatures much higher than 130°F would risk expulsion of somewhat volatile anti-oxidation additives in the oil and should be avoided.

A comparison of effluent H_2S concentrations at various sparging rates for single stage and two-stage operation is shown in Fig. 3. These graphs were plotted for seal oil entering the disengagement vessel with an H_2S concentration of 2100 wppm. The far more effective disengagement capability of two-stage vs. single stage operation is immediately recognized from the graphs.

The effluent composition scale factors shown in Fig. 4 were developed to prorate sparger vessel performance to conditions other than those given on Fig. 3. This takes into account the effects of liquid rate, pressure, temperature and inlet condition. Note that the scale factor is 1.0 in each case for the given conditions of Fig. 4. To use these factors, a number picked off any of the plots in Fig. 4 is to be multiplied by the effluent H_2S concentration given for two stages in Fig. 3. When two or more of the conditions are different, the scale factors should be multiplied together to produce a combined factor. These factors take into account both equilibrium considerations and mass transfer fundamentals. They should be considered valid for collecting drums having a diameter range from 2 to roughly 4 feet.

Application summary. H_2S can be expelled from compressor seal oil by bubbling air or N_2 through the oil at flow rates which are to be governed by the volume of contaminated oil and time available to perform this reclamation procedure. Both batch and continuous processing steps are feasible. Batch processing lends itself to the purification of centrally collected volumes of sour seal oil whereas continuous processing is generally recommended for individual collecting points. Continuous processing, if used, should employ two processing stages because this method will reduce the H_2S content in the reclaimed seal oil to values well below 0.5 wppm even if processing conditions (temperature, air flow rate, pressures) should fluctuate.

Safety considerations should govern the selection of N_2 versus air as a sparging medium. While the majority of sour gas compression services will produce H_2S concentrations in the seal oil outside of explosive limits, we cannot overlook another very important fact; namely, that there are possibly other light hydrocarbon gases or hydrogen dissolved in the sour seal oil. These gases may cause a significant lowering of the flash point of the seal oil and thus materially alter the explosivity limits. Moreover, air bubbling through liquid has been shown to generate static electricity. The more violent the agitation, the greater the electric charge. As a consequence of these concerns, the gas composition of a given sour gas compression service should be established and the solubility of light hydrocarbon and hydrogen constituents determined. These data should then be reviewed to determine if air sparging is acceptable for the service under consideration. If air sparging is judged a safety risk, nitrogen sparging should be employed.

ACKNOWLEDGMENT

The writer is indebted to Dr. C. J. Colwell and Dr. M. F. Randrup of ER&E's Fractionation and Thermodynamics Section for valuable assistance in mass transfer rate analysis and multistage stripping efficiency determinations.

LITERATURE CITED

[1] Reamer, Sage and Lacey, "Experimental Data for the Solubility of H_2S in Pentane and Decane," *Ind. Eng. Chem.*, Vol. 45 (1953), pp. 1805-1812.

How to Troubleshoot Large Industrial Fans

Industrial fans, or blowers, are vital to many processes in the HPI. If a problem occurs, it must be diagnosed and solved without delay. Here are some proven new diagnostic techniques to consider.

J. W. Martz and **R. R. Pfahler,** Structural Dynamics Research Corp., Cincinnati, Ohio

THE USE OF large fans, or blowers, in chemical processing, petrochemical, power plant and other industry applications has experienced a tremendous growth in the past ten years. Much of this growth has been brought about by demand for cleaner air. Associated boilers, furnaces and related systems have produced increasingly hotter, more corrosive and particulate-matter-laden gases to be moved. With the handling of more demanding gases in greater volumes have come larger, faster fans which have been accompanied by certain design and operating problems. Structural Dynamics Research Corp. (SDRC) has been involved in some 40 troubleshooting and/or design projects on large fans and fan systems during this period. Testing and analysis procedures developed by SDRC under sponsorship of United States Steel Corp., Pittsburgh, Pa., have widespread application among designers, manufacturers and users both in improving designs and in correcting malfunction and failure problems of in-service equipment.

Fans dealt with are generally used in two major functional applications: forced draft or induced draft. Forced draft fans are used to push air through a furnace, boiler or process apparatus. Induced draft fans are exhaust fans —drawing gases from the process—usually through pollution control equipment. Both types of fans are usually found with one of four common impeller arrangements—paddle wheel, radial tip, double inlet airfoil and single inlet fans (Fig. 1). Fans to be discussed have impeller diameters up to 166 inches and may weigh up to 10,000 pounds. Operating speeds range from 600-1,800 rpm, depending upon the system and its use. Fig. 2 shows the design of a typical system, listing important components. Generally, these units are driven by large AC motors although some recent designs are turbine driven. Nearly 90 percent of these systems use direct gear coupling, with the remainder having hydraulic fluid couplings for variable speed applications. These systems can develop problems that are manifested in different ways.

PROBLEMS IN FAN DRIVE SYSTEMS

There are many variables in fan design and, hence, potential problems. One problem area deals with design of fans for flow, efficiency and performance. A second group is concerned with noise associated with installations. A third is expressed in terms of ductwork vibration and panel weld cracking. And still another, most important to this article, relates to the structural integrity of the fan system and components.

Most structural problems appear in two specific areas: (1) system vibration and, (2) blade or wheel cracking and failure. System vibration is concerned with axial, lateral or vertical vibration of fan rotor and impeller. Blade failure relates to steady state stresses, upon which are sometimes superimposed oscillating stresses in excess of yield or fatigue limits of the material.

System vibration and its causes. In a normal fan system there are two factors that bear on vibration. These are the dynamic forces and system dynamic characteristics.

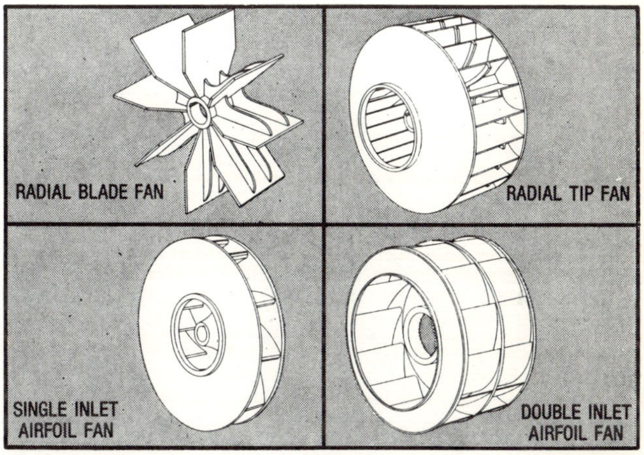

Fig. 1—Common impeller arrangements for large industrial fans and blowers.

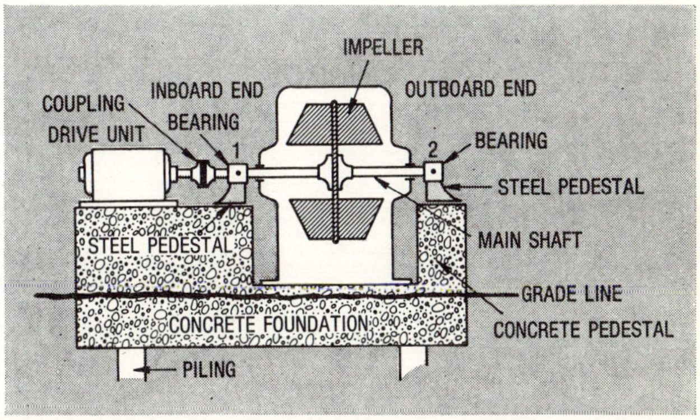

Fig. 2—Layout of a typical industrial fan system.

- Motor torque transients
- Motor rotor imbalance
- Coupling misalignment
- Bearing self-induced fluid film forces
- Fan impeller imbalance
- Aerodynamic forces
- Forces from nearby machinery

Fig. 3

Dynamic forces, as shown in Fig. 3, are those forces that are time varying or frequency dependent, as opposed to static or deadweight forces. Numerous sources of dynamic force inputs affect a fan drive system such as motor torque transients (particularly at start-up), imbalance forces from the motor rotor, misalignment or improper lubrication of coupling, fluid film self-excited forces from bearings (particularly in higher speed units), impeller imbalance, aerodynamic forces and inputs from nearby machinery.

The most predominant dynamic forcing function is imbalance of the fan impeller. This forcing appears once per revolution at operating speed, but sometimes exhibits harmonics at the two or three-per-rev frequency. This can be caused by a shift of the impeller on the shaft or, more commonly, from a buildup of waste material on the impeller.

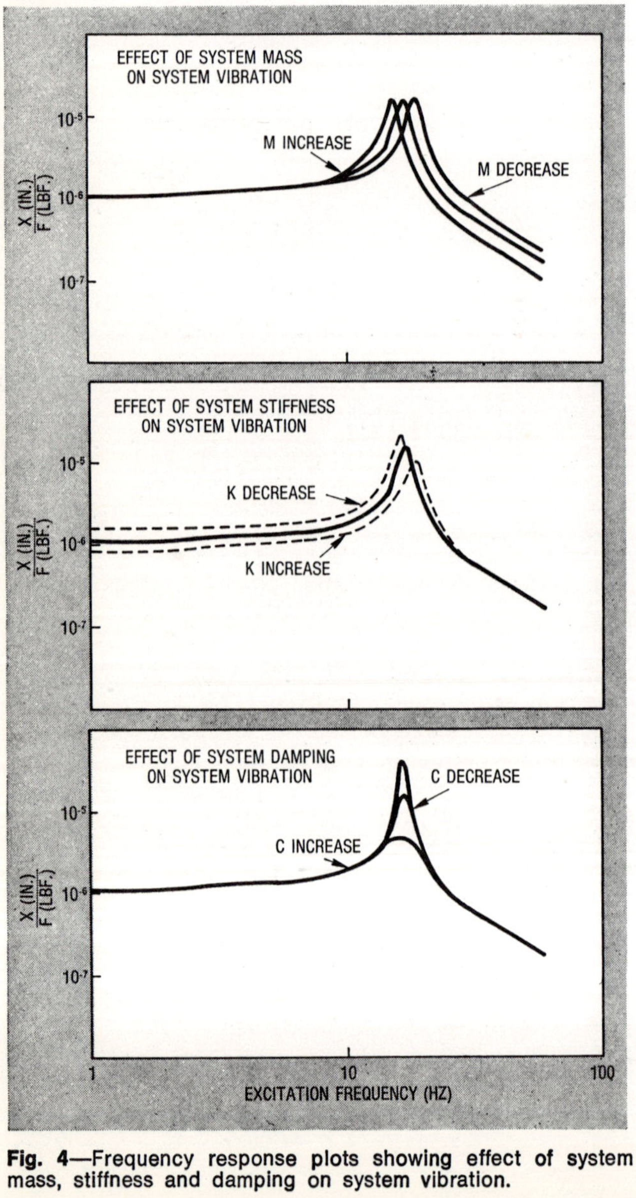

Fig. 4—Frequency response plots showing effect of system mass, stiffness and damping on system vibration.

Aerodynamic forces are important in looking at axial vibration, particularly in double inlet fans. Here, aerodynamic forces can be severe enough to rock the entire system in the axial direction. These dynamic force inputs, acting in combination with dynamic characteristics, will determine resultant vibration of the fan system.

The system's dynamic characteristics are determined by its stiffness, mass and damping. These are known properties in a system and are determined at the design stage.

A part of the dynamic characteristics are attributable to the system's natural frequencies. Fig. 4 graphically illustrates this concept. This illustration represents measured lateral vibration divided by a known input force. Measurements are from a fan bearing block observed during an onsite test of a large fan system.

Practical aspects of stiffness, mass, damping. Most large fans are designed to possess a "stiff-shaft system." In such a system the unit operates below the first natural frequency. A fan system does not have a single natural frequency, obviously. Rather, it has an infinite number of them. In most systems the first mode of vibration is the only one of great concern, but other modes can be significant. Considering this kind of system, there are numerous sources of stiffness, mass and damping acted on by the input force.

The first major stiffness is fan shaft and hub lateral bending as shown in Fig. 5. Next would be the bearing fluid film stiffness, affected by bearing length and diameter, oil viscosity, temperature and clearances. Moving down, the next stiffness is the steel pedestal and sole plate. Next is an area found to be extremely troublesome in fan system stiffness—the interface stiffness of the steel pedestal and concrete pedestal. This is affected by movement in the sole plate, anchor bolts or wear at this point. The same situation has been noted between the bearing block and steel pedestal. Finally, there is bending stiffness of the concrete pedestal and foundation and its relative position and strength in the soil (Fig. 5). Rotor-impeller and one-half the coupling are the only principal masses. Foundation mass is important only in the combination of a stiff shaft and weak foundation that exhibits considerable motion. Major damping is in the bearings' fluid film. Past that point there is damping from friction between moving parts in the pedestal-sole plate-concrete juncture and, finally, damping of soil interface with the foundation. All these sources of stiffness, mass and damping can now be applied to a practical "fix" of the problem.

Using stiffness, mass, and damping for problem solving. Fig. 6 represents the same base plot as Fig. 4. An operating speed in Zone 1 would dictate attacking system stiffness as a means of reducing vibration. This may be accomplished by changing factors that provide the stiffness. The hub and shaft may be difficult to modify effectively in an existing system without computer modeling, but bearings present a ready source of improving stiffness. A larger bearing surface could be incorporated or the clearances reduced, or, simply, any other change to arrive at a stiffer bearing could be made. Analysis may show a weak pedestal that needs stiffening or looseness at the sole plate indicating need for replacing shoe bolts or regrouting. Foundation stiffening may also be needed.

Zone 2 is the area where all three factors affect compliance. Natural frequency is important here. Particularly if the fan is operating at a speed just below natural frequency, a combination of decreasing mass and increasing stiffness should be considered to help push natural frequency above operating speed. Here, state-of-the-art damping treatments, including optimized constrained-layer material and tunable dynamic absorbers, have been used successfully on various kinds of smaller fan systems and other rotating equipment to reduce system vibration.

In Zone 3, operation is above natural frequency and thereby is mass dependent. An increase in mass is warranted to lower both compliance and natural frequency. Because of the possibility of overstressing blades through mass added to the impeller, Zone 3 operation warrants strain gage testing and possible finite element computer analysis. Zone 3 operation is generally unfavorable to performance, since the unit must pass through natural frequency to reach operating speed. Such operating parameters can best be determined through field testing of the system.

Field testing techniques. Numerous factors, many of which have been discussed, can cause system vibration problems. Thus, the first step in field testing is to define the cause of vibration. The second objective of field testing is to analyze data in order to recommend a solution.

The initial step is to determine the operating zone of the fan. This requires mounting a series of accelerometers and proximity probes on fan system components. Accelerometers measure vibration of the pedestals and foundation in various directions, while proximity probes monitor the action of the shaft across the bearing fluid film. Fig. 7 shows typical instrumentation of an inboard bearing block and shaft for deflection measurements.

Output from these transducers can be wired to several different signal analyzers for evaluation. For example, a real time analyzer (RTA) converts raw vibration data from a vibration versus time signal to a vibration versus frequency signal. A signature ratio adapter (SRA), used in conjunction with the RTA, locks in on one component of the vibration signal and tracks it across the frequency spectrum. Using the RTA signal along with a tachometer signal, the SRA can lock in on any vibration component and follow its amplitude as the fan is brought to operating speed. The key factor in testing is a narrow-band frequency analysis. This analysis determines what distinct frequency components make up total system vibration. Readout from the RTA is a series of peaks of different amplitude at each frequency comprising the signal. These frequencies are the ones the SRA will track across the spectrum. A new testing tool even more powerful than the RTA is the digital fourier analyzer (DFA). Using its own built-in mini-computer, the DFA can be used even more effectively than the RTA by being programmed for additional analyses. This programming can be done onsite as data are observed by the test engineer. It takes only a few minutes to complete. The DFA is also used for determining vibratory mode shapes, which will be discussed later.

The next step is to perform speed sweeps, starting and

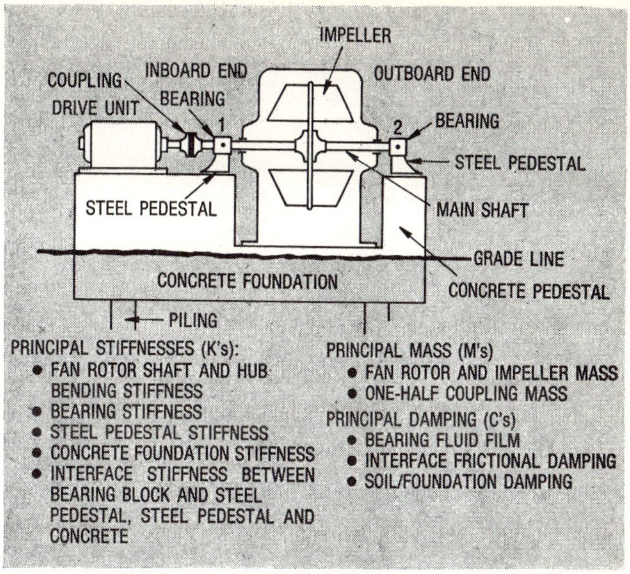

Fig. 5—Principal stiffness, mass and damping elements in a fan system.

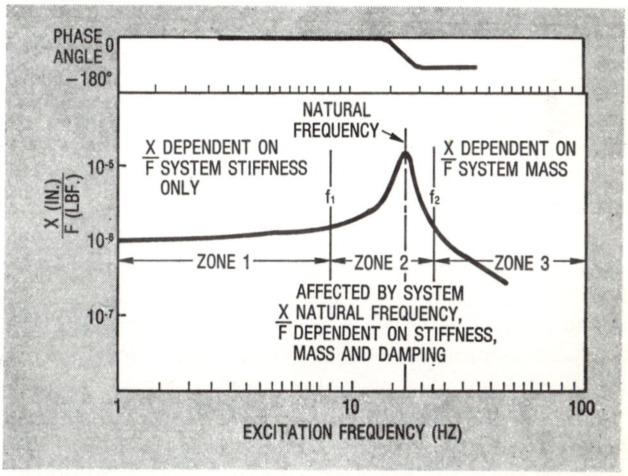

Fig. 6—Frequency response plot showing three operating speed zones. Each zone exhibits its own dependencies for affecting vibration.

stopping the fan several times, while watching frequency components already identified. The SRA performs order tracking in which it locks in on each frequency of interest and calculates component amplitude of deflection as speed changes. In other words, it tracks each vibration source over the entire speed spectrum. This technique determines natural frequency and fan operating zone.

Once the operating zone and natural frequency have been identified, the system can be analyzed to determine an acceptable modification. Stiffness, mass and damping must be thoroughly evaluated. At this point, operating deflection shape, which shows the relative stiffness of each component and helps decide which stiffness to change, is determined.

Data are taken with the fan operating normally. If the fan speed is fairly close to the first natural frequency, deflection shape assumed by the system is representative of the first mode shape. Test setup consists of accelerometers placed at various fixed points on the structure and proximity probes on moving components to measure relative deflection of shaft to bearing.

Resultant signals are processed in a mini-computer to

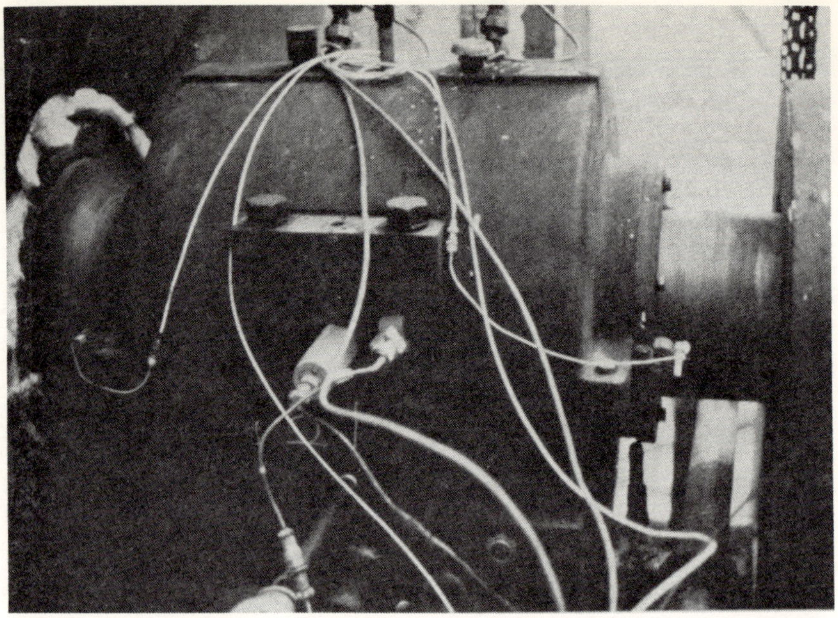

Fig. 7—Typical inboard fan bearing block instrumented for vibration and deflection measurements.

plot the deflection shape which may reveal certain obvious system weaknesses. These weaknesses are the elastic members used to stiffen or to move the first natural frequency away from the operating speed (see Fig. 8).

As stated earlier, data could reveal a large amplitude across the bearing fluid film, and the bearing might be modified or replaced. Farther down the structural path, the test could indicate a discontinuity across the sole plate, pinpointing a location to check for looseness or design inadequacy. Finally, the foundation may be rocking, where changing the bearing or stiffening the pedestal would not help.

Another type of testing is forced response measurement. This involves putting a known force into the system and measuring the response of each component. A reliable technique for gathering such data is impact testing, but this method requires sophisticated instrumentation to make sense of the data. However, this is offset by using recorders so data can be analyzed later under ideal conditions. There are inherent advantages to this method because the test can be performed in the field without using large electro-mechanical or hydraulic exciters. This method is far superior to others for cost, accuracy and speed.

The most common impact testing technique uses an instrumented heavy hammer to provide input force. A load cell in the hammer measures input force, and an accelerometer on the pedestal measures transient response of the system. Fed to a digital fourier analyzer, these signals show force level across a certain frequency band and displacement as a function of frequency. Force divided by displacement is the stiffness.

After the field test, observed conditions and operating parameters are more thoroughly examined by further analysis of measured data. The objective, of course, is to be able to make intelligent recommendations for a redesign or "fix" without trial-and-error modifications to the actual system. Testing is directed to this end.

WHEEL AND BLADE FAILURES

Wheel and blade failures fall into three major areas: material properties, quality control and engineering design problems.

Basically, material properties concern ability of the fan to resist erosion, corrosion and effects of expansion. Quality control problems are related to weld defects, welding procedures and materials. Engineering design problems have two components: (1) steady state conditions and, (2) combined steady state and dynamic or alternating stresses.

Fatigue stress problems can be divided into two classes: low-cycle fatigue and high-cycle fatigue. Low-cycle fatigue is caused by a high, static stress existing in the design. This occurs only in a relatively small number of cycles in the life of the structure and can occur in systems subjected to an unusually high number of starts and stops.

High-cycle fatigue occurs when a high steady-state stress (at or near yield point of the material) exists in combination with low-amplitude vibratory stresses. Because these vibrations occur at high frequencies, the wheel experiences several million cycles in a short time. This also can occur when steady-state stress is well below yield, but vibratory stresses have high amplitude.

Blade failure analysis. Analysis for blade failures follows generally the same approach as that for system vibration but there are notable additions. The first step in determining the cause of a wheel or blade failure is to perform an operating strain gage test. Purpose of the test is to measure steady-state and oscillating strain values at discrete locations on the blade.

A strain gage converts the mechanical parameter to an electrical signal that is proportional to strain in the metal and then transmits it to a recorder and an RTA. On fan systems this is accomplished by use of miniature FM radio transmitters mounted on the rotating element (see Fig. 9). These transmit an accurate signal to a receiver that relays it to a tape recorder, strip chart

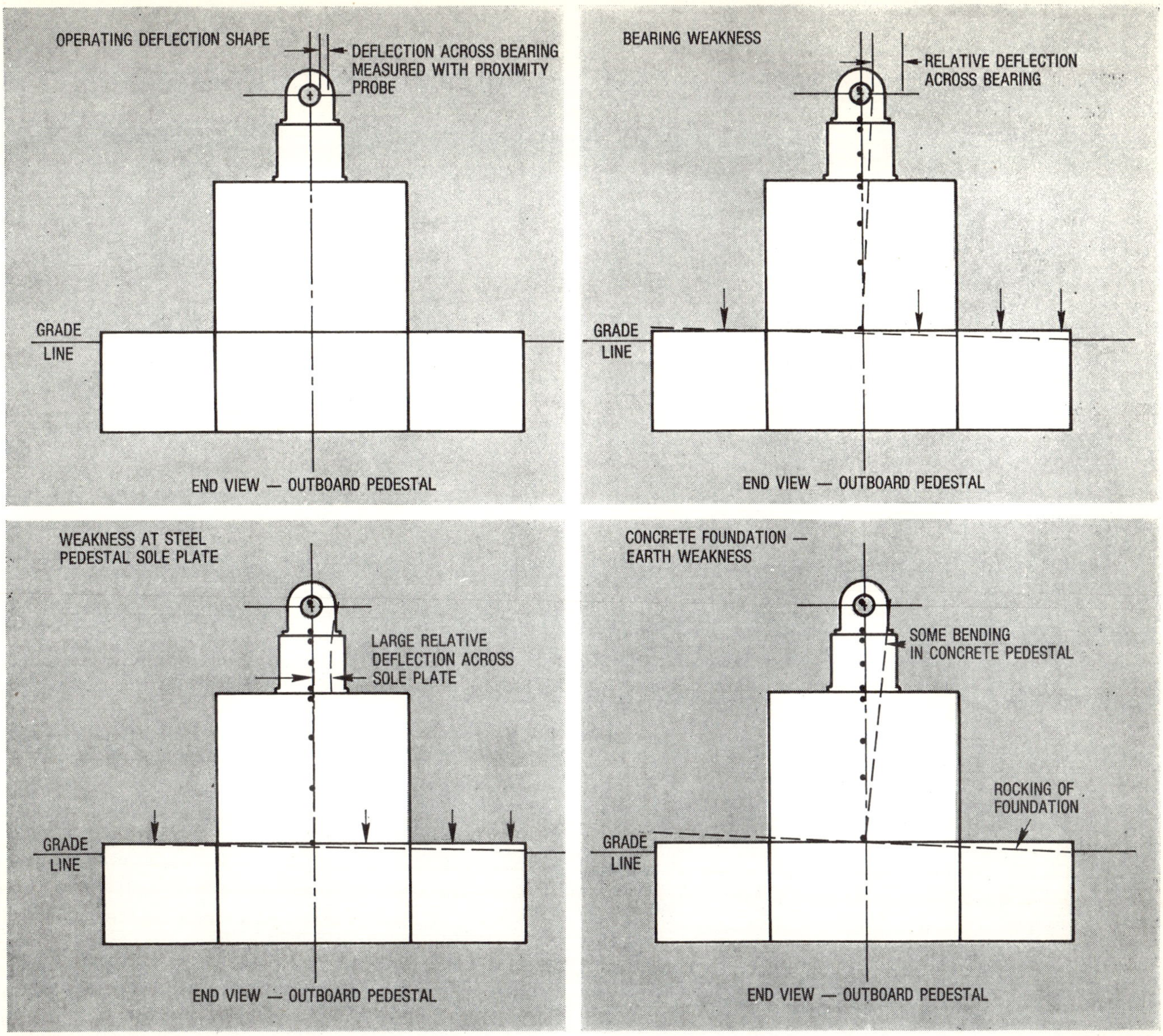

Fig. 8—Various component weaknesses revealed by testing.

recorder or RTA. Strain gages are placed in areas of high stress, normally where cracking has occurred. This procedure is an exacting process, particularly because of pressures, temperatures and general environment inside the fan system. Strain is monitored through a start-up. Oscillation is observed as well as evidence of dynamic strain (usually peaks coming into and out of the signal). If these are seen, analysis for frequency content is made with the RTA. This information permits an accurate analysis at minimum cost and in the shortest possible time.

The next step is impact testing of the wheel to find its natural frequencies. Mode shape measurements are taken to see how the blade acts at those frequencies. Mode shapes are actual shapes the wheel components take on at their natural frequencies. These shapes help determine causes of high stresses that have contributed to failure. Fig. 10 represents various typical mode shapes.

These mode shapes are developed by mounting accelerometers at various points on the blade sufficient to represent its shape. Relative displacements at each of these points, in response to a known input force at frequencies of interest, are used to define mode shapes relative to the original shape.

With all of this information, a finite element computer model of the wheel can be constructed and computer

Fig. 9—Miniature FM transmitters used to send strain signals from an operating fan. Sending antenna is behind fixturing ring.

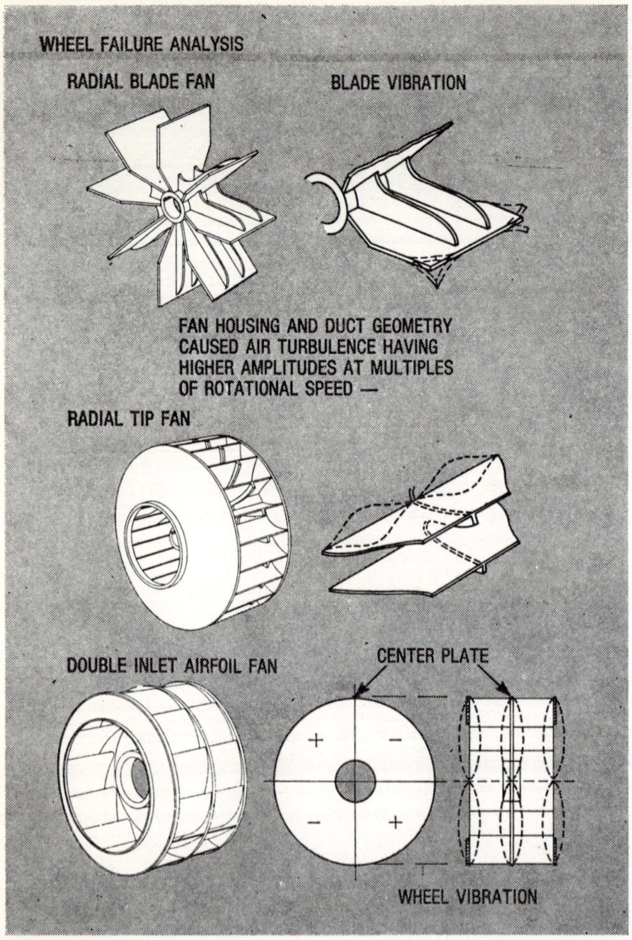

Fig. 10—Various mode shapes, observed during impact testing for blade and wheel failure, can occur at the first and subsequent natural frequencies of the component. Different mode shapes of the same component will occur at different natural frequencies.

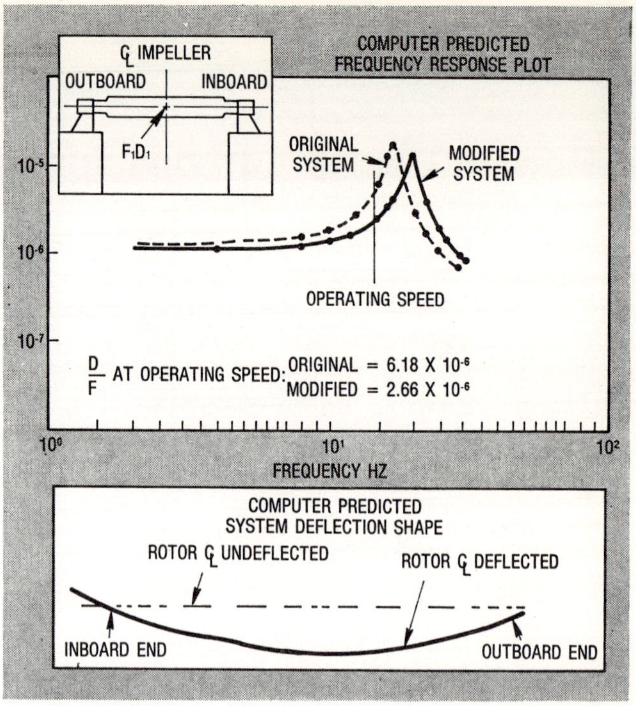

Fig. 11—Results of computer vibration analysis of a fan system. When correlated with actual test results, this provides a powerful tool for modification of design.

output matched to measured steady-state stresses. Once the model is verified, it can be used to evaluate potential modifications.

Computer aided analysis. Finite element analysis is a valuable tool in modeling a fan wheel and in predicting its response. The first step in constructing the model is to determine a representative section of the wheel. The analyst must decide how little of the wheel can be studied while still being able to predict action of the whole wheel. Usually, this means a single blade and its support structure must be modeled.

Next, this representative sector is divided into a grid of small, plate-like finite elements. Each particular element is defined to the computer by its material properties, stiffness and geometry. By setting certain boundary conditions on the model, it can be made to simulate the entire wheel. Computer analysis of fan component vibration is shown in Fig. 11.

Next is a computer analysis. The computer calculates area, volume and mass of each element. It determines the radius of each element and applies centrifugal and other loads to the center of each. The computer calculates stresses and plots stress patterns automatically. It then draws an incremental plot showing how the section of the wheel is deformed by loadings imposed. This plot points out areas where the structure should be modified to relieve stress. Finally, the computer analysis is rerun with recommended changes incorporated to see if conditions improve. Only when the paper model acts satisfactorily is the wheel physically modified.

Computer analysis provides accurate predictions of stresses and deflections of components. These are both time and cost effective in evaluating changes in design, whether the purpose is to eliminate a problem of vibration or stress or to modify the unit for greater output or more severe conditions.

How to Uprate Turboequipment by Optimized Coupling Selection

Compressor and driver shafts often prove to be overstressed in equipment uprate situations. Here's how a change from conventional gear type couplings to diaphragm couplings can lower shaft stresses sufficient to avoid shaft replacement during power uprates of turboequipment.

Heinz P. Bloch, Exxon Chemical Co., Baytown, Texas

EQUIPMENT uprate projects involving turbomachinery often *appear* to require time consuming and costly shaft replacements. However, closer examination of how the equipment vendor arrived at his maximum allowable stress levels may frequently show that such shaft replacements can be avoided without undue risk if the coupling selection is optimized. This conclusion is based on the fact that gear-type couplings have the potential of inducing in a shaft both torsional stresses *and* bending stresses, whereas diaphragm couplings tend to primarily induce torsional stresses and insignificant bending stresses at best.

The economic incentives of finding ways of salvaging major rotating equipment shafts are illustrated on a steam turbine shaft originally rated to transmit 17,600 hp maximum at 6,400 rpm. When uprating the turbine to 19,600 hp was being considered, the required changeout of stationary steam path components was to cost around $60,000, but a combined replacement cost of about $500,000 was quoted for the main and spare rotor shafts. A rigorous calculation of shaft stresses showed the shaft factor of safety to be greater at 19,600 hp using a diaphragm coupling than at 17,600 hp using a conventional gear coupling.

MAXIMUM SHAFT STRESS CAN BE CALCULATED

Efforts to determine whether or not this equipment power uprate required shaft replacement were preceded by shaft stress calculations. The coupling end of the steam turbine shaft used in our example incorporated the dimensions shown in Fig. 1.

With the shaft material ASTM A-293 Class A steel, heat treatment and stabilization at 1,000° F had resulted in the following properties:

Ultimate strength in tension .... $\sigma_{ut} =$ 105 000 psi

Ultimate strength in shear .... $\tau_{ut} =$ 60,600 psi

Endurance limit in shear .... $\tau_E =$ 30,300 psi

Minimum yield strength in tension ..... $\sigma_{yp} =$ 80,000 psi

Minimum yield strength in shear ...... $\tau_{yp} =$ 40,000 psi

Endurance limit in tension .... $\sigma_E =$ 52,500 psi.

Most of these properties are used in a Soderberg diagram[1] similar to Fig. 2.

In addition to some of the nomenclature given earlier for the shaft properties, the diagram uses σ_m (= steady tensile stress component), σ_a (= alternating tensile stress component), and k_f, the stress concentration factor for the particular keyway dimensions shown in Fig. 1. The stress concentration factor for our sample keyway is 2.9.[2]

Line B_1A_1 is considered to define the limiting stress condition for a specimen with stress concentrations. If the steady stress on the specimen is given by the absicca OD, then the limiting amplitude of the alternating stress is given by the ordinate DC, and point C will represent the limiting stress condition. The corresponding safe condition will be represented by point F with coordinates σ_m and σ_a (or τ_m and τ_a). These coordinates are obtained by dividing the coordinates of point C by the factor of safety n. From the similarity of triangles we have

$$OB_1A_1 \mathrel{\hat{=}} DCA_1, \therefore \frac{\overline{CD}}{\sigma_E/k_f} = \frac{\overline{DA}}{\sigma_{y.p.}} = \frac{\sigma_{y.p.} - \overline{OD}}{\sigma_{y.p.}},$$

$$\text{or } \frac{\overline{CD}}{\sigma_E/k_f} + \frac{\overline{OD}}{\sigma_{y.p.}} = 1$$

Dividing this equation by n, we obtain for the safe stress condition (Point F)

$$\frac{\sigma_a}{\sigma_E/k_f} + \frac{\sigma_m}{\sigma_{y.p.}} = \frac{1}{n}$$

$$n = \frac{1}{k_f\dfrac{\sigma_a}{\sigma_E} + \dfrac{\sigma_m}{\sigma_{y.p.}}} \quad (1)$$

This is the expression for the factor of safety in the case of **uniaxial stresses.**

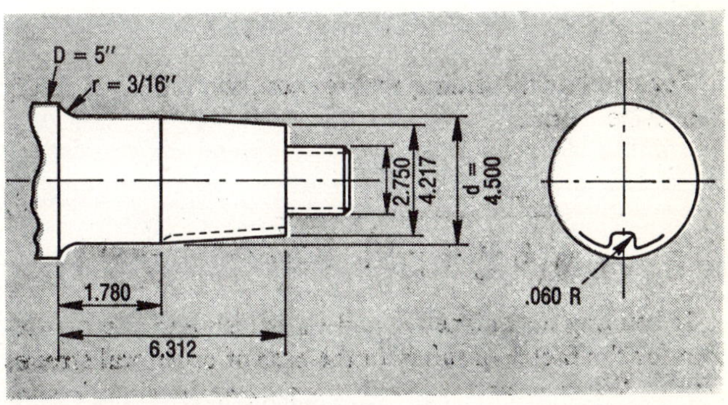

Fig. 1—Shaft dimensions.

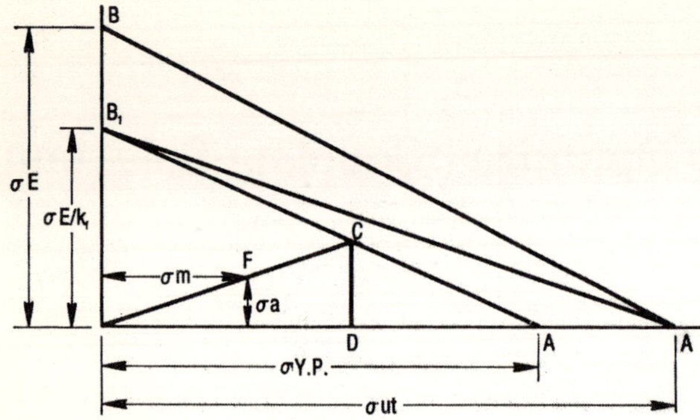

Fig. 2—Soderberg diagram.

In the case of combined stresses, the equivalent stresses for σ_a and σ_m should be substituted in the previous equation and literature citation 1 shows how the equation for **torsion** reduces to Equation 2. Using $k_{f'}$, instead of k_f

$$n = \frac{1}{\sqrt{3}\left(k_{f'}\frac{\tau_a}{\sigma_E} + \frac{\tau_m}{\sigma_{y.p.}}\right)} \tag{2}$$

For **combined bending and torsion,** the factor of safety is obtained from

$$n = \frac{1}{\sqrt{\left(k_f\frac{\sigma_a}{\sigma_E} + \frac{\sigma_m}{\sigma_{y\cdot p.}}\right)^2 + 3\left(k_{f'}\frac{\tau_a}{\sigma_E} + \frac{\tau_m}{\sigma_{y.p.}}\right)^2}} \tag{3}$$

If bending acts alone, τ_a and τ_m vanish and the expression for the factor of safety in the case of unilateral stresses results. When σ_a and σ_m vanish, we have the equation for n in torsion.

TORSIONAL STRESSES CAN BE READILY CALCULATED

For the desired uprate conditions given earlier, we calculated shaft torque output

$$T = \frac{(63{,}000)\ (19{,}600\text{ hp})}{6{,}400\text{ rpm}} = 193{,}000\text{ lb.-in.}$$

The steady torsional stress is, therefore,

$$\tau_m = \frac{16T}{\pi d^3} = \frac{(16)\ (193{,}000)}{\pi(4.5)^3} = 10{,}900\text{ psi.}$$

Making the generally accepted assumption that the alternating torsional stress will not exceed 20 percent of the steady stress, we obtained

$$\tau_a = (0.2)\ (10{,}900) = 2{,}180\text{ psi.}$$

BENDING MOMENTS ASSESSED FOR GEAR COUPLING

There are three relevant bending moments caused by a gear coupling when transmitting torque with angular or parallel misalignment (an additional moment, due to overhung weight of coupling, is usually negligibly small and was omitted from the bending moment analysis for both gear and diaphragm couplings):

• Moment caused by shift of contact point. This moment acts in the plane of angular misalignment and tends to straighten the coupling. It can be expressed as

$$M_c = \frac{T}{D_p/2}\left(\frac{X}{2}\right)$$

Where T is the shaft torque, D_p is the gear coupling pitch diameter and X is the length of tooth face (see Fig. 3).

• Moment caused by coupling friction. This moment acts in a plane at right angle relative to the angular misalignment. It has the magnitude

$$M_F = T\mu$$

where μ is the coefficient of friction.

• Moment caused by turning torque through misalignment angle α. It acts in the same direction as the friction moment M_F and can be expressed as

$$M_T = T\sin\alpha$$

The total moment is the vector sum of the individual moments;

$$M_{\text{TOTAL}} = \sqrt{M_c^2 + (M_F + M_T)^2}$$

A gear coupling suitable to transmit 19,600 hp at 6,400 rpm was assumed to have a pitch diameter $D_p = 9$ inches and a face width of 1.3 inches. Using a coefficient of friction $\mu = 0.3$ and a misalignment angle $\alpha = 0.057°$ (approximately 0.001 in./in. parallel offset), we calculated

$$M_c = \frac{193{,}000}{4.5} \times \frac{1.3}{2} = 27{,}878\text{ lb.-in.}$$

$$M_F = (193{,}000)\ (0.3) = 57{,}900\text{ lb.-in.}$$

$$M_T = (193{,}000)\ (\sin 0.057) = 193\text{ lb.-in.}$$

$$M_{\text{TOTAL}} = \sqrt{(27{,}878)^2 + (57{,}900 + 193)^2} = 64{,}435\text{ lb.-in.}$$

This moment is fixed relative to the actual angular misalignment α and is seen by the shaft as a rotating bending moment at a once-per-revolution frequency.

BENDING MOMENTS CALCULATED FOR CONTOURED DIAPHRAGM COUPLING

The contoured diaphragm coupling causes two bending moments:

• Moment caused by angular misalignment which results in bending the diaphragm,

$$M_B = k_B\alpha$$

In this expression, k_B equals the angular spring rate of the diaphragm (lb.-in./degree) and α is the misalignment angle. This moment acts in the plane of angular misalignment, as did M_c in the gear coupling analysis.

- Moment caused by turning torque through a misalignment angle α. It can be expressed as

$$M_T = T \sin \alpha$$

The total moment is now

$$M_{TOTAL} = \sqrt{M_B^2 + M_T^2}$$

A suitable contoured diaphragm coupling has a diaphragm diameter of 16.5 inches and an angular spring rate $k_B = 18{,}800$ lb.-in./degree. Thus, for the misalignment angle used earlier,

$$M_B = (18{,}800)\ (0.057) = 1{,}080 \text{ lb.-in.}$$

$$M_T = (193{,}000)\ (\sin 0.057) = 193 \text{ lb.-in.}$$

$$M_{TOTAL} = \sqrt{(1{,}080)^2 + (193)^2} = 1{,}095 \text{ lb.-in.}$$

ALTERNATING STRESSES COMPARED

Comparing the bending moments causd by gear couplings with those resulting from contoured diaphragm couplings showed the former to be significant and the latter virtually negligible in comparison.

Cyclic bending stress imposed on a gear coupling-equipped shaft can be computed from

$$\sigma_a = \frac{M_{TOTAL} \times C}{I},$$

where C and I are the shaft radius and shaft area moments of inertia, respectively. Thus

$$\sigma_a = \frac{(64{,}435)\ (2.25)}{\pi d^4/64}$$

$$= \frac{(64{,}435)\ (2.25)\ (64)}{\pi(4.5)^4} = 7{,}206 \text{ psi.}$$

In addition, there is a mean tensile stress acting on the shaft cross-sectional area. This mean stress equates to

$$\sigma_m = \frac{T\mu}{(D_p/2)\ (\pi C^2)\ \cos\theta}$$

$$= \frac{(193{,}000)\ (0.3)}{(4.5)\pi(2.25)^2(0.94)} = 860 \text{ psi.}$$

where $\theta = 20°$, the pressure angle assumed for the gear teeth.

The cyclic bending stress seen by the diaphragm coupling—equipped shaft was obtained by a rapid ratio calculation:

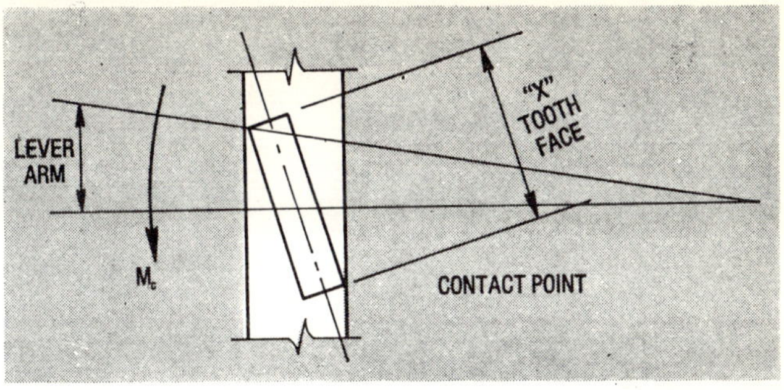

Fig. 3—Shift in contact point.

$$\frac{\sigma_a(\text{Diaphragm coupling})}{\sigma_a(\text{Gear coupling})} = \frac{M_{TOTAL}(\text{Diaphragm coupling})}{M_{TOTAL}(\text{Gear coupling})}$$

$$\sigma_a\ (\text{Diaphragm coupling}) = \frac{(1{,}095)\ (7{,}026)}{64{,}435} = 122 \text{ psi.}$$

The mean tensile stress acting on the cross-sectional area of a diaphragm coupling-equipped shaft depends on how far the diaphragm is displaced axially from its neutral rest position and the axial spring rate of the diaphragm. Assuming the diaphragm in this case were displaced by its maximum permissible distance of 0.100 inches, it would exert a force of 1,950 pounds on the shaft cross-sectional area.[4] This would cause a mean stress

$$\sigma_m = \frac{1{,}950 \text{ lbs.}}{\pi C^2} = \frac{1{,}950}{\pi(2.25)^2} = 125 \text{ psi.}$$

SHAFT FACTOR OF SAFETY CAN NOW BE EVALUATED

Before actually calculating the shaft factors of safety for torque transmission with either coupling type, one must determine the stress concentration factors k_f and $k_{f'}$. Using values of r/D, D/d and R/d from Fig. 1, literature 3 gives stress concentration factors $k_f = 1.95$ and citation 2 gives $k_{f'} = 2.9$. The stress concentration factor $k_{f'}$ results from the keyway and must be used in torsional stress calculations. Factor k_f takes into account the shaft step going from 4.5 to 5-inch diameters. It must be used in the bending stress calculation.

Since the gear coupling-equipped shaft is loaded in torsion *and* (potentially) bending, we employed the expression for combined torsion and bending and obtained the factor of safety (from Equation 3):

$$n = \frac{1}{\sqrt{\left(k_f\frac{\sigma_a}{\sigma_E} + \frac{\sigma_m}{\sigma_{y.p.}}\right)^2 + 3\left(k_{f'}\frac{\tau_a}{\sigma_E} + \frac{\tau_m}{\sigma_{y.p.}}\right)^2}}$$

$$= \frac{1}{\sqrt{\left(1.95\frac{7{,}206}{52{,}500} + \frac{860}{80{,}000}\right)^2 + 3\left(2.9\frac{2{,}180}{52{,}500} + \frac{10{,}900}{80{,}000}\right)^2}}$$

$$= 1.90$$

Using the same mathematical expression to calculate the factor of safety for the shaft equipped with the contoured diaphragm coupling we obtained

$$n = \frac{1}{\sqrt{\left(1.95\frac{122}{52{,}500} + \frac{125}{80{,}000}\right)^2 + 3\left(2.9\frac{2{,}180}{52{,}500} + \frac{10{,}900}{80{,}000}\right)^2}}$$

$$= 2.25$$

The value $\left(1.95\frac{122}{52{,}500} + \frac{125}{80{,}000}\right)^2$ in the above expression is so small that application of the operation for pure torsion (Equation 2) would have given the same factor of safety, $n = 2.25$.

SHAFT STILL ADEQUATE UNDER UPRATE HORSEPOWER CONDITIONS

At this point, we may want to recall that the above analysis was made to determine if the steam turbine shaft would require replacement should the output horsepower be increased beyond the manufacturer's maximum permissible rating of 17,600 hp. Rather than engaging in a debate as to safe maximum absolute values of stress in turbine shafts, we merely investigated what factor of safety the shaft design embodied while equipped with a suitable gear coupling transmitting 17,600 hp at 6,400 rpm:

$$\text{Shaft torque } T = \frac{(63{,}000)\ (17{,}600)}{6{,}400} = 173{,}250 \text{ lb.-in.}$$

$$\text{Steady torsional stress } \tau_m = \frac{16T}{\pi d^3} = \frac{(16)\ (173{,}250)}{\pi(4.5)^3}$$

$$= 9{,}680 \text{ psi.}$$

$$\text{Alternating torsional stress } \tau_a = (0.2)\ (9{,}680) = 1{,}940 \text{ psi.}$$

$$M_C = \frac{173{,}250}{4.5}\left(\frac{1.3}{2}\right) = 25{,}025 \text{ lb.-in.}$$

$$M_F = (173{,}250)\ (0.3) = 52{,}000 \text{ lb.-in.}$$

$$M_T = (173{,}250)\ (\sin 0.057) = 173 \text{ lb.-in.}$$

$$M_{\text{TOTAL}} = \sqrt{(25{,}025)^2 + (52{,}000 + 173)^2}$$

$$= 57{,}864 \text{ lb.-in.}$$

$$\sigma_a = \frac{(57{,}864)\ (2.25)\ 64}{(\pi)\ (4.5)^4} = 6{,}472 \text{ psi.}$$

$$\sigma_m = \frac{(173{,}250)\ (0.3)}{(4.5)\ \pi\ (2.25)^2 0.94} = 770 \text{ psi.}$$

$$n = \frac{1}{\sqrt{\left(1.95\frac{6{,}472}{52{,}500} + \frac{770}{80{,}000}\right)^2 + 3\left(2.9\frac{1{,}940}{52{,}500} + \frac{9{,}680}{80{,}000}\right)^2}}$$

$$= 2.14$$

The shaft factor of safety, while certainly adequate for operation at 17,600 hp with gear-type couplings, was thus shown to be only 2.14 in the original design case with conventional gear couplings. Equipping the same shaft with contoured diaphragm couplings makes operation at 19,600 hp not only possible, but will actually increase the factor of safety to 2.25.

WHICH DIAPHRAGM COUPLING?

Several manufacturers have both the capability and experience to provide diaphragm couplings for critical petrochemical machinery. In reviewing this experience, machinery engineers may want to compare the combined stresses acting in diaphragms of operating couplings with the stresses that would act in the diaphragms of couplings being considered, since this, in essence, establishes prior experience at a given stress level. Proper installation reviews and an understanding of axial shaft growth of coupled machines should serve as verification that the manufacturer's maximum permissible axial diaphragm deflection will not be exceeded. Noncontacting proximity probes can be adapted for continuous monitoring of diaphragm displacement and have been installed on process compressor trains where the owner wanted to be sure beyond any doubt.[5]

ACKNOWLEDGMENT

Originally presented at the 4th Turbomachinery Symposium, College Station, Texas, October 1975.

LITERATURE CITED

[1] Timoshenko, F. P., "Strength of Materials—Advanced Theory and Problems," Van Nostrandt-Reinhold Publishing, N.Y. 3rd Edition, 1956.
[2] Peterson, R. E., "Stress Concentration Design Factors," John Wiley and Sons, Inc., New York, 1953.
[3] Wilson, C. E., Jr., "Mechanisms—Design Oriented Kinematics." American Technical Society, Chicago, 1969.
[4] Bendix Fluid Power Corp., Utica, N.Y. Technical Bulletin, "Contoured Diaphragm Couplings," 1973.
[5] Bently-Nevada Co., Minden, Nev. Technical Bulletin on Proximitor Alignment Systems, 1974.

Part V.

Reciprocating Compressors

Reciprocating Compressors—An Introduction

The reciprocating compressor is fundamentally a fixed-capacity machine with variable capacity obtained by altering the speed of the prime mover. Capacity may be varied on multiple-cylinder machines by making the head or outer ends and the crank or inner ends inoperative by using devices that lift suction valves open manually or automatically. Additional clearance volume may be provided, which can bé cut in or cut out to vary the compressor capacity as required.

For some applications, it is often desirable to include a large clearance volume to maintain approximately constant horsepower over a wide range of pressure conditions. Thus, economical operation is attained by operating the prime mover at its rated load for a broad range of operating pressures.

Speeds

Reciprocating compressor speeds may range from 125-514 rpm. Piston speeds range from 500-950 fpm, the majority being 700-850 fpm. The nominal gas velocity through the valving is usually in the range of 4,500-8,000 fpm. Discharge pressures may range from vacuum pumps to 30,000 psi.

Frames and Cylinders

A reciprocating compressor consists essentially of a frame (or frames) to which are attached the necessary compressor cylinders and driver, plus the necessary auxiliaries such as intercoolers where required (either machine mounted or remotely located), interconnecting gas piping, jacket water piping, et cetera. Every compressor manufacturer has a group of standard frames, each with definite design limitations on speed and load carrying capacity. Load carrying capacity involves two considerations: horsepower, and forces created by pressure differential across the pistons. The latter is generally referred to as frame load.

Horsepower limitations are determined by the ability of the crankshaft and crank webs (or crank discs) to withstand the necessary torque or turning effort without over-stress, and by the ability of the bearings to dissipate the heat of frictional loading.

Frame load limitations are established to prevent stresses in the frame, piston rods, connecting rods, bolting and other frame parts from exceeding safe conservative values, and to eliminate the imposition of excessive pressure per projected unit bearing area. The frame load for each compressor crank is equivalent to the sum of the differential loads of all pistons attached to that crank. The load for each piston is the difference in the products of net piston areas and pressures exerted on the piston faces. Loadings should be determined for both directions of piston movement. The loadings are not the same in each direction and where the ratio of piston rod area to piston area is high, they may be vastly different. An undesirable negative loading may sometimes result and provision must be made, such as a tail rod addition, to remove the unbalance.

Compressor Selection

The inquiry data determines the types and sizes of compressor and driver needed. The procedure is briefly this:

1. The compression characteristics of the gas are determined. For single-component gases the necessary information can be found in published tables. For mixed gases these characteristics are determined from the gas analysis. The physical constants to be determined are the k value (C_p/C_v), the molecular weight, the critical pressure (P_c) and the critical temperature (T_c). The critical pressure and temperature as determined from the analysis of mixed gases are sometimes termed pseudo-criticals.

2. The total ratio of compression is determined. This is the ratio of final absolute initial suction pressure. It must now be decided whether the compressor is to have one, two, three or more stages of compression. Experience will indicate quite readily how many are needed. The primary consideration is the expected discharge temperature. A sufficient number of stages, with cooling between them, must be used to prevent the discharge temperature of any of them from exceeding practical limits.

Another consideration is the relationship between volumetric efficiency and the ratio of compression. Gases having the higher k values will show a higher volumetric efficiency for the same cylinder clearance and ratio of compression than lower k value gases. An extremely low volumetric efficiency for normal service is not desirable since an excessively larger cylinder is needed to obtain the desired capacity. This results in a heavy differential load (frame load) and may require a heavier frame than would otherwise be needed.

It is customary in the refinery and chemical process industries for the project engineer to establish the number of stages of compression, and for each, the suction pressure, suction temperature and discharge pressure, with due consideration given to pressure drop and cooling between stages. If this information is not contained in the inquiry, the manufacturer will ordinarily assume nearly equal ratios per stage to be satisfactory and will assume appropriate intercooler and interconnecting piping pressure losses between stages. He will likewise assume suction temperatures for the second, third, etc., stages consistent with the available cooling water temperatures as shown in the inquiry.

This part of the prodecure shows why it is essential that not only the suction pressure, but all interstage pressures, if given, as well as the final discharge pressure, be fully identified as either gage or absolute. Where gage readings are specified, the feet elevation above sea level at the plant site must be known so that absolute pressures, mandatory in selection procedure, may be determined.

3. The required capacity per compressor is determined and is converted to cfm at inlet conditions of the respective stages. The proper corrections for temperature, pressure and compressibility must be made. Where the desired capacity is stated as dry gas, the inlet volume of each stage must be increased by the ratio of the total absolute pressure to absolute partial pressure of the dry gas at stage inlet. This step of the procedure again emphasizes the necessity for clarity and completeness in the inquiry data. The desired capacity rate must be explicit, and the conditions of pressure and temperature at which it is measured must be given.

4. A rough check is made to determine the horsepower required. For this purpose generalized charts are used. The factors involved are the total actual capacity of each stage (cfm at inlet), the ratio of compression per stage, the specific gravity and the k value of the gas. Again this emphasizes the requirement for complete gas characteristics, clearly stated pressures, temperatures, and capacities.

The total horsepower required for the compressor (sum of horsepowers per stage) determines the required frame size. The kind of drive, whether steam, motor, gas, etc., as stated in the inquiry will dictate the type of frame best suited. There is usually an established length stroke and speed for the frame selected. However, for a motor or steam turbine drive it may be necessary or advantageous to modify one of both to better match driver characteristics. Having established these details, the piston speed, which in feet per minute is equal to [(stroke, in.)/12](2)(rmp), as needed in Step 5, can be calaculated.

5. The approximate compressor cylinder sizes can now be determined. The required capacity of each has already been established (Step 3). From appropriate charts or formulae the volumetric efficiency of each stage of compression is developed. This involves the ratio of compression, the k value of the gas and the compressibility factors at inlet and discharge conditions. A trial selection must be made, assuming cylinder clearance, and the size and clearance readjusted to match cylinder patterns and/or designs available.

This procedure again emphasizes the requirement for exacting inquiry data. If facts for establishing critical constants are missing the cylinders could be sized incorrectly, sometimes appreciably.

6. The frame loadings can now be checked to be sure that all are within the design limits established for the selected frame. If all are too high, a heavier frame must be used. If one of several are high it may be possible to readjust the ratios of compression to bring all loadings within the rating. Failing this, a heavier frame becomes necessary.

7. Assuming the frame requirements have been met, the selection so far as stroke, speed, number of stages and sizes of cylinders is now complete. One more detail remains—a more accurate check of the horsepower required. In Step 4 generalized horsepower charts were used. Such charts are termed generalized because they are based on average values of mechanical and compression efficiencies. It should be emphasized here that it is impossible to include in any one horsepower chart, such as the widely used "bhp per million" chart, values other than average for these two factors. Therefore, after the selection has been made and all factors involved in determining the compression and mechanical efficiencies are known, the horsepower can be quite accurately established.

Selection is now complete with full performance characteristics known for the conditions of service.

Multiple Service

Quite often an inquiry will list one or more alternate service conditions, such as a range in suction pressure or discharge pressure or both. Or perhaps, operation on alternate gas streams may be desirable, where pressures and capacities may or may not be identical. In such applications, the requirements must be analyzed and the selection based upon the most severe conditions. It may be necessary to select cylinder sizes for one service, to obtain the desired capacity, and to base frame selection on the alternate service because of high horsepower and/or frame loadings required by the alternate.

Performance Calculations

Table 1 lists some useful equations used in compressor selection and performance calculations.

The equations given are all theoretical. A certain modification of results obtained must be made to arrive at actual results. For instance, the actual volumetric efficiency is always lower than that found from the equation shown because allowance must be made for entrance losses caused by wire

Table 1

Performance Equations for Reciprocating Compressors

1. Ratio of Compression = P_2/P_1

2. Theoretical Volumetric Efficiency =

$$1.00 - \left(\frac{R^{1/k}}{f} - 1\right) C$$

3. Theoretical Mean Effective Pressure =

$$\frac{kP_1}{k-1}\left(R^{(k-1)/k} - 1\right)$$

4. Theoretical (isentropic) Horsepower = $.004364\ Q_1$ (M.E.P.)

5. Theoretical Discharge Temperature = $T_1\ r^{(k-1)/k}$

6. Theoretical Terminal Volume Expressed as Percent of Vs = $\frac{f}{R^{1/k}}$

Note: All pressures and temperatures are in absolute units.

where:

f	=	Ratio of compressibility factors, F_2/F_1
C	=	Cylinder clearance volume (volume in one cylinder end in excess of the piston displacement per stroke), usually expressed as a percentage of stroke displacement
$M.E.P.$	=	Mean Effective Pressure
Q_1	=	Capacity, cfm.

drawing through the cylinder ports and valves and for increase in volume caused by heating during the inlet period. In addition, compensation must be made for leakage past piston rings and cylinder valves and for packing vent losses. Obviously, leakage is primarily a function of pressure differential and the type of cylinder construction. Nonlubricated cylinders (carbon piston rings, etc.) can be expected to develop greater slippage than those receiving lubrication, as the oil effectively seals surface contacts in most instances. However, also of importance is the type of gas being handled. Certain gases are simply harder to retain than others and the problem is closely connected with molecular weight and viscosity. There is no hard and fast compensating factor that can be given to cover all these variables. It is customary to allow 3 percentage points for wire drawing and heating entrance losses, so that 0.97 may be substituted in for formula for 1.00. Slippage losses may vary from 2 to probably a maximum of 5 percentage points for ordinary applications involving oil-lubricated cylinders and from perhaps 4 to perhaps 10 percentage points for nonlubricated construction. Thus a more practical formula might be

$$\text{Volumetric Efficiency} = 0.97 - (r^{1/k}/f - 1)\ C - L$$

Where L is the slippage loss just discussed.

Brake Horsepower

The theoretical horsepower naturally does not include fluid or mechanical friction losses. The fluid losses are grouped in the term Compression Efficiency, while the mechanical losses are covered by the term Mechanical Efficiency.

The major factor involved in determining the compression efficiency is the valve loss or pressure drop through the inlet and discharge valves. These fluid losses are a function of gas density and valve velocity. The suction and discharge pressures and the molecular weight establish the density. The valve velocity is fixed by the valve area available in the selected cylinders and the piston speed. Valve velocity is normally stated in feet per minute and is the ratio of piston area to valve area per cylinder corner multiplied by the feet per minute piston speed. It is likewise equal to $144d/A$ where d is the piston displacement of one cylinder in cubic feet per minute, and A is the valve area in square inches per cylinder corner. Valve area per corner is the area of the suction or the discharge valves in one end of the cylinder.

Valve losses are usually read from graphs plotted from research data. Applicable corrections for density are designated. There is no precise general equation available for calculating these losses because of the numerous variables in valve designs in use, such as spring loadings, unbalanced areas and flow patterns.

There are other considerations. During compression, a small portion of the gas continually slips past the piston rings and suction valves. Work has been done on this gas, yet it is not delivered to the discharge system. Also slippage past the discharge valves allows gas which has already been delivered to the discharge system to return to the cylinder. Recompression and redelivery take place.

The cooling effect of cylinder jacketing is still another factor. Removal of heat by the jacket water would shrink the volume during compression. This is a true saving in power expended. Unfortunately, it is not of any great significance, except in small cylinders handling rather low density gas through high compression ratios where the jacket surface is large in proportion to the amount of work performed and heat generated. Further, cold jacket water should be avoided when the compressor is handling readily condensible gases which enter the cylinders at or near the dew point. In these instances it is actually desirable to superheat the gases during the inlet and early compression portions of the cycle by hot jacket water to prevent condensation within the cylinders. Condensation, liquid of any kind, and wet saturated gases are extremely detrimental to lubrication. Many times the same factors adversely affect nonlubricated construction, especially valve service life.

Horsepower Calculations

The following equation may be used to calculate the energy required for a reciprocating compressor:

$$E = 144\ (k/k - 1)\ (P_1 Q_1)\ (r^{(k-1)/k} - 1)$$

where:

E	=	Foot pounds of energy per minute
k	=	C_p/C_v
P_1	=	suction pressure, psia
Q_1	=	Actual capacity, cfm, measured at inlet condition
r	=	Ratio of compression = P_2/P_1

The theoretical horsepower then becomes

$$HP = (144/33{,}000)\ [(k/k - 1)]\ (P_1 Q_1)\ (r^{k-1/k} - 1).$$

Other factors remaining constant, the horsepower change is directly proportional to the change in suction pressure. For instance, if the suction pressure is increased 50%, the horsepower increases in the same proportion.

Other factors remaining constant, the horsepower required changes in direct proportion to the change in Q_1. For example if the capacity is reduced 50%, horsepower is likewise reduced 50%.

The variation brought about by change in k is three-fold. First, the capacity is altered. Secondly, the value of $k/(k-1)$ differs and thirdly, the value of $(r^{(k-1)/k} - 1)$ changes.

Suppose for example the k value change is such that Q_1 increases 5.13%. This would mean a corresponding decrease in $k/(k-1)$ of 33.3% but an increase in $(r^{(k-1)/k}-1)$ of 60.7%. The ratio of altered to original horsepower then is 1.0513 (0.667) (1.607) = 1.125. Thus there is a net change of +12½% in power required.

This change would occur if a compressor with a 10% cylinder clearance compressing a gas of k value 1.24 through 4 ratios of compression were changed to air service (k = 1.41), suction pressure unchanged.

The cylinder clearance does not appear in the horsepower formula but a change in clearance results in altered volumetric efficiency and capacity. The effect of clearance change on horsepower is in direct proportion to the change in capacity brought about by the change in clearance. Should

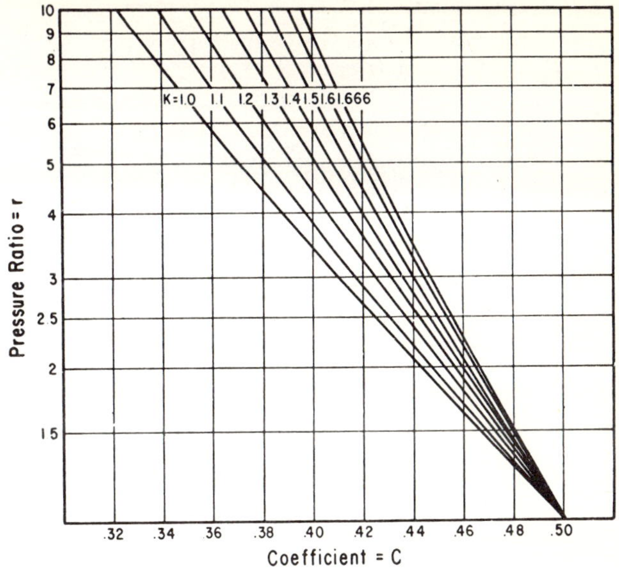

Fig. 1—Horsepower deviation coefficient. Procedure: (1) Determine compressibility factors F_1 and F_2 from compressibility charts; (2) Calculator $f = F_2F_1$; (3) From given r and k determine coefficient (c) from this chart; (4) Change in horsepower (deviation) = $C(F - 1)\ hp$ (Ideal Isentropic hp); (5) Corrected horsepower is algebraic sum of Ideal bhp and deviation (item 4).

the clearance change reduce the capacity 25%, the horsepower likewise would be 25% lower.

Should the compression ratio vary, so will the capacity. The value of $(r^{(k-1)/k}-1)$ would also vary.

Assume the compressors described in the preceding paragraph were required to operate at 5 ratios of compression instead of at 4 ratios. The k value is 1.24 and the suction pressure and cylinder clearance are unaltered. The decrease in volumetric efficiency would be approximately 8.0%. The change in $(r^{(k-1)/k} - 1)$ would be an increase of 18.3%. The net change therefore would be (0.920) (1.183) or an 8.8% increase in the power required.

The effect of compressibility on the horsepower required per cubic foot capacity measured at intake conditions does not appear in the equation. It is small for average operating conditions. In high-pressure units, however, effects of sizeable proportions are sometimes found. For example, consider a cylinder handling carbon dioxide for a urea plant. The suction pressure is 1,720 psia at 115° F. The discharge pressure is 3,440 psia. From published data we find the critical pressure of carbon dioxide to be 1,073 psia and the critical temperature to be 548°R. At suction conditions the reduced pressure, P_1/P_c = 1,720/1,073 = 1.6, the reduced temperature, T_1/T_c = 575/548 = 1.05. From a standard compressibility chart (1 psia basis), the compressibility factor, F_1, at intake conditions is 0.312.

The theoretical discharge temperature, using the equation in Table 1, is

$575(3{,}440/1{,}720)^{(1.3-1)1.3} = 215°F$. Therefore, at discharge conditions of 3,440 psia and 215°F; $P_2/P_c = 3{,}440/1{,}073 = 3.2$; $T_2/T_c = 675/548 = 1.232$. Referring again to a standard compressibility chart, the compressibility factor, F_2, at discharge conditions = 0.575. Therefore $F_2/F_1 = f = 1.842$.

Referring now to Figure 1 and following horizontally from a Ratio of Compression of 2 to the intersection with the diagonal line representing $k = 1.3$ (C_p/C_v for carbon dioxide), and reading vertically downward the coefficient C equal to 0.456 is found. Using the explanation given below the figure, the increase in the theoretical horsepower will be $C(f - 1) = 0.456\ (8.42) = 0.384$ which means an increase of 38.4% in theoretical horsepower caused by the compressibility effect.

This example is probably one of the most dramatic which might have been selected, but it has been used to illustrate the possibility of incurring an overload unless all conditions surrounding the inquiry are known and evaluated. The converse is sometimes true and the driver may be appreciably underloaded. However, driver underload does not prevent operation, whereas compressor overload may. The possibilities must be checked to be reasonably certain. Compressibility factors are important.

Use Compressors Safely

This practical checklist can help persons who buy, install, inspect, maintain or operate compressors. Special attention should be paid to personnel hazards, piping, oil, the building, fire protection and maintenance.

Robert J. Cordes, Marathon Oil Co., Findlay, Ohio

COMPRESSORS CAN BE dangerous if not designed, installed, operated and maintained properly. Consequently, a periodic review of common problems that could lead to a loss incident is helpful to persons with responsibility for fire protection and safety in a hydrocarbon processing plant. With such a review in mind, I prepared guidelines for eight key areas of concern: Compressor engine, electrical, hazards, piping, oil, building, fire protection and maintenance. These guidelines can be used as a checklist to aid in more compressor-operation reliability. The goal is to help you prevent loss of life, money and gas.

I will deal only with reciprocating gas compressors, but much of the material applies to *any* type of compressor. While I have tried to categorize the various subject matter, you will find information that can fit well under several categories.

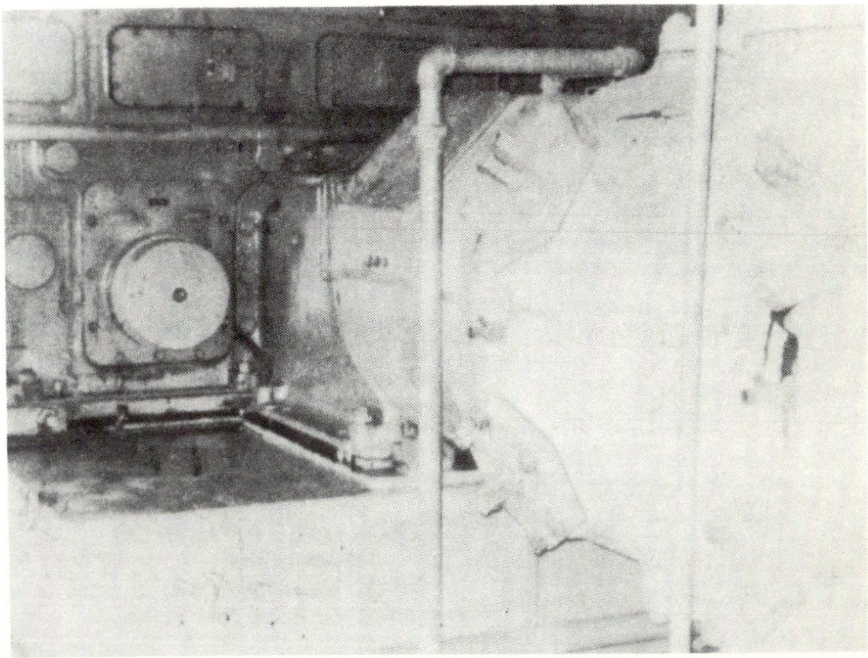

Fig. 1—An explosion vent on an engine can prevent overpressure damage.

COMPRESSOR—ENGINE

☐ **Explosion vents should be installed** on the engine. A crank case explosion will then be vented to atmosphere rather than overpressure the engine thereby causing equipment damage and possible personal injury. See Fig. 1.

☐ **A high temperature on the engine cylinder** can be predicted if the copper tubing to the fuel gas injection valve becomes black. It becomes black because the injector valve has burned out causing heat from the cylinder to travel back into the tubing.

Possibly there is also some burning in the gas supply line to the engine cylinder.

☐ **The compressor valve cover plates are a source of gas leakage,** as is packing on the compressor piston rod. There are several methods of handling a leak at the piston rod packing to keep gas out of the compressor room.

One method is to connect piping or tubing to the packing housing and vent gases out of the building to an elevated atmospheric vent.

A less efficient way of removing the gases is by allowing them to flow into the distance piece. The distance piece is then vented to atmosphere. If this method is used, the inspection plates must be bolted tightly on the distance piece.

When handling toxic gases, it is sometimes desirable to fit the packing housing with an exhauster. A vacuum of a couple of inches of water prevents toxic gases from traveling into the distance piece and possibly escaping into the room.

☐ **Experienced operators know gas compressor valves are not seating when the suction and discharge valve covers become hot** (Fig. 2). Two common causes are broken valve springs and deposits. Be sure to examine discharge piping for broken springs and deposits when the compressor is down for maintenance.

Air compressors are frequently located near gas compressors. Hot carbon deposits in an oil-lubricated air

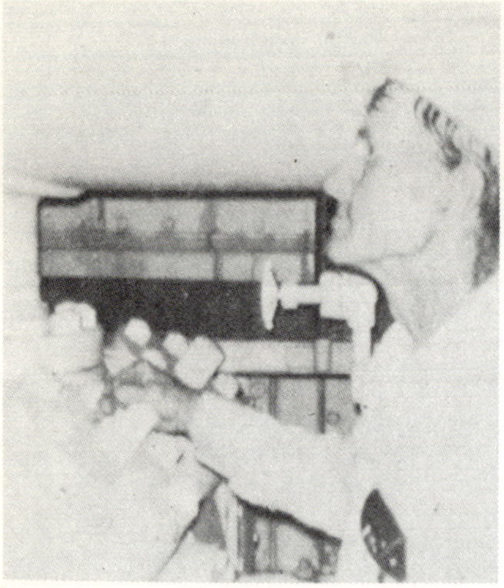

Fig. 2—By putting his hand on a suction-valve cover, an operator can determine if it is hot.

Fig. 3—These shutoff valves, upstream of the relief valves, should be locked or car sealed in the open position. Car sealing is preferred because it enables quick closing of the valve if conditions warrant it.

compressor can cause an explosion because all of the necessary parts of the fire triangle are present. Synthetic lubricants will do away with the deposit problem and the possibility of a fire. But a word of caution. They should be put into the system initially because of their solvent ability to remove paint and because they are not compatible with a few materials. Paint and other loosened materials will flow downstream and can plug small ports and channels. A line strainer is needed if the air compressor is converted to a synthetic lubricant.

☐ A properly sized relief valve should be installed on the reciprocating compressor discharge piping between the final cylinder and the first shut-off valve downstream to prevent overpressuring the equipment. Shut-off valves placed within the piping between stages will also require relief valves (Fig. 3). Here is an example:

If a multistage machine has separate coolers and receivers with isolation valves for each stage of compression, install relief valves on the discharge piping of each cylinder. At some locations an engine-driven four cylinder compressor has, as an example, three cylinders in one particular service and the other cylinder in another service. In this instance, be sure that *both* process streams can be properly relieved.

The relief valve should discharge to a lower pressure system, such as the first-stage suction line in the same service. Shut-off valves upstream or downstream of the relief valve should be locked or car-sealed in the "open" position. Car sealing is preferred because the valve could quickly be closed if conditions warrant it.

The relief valve discharge line should be of pipe. A short piece of steel braided hose can be attached to the piping near the discharge of the relief valve. Hose will help stop vibration and will provide for thermal expansion *but* it must be rated for the pressure expected, i.e. in excess of the relief valve setting. Hose is placed downstream of the relief valve which normally places it in a lower pressure and temperature service.

Plain rubber hose should not be used because of its tendency to weather, the likelihood it will fail in a fire and it probably cannot withstand high pressures. A hose which fails will spray compressed gas in the area of the engine-compressor unit.

☐ **Compressor valve covers should be checked to make sure that they are fully bolted** rather than have only some of the threads of the nut in contact with the stud bolt (short bolted). Full thread contact is necessary to reduce the chance of the compressor valve cover being blown off and a fire occurring.

There is a remote possibility that a longer nut has been substituted which gives the appearance of short-bolting. But this is unlikely. Look at the other nuts to see what size they are. Stud bolts should be replaced when their threads become worn, corroded or split because the holding ability of the nut-bolt connection is reduced. Be certain that the new stud is long enough to engage all of the nut threads.

Another problem occurs if the replacement bolts and nuts are of a different hardness and steel composition. Compressor manufacturers specify nut and bolt properties. They also specify a torque when putting the nuts on, so that the load is evenly distributed on all nut-bolt connections. Impact wrenches have exerted enough tension on the bolts to cause the cast iron casing to crack resulting in bolt modification before the compressor can be put in service.

☐ **Beware of home-made clearance pockets or clearance bottles to reduce compressor capacity.** Some are of questionable design. They should be fabricated to ASME specifications. Ideally a bottle which can be bolted in place is best. Threaded clearance bottles can develop cracks at the threads due to gas pulsation as the

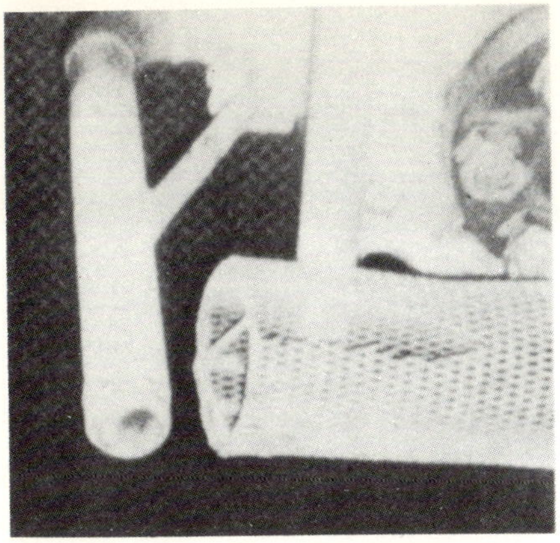

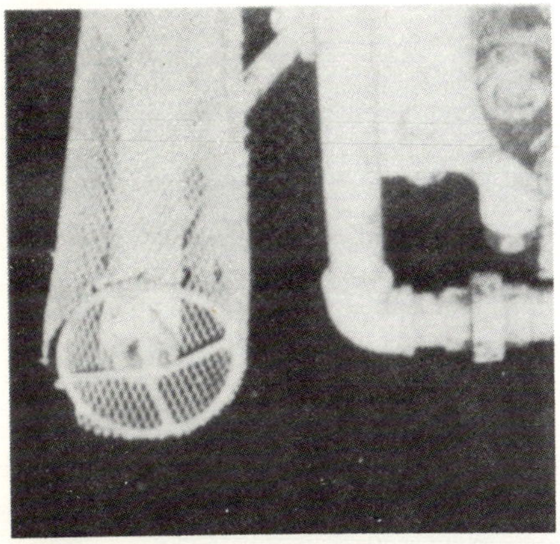

Fig. 5—The guard is in place on this hot home-made clearance bottle. Wear is not evident.

pressure changes in the cylinder. See Figs. 4 and 5.

☐ **Monitors can shut down a compressor if physical operating conditions change.** Vibration switches are one type. Some others are high lube-oil temperature, low lube-oil level, low oil pressure, high engine-exhaust temperature, low first-stage suction pressure, low or high-discharge pressure, overspeed, high-engine temperature and high liquid level in the gas scrubber ahead of the compressor first stage.

Compressors should have alarm and/or shutdown devices. The shutdown device is better. It protects the compressor but it might throw operations into an emergency situation.

At manned installations, when possible, an alarm should precede the compressor shutdown. Operators may be able to keep it running by taking

Fig 6.—Grout has separated from the foundation concrete, reducing the support area under the machine.

some necessary action if they receive an alarm.

Operators usually have no problem in knowing when a compressor goes down. They know by the change in sound level in the plant and control-board indication.

In remote locations where compressors are part of a distribution system, a signal should be immediately sent to a manned control station.

☐ **Watch for cooling problems.** Engine and compressor cylinders are usually water or antifreeze cooled. Water is used for the intercooler between stages too. The water is supplied from cooling towers, fin fan coolers or vertical radiators at one end of the compressor-engine block. Poor cooling water condition and quality can result in poor heat transfer, plugging of small ports, corrosion and increased cylinder operating temperatures. Cooling towers, because they are not a closed system, will accumulate bugs, leaves, etc., plus pieces of redwood which are circulated through the equipment and create heat transfer problems.

☐ **Observe forces that influence compressor operation.** The machine has forces going in many directions. These can cause damaging vibration plus excessive deflection of parts of the equipment. For that reason, grout is added between the machinery and its foundation bolts—it helps the foundation bolts to do their job.

Oil, vibration and heavy stress loading frequently cause grout to fault or break up (Fig. 6). Epoxy grouts are popular, having replaced cement grouts both at the initial setting of the machinery and when there is no recourse but to regrout. Be certain grout is in good condition.

Foundation bolts should be in place with nuts tightly in place. Their purpose is to secure the machinery to the heavy concrete foundation. The foundation should be physically separated from the compressor building floor to reduce vibration transmission. Sometimes machine movement becomes intense enough to give the appearance of lube oil being pumped from under the machine.

☐ **Cold climates and/or long suction-piping runs can cause some of the gas to condense—an undesirable condition** because liquids cannot be compressed. A large quantity of liquid could cause the compressor cylinder

Fig. 7—Gas knockout drums should be checked to be certain that valves are open from the drum to the Hi-Lo liquid-level controller.

to be destroyed when the piston attempts compression.

Gas knock-out drums, heaters such as steam coils inside or outside of the suction drum or piping and centrifuge separators in the suction piping will eliminate condensate from the stream. Gas knock-out drums should be checked to be certain valves from the drum to the Hi-Lo liquid level controller are open (Fig. 7).

The alarm, if one is installed, should be tested routinely.

Valves on the automatic liquid draw should be correctly lined up.

Auxiliary heaters, such as steam coils, should be in service. Therefore, check the valves on that system.

Check centrifuge separators to be certain the liquid drain valve is open. Severe cold weather should be anticipated because, despite valves being in the open position, the line could become blocked due to freezing.

☐ **The engine air intake pipe should be upwind so that if there is a gas leak at the compressor, gas will not enter the combustion air piping** (Fig. 8). Intake air should not be located where it can come in contact with cooling-tower fog. This could freeze off intake screens and filters in winter.

Fig. 8—Engine air intake is upwind of the compressor engine. Note the well-ventilated building.

The spring relief plate at the engine air intake should be checked to determine if it is jammed open or if the spring has failed. An open plate allows air to bypass the air filter causing insects and dirt to be drawn into the engine.

☐ **The air-starting valve mounted on the power cylinder head should be inspected whenever the engine is opened up** (Fig. 9). If the valve sticks open, high pressures can be built up in the air piping because of the direct connection to the engine cylinder. A rupture disc or relief valve between the cylinder and first valve on the air line can prevent it from violently failing. After the engine is running, a small vent valve on the air line should be opened to provide a means of detecting a leaking air-starting valve.

The leakage can allow blow-by combustion gases to damage the valve or cylinder head, and warrants immediate attention. Faulty air valves can allow fuel gas to leak into the starting air line and be ignited from the power cylinder resulting in a fire, explosion or detonation.

ELECTRICAL

☐ **Ignition systems should be in explosion-proof conduit as much as practical** (Fig. 10). Conduit will protect the wiring from heat, fire and abrasive wear on the engine. All of these can cause a short circuit and an engine shutdown.

Fig. 9—The air-starting valve should be inspected whenever the engine is opened up, to be certain that the spring, guide and seating surface are in good-enough condition to prevent overpressuring the air line to the engine.

Fig. 10—Ignition systems should be in explosion-proof conduit, to the extent that this is practical.

☐ **All electric motors, whether the compressor driver or for the water cooling pump, should be grounded in accordance with the National Electrical Code.**

☐ **High tension coils (transformers) produce high voltages. Avoid contact with them during operations to prevent serious or fatal injuries.**

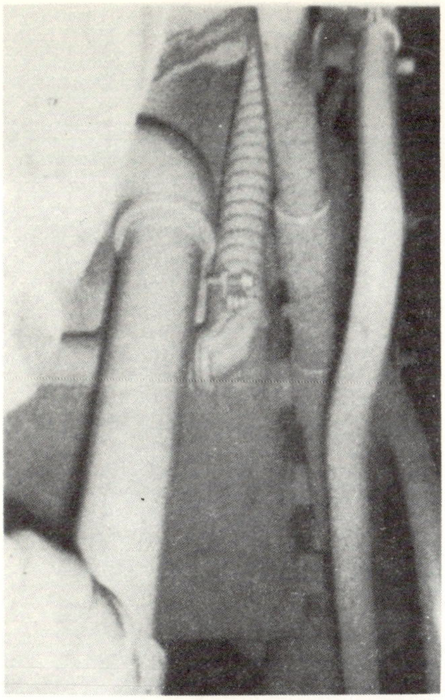

Fig. 11—This split-ell electrical connection is flanged up tight, as intended. Vibration has not affected it at this point.

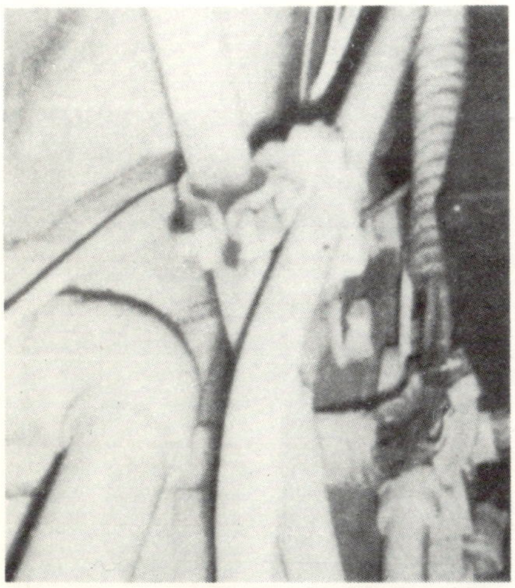

Fig. 12—A neglected split-ell connection separated by vibration. All but one bolted connection loosened, the enclosure slipped from the wire, and continuing vibration poses the danger of wiring-insulation wear-through. A short could cause a shutdown, not to speak of electrical-shock hazard to personnel.

HAZARDS

☐ **Check bolted stairways, walkways and handrails for loose connections** which might cause the fixture to move when someone uses it. Vibration can loosen bolts on a step so that it will move a few inches when stepped on, possibly causing a personnel injury. In extreme cases, welded connections might fail.

Fig. 13—The white hard hat rests on the building floor. An operator must step from the floor to the metal floor of this skid-mounted compressor. Note the excessive height for only one step. Moreover, the metal floor at the base of the ladder is slick because of a fine film of oil.

☐ **Check the means by which an operator is expected to climb up on the compressor-engine unit** (Fig. 13). Many compressors seem to have a very narrow ladder made of ¾-inch-diameter bar. An operator is expected to swing up or climb up to the engine-head platform without burning himself or slipping off the ladder. And yet the handrail alongside the ladder is often as poorly designed as the ladder.

The ¾-inch bar stock gives the operator a slick piece of steel to put his foot on. Stair treads would be much better (Fig. 14). The platform alongside the engine heads is often slick and narrow. This condition can sometimes be improved in the field. There should be a handrail along the platform.

☐ **Hot exhaust pipes can cause personnel burns.** A means of protecting personnel—such as a noncombustible blanket, insulation or shield may be needed.

☐ **Be sure that gaskets are serviceable.** Some exhaust pipes have been installed with a quick-coupling device containing gaskets of rubber and other materials. The gaskets have

Fig. 14—A well-designed ladder and platform will include hand rail, mid rail and kick plate.

temperature limitations. Improper gaskets will eventually fail and cause carbon monoxide to enter the room. Be alert also for cracks or leaks in the exhaust manifold and exhaust pipe at flanged connections and expansion joints.

☐ **Motor-driven gas and air compressors should have a sturdy guard over the belt and pulley or coupling.** A mesh is preferred rather than a solid guard. Mesh allows visual inspection. The mesh should conform to existing applicable codes.

PIPING

☐ **Check for problems caused by vibration** (Fig. 15). Clamps on piping, especially U-bolts, will wear on the pipe creating a thin section where a failure can occur. Pipe shoes will support compressor piping and eliminate both erosion and corrosion. Vibration dampeners and gussets might be necessary to stop or reduce the destructive vibration.

☐ **Pipe flanges and threaded connections can be sources of gas leaks.** Union fittings on the fuel injection line at the engine head are often a source. It is difficult to find the location of a leak. Usually, you cannot use your sense of hearing to locate it because the leak is small and the compressor room is noisy. An ultrasonic detector can pick up the leak

Fig. 15—Wood blocks are used here to reduce vibration.

Fig. 16—Gage glass valves should be fully opened on gas scrubbers.

Fig. 17—This vent valve is properly located, but the drain valve should be adjacent to the bottom of the horizontal suction line. As it is, if the drain hose leaks, the compressor will need to be shut down, because the drain valve was not installed in the right location.

or you can use a portable combustible gas detector.

☐ **Pressure gages on vibrating piping should be checked.** The valve below the gage should be partially closed to reduce needle swing due to gas pulsation and to reduce the size of a leak if the gage or its screwed piping should fail.

☐ **Brass valves should not be used on steel piping** because of the galvanic action which takes place between brass and steel. Also, brass fails (melts) at a lower temperature than steel under fire conditions. Moreover, brass is susceptible to attack by some materials.

☐ **Gas scrubbers (or knockout drums) should have their gage-glass valves fully opened** (Fig. 16). If they are not fully opened, in many cases the ball check valve cannot seat if the glass breaks. This is important regardless of the location. But it takes on added importance at unmanned remote locations where the compressor is skid mounted and the knockout drums are very close to the compressor and engine.

☐ **Engineers and maintenance men should question the design anytime a pipe is run less than seven feet above the piston head walkway and the ladder to that walkway.** Operators should not need to climb over hot or cold pipes while performing routine operations.

☐ **Vent and drain valves should be located as close as possible to the piping they are servicing** (Fig. 17). This is especially important on compressor piping because it is susceptible to vibration. If the vent or drain valve is installed far from the piping it is servicing, a larger moment force will exist. The pendulum effect due to vibration can cause the connecting pipe to crack or fail at a welded or threaded connection.

Threaded connections should be back welded for extra strength. Along the same vein, heavy valves should not be hung at the end of a horizontal small-diameter line.

All valves should be plugged unless they are frequently used. The plug protects the threads from corrosion and paint jobs. It also prevents a hydrocarbon leak should the valve fail because of fire or vibration.

☐ **All fuel-gas piping should be welded or, if screwed piping, it should be back welded.** Screwed connections on the compressor suction and discharge piping should be back welded to and including half of the nearest valve from the compressor piping. This will provide a tight system with less chance of leakage.

OIL

☐ **A common problem around many compressors is spilled oil.** It can be found as you walk (or slip) in the door, on the floor around the compressor-engine block and on the working platform at the compressor. The operator usually insists he is doing his best to keep it wiped up.

Some oil comes from operators who overflow those tiny lubricators and some drains from the hose attached to lube-oil barrel pumps which are wheeled from compressor to compressor.

Oil and grease should be wiped up immediately to prevent slips which are easy to come by on a smooth compressor floor. A person whose head strikes the floor can and probably will suffer a serious injury.

☐ **At some remote unattended locations** especially, you will find an elevated lube oil tank connected to an engine with a rubber hose. The tank valve is left in the open position to gravity fill the crank case. The rubber hose should be replaced with steel pipe (Fig. 18). A fire would cause the rubber hose to fail. Contents of the lube oil tank would then feed the fire.

☐ **Check the lubricator to be certain all lubricator cylinders are filled and**

Fig. 18—Appropriately, this elevated lube oil tank has a steel-pipe fill line to the engine, rather than a rubber hose or other high-risk material.

are pumping oil. Check also for cracked lubricators. Most lube oil systems have rupture discs which relieve when the oil pressure becomes high due to flow stoppage. The release of system pressure causes a no-flow condition and engine shutdown.

BUILDING

☐ **Fresh air paks should be considered for use when the building or immediate area is gassy and needs to be entered,** as an example, to close a valve. Of course, if there is a chance of ignition, no one would be expected to enter the gassy area. The fresh air paks might be needed also if a fire developed and someone needed to enter a smoky area.

☐ **Vibration effects equipment near the compressor.** As an example, you might find pipe rubbing on floor grating, eventually causing a thinner pipe wall at one place.

Vibration caused the small nuts and bolts to drop off of a shield over electrical wires leaving one nut and bolt in place which caused the metal electrical shield to rub on some of the wires. Vibration will cause the metal shield to cut into the insulation of the electrical wires creating a short circuit (Figs. 11-12).

The building can be weakened structurally because of loosened bolts and cracked welds. All materials and debris left from an engine overhaul should be removed because vibration might cause it to fall later.

FIRE PROTECTION

☐ **Remote shut-off valves should be placed on the compressor suction line a considerable distance from the compressor** to stop flow should a fire occur. In many cases, check valves or shut-off valves are justified on discharge piping also because the gas can flow backward from pressurized processing equipment and continue to feed the fire. The location of the compressor may dictate locking the valves open.

The valves should be identified by color coding or a sign to indicate their service.

☐ **Metal thickness tests should be considered for compressor piping susceptible to external corrosion or internal corrosion and/or erosion** because of the atmospheric conditions and chemical analysis of the gas handled. Thickness tests should especially be made on ells where there can be erosion due to flow.

The metal will not be worn uniformly in the ell section, nor is there any reason to expect that thickness readings taken 90° apart will reveal appreciable wear. Readings may need to be taken every 45° or even 30° on large diameter pipe.

☐ **Gas detectors should be located in the sump area under compressors and also above each compressor unit.** They should not be placed too near the machine. Rather, they should be high enough so they will not interfere with maintenance and crane travel. They should be set to alarm at about 15 percent LEL.

In some cases, instrumentation will shut down the engine if 60 percent LEL is reached. A machine shutdown may create many problems. Therefore, a plant might be satisfied with only an alarm.

☐ **The compressor building at manned locations should have fire protection available.** Extinguishers containing a material suitable for flammable liquids and electrical fires should be placed near doorways and compressors.

Sufficient fire hose should be available from nearby hydrants to permit spraying cooling water on the area near the fire or to extinguish a lube-oil fire with a fine spray pattern. Cooling water is not expected to cause equipment warpage. Instead, it will prevent damage to conduit, monorails, roof and support columns until the fire dies due to lack of fuel. Hopefully, fuel feeding the fire will have been shut off earlier.

MAINTENANCE

☐ **Precautions should be taken by maintenance men when they work on compressors.** They should be certain the pressure has been relieved from the compressor cylinder prior to their work regardless of whether the work is to be done at the compressor or engine. The trapped pressure might cause the engine crankshaft to move. A gas test with a portable detector should be made before welding. Welders in remote locations should not make gas tests by passing a lit torch over the open flange of a vessel. In some cases, they will get by with that. However, if a large drum with a small opening has a flammable mixture, a serious accident will occur because of insufficient pressure-relieving area.

☐ **The flywheel should be locked in place when the engine or compressor is opened** to prevent injury to employees who must put their hands and arms inside the engine or compressor cylinder (Fig. 19). The guard should be on the flywheel while the machine is in service to prevent personnel or equipment from contacting the wheel.

☐ **Engine analyzers are good preventative maintenance tools** which allow you to safely operate over a longer period of time. Analyzers yield funny looking lines which tell an experienced analyzer operator of all types of problems inside the machine. They tell of bearing troubles, spark plug troubles and piston blow-out just to name a few.

☐ **If the water is drained from the engine radiator during a compressor shutdown because of repair work, fill**

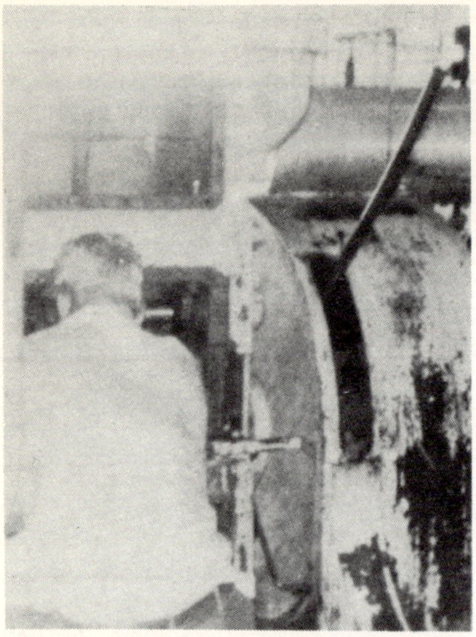

Fig. 19—This machine's flywheel bar is not locked out, and an employe has his arm inside the engine—a dangerous situation.

the radiator as soon as repairs are completed. If you don't, the deposits inside the radiator will dry. Then, when the engine is put back into service, deposits will circulate through the system. A filter ahead of the water pump will catch those materials. Outwardly the water pump will appear to be doing its job but its filter could be nearly plugged off resulting in a low circulation rate and a hot engine.

☐ **A means of preventing gas from leaking into the engine or compressor during overhaul is necessary.** That gas can leak into the room when the equipment is opened up. Before opening up the engine, vent the fuel gas line besides closing two block valves on the line. Some fuel gas shut-off valves automatically provide the double block and bleed; shutting the fuel-gas valve causes a vent line valve to open. If a positive shut-off such as a plug is inserted in a line which is not vented, pressure might build up behind the plug because of leaky block valves. Pressure must be bled off before attempting to remove the plug. A ¼-inch valve fitted on the plug is an easy reminder to employees that it is hazardous to remove the plug if it is under pressure. It is better to have a double block and bleed.

If opening up the compressor part of the machine, pancake/skillet blinds should be placed between flanges or a blind flange installed on existing piping. Those positive shutoffs are easier said than done at most installations. So a double block and bleed is satisfactory if the bleed line (i.e. a vent line to atmosphere) consists of more than a small-diameter tubing. Remember, very high pressures are held back by the block valve. If there is leakage around a gate or plug in a valve, a small vent line may not be able to fully bleed down the line. This can result in gas leaking through the second block valve into the compressor room.

It is possible, in some compressors, to install suction valves in the discharge pockets and vice versa. Improper installation could result in excessive internal cylinder pressures creating an extremely hazardous condition. Maintenance men should take precautions to properly identify the valves before installing them. Chalk-marking a valve which will be reinserted is a good practice.

It's easier to select and operate a compressor if you know its design and characteristics.

T. O. Kuivinen
The Cooper-Bessemer Corporation
Mount Vernon, Ohio

THE RECIPROCATING compressor has been well established for a wide range of applications. Speeds may range from 125 to 514 revolutions per minute. Piston speeds range from 500 to 950 feet per minute, the majority being 700 to 850 feet per minute. The nominal gas velocity through valving is usually in the range of 4500 to 8000 feet per minute. Discharge pressures may range from vacuum pumps to 30,000 pounds per square inch.

The reciprocating compressor fundamentally is a fixed capacity machine with variable capacity obtained by altering the speed of the prime mover. Capacity may be varied on multiple cylinder machines by making head or outer ends and crank or inner ends inoperative by utilizing devices that lift suction valves open manually or automatically. Additional clearance volume may be provided, which can be cut in or cut out, to vary the compressor capacity as required.

For some applications, it is often desirable to include a large clearance volume to maintain approximately constant horsepower over a wide range of pressure conditions. Thus, economical operation is attained by operating the prime mover at its rated load for a broad range of operating pressures.

During design, all parts are given thorough stress analysis to determine that operating stresses are within the endurance strengths of the materials used. Adequate factors of safety are incorporated to cover expected variations in manufacture, materials, and customer usage of the compressor.

Many of these designs are fatigue tested by submitting the actual cylinder assembly to pulsing pressure supplied by a hydraulic pump. Tests were carried out in increments involving at least 6 million cycles of pressure pulses at each pressure until some part failed. From these tests the maximum allowable working pressure for a suitable factor of safety is determined.

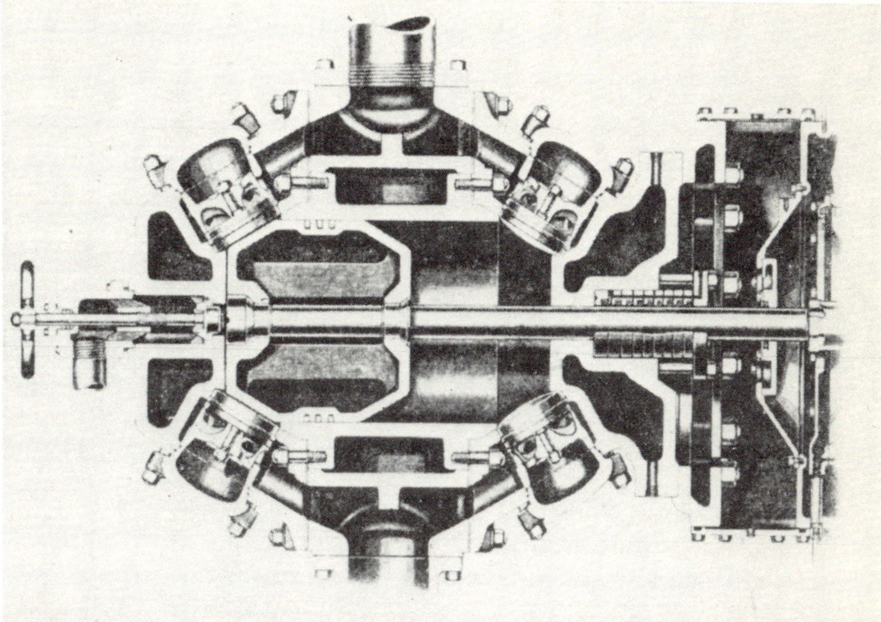

FIGURE 1—Compressor cylinder for medium pressure service. Multiple valves are used to keep clearance volume at a minimum.

Construction—Larger diameter lower pressure cylinders are usually constructed as shown in Figure 1. Pistons are of medium strength specification cast iron or of aluminum alloy. Either type are light construction with ribbing to maintain minimum reciprocating weight. Cylinder heads, cylinders, and cylinder liners are made of specification cast iron. Cylinder heads are fitted with multiple valves to keep clearance volume at a minimum. Piston rods are usually of medium carbon steel. The area on which the packing runs is hardened for long wear life (usually by induction heating and water quenching).

Piston rings are usually cast iron but may be special bronze, multiple piece cast iron, or multiple piece thermosetting plastic. Selection of this material is based on the gas and the type of the compressor service. Piston rod packing is usually multi-stage metallic using materials similar to the piston rings. Practically all compressor cylinder manufacturers supply piston rod packing engineered and manufactured by specialists in that field.

Design—Suction and discharge valves cover a greater area of individuality in design than other parts of a compressor cylinder. In general multiple disc plate valves or strip valves seating on multiported cast iron seats are used. The valve plates open against very light springs in a valve guard. Narrow seating areas and light springs minimize pressure differentials required to open the valves, to keep compression work loss to an absolute minimum. Valves are held in place by cages and caps that in turn are restrained by a circle of studs or bolts.

The outer head of a compressor cylinder is usually equipped with an additional clearance volume chamber that becomes effective by manually opening a built-in valve. In this way clearance space of the outer compression space is altered to reduce the capacity. The volume is usually determined while engineering the compressor cylinder application. Compressor cylinders are sometimes fitted with a fixed clearance pocket cast right in the head. Sometimes a selective clearance volume bottle is attached to the head. Sometimes a variable volume, controlled by manual or hydraulic positioning of an auxiliary piston within the clearance chamber, is provided.

For services of around 1000 psi, a cylinder design as shown in Figure 2 is often used. The higher pressures naturally require heavier walls in the body and head castings. Valves are arranged at right angles to the main bore.

Cylinders as shown in Figure 2 are designed with greater built-in clearance. Added clearance is obtained by using the unique double deck valve construction. When pressure ratios are low, cylinders of this type (equipped with double deck valves and large clearance volumes) are self unloading for a wide range of pressure variations. Thus the compressor driver remains at or near full load through a larger range of pressures and capacities with good economy in fuel consumption or motor efficiency.

For higher pressure service, the designer soon runs out of space for all components when he attempts the design with even the best grades of specification cast irons. A few years ago a new material, nodular iron, was developed that fills this need. Nodular iron is produced in the foundry similarly to high strength iron. Patented processing produces an iron having the free graphite uniformly distributed in spherical nodules. The nodular iron material is stronger since the notch effect caused by graphite flakes has been eliminated. Furthermore, nodular iron material is ductile.

Ductile iron's tensile strength compares with mild steel castings. Its ductility, though less than steel, runs from three to ten per cent elongation depending on whether the metal is used in the as-cast condition or annealed by subsequent heat treatment. Many designs exactly like Figure 2 have been giving excellent service up to 1500 psi using nodular iron castings. The versatility of this new high strength ductile foundry product should continue to promote usage at even higher pressures as acceptance increases.

Steel castings are used quite universally in higher pressure cylinders.

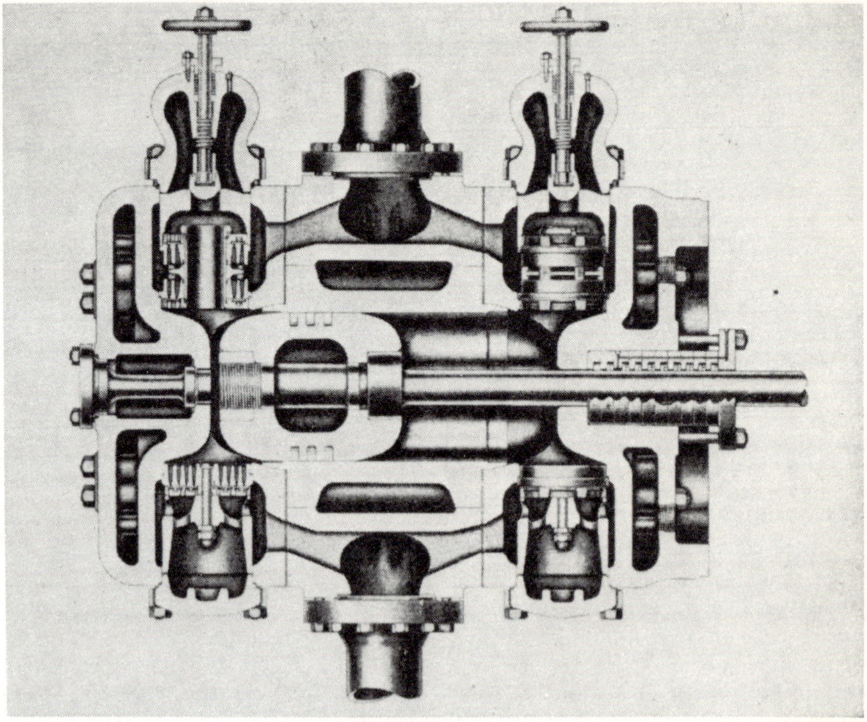

FIGURE 2—Compressor cylinder for 1000 to 1500 psi service. Heavy walls are required and valves are at right angles to the bore.

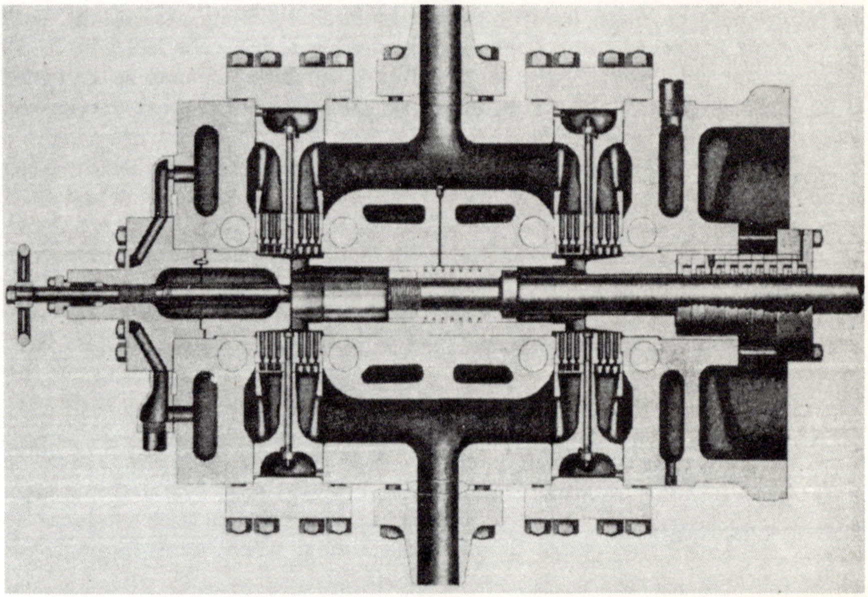

FIGURE 3—Compressor cylinder for 3500 psi service. Tie-in bolts help carry the load.

Since steel castings are more difficult to produce than iron castings compressor cylinder designs are limited to those employing fewer valves.

As the designer attempts to produce designs to withstand higher pressures he discoveres the maximum stress occurs at the junction of the bore for the valve and the main bore of the cylinder. As it becomes impossible to increase wall thickness at this point in an effort to reduce the stress, the design shown in Figure 3 has been produced. The cylinder body is a steel casting fitted with a shrunk in cast iron liner. Eight alloy steel tie bolts are installed at the junction points of valve pocket bores and cylinder bore. Controlled preset of these tie bolts at assembly places the material in compression stress at the junction point. Thus, the internal pressure, when in service, starts to stress the cylinder from a negative level so that maximum resultant stress is reduced. The bolts help to carry the load providing greater endurance strength to handle higher working pressures.

Suction and discharge valves, of heavier design for pressures involved, are mounted in the cylinder body. The cylinder heads are inserted within the body and firmly seated on metal ring gaskets by the cylinder head studs. Flat head gaskets are not suitable for the higher pressures. The usual clearance volume unloaders are provided when required.

Many of the high pressure chemical processes require recirculation in the system. This often means handling relatively small volumes at high pressures and extremely low pressure ratios. Single acting cylinders result in heavy bearing loads on the driving frame. Double acting cylinders also give high unbalance between the outer end and the inner end because the piston is hardly larger in diameter than the piston rod. As a result, recirculator cylinders are usually built as double acting cylinders with a tail rod of the same diameter as the piston rod. Extra crossheads are unnecessary because the piston and rod weight is low. The rod is easily guided by using bronze bushings in the inner and outer cylinder heads. Such a design is shown in Figure 4. Usually the pressures in this design are in the range where steel forgings are required in the cylinder body and in the heads. Steel forgings can be produced in materials of greater tensile and endurance strength than obtainable in steel castings. Higher cost, however, is involved because the shapes required in a compressor cylinder must be produced by machining when steel forgings are used.

When extreme pressures are encountered, the design choice of necessity becomes the single-acting plunger-type of cylinder. Special materials are involved to get plunger surfaces hard enough to endure packing pressure. Nitrided steels are most commonly used. Cylinders are forged steel with iron liners with extremely careful attention paid to design details in order to minimize stress concentrations. Special valve designs, usually operating in line with the cylinder axis, are required to endure the dense gas being handled. All parts are more heavily constructed, and many parts are made of very high strength steels.

Figure 5 shows a design for 15,000 psi service. The plunger attaches to a spherical seated aligning device driven from the driving gear crosshead. Without such an aligning means, it would be difficult to maintain the operation of the metallic packing that seals the plunger.

Many of the gases being compressed in various processes react with some of the materials ordinarily used in compressors. The engineering of such applications begins with consultation and determination of corrosive and other effects. Valves may be made of stainless steels, piston rods may be stainless steel coated or nitrided. Liners may be of special alloys like Ni-Resist iron. Piston rings and piston rod packings might be ordinary cast iron, bronze, thermosetting phenolic plastic, or babbit-faced iron. Sometimes it is necessary to plate gas passages with a suitable coating, but this is quite rare. The combination metallurgist-designer team is essential to solution of such special applications.

Certain processes require the delivery of compressed gas or air that is absolutely devoid of any lubricant. If such is the case, it is often more expedient to construct the compressor for non-lubricated operation rather

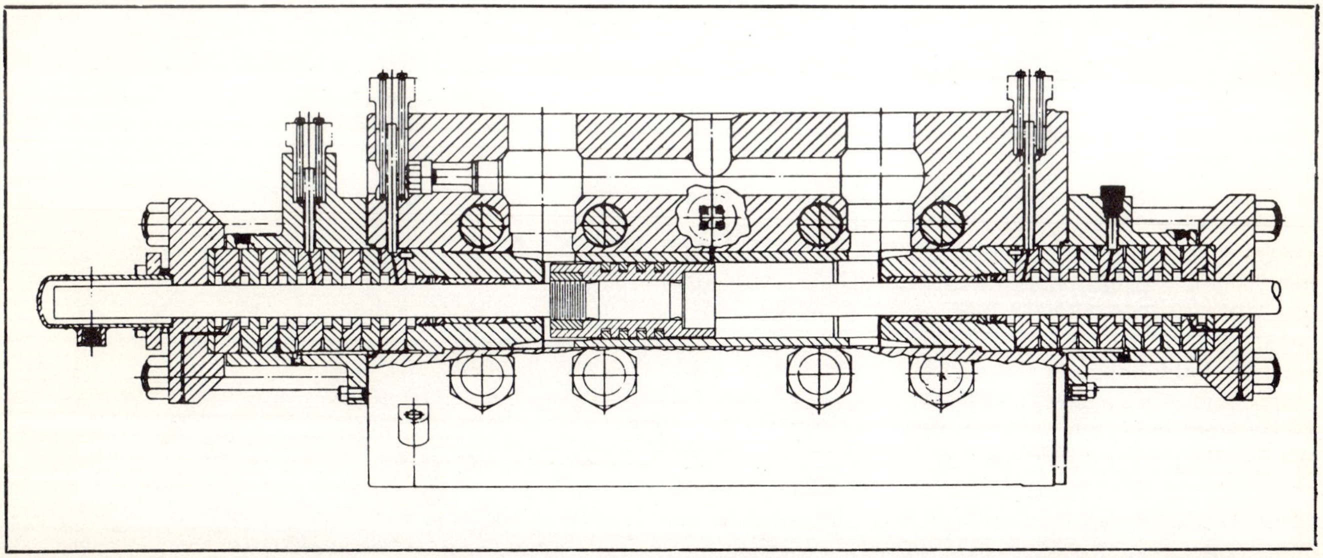

FIGURE 4—Recirculator cylinder for high pressure service. Double acting cylinder with a tail rod of the same diameter as the piston rod.

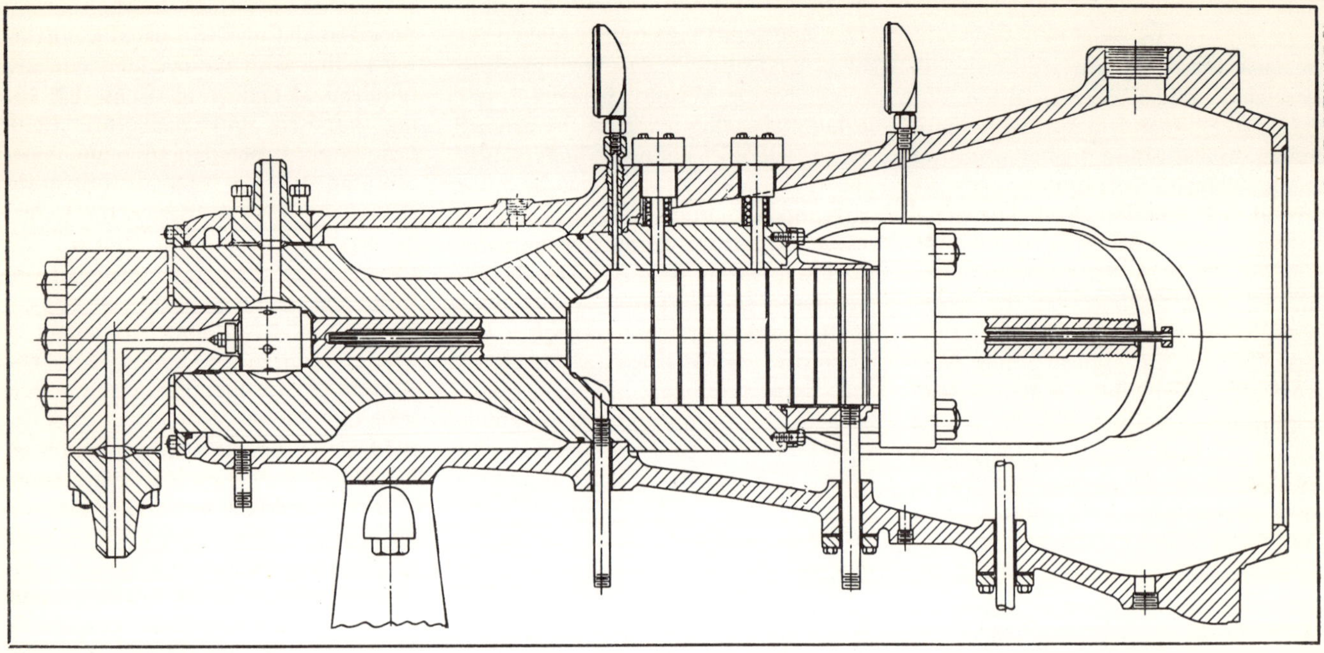

FIGURE 5—Single-acting cylinder for 15,000 psi service. Plunger attaches to a spherical seated aligning device.

than attempt to deliver absolutely dry gas by extraction of lubricant carried over in the gas after compression.

The usual solution to this problem is to construct the piston of multiple disc spacers of dense fine grain graphite. Sectional piston rings, backed up by an expander spring, of the same type of graphite operate in the ring grooves formed between the piston pieces. The piston rod packing is also made of this same carbon material. Piston rods, hardened as usual, and cylinder liners are machined, ground, and honed to a very fine micro-finish. A very close grained iron is used for the cylinder liner.

Best wear life of carbon parts is attained when there is present in the gas a very slight amount of vapor. The well known platforming process is served with such designs as shown in Figure 6.

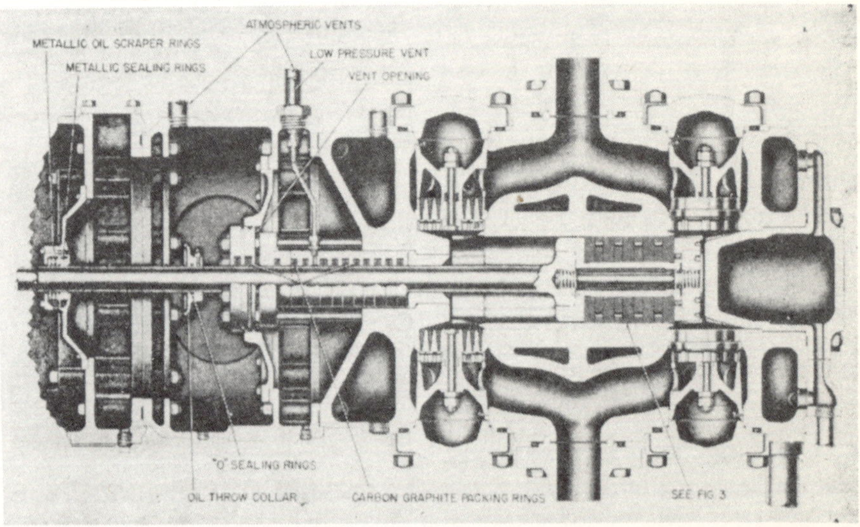

FIGURE 6—Non-lubricated compressor cylinder for moderate pressure. Best wear life of carbon parts if the gas has some vapor.

Lubrication—Reciprocating compressor cylinders are lubricated by the force feed mechanical lubricator. The lubricator is equipped with multiple units of individual plunger pumps with the unit having a built-in oil supply sump. A metered quantity of oil is forced into the cylinder at one or more points through check valves and to each packing as required. Such lubricators may be mechanically driven by the compressor or by an individual electric motor. The electric motor is more often used with motor-driven compressors to provide positive lubrication before the compressor motor is started.

Correct lubrication is vital to efficient operation of the compressor cylinder. In many cases the cylinders may be lubricated with the same oil as used to lubricate the engine or the compressor shaft and connecting rod bearings.

There are many cases where special consideration of lubrication problems is required. When the discharge temperature is high, the lubricant must have a high flash point when compressing air. Oxygen compressors must not use petroleum products nor vegetable oils for lubricants. Certain synthetic lubricants are suitable. When the compressor cylinder is handling gases carrying gasoline vapors or other condensible hydrocarbon vapors, a vegetable oil, or some combination of vegetable and petroleum oils may be used. A soap lubricant is sometimes recommended.

If gases are highly dehydrated, there

is a tendency to absorb the lubricant. A heavier bodied oil must, therefore, be used. Some gases have a chemical affinity for petroleum oil. In this case a special lubricant must be used, depending upon the gas.

In some processes, gases are involved which cannot be contaminated with lubricating oil. A lubricant must be used which, if carried into the gas, will not produce any ill effects. For example, an alcoholic soap lubricant has been found successful in compressing propane and carbon dioxide.

Compressor Drivers—The particular gas being compressed may not necessarily dictate the method used to drive the compressor. If the plant arrangement and availability of fuel dictates using an internal combustion engine, then compressor cylinders are generally integral as a part of an engine-compressor unit. Electric motor drives are frequently used. Some installations have available process steam; so, geared steam turbines become the logical choice. In the latter two cases, the compressor cylinders are attached to a compressor frame carrying its own crankshaft, connecting rods, and crossheads.

Quite a range of unit arrangements have been installed. These seem to run in all combinations from single cylinder compressor units to eight cylinder units and from one to six stages. The compressor manufacturers generally engineer each application to have an individual crank, connecting rod, and cross head for each compressor cylinder. Sometimes the best solution for multi-stage units requires placing some of the cylinders in tandem on a single connecting rod and crosshead.

It becomes apparent that the number of cranks, size of compressor pistons and phase relationship of multiple crank units influences the nature and magnitude of inertia forces resulting from the reciprocating motion. The compressor manufacturers keep these forces minimized by suitable arrangement of angles between the various cranks, by counterweighting, and by matching piston weights as near as possible.

Foundations—When engine-compressor units are installed, foundation requirements are controlled by the engine. A foundation designed as wide as possible and with sufficient mass will operate satisfactorily without excessive transmission of inertia forces to surrounding equipment and structures.

Reinforced concrete is the most universally used foundation construction. If the proposed location requires consideration of using friction piling, then the foundation construction must be carefully engineered to cope with the nature of the inertia forces involved. Piling structures must avoid resonance with the frequency of the reciprocating inertia forces and be of ample strength and suitable distribution to avoid forced vibration from these forces.

Motor or turbine driven compressor unit foundations are engineered to suit the compressor requirements. Breadth rather than depth is desired in the concrete mass. High foundations should be avoided so as to avoid rocking frequencies of the foundation that might be in resonance with inertia force frequencies.

A well engineered installation contributes greatly to successful long life performance of reciprocating compressors. A poorly engineered installation invariably results in many problems that preclude good performance no matter how well engineered the compressor unit itself might be.

Vibration—With multiple crank compressor units there are the problems of torsional vibration resulting from the harmonics of the varying pressures within the various compressor cylinders. The natural frequency of the compressor shafting may be controlled by suitable engineering of the coupling between the compressor and the driver. It can be controlled by selection of counterweights, and by adjustment of the mass of the driver or compressor pistons. The engineer correlates these factors to obtain a desired natural frequency to keep serious critical speeds removed from the operating speed range. Compromise enters the picture if the desired speed range conflicts with a feasible solution of the problem.

Along with the torsional vibration problem, it is necessary to investigate torque variation, caused by the compressor cylinder pressures, when motor drives or geared steam turbine drives are used. This must be done for all conditions of unloading.

When alternating current motors are used as drivers, suitable WR^2 (flywheel effect) must be incorporated to keep current pulsation (caused by torque variation) within the tolerable limits established by motor manufacturers. Otherwise overheating and inefficient operation occurs. In the case of synchronous motors, the pulsation may be large enough to pull the motor out of synchronism and trip it off the line.

When geared turbine or other geared drives are used, torque variation resulting from compressor pressures or harmonics thereof may exceed the tolerance prescribed by the gear manufacturer. Excessive torque variation results in high tooth stresses and runs the risk of gear failure. Any torque reversal is intolerable because of gear tooth separation resulting in severe impact stress and impact noise. Short gear life in such cases is inevitable. High maintenance costs and unwanted down time are the penalties.

Performance Characteristics of Reciprocating Compressors

Check this to be sure that the compressor will operate according to its design.

Howard M. Boteler
Worthington Corporation
Houston

THROUGHOUT the various branches of the petroleum and petrochemical industries one finds a great variety of reciprocating compressors performing a myriad of chores, each essential to the particular process to which applied. Some of the newer units may still be performing the jobs originally assigned them, operating under original contract conditions. Many others have been assigned new duties, some of them perhaps several times.

Frames and Cylinders—A reciprocating compressor consists essentially of a frame (or frames) to which are attached the necessary compressor cylinders and driver, plus the necessary auxiliaries such as intercoolers where required (either machine mounted or remotely located) interconnecting gas piping, jacket water piping, etc. Every compressor manufacturer has a group of standard frames, each having definite design limitations on speed and load carrying capacity. Load carrying capacity involves two considerations, namely horsepower and forces created by pressure differential across the pistons. The latter is generally referred to as frame load.

Horsepower limitations are determined by the ability of the crankshaft and crank webs (or crank discs) to withstand the necessary torque or turning effort without overstress, and by the ability of the bearings to dissipate the heat of frictional loading.

Frame load limitations are established to prevent the stresses in the frame, piston rods, connecting rods, bolting and other frame parts from exceeding safe conservative values, and to eliminate the imposition of excessive pressure per projected unit bearing area. Frame load for each compressor crank is equivalent to the sum of the differential loads of all pistons attached to that crank. The load for each piston is the difference in the products of net piston areas and pressures exerted on the piston faces. Loadings should be determined for both directions of piston movement. The loadings are not the same in each direction and where the ratio of piston rod area to piston area is high, they may be vastly different. An undesirable negative loading may sometimes result and provision must be made, such as a tail rod addition, to remove the unbalance.

Essential Information—Before a truly complete analysis of a compressor performance problem can be made certain essential data concerning the problem must be available. Such information must come from the prospective user in the inquiry. Figure 1 is an inquiry data sheet indicating facts which must be known for a full analysis. If the prospective user will furnish all of the information requested in this form, a much more intelligent analysis of the problem can be made than would otherwise be possible. A glance at the form will show numerous essential and revelant data. How essential the information is can perhaps best be realized by understanding how it is used.

To better appreciate this, however, it is necessary to become familiar with certain terms and definitions used in the industry. Figure 2 sets forth the more common ones. These have perhaps appeared in print on many previous occasions, but are repeated here for convenience so that their true significance may be better understood.

The inquiry data determines the types and size of compressor and driver needed if new equipment is requested or if desired, whether an existing compressor and its driver are suitable for the stated service conditions. Or whether modification can be made to an existing machine to permit its use safely under the stipulated conditions. The procedure is briefly this:

1. The compression characteristics of the gas are determined. For single component gases the necessary information can be found in published tables. For mixed gases these characteristics are determined from the gas analysis. The physical constants to be determined are the K value (C_p/C_v) the molecular weight, the critical pressure (Pc) and the critical temperature (Tc). The critical pressure and temperature as determined from the analysis of mixed gases are sometimes termed pseudo-criticals.

2. The total ratio of compression is determined. This is the ratio of final absolute discharge pressure to absolute initial suction pressure. It must now be decided whether the compressor is to have 1, 2, 3 or more stages of compression. Experience will indicate quite readily how many are needed. The primary consideration is the expected discharge temperature. A sufficient number of stages, with cooling between them, must be used to prevent the discharge temperature of any of them exceeding practical limits.

Another consideration is the relationship of volumetric efficiency and ratio of compression. Gases having the higher K values will show a higher volumetric efficiency for the same cylinder clearance and ratio of compression than low K value gases. An extremely low volumetric efficiency for normal service is not desirable since an excessively larger cylinder is

INQUIRY DATA SHEET

1. User__________ Location__________
2. Contractor or Agent__________ Location__________
3. If this inquiry is for revised service conditions for Existing Compressor(s) give size_____ type_____ Serial Number(s)_____
4. Type of gas_____ If mixed gases list on the reverse side each constituent defined as % volume or % weight; mols or pounds (state which).
5. Capacity_____ Mols._____ Lbs._____ cubic feet_____ per minute_____ per hour_____ per 24 hours_____ measured at_____ PSIA,_____ (°F)_____ (°C)_____ water vapor included,_____ excluded_____.
6. Number of compressors_____, each to handle_____% of capacity stated in (5).
7. Rela. Humidity % (for air comps)_____; water vapor content_____% saturation.
8. Customer's statement of molecular weight_____, K value (CP/CV)_____, critical pressure (Pc)_____ PSIA, critical temperature_____ (oR)(°F)(°C).
9. Customer's statement of (a) corrosive constituent_____
   (b) Polymer Formation_____
   (c) Regards maximum discharge temperature permissible_____
   (d) Catalyst carry over_____
   (e) Liquid entrainment_____
10. Compressor suction conditions: Temp._____ (°F)(°C); Pressure_____ (psia)(psig)
11. Compressor discharge pressure:_____ (psia)(psig)
12. Altitude at plant site_____ ft. above sea level.
13. Ambient air conditions (for filtering recommendations)_____
14. Cooling water temp., max._____ Min._____ (°F)(°C)
15. Drive:
    Motor: Synchronous_____; Induction_____; Volts_____; Phase_____; Cycle_____;
    Power Factor_____ Enclosure_____ Explosion proof_____
    MVD_____; Coupled_____; Direct conn._____
    Steam: Steam at throttle_____ (psia)(psig) and_____ (°F)(°C) Total Temp.
    Exhaust pressure_____ (Inches HG abs)_____ (Inches HG Vac)_____ (psia)(psig)
    Gas Engine: Fuel gas Btu/cu.ft._____ (LHV)(HHV) Pressure_____ psig
    Free hydrogen_____ % volume. Type gas fuel_____
    Oil Engine: Type oil_____ Cetane rating_____
16. Compressor Control:_____; From suction_____; From disch._____;
    Manual_____; Automatic_____; Capacity reduction, total_____%
    No. steps_____ Air pressure available for control operation_____ psig
17. Space limitations; Width_____ Length_____ Height_____
18. Date

FIGURE 1—Inquiry Data Sheet—List of facts to be determined for analysis of compressor performance.

needed to obtain the desired capacity. This results in a heavy differential load (frame load) and may require a heavier frame than would otherwise be needed.

It is customary in the refinery and chemical process industries for the project engineer, for process reasons or otherwise, to establish the number of stages of compression, and for each, the suction pressure, suction temperature and discharge pressure with due consideration given to pressure drop and cooling between stages. If this information is not contained in the inquiry the manufacturer will ordinarily assume nearly equal ratios per stage to be satisfactory and will assume appropriate intercooler and inter-connecting piping pressure losses between stages. He will likewise assume suction temperatures for the second, third, etc. stages consistent with the available cooling water temperatures as shown in the inquiry.

This part of the procedure shows why it is essential that not only the suction pressure, but all interstage pressures, if given, as well as final discharge pressure be fully identified as either gage or absolute. Where gage readings are specified the feet elevation above sea level at the plant site must be known so that absolute pressures, mandatory in selection procedure, might be determined.

3. The required capacity per compressor is determined and is converted to cubic feet per minute at inlet conditions of the respective stages. The proper corrections for temperature, pressure and compressibility must be made. Where the desired capacity is stated as dry gas, the inlet volume of each stage must be increased by the ratio of the total absolute pressure to absolute partial pressure of the dry gas at stage inlet. This step of the procedure again emphasizes the necessity for clarity and completeness in the inquiry data. The desired capacity rate must be explicit and the conditions of pressure and temperature at which measured *must* be given.

4. A rough check is made to determine the horsepower required. For this purpose generalized charts are used. The factors involved are the total actual capacity of each stage (*cfm* at inlet), the ratio of compression per stage, the specific gravity and the *K* value (C_p/C_v) of the gas. Again this emphasizes the requirement for complete gas characteristics, clearly stated pressures, temperatures, and capacities.

The total horsepower required for the compressor (sum of horsepowers per stage) determines the required frame size. The kind of drive, whether steam, motor, gas, etc. as stated in the inquiry will dictate the type of frame best suited. There is usually an established length stroke and speed for the frame selected. However, for motor or steam turbine drive it may be necessary or advantageous to modify one or both to better match driver characteristics. Having established these details, the piston speed, which in feet per minute is equal to $\frac{\text{stroke in inches}}{12} \times 2 \times \text{rpm}$, as needed in Step 5, can be calculated.

5. The approximate compressor cylinder sizes can now be determined.

TERMS AND DEFINITIONS

1. D = Piston displacement, expressed as cubic feet per minute, is equal to square feet of net piston area × length of stroke in feet × number of compression strokes per minute.
2. V_s = Actual capacity, cubic feet per minute, compressed and delivered, expressed at the conditions of total temperature and total pressure prevailing at compressor inlet.
3. VE = Volumetric Efficiency = $\frac{V_s}{D}$
4. THP = theoretical horsepower is the horsepower required to compress isentropically the gas delivered by the compressor through the specified pressure range. This is often referred to as the adiabatic horsepower.
5. CIHP = Compressor indicated horsepower is the actual work of compression developed in the compressor cylinder(s) as determined from the indicator card.
6. CE = Compression efficiency is the ratio of theoretical horsepower to compressor cylinder indicated horsepower = $\frac{THP}{CIHP}$
7. BHP = Brake horsepower is the measured horsepower input at the crank shaft.
8. SIHP = Steam indicator horsepower is the power developed in the steam cylinder(s) of an integral or coupled steam engine driven unit, as determined from the steam cylinder(s) indicator card, to perform the required gas compression duty.
9. ME = Mechanical efficiency is the ratio of compressor cylinder indicated horsepower to brake horsepower or to steam indicated horsepower = $\frac{CIHP}{BHP}$ or $\frac{CIHP}{SIHP}$
10. OE = Over-all efficiency is equal to (CE) × (ME) or $\frac{THP}{BHP}$ or $\frac{THP}{SIHP}$
11. P_s = Suction pressure, pounds/square in absolute.
12. T_s = Suction temperature, degrees Rankine = °F + 459.6.
13. P_d = Discharge pressure, pounds/square in absolute.
14. T_d = Discharge temperature, degrees Rankine = °F + 459.6.
15. R = Ratio of compression = P_d/P_s.
16. C_p = Specific heat at constant pressure is the number of BTU required to raise the temperature of 1 pound of perfect gaseous fluid one degree Fahrenheit, pressure remaining constant.
17. C_v = Specific heat at constant volume is the number of BTU required to raise the temperature of 1 pound of perfect gaseous fluid one degree Fahrenheit, volume remaining constant.
18. K = The ratio of specific heats = C_p/C_v.
19. T_c = Critical temperature is the temperature, degrees R, of a gas above which it cannot be liquefied.
20. P_c = Critical pressure and is the absolute pressure per square inch required to liquefy a gas at its critical temperature.
21. Pr = Reduced pressure is equal to P/P_c.
    Prs = Reduced pressure at suction pressure = P_s/P_c.
    Prd = Reduced pressure at discharge pressure = P_d/P_c.
22. Tr = Reduced temperature is equal to T/T_c.
    Trs = Reduced temperature at suction temperature = T_s/T_c.
    Trd = Reduced temperature at discharge temperature = T_d/T_c.
23. F = Compressibility factor which expresses the deviation from the perfect gas law. This is sometimes given the symbol "Z".
24. F_s = Compressibility factor at compressor inlet pressure and temperature.
25. F_d = Compressibility factor at compressor discharge pressure and temperature.
26. f = Ratio of compressibility factors = F_d/F_s.
27. C = Cylinder clearance volume and is the volume in one cylinder end in excess of the piston displacement per stroke. It is usually expressed as a percentage of stroke displacement.
28. TV = Terminal volume is the space occupied at discharge pressure and temperature of a unit volume of gas entering the cylinder at suction pressure and temperature.

FIGURE 2—Terms and definitions used in this article as applying to reciprocating compressors.

The required capacity of each has already been established (Step 3). From appropriate charts or formulae the volumetric efficiency of each stage of compression is developed. This involves the ratio of compression, the *K* value of the gas and the compressibility factors at inlet and discharge conditions. A trial selection must be made, assuming cylinder clearance, and the size and clearance readjusted to match cylinder patterns and/or designs available.

This procedure again emphasizes the requirement for exacting inquiry data. If facts for establishing critical constants are missing the cylinders could be sized incorrectly, sometimes appreciably.

6. The frame loadings can now be checked to be sure that all are within the design limits established for the selected frame. If all are too high a heavier frame must be used. If one of several are high it may be possible to readjust the ratios of compression to bring all loadings within the rating. Failing this, a heavier frame becomes necessary.

7. Assuming the frame requirements have been met, the selection, so far as stroke, speed, number of stages and sizes of cylinders is now complete. There remains one more detail—a more accurate check of the horsepower required. In Figure 2 will be noted the terms "compression efficiency" and "mechanical efficiency." In Step 4 generalized horsepower charts were used. Such charts are termed generalized because they are based on average values of mechanical and compression efficiencies. It should be emphasized here that it is impossible to include in any one horsepower chart, such as the widely used "bhp per million" chart, other than average values for these two factors. Therefore, after the selection has been made and all factors involved in determining the compression and mechanical efficiencies are known, the horsepower can be quite accurately established.

The selection is now complete with full performance characteristics known for the conditions of service for which selected.

Multiple Service—Quite often an inquiry will list one or more alternate service conditions, such as a range in suction pressure or discharge pressure or both. Or perhaps, operation on alternate gas streams may be desirable, where pressures and capacities may or may not be identical. In such cases the requirements must be analyzed and the selection based upon the most severe conditions. It may be necessary to select cylinder sizes for one service, to obtain the desired capacity, and to base frame selection on the alternate service because of high horsepower and/or frame loadings required by the alternate.

PERFORMANCE CALCULATIONS

The Indicator Card—Figure 3 lists some useful formulae used in compressor selection and performance calculations. From these formulae the variables involved can readily be seen.

PERFORMANCE FORMULAE RECIPROCATING COMPRESSORS

1. **Ratio of Compression** = ... $\frac{P_d}{P_s}$

2. **Theoretical Volumetric Efficiency** = ... $1.00 - \left(\frac{R^{\frac{1}{k}}}{f} - 1\right)C$

3. **Theoretical Means Effective Pressure** = ... $\frac{KP_s}{K-1}\left(R^{\frac{k-1}{k}} - 1\right)$

4. **Theoretical (isentropic) Horsepower** = ... $.004364 \times V_s \times (\text{M.E.P.})$

5. **Theoretical Discharge Temperature** = ... $T_s\left(R^{\frac{k-1}{k}}\right)$

6. **Theoretical Terminal Volume Expressed as Percent of V_s** = ... $\frac{f}{R^{\frac{1}{k}}}$

Note: All pressures and temperatures are in obsolute units. For nomenclature refer to Figure 2.

FIGURE 3—Performance Formulae—Used to determine compressor characteristics.

The formulae given are all theoretical. Certain modification of results obtained must be made to arrive at actual results. For instance, the actual volumetric efficiency is always lower than that derived from the formula shown because allowance must be made for entrance losses due to wire drawing through the cylinder ports and valves and for increase in volume due to heating during the inlet period. In addition, compensation must be made for leakage past piston rings and cylinder valves and for packing vent losses. Obviously, leakage is primarily a function of pressure differential and type of cylinder construction. Non-lubricated cylinders (carbon piston rings, etc.) can be expected to develop greater slippage than those receiving lubrication, as the oil effectively seals surface contacts in most cases. However, also of importance is the type of gas being handled. Certain gases are simply harder to retain than others and the problem is closely connected with molecular weight and viscosity. There are no hard and fast compensating factor that can be given to cover all these variables. It is customary to allow 3 percentage points for wire drawing and heating entrance losses, so that 0.97 may be substituted in the formula for 1.00. Slippage losses may vary from 2 to probably a maximum of 5 percentage points for ordinary applications involving oil lubricated cylinders and from perhaps 4 to perhaps 10 percentage points for non-lubricated construction. Thus a more practical formula might be:

Volumetric Efficiency = $.97 - \left(\frac{R^{1/k}}{f} - 1\right)C - L$, where L is the slippage loss just discussed.

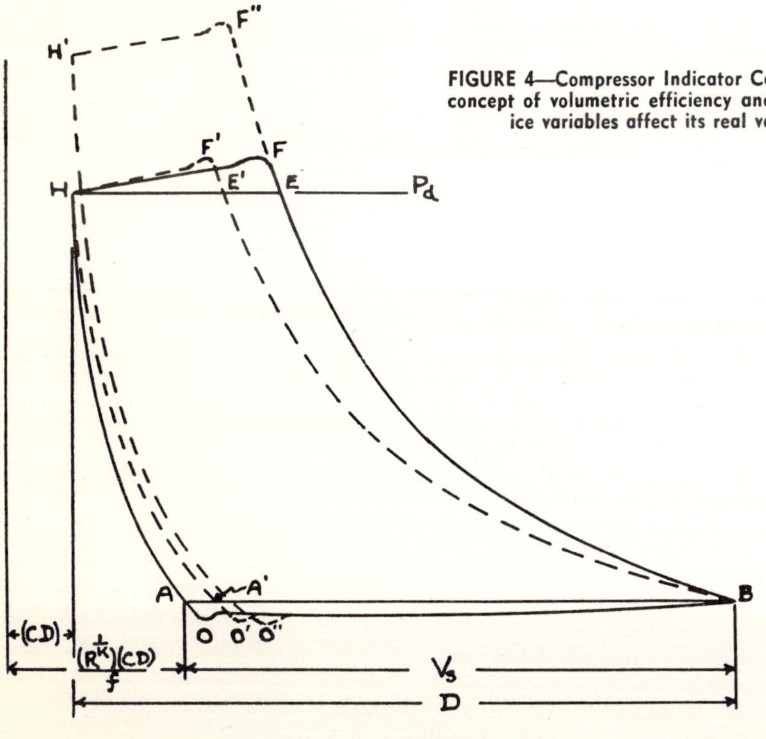

FIGURE 4—Compressor Indicator Card—Visual concept of volumetric efficiency and how service variables affect its real value.

A visual concept of volumetric efficiency can be had by reference to Figure 4. The indicator card shows what takes place within the compressor cylinder. The shape or appearance of the card varies with pressure conditions, cylinder clearance, gas characteristic and other factors.

In Figure 4, diagram ABEH represents behavior of a perfect gas under ideal conditions. P_s and P_d represent the suction and discharge pressures at the inlet and discharge flanges of the cylinder respectively. Movement of the piston from left to right represents the suction stroke. Point B represents the extreme point of piston travel on the suction stroke.

At B, the direction of piston travel is reversed and as the piston moves from right to left on the discharge stroke, compression of the gas trapped within the cylinder proceeds along line BE. At point E, pressure within the cylinder is equal to discharge pressure. Point E, therefore represents effortless opening of the discharge valve. From this point to the end of the discharge stroke at H, the discharge valve remains open while gas is being delivered to the discharge system.

It will be noted, however that some of the gas has been compressed into the clearance spaces within the cylinder and remains in the cylinder when the discharge valve closes. These clearance spaces are found between the valves and the piston, in the passageways in the valves themselves, in head cutouts under the valves for gas passages, linear clearance between the piston and head faces, etc.

As the piston reverses direction at H and again starts on the suction stroke, the gas at discharge pressure in the clearance space begins to expand. Volume increase and pressure reduction continue as the piston moves. Finally, the pressure in the cylinder is equal to suction pressure, P_s, and the effortless suction valve opens at A, permitting the inflow of fresh gas for that portion, AB, of the suction stroke.

Since the displacement is a function of the length of stroke and since the length of the indicator card is proportional to the length of stroke, we can in the diagram, substitute displacement, D, for length of indicator diagram. We can also represent that portion AB of the suction stroke where in gas is flowing into the cylinder by V_s, the actual capacity at intake conditions. $\frac{V_s}{D}$, then is the apparent volumetric efficiency. The value indicated here is usually higher than the actual because the reduction caused by leakage loss is not fully effective at the point of valve opening. There is some leakage from the clearance spaces past the rings and suction valves, as the gas expands. This creates a faster than normal pressure decline and would tend to move point A to the left, for an apparent increase in volumetric efficiency. On the other hand, the expansion from H to A is not completely isentropic and therefore the temperature and volume are greater than theoretical. This slows the actual pressure drop and would tend to move point A to the right for a decrease in volumetric efficiency.

The increase in volume during the intake, AB, because of heat picked up by the gas, would slow the rate of flow from the suction system. In addition, leakage past the rings and valves during the discharge stroke reduces the volume delivered to the discharge system. Since the definition of actual capacity takes these losses into consideration, so must the volumetric efficiency. Further, the actual valve opening does not occur at A. The pressure in the cylinder must be sufficiently below the pressure at the inlet flange to overcome the effect of unbalanced valve area and initial spring resistance before the valve can open. The actual opening is, therefore at point O rather than at point A. Since none of these losses are reflected in the apparent or indicated volumetric efficiency, it is usually higher than the actual. The deviation will vary, however five percentage points might be considered the expected average.

The equation for theoretical volumetric efficiency shown in Figure 3 can be explained also by reference to the indicator diagram; Figure 4. For the expansion we can write the equation

$$\frac{P_1V_1}{T_1F_1} = \frac{P_2V_2}{T_2F_2}$$

Rearranging

$$V_2 = \frac{T_2F_2P_1V_1}{T_1F_1P_2}$$

Substituting subscripts d (Discharge) and s (Suction) for subscripts 1 and 2 respectively.

$$V_2 = \frac{T_sF_sP_dV_d}{T_dF_dP_s}$$

But:

$V_2 = V_{sc} =$ the volume of the clearance when expanded isentropically from pressure P_d to pressure P_s.

Also:

$$\frac{T_s}{T_d} = \frac{1}{R^{\frac{k-1}{k}}}$$

$$\frac{F_s}{F_d} = \frac{1}{f}$$

$$\frac{P_d}{P_s} = R$$

$$V_d = CD$$

Substituting:

$$V_{sc} = \frac{1}{R^{\frac{k-1}{k}}} \times \frac{1}{f} \times R \times CD = \frac{\left(R^{\frac{1}{k}}\right)\left(CD\right)}{f}$$

$$\text{Volumetric Efficiency} = \frac{V_s}{D}$$

$$\text{But } V_s = (D + CD) - \frac{(R^{1/k})\,(CD)}{f}$$

Substituting

$$\text{Volumetric Efficiency} = \frac{D + CD - \frac{(R^{1/k})\,(CD)}{f}}{D}$$

$$= 1 + C - \frac{(R^{1/k})\,(C)}{f} = 1 - \left(\frac{R^{1/k}}{f} - 1\right)C$$

Brake Horsepower—The theoretical horsepower naturally does not include fluid or mechanical friction losses. The fluid losses are grouped in the term Compression Efficiency while the mechanical ones are covered by the term Mechanical Efficiency.

The major factor involved in determining the compression efficiency is the valve loss or pressure drop through the inlet and discharge valves. These fluid losses are a function of gas density and valve velocity. The suction and discharge pressures and the molecular weight establish the density. The valve velocity is fixed by the valve area available in the selected cylinders and the piston speed. Valve velocity is normally stated in feet per minute and is the ratio of piston area to valve area per cylinder corner multiplied by the feet per minute piston speed. It is likewise equal to $\frac{144D}{A}$ where D is the piston displacement of one cylinder, in cubic feet per minute, and A is the valve area in square inches per cylinder corner. By valve area per corner is meant the area of the suction *or* the discharge valves in one end of the cylinder.

Valve losses are usually read from graphs plotted from research data. Applicable corrections for density are designated. There is at present, so far as is known, no precise general formula available for calculating these losses because of the numerous variables in valve designs in use, such as spring loadings, unbalanced areas and flow patterns.

A better understanding of the losses involved in compression efficiency may also be obtained by reference to the indicator diagram. If it were possible to get the gas into and out of the cylinder without fluid losses, the indicator card ABEH could be realized. This card may be said to represent the ideal or theoretical horsepower requirements. But fluid losses are present. Therefore the actual inlet pressure in the cylinder is below that at the cylinder inlet flange. Likewise the pressure in the cylinder during the delivery interval EH is above that at the cylinder discharge flange.

A more practical card would be AOBFH. Here the area AOB, fluid losses through the inlet ports and valves, and the area EFH fluid losses through the discharge valves and ports are included in the card area. Since the horsepower is proportional to the card area, the larger area by reason of fluid loss inclusion, means greater horsepower.

There are other considerations. During compression, BF, a small portion of the gas continually slips past the piston rings and suction valves. Work has been done on this gas, yet it is not delivered to the discharge system. Also slippage past the discharge valves allows gas which has already been delivered to the discharge system to return to the cylinder. Recompression and redelivery take place.

Unless leakage is abnormal the theoretical location of point E is not

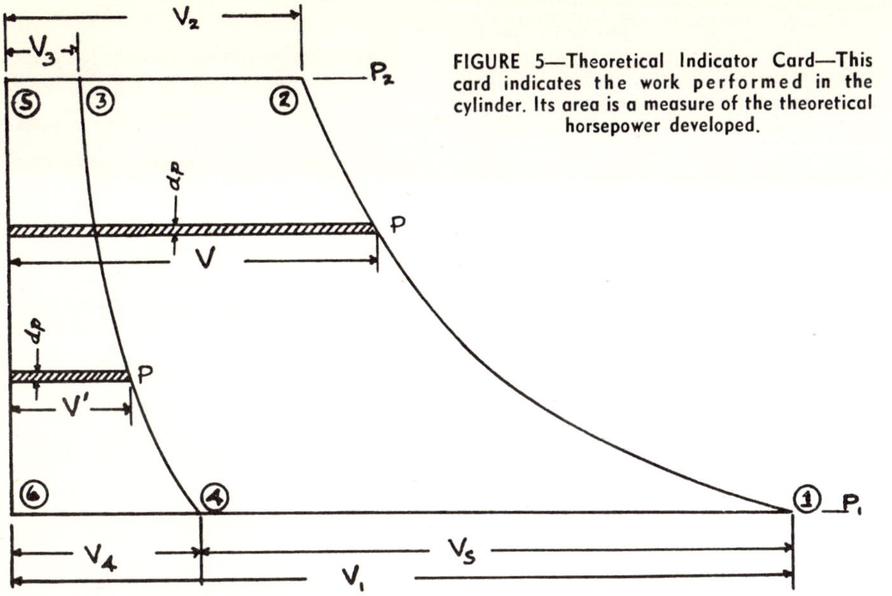

FIGURE 5—Theoretical Indicator Card—This card indicates the work performed in the cylinder. Its area is a measure of the theoretical horsepower developed.

greatly altered. Yet an over-all loss has occurred, first because more gas must be taken into the cylinder to compensate for piston ring and suction valve slip, secondly because work is performed on this lost capacity, and finally because leakages back through the discharge valves must be recompressed and redelivered to the discharge system.

There is still another factor, the cooling effect of cylinder jacketing. Removal of heat by the jacket water would shrink the volume during compression and would tend to move point F to the left reducing the power required. This is a true saving in power expended. Unfortunately it is not of any great significance, except in small cylinders handling rather low density gas through high ratios of compression where the jacket surface is large in proportion to the amount of work performed and heat generated. Further, cold jacket water should be avoided where the compressor is handling readily condensible gases which enter the cylinders at or near the dew point. In these cases it is actually desirable to superheat the gases during the inlet and early compression portions of the cycle, by hot jacket water, to prevent condensation within the cylinders. Condensation, liquid of any kind, and wet saturated gases are extremely detrimental to lubrication. In many cases the same factors adversely affect non-lubricated construction, especially valve service life.

The compression efficiency can be said to equal the ratio of the area of the theoretical card to that of the actual card, the actual card including the fluid entrance and exit losses as above outlined. A reasonable allowance must also be made for the slippage losses. These losses may ordinarily be considered as a reduction in compression efficiency equivalent to ½ to ¾ of the percentage point losses indicated in the volumetric efficiency formula. Again these losses must be predicated on experience and dictated by cylinder construction features, type of gas handled and operating conditions.

Horsepower Formula Development—The indicator diagram illutrates excellently the derivation of the formula shown herein for theoretical horsepower. To prevent confusion a second diagram for this purpose is shown in Figure 5. Path 1-2-5-6 is the compression cycle of a single stage compressor with zero clearance, while path 1-2-3-4 represents single stage compression with cylinder clearance. The energy represented by the area within the paths 1-2-3-4, compression with clearance, is the energy bounded by paths 1-2-5-6 minus the energy bounded by paths 3-5-6-4. This latter is the energy recovered from the expansion of the gas in the clearance space.

The energy of the zero clearance cycle is

$$\int_{P_1}^{P_2} V dp \quad \ldots\ldots (1)$$

The energy from expanding clearance gas is

$$\int_{P_1}^{P_2} V^1 dp \quad \ldots\ldots (2)$$

The net energy therefore $= \int_{P_1}^{P_2} V dp -$

$$\int_{P_1}^{P_2} V^1 dp \quad \ldots\ldots (3)$$

For isentropic compression and expansion

$P_1 V_1^k = PV^k$ from which

$$V = P_1^{\frac{1}{k}} V_1 \left(P^{-\frac{1}{k}} \right) \quad \ldots\ldots (4)$$

$$\text{and } V^1 = P_1^{\frac{1}{k}} V_4 \left(P^{-\frac{1}{k}} \right) \quad \ldots\ldots (5)$$

Substituting valves of V and V^1 from Equations 4 and 5 into Equation 3

$$\text{Net E} = P_1^{\frac{1}{k}} V_1 \int_{P_1}^{P_2} P^{-\frac{1}{k}} dp - P_1^{\frac{1}{k}} V_4 \int_{P_1}^{P_2} P^{-\frac{1}{k}} dp \quad \ldots\ldots (6)$$

$$= P_1^{\frac{1}{k}} V_1 \left[\frac{P^{\frac{k-1}{k}}}{\frac{k-1}{k}} \right]_{P_1}^{P_2} - P_1^{\frac{1}{k}} V_4 \times \left[\frac{P^{\frac{k-1}{k}}}{\frac{k-1}{k}} \right]_{P_1}^{P_2}$$

$$= \left(\frac{k}{k-1} \right) \left(P_1^{\frac{1}{k}} V_1 \right) \left[P_2^{\frac{k-1}{k}} - P_1^{\frac{k-1}{k}} \right] - \left(\frac{k}{k-1} \right) \left(P_1^{\frac{1}{k}} V_4 \right) \left[P_2^{\frac{k-1}{k}} - P_1^{\frac{k-1}{k}} \right]$$

$$= \left(\frac{k}{k-1} \right) P_1^{\frac{1}{k}} \Big) \left(V_1 - V_4 \right) \times \left[P_2^{\frac{k-1}{k}} - P_1^{\frac{k-1}{k}} \right] \quad \ldots\ldots (7)$$

But $(V_1 - V_4) = V_s$, which is the actual capacity, CFM, measured at intake conditions. Substituting $V_s = (V_1 - V_4)$ in Equation 7.

Net E =

$$\left(\frac{k}{k-1} \right) \left(P_1^{\frac{1}{k}} \right) \left(V_s \right) \left[P_2^{\frac{k-1}{k}} - P_1^{\frac{k-1}{k}} \right] \quad \ldots\ldots (8)$$

Multiplying each side of equation 8 by

$$\frac{P_1^{\frac{k-1}{k}}}{P_1^{\frac{k-1}{k}}}$$

Net Energy =

$$\left(\frac{k}{k-1} \right) \left(P_1 V_s \right) \left[\frac{P_2^{\frac{k-1}{k}}}{P_1^{\frac{k-1}{k}}} - 1 \right] \quad \ldots\ldots (9)$$

But $\frac{P_2}{P_1} = R$: and substituting in Equation 9

Net Energy =

$$\left(\frac{k}{k-1}\right)\left(P_1 V_s\right)\left(R^{\frac{k-1}{k}}-1\right) \quad \ldots\ldots (10)$$

The energy obtained from solving Equation 10 is in foot pounds per minute provided P is expressed as pounds per square foot absolute. Since in compressor work it is customary to use pounds per square inch as normal units, it is necessary to multiply by 144 and the energy formula then becomes:

$$W = 144\left(\frac{k}{k-1}\right)\left(P_1 V_s\right)\left(R^{\frac{k-1}{k}}-1\right) \quad \ldots\ldots (11)$$

where

W = Foot pounds of energy per miinute.
$K = C_p/C_v$
$P_1 = P_s$ = PISA suction pressure
V_s = Actual capacity, CFM,/minute measured at inlet condition
R = Ratio of compression = $P_2/P_1 = P_d/P_s$

The theoretical horsepower then becomes:

$$HP = \left(\frac{144}{33000}\right)\left(\frac{k}{k-1}\right)\left(P_s V_s\right) \times \left[R^{\frac{k-1}{k}}-1\right] \quad \ldots\ldots (12)$$

No attempt will be made here to explain the development of the horsepower formula including compressibility factors as it is felt that such is beyond the intended scope of this paper.

Variables of Performance—Volumetric Efficiency as represented by formula shown in Figure 3 and also by the relationship V_s/D, Figure 4, the indicator card, is obviously affected by:

(a) The cylinder clearance.
(b) The ratio of compressibility at discharge pressure and temperature to that at suction pressure and temperature.
(c) The K value of the gas being compressed.
(d) The ratio of compression.

By inspection of the formula it is noted that

(a-1) Increase in clearance lowers the volumetric efficiency. Thus it is common practice to use clearance pockets (additional clearance) to reduce capacities.
(b-1) Decrease in f lowers the volumetric efficiency. For perfect gases $f = 1$. For most hydrocarbon mixtures $f < 1$ while for certain diatonic gases especially hydrogen and nitrogen $f > 1$ under certain conditions of pressure and temperature.
(c-1) Decrease in K value lowers the volumetric efficiency. Thus, the volumetric efficiency of a compressor handling propane (K = 1.15) would be lower than were the same cylinder compressing hydrogen (K = 1.40).
(d-1) Increase in the ratio of compression lowers the volumetric efficiency.

A change in any of the above variables opposite to that indicated results in a higher volumetric.

The same phenomena can be visualized from the indicator card, figure 4.

(a-2) When clearance is increased there is more gas trapped in the clearance spaces. The numerical value of $\frac{(R^{1/k})\ (CD)}{f}$ is therefore increased and V_s, actual intake capacity is decreased as is V_s/D, the apparent volumetric efficiency. This would result in the expansion line taking some path, such as HA'O' resulting in delayed opening of the suction valves. The compression line would be less steep and might be represented by BE'F'. The delivery interval F'H is also shorter. This is a further indication of reduced capacity.
(b-2) A decrease in f would have the same direction tendency upon the location of O and E as does an increase in clearance (a-2).
(c-2) Same tendency as (a-2) and (b-2).
(d-2) Here P_d would be higher than shown on the diagram or P_s would be lower, or both.

In either case the numerical value of $\frac{(R^{1/k})\ (CD)}{f}$ would increase. For instance if suction pressure remained constant and discharge pressure increased the compression would be carried to F''. Thus, a shorter delivery interval F''H' would prevail. Expansion of clearance gases would begin at H' and end at O'', where inlet would begin. Thus V_s, V_s/D and F''H' would be reduced below similar values prevailing at the lower P_d.

THEORETICAL HORSEPOWER as represented by the formula in Figure 3 and by the area of the indicator diagram Figure 5, is affected by

a) The suction pressure, P_s.
b) The capacity, V_s.
c) The K value of the gas.
d) Cylinder clearance.
e) The ratio of compression.
f) The ratio, f, of compressibility factors.

By reference to formula.

a) Other factors remaining constant the horsepower change is directly proportional to the change in suction pressure. For instance if the suction pressure is increased 50 percent the horsepower increases in the same proportion.
b) Other factors remaining constant the horsepower required changes in direct proportion to the change in V_s. For example if the capacity is reduced 50 percent, horsepower is likewise reduced 50 percent.
c) The variation brought about by change in K is three-fold. First, the capacity is altered. Secondly, the value of $\frac{K}{K-1}$ differs and thirdly, the value of $R^{\frac{k-1}{k}}-1$ changes. The net proportionate increase or decrease is the product of the ratio of altered to original values of the affected phrases in the formula. Suppose for example the K value change is such that V_s increases 5.13 percent. This would mean a corresponding decrease in $\frac{K}{K-1}$ of 33.3 percent but an increase in $R^{\frac{k-1}{k}}-1$ of 60.7 percent. The ratio of altered to original horsepower then is $1.0513 \times .667 \times 1.607 = 1.125$. Thus there is a net change of $+12\frac{1}{2}$ percent in power required.

This change would result if a compressor with 10 percent cylinder clearance compressing a gas of K value 1.24 through 4 ratios of compression were changed to air service (K = 1.41), suction pressure unchanged.
d) The cylinder clearance does not appear in the horsepower formula but as noted under volumetric efficiency, item (a), a change in clearance results in altered volumetric efficiency and capacity. The effect of clearance change on horsepower is in direct proportion to the change in capacity brought about by the change in clearance. Should the clearance change reduce the capacity 25 percent the horsepower likewise would be 25 percent lower.
e) Should the ratio of compression vary, so will the capacity (see volumetric efficiency, item (d). Likewise, formula phrase $R^{\frac{k-1}{k}}-1$ would vary. Assume the compressors described in the preceding paragraph were required to operate at 5 ratios of compression instead of at 4 ratios. The K value is 1.24 and the suction pressure and cylinder clearance are unaltered. The decrease in volumetric efficiency would be approximately 8.0 percent. The change in $R^{\frac{k-1}{k}}-1$) would be an increase of 18.3 percent. The net change therefor would be (.920 × 1.183) or 8.8 percent increase in power required.
f) The effect of compressibility on the horsepower required per cubic foot capacity measured at intake conditions does not appear in the formula. It is small for average operating conditions. In high pressure units, however effects of sizeable proportions are sometimes found. For example, consider a cylinder handling carbon dioxide for a urea plant. Suction pressure is 1720 PSIA at 115 F. The discharge pressure is 3440 PSIA. From published data we find the critical pressure of carbon dioxide to be 1073 PSIA and the critical temperature to be 548° R. At suction conditions the reduced pressure, $Prs = \frac{1720}{1073} = 1.6$. The reduced temperature, $Trs = \frac{575}{548} = 1.05$. From a standard compressibility chart (1 psia basis), the compressibility factor, F_s, at intake conditions is .312.

The theoretical discharge temperature, using the formula from Figure 3, is 675 R(215 F.). Therefore, at discharge conditions of 3440 PSIA and 215 F.; $Prd = \frac{3440}{1073} = 3.2$. $Trd = \frac{675}{548} = 1.232$. Referring again to a standard compressibility chart, the compressibility factor, F_d, at discharge conditions = .575. Therefore $F_d/F_s = f = 1.842$.

Referring now to Figure 6 and following horizontally from a Ratio of

Compression of 2, to the intersection with the diagonal line representing K = 1.3 (C_p/C_v for carbon dioxide) and reading vertically downward the coefficient "C" equal to .456 is found. Using the explanation given below the chart in this figure, the increase in theoretical horsepower will be C (f — 1) = .456 × 8.42 = .384 which means an increase of 38.4 percent in theoretical horsepower due to the effect of compressibility.

This example is probably one of the most dramatic which might have been selected but it has been used to exemplify the possibility of incurring overload unless all conditions surrounding the inquiry are known and evaluated. The converse is sometimes true and the driver may be appreciably underloaded. However, driver underload does not preclude operation whereas overload may. The possibilities must be checked to be reasonably certain. Compressibility factors are important.

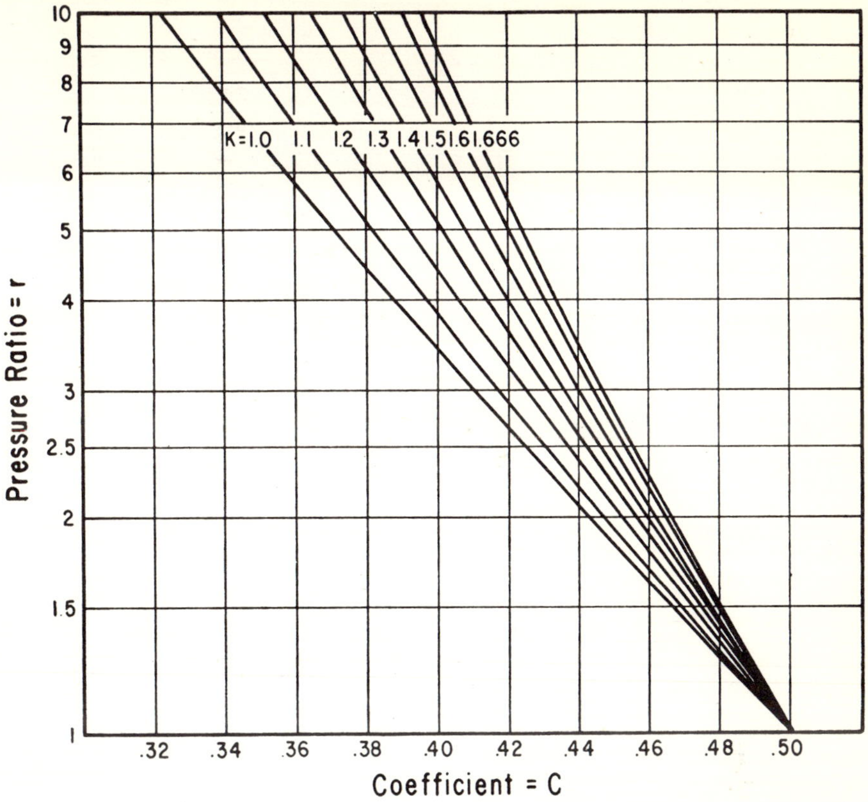

Procedure:

1. Determine compressibility factors F_1 and F_2 from compressibility charts
2. Calculate $f = \frac{F_2}{F_1}$
3. From given r and k determine coefficient (c) from this chart
4. Change in horsepower (deviation) = C (f — 1) HP (Ideal Isentropic HP)
5. Corrected horsepower is algebraic sum of Ideal BHP and deviation, Item 4 above.

FIGURE 6—Horsepower Deviation Coefficient.

Dirty Corrosive Gases and Wet Gases—The user should not hesitate to give the manufacturer a complete and frank statement of the expected qualities of the gas to be compressed. If the gases have a tendency to polymerize when subjected to normal discharge temperatures (250 F. to 350 F.) such fact should be made known. Additional stages of compression, beyond ordinary practice, may be needed. The user should attempt to give his version of temperature limitation to assist the manufacturer in his selection.

Every precaution should be taken to prevent catalyst dust, or dirt of any kind, entering the compressor. However, if previous experience indicates, regardless of practical safeguards used, that dust inclusion cannot be prevented this fact should be made known. While no one can guarantee 100 percent satisfactory compressor service under such conditions, yet a true statement and discussion of the facts could aid in the selection of more suitable and longer lasting materials for such adverse operating conditions.

The same might be said about corrosive constituents in the gases. Here the manufacturer's suggestions on predrying and preheating the inlet gas may prove invaluable. Likewise, provisions might be made in the cylinder jacketing to keep the gas dry while it is in the cylinder. Many troublesome corrosive problems have been cured in this manner.

Entrained liquid should never be permitted to enter a compressor cylinder. Neither should the jacketing arrangement and jacket temperature cause condensation within the cylinder. Adequate separators or scrubbers with reliable and positive drain features must be used before successful operation can be expected. This liquid removal equipment should be adjacent to the cylinder inlet to preclude post scrubber condensation. Once in the cylinder, readily condensable gases should be heated, rather than cooled, by the cylinder jackets. Hot jacket water is preferable to no water because its use does tend to maintain more uniform cylinder casting temperatures, thus minimizing distortion. A simple thermosyphon jacket system may suffice in many cases when provided with a coolant having a suitable boiling point.

Installation, Operation and Maintenance

These experience-proven tips will save costly errors.

Robert S. Ridgway
Standard Oil Company of California
Whittier, Calif.

FIGURE 1—Gas headers outside bridged by separate walkway from each other.

PROPER INSTALLATION and handling of compressors is not acquired so much from a text as it is from experience. The sharp operator can profit considerably from the mistakes and experiences of others.

This discussion will be confined to the compressor only and not its prime mover except where the relationship cannot be disregarded. The article will also dwell lightly upon or omit entirely those things which are commonly accepted, and those installation practices which every manufacturer's service man will follow. Thus the presentation will be devoted principally to pointing out those things which are often omitted or neglected in many installations.

INSTALLATION

Selection of Site—The proper location for the installation involves such factors as safety, good soil, accessibility and proximity to cooling water. The compressor house location must be influenced by safety, particularly if the prime movers are internal combustion engines. Convenience to the gas being compressed, good soil for foundations, accessibility to a roadway for transport, and proximity to cooling water are influencing factors, although fuel and water for engines or electric power for motors can be brought to the site rather readily.

Foundations have a bearing upon the selection of the site. If new equipment is being installed, the foundation problem will probably be much less than with old equipment. Most manufacturers are giving more and more attention to balance of reciprocating machines, and care should be exercised when compressor equipment is being selected, to favor better balance. The site probably will be required to furnish a better foundation if gas engines are used for prime movers than if electric motors are supplied.

Header Location—After the site has been selected the arrangement of plant and equipment must be planned. All of these items are so interrelated that it will be impossible to start with one factor and progress logically through the sequence of items involved in the plant. Piping will probably be the major factor at the site, the location of the headers being all important. This location must be planned with considerable care. It must be convenient to the area of gas supply, and provide ample room for the necessary scrubbers. Convenience seems to dictate running the headers parallel to the long axis of the plant (Figure 1). If pulsation dampeners are used, the tuned length of the resistance tubes will frequently dictate a minimum distance out from the plant to the headers.

The gas headers were usually placed in a trench in the earlier plants, but more recently either the concrete trench has been eliminated entirely, or placed above grade so that drainage can be obtained. Height of the plant floor is thus dictated by

FIGURE 2—Longitudinal walkway for access to stop valves.

the location of the headers, or by drainage of the subfloor under the gas surge chambers which are usually placed directly under the compressor cylinders, inasmuch as short connections between the cylinders and surge chambers are essential.

Arrangement of Units—The arrangement of units within the building is often the subject of considerable controversy. Engine-driven, direct-connected units are usually found to be most convenient when they are arranged with their crankshafts forming a single line, end to end. Occasionally to save space machines are set in a double row with the engine ends together, back to back. The compressor cylinders face the outside of the plant, and headers are run on both sides of the building. Operators usually regret this decision because the runway between engine sides is consequently narrow, dark and hot, and maintenance is difficult.

Some operators prefer to install their units crosswise of the plant, carrying the compressor piping to the building side in a trench. This arrangement was designed to permit crankshaft removal through the ends of the crankcase. However, such a design is questionable as shafts are never removed in many instances, and if this is ever necessary, such infrequent engine dismantling will not pay for the many other inconveniences. Where electric driven opposed compressors are used, with cylinders on both sides of the frame, setting the units with their crankshafts crosswise of the plant will be the preferred method. The piping from the cylinders on both sides of the unit can then be carried in basement trenches to the headers on one side of the building only.

Regardless of how the units are arranged, however, the principal item to remember is not to attempt to make a saving on the installation by skimping on the floor area, or getting the units too close together. On an eight-cylinder engine with a rail overhead, four pistons must be set on the space at each end of the engine during overhaul and worked upon, hence it will pay good dividends to allow plenty of room between units. Also, plenty of space should be allowed both in back of the machine for crankcase work, and on the compressor side for pulling pistons and rods.

Thought should be given during layout time for the possible addition of future units, whether or not it

FIGURE 3—Novel method of installing high pressure stop valves. Safety valves securely anchored and very accessible.

seems unlikely at the outset. This must include space for the headers also. If auxiliaries are located in the building they must be either at the end where no expansion is contemplated, or in the center of the plant for easier access.

Compressor Building—Occasionally, in mild climates, someone will question the necessity of a building. This structure should never be omitted since, although rains may be infrequent, the water will find its way into the crankcases and distance pieces, and both rains and winds will seriously hamper the proper attention necessary for good operation and maintenance. Then, too, it is essential that a rail for a crane be placed over the compressor cylinders and another over the engine heads. Or a bridge crane can straddle the full width of the plant for only a slight increase in cost, and with much more usefulness. These can be incorporated into the structure. Care should be taken to see that the craneways extend through the end of the building for loading and unloading compressor equipment onto trucks.

The building should be well provided with windows for good lighting, as this will pay out in better maintenance and operation. Enough of the windows should open so that good ventilation is possible. Doors should be provided opposite each compressor for quick access to the valves located outside at the sides of the walkway.

It will be found much more satisfactory and convenient to make the floor of open grating wherever a trench or basement is underneath, rather than of floor plate. The grating is so much lighter that one man can remove it with ease, whereas floor plate is a back-breaker. Further, it has been found that the possibility of constant observation of surge chambers and piping underneath is quite advantageous. When this area is visible it never accumulates trash. The safety men should be called in to spot their fire protection in the early stages of design. An office, or room for records to be kept, also wash and change rooms, should be located as convenient to the building site as possible.

Piping—The compressor and service piping will constitute a major part of the installation. We have briefly noted the advisable location of the headers, but have not discussed their arrangement. Considerable study should be made of the header anchorages to care for stresses and thermal expansion of the lines. For this purpose piping contractors have developed pipe "hold downs" which permit either side or longitudinal deflection while restraining the other. These are used in conjunction with solid anchorages. Where side deflection of headers by lead-out lines will permit, solid anchorages can be installed near the center of each longitudinal header, and spring hold-downs may be used on either side. It is often advisable to "cold spring" headers and lead-out lines so that the minimum stress will be imposed when the lines are up to temperature. The pipe will expand .001 inches per inch for every 150 degrees of temperature rise, and this can add up to an appreciable distance in the length of a long compressor house. We note these items merely to call the reader's attention to the fact that this is a problem for an expert, and due recognition should be given to it.

The stop valves are normally placed in the lead-in and lead-out lines between the headers and the cylinder surge chambers, and these valves will usually be found to be most convenient when arranged close to the building wall. For access to them, a walkway is either run for the length of the building (Figure 2), or a platform and steps are provided opposite each compressor, with the valve extension handles being brought to the walkway railing. The safety valves should be located at an accessible spot close to the building so that their discharge lines can be carried individually above the building eaves. These vent lines must be quite firmly anchored if the valve is in high pressure service (Figure 3). When it is not possible to discharge the "pops"

FIGURE 4—Vertical suction surge chambers.

in this manner, and vents must go to a common vent line, it must be amply sized for the worst possible condition.

Another point to watch in the compressor piping is the cylinder suction connections. Suction flanges are often rated at a lower working pressure than the discharge flanges on any cylinder, hence, with the compressor down, the suction stop valve closed, and the discharge open, that section of intake line between the stop valve and the cylinder can conceivably be subjected to discharge pressure by leakage of one suction and one discharge compressor valve. To protect against this possibility, a safety valve must be installed on the suction line, or the suction flange and piping must be adequate to safely handle the discharge pressure.

The point of having suction headers and piping extremely clean cannot be over-emphasized, as it will make so much difference in the amount of compressor valve trouble which will be encountered. Anyone who has experienced the incessant valve trouble caused by dirty lines will agree with the foregoing statement. We have found it wise to go to the extent of sand blasting all of the pipe used in fabricating headers, lead-in lines and surge chambers, and sealing them to protect against corrosion during construction. After the pipe work is completed, and just prior to startup, all lines should be blown down with 100 psig air in an adequate volume to give a good blow, with piping disconnected at the outer ends. If any of the piping or surge chambers have lain idle through inclement weather, they should be treated with great suspicion and thoroughly cleaned before using.

Heavy hardware cloth should be used to back up screen wire and made into steel ringed plates which can be inserted in the suction lines between flanges near the cylinders. In addition, screens should be placed over the compressor suction valves to remove all small particles. These screens should be removed after 2 or 3 weeks of operation.

Pulsation—Pulsation filters are commonly used today on suction and discharge lines adjacent to the compressor cylinders for two reasons: to remove disturbing pulsations from the plant pipe lines, and to prevent starvation and hence capacity loss in the case of the suction lines, and excessive pressures, resulting in high horsepower and some capacity loss, in the case of the discharge lines. Such filters are offered by several concerns and are readily obtainable. Some operators have been lucky about their pipe line pulsations and are still not sold on the advisability of installing filters. On high-pressure installations where the gas density is high, there seems to be no alternative, and the early installations without them usually experienced considerable trouble at the welded connections to the surge chambers.

If pulsation filters are to be omitted on either high or low pressure lines, then adequate surge chambers must be provided on both the suction and the discharge. The surge chambers will tend to prevent starvation of the cylinder on the suction side. There have been instances where cylinder capacity has been reduced by as much as 25 percent by operating without a suction surge chamber. Similarly, the discharge surge chamber will prevent excessive momentary discharge pressures which can result in severe overloads and also somewhat reduce cylinder capacities. These surge chambers may serve several cylinders on the same machine. Their capacity must not be less than seven times the cylinder displacement (bore × stroke) for every cylinder served. They must be relatively close to the cylinder which they serve, else, in combination with the "gas jacket" around the cylinder, and the "too long" connecting line, they may constitute a filter themselves which may be improperly tuned, and some disturbing pulsations may result. However, don't let the fear of this unusual situation tend to favor omitting the surge chambers. Their omission will almost invariably cause shortage or overload conditions, even though it may not be recognized as such.

Discharge surge chambers normally are located on sleepers under the floor. Most high-pressure suction surge chambers are in a similar location for convenience in connecting to cylinder manifolds. Low-pressure suction surge chambers usually are placed across the top of the cylinders so that the cylinder connections can be extremely short. This also permits easy removal. Suction surge chambers may be placed vertically where space is limited (Figure 4). All surge chambers should be equipped with drains and pressure taps for checking pressure losses. Last but not least—don't place them where they cannot be removed for inspection or possible repair!

Scrubbers should be placed on all suction lines. These vessels should be equipped with a separating element to remove liquid, and should not be merely a "knockout drum." The latter does not do a satisfactory job of separating entrained vapor. Scrubbers should be equipped with a gage glass, automatic drain, high level alarm and a shut down device to shut the compressors down on high liquid level. Scrubbers should be located at an accessible spot convenient to the gas headers.

Unless the engines are supercharged, it may be wise to silence the air intakes to the scavenging cylinders of two cycle engines if noise is objectionable. This can be done with pulsation filters. It will be wise to locate the air intakes on such units at some distance from the building wall.

FIGURE 5—Special flange fitting below cylinders to permit cold supports.

Unloading—On each compressor discharge (or for several cylinders in parallel on the same machine) a vent to atmosphere or a vent line should be provided for unloading. The safety valve usually discharges to this same vent. Unless some unusual situation demands that no gas be lost from the system, this type of unloading has some advantages over the bypass system. If, for any reason it cannot be used, then a bypass line and valve must connect the discharge to the intake, inside of the stop valves. Either system permits unloading during the startup and shutdown operations of the compressor.

Grout—In installing the compressor, the manufacturer will supply instructions for grouting and levelling, and they need not be repeated here. We would note, however, that if the crankcase is equipped with an oil pan or sump which projects below the rest of the crankcase, it is usually better not to grout around this pan. This can be worked out either by installing a piece of fiber board between the pan and foundation around the top rim, or placing a piece of fire hose around this section which is pressured with water during the grouting operation, then removed. The more recent practice of following the wet grout with a dry grout to prevent shrinkage has worked out excellently. Any wedges placed for levelling should be removed after 24 hours, else the frame will ride the wedges later on.

Cylinder Supports—The manufacturer will furnish full instructions on assembling the machine and installing cylinders. However, close attention should be paid to the method of supporting the compressor cylinders. It is imperative that the hot discharge line should not go downward to a rigid base, or base ell, which will result in an upward thrust on the cylinder when the line heats up. Cylinders are forced out of line, and occasionally distance pieces are cracked by such methods of installation. In earlier periods, the base ell was the standard method of supporting cylinders. They were usually held up by jackscrews, and manufacturer's service men were careful not to run these jackscrews down until after the discharge line had come up to temperature. Later, however, when the unit was down and cold, the fact that the jackscrews did not reach the support plate was usually noticed by some attentive mechanic, and the damage was done.

Customer pressure has caused the manufacturers to correct this trouble in various ways. Some have put lugs on the sides of the cylinders to which cast legs fit which support the cylin-

FIGURE 6—Method of locating thermometer in discharge line.

der, straddling the discharge trench. Others have made a wide flange fitting for the discharge, the projections on which are supported by 3-inch or 4-inch pipes which serve no other purpose but to support the cylinder, and are therefore always at room temperature (Figure 5). These methods serve the purpose excellently, but are completely nullified if the contractor fails to recognize the principles involved, and runs the discharge lines directly downward into base ells or surge chambers or any fittings which are rigid. Discharge lines should drop down into an ell which floats free and directs the line into a surge chamber. The ell can then bend under thermal stresses and relieve the cylinder of this strain. Within the last month we had occasion to look over a new installation where the contractor had defeated the provisions for good alignment provided by "cold supports" by installing rigid discharge spools directly into firmly anchored surge chambers directly below the cylinders on more than half of the units.

While on the subject of installation we would warn against the use of an impact wrench in running nuts. The impact wrench is very convenient when dismantling machines, but should never be used for setting up, due to the ease with which bolts and studs are overstressed. The new torque-limiting air wrenches are a different matter, and bid fair to combine a number of desirable qualities into one. Overstressing nuts has caused so much distortion in the past that we have found it wise to supply 1000 foot-pound torque wrenches for assembly of machine parts.

Jacket Water—In arranging to "service pipe" the jacket water, care should be taken to insure that the outlet water flows upward to the outlet header. If the outlet line from any cylinder drops downward it will form an air or gas trap which usually results in blocking off that cylinder, and diverting its cooling water through the others. If it becomes necessary to drop lines or headers downward prior to reaching the surge drum, then air expelling traps must be installed at all of these points to remove the gas which occurs whenever a gasket leaks, or the air which is sucked in by the centrifugal pump or trapped in the system after any dismantling. We have found it most convenient to install a vertical standpipe or surge drum of 20- or 24-inch pipe sufficiently high to collect all jacket waters, and provide an ample head for the centrifugal jacket-water pump. A permanent ladder is provided for this open top tank, and all outlet waters are directed into it at any elevation. The operators make a daily routine of climbing the ladder and checking the bubbles. This normally permits spotting gasket leaks and correcting them before they become serious. In piping the compressors it is good practice to install an industrial thermometer at each water outlet, both to be able to regulate the volume throughput, and to indicate trouble due to overheating.

Compressor manufacturers usually overestimate the heat rejection rate of the cylinder to the jacket water. Rates of 500 Btu per bhp-hr., and 175 times the ratio of compression are two of the most common factors supplied for this heat rejection. In our observation, both are considerably higher than any cylinder rejection rate we have noticed. In fact, it would seem that 100 times the compression ratio would be ample and much closer to the expected rate than the figures noted above. A few of the gas transmission line companies have installed cylinders with no water jackets at all, and put the major portion of the heat into the gas stream, radiating a small portion to the air. Such cylinders fit in well with their program, for normally the ratio of compressions on such stations is below 3. We know of instances where operators have filled the water jackets with oil and left the top openings unplugged. This equalizes the jacket temperature around the cylinder, but forces the rejected heat into the gas being compressed. It causes a small increase in the horsepower of compression, but could make the occasional gasket leaks difficult to find. Such practices are usually confined to ratios of compression below 3.

In water cooling the compressor cylinders, a temperature rise of 10 degrees will be found quite satisfactory. The temperature level may be kept around 120 to 130 degrees, and even higher if the gas is particularly wet, in order to prevent condensation in the cylinder and consequent high ring and wall wear. Such temperatures are quite favorable to dry air cooling of the jacket water, particularly in high ambient temperatures.

Dial Thermometer—During installation, a minimum of instrumentation will include several items which are often omitted. Foremost of these is a heavy duty dial thermometer in each compressor discharge line, adjacent to the cylinder (Figure 6). This is for the purpose of supplying a ready index of the performance of the compressor valves and rings. Those who have not used such indicating thermometers have no idea of their usefulness in pointing out the situation when valves are going bad or valve gaskets are beginning to leak. This is particularly valuable on high pressure cylinders where valve or gasket leaks for only a short period of time either damage the valve beyond repair, or make the repair difficult. Similarly, when cylinder gasket recesses become slotted by high-pressure gas blowing through, it is a long and tedious job resurfacing them. A slight increase in temperature indicates trouble long before such trouble can be detected by the feel of the valve covers, and operators are usually enthusiastic about such a "trouble detector." A heavy duty dial thermometer without gears should be used to prevent destruction from vibration. Such equipment stands up well in this service.

The metallic packer should also be equipped with a thermometer to indicate its performance. If the manufacturer has not done so, the packer flange should be drilled and tapped, radially, for the installation of either a dial thermometer, or the bulb for a remote indicating instrument. The latter is required if the gas being compressed is poisonous, requiring the distance piece to be normally closed. This makes a very neat and convenient installation, as the temperature gages may all be mounted together on a single instrument panel, and any packer trouble will be pointed out rather forcefully. There have been instances where we have placed two such tapped wells in the packer flange and have placed a thermally-operated engine shut-down device in the second one. Such installations are for plants which operate unattended and we do not care to burn up the packer and ruin the piston rod in the event of packer trouble. We have seen instances where other operators have provided such protection in attended plants.

FIGURE 7—Single section three-fan unit for cooling jacket water.

Packers—While on the subject of packers, it has proven quite profitable to install a grease gun fitting and check valve on a tee fitting to the oil line leading to the inner lubrication point on the packer. This is for the purpose of breaking-in a new packer. During startup the grease gun can be filled with a heavy oil, and supplementary injections should be made frequently until both the thermometer in the packer flange and lack of audible blow-by indicate that the packer is seating properly.

Double acting wipers should always be used around the piston rod. Originally, single acting wipers were installed with the thought of stripping the crankcase oil from the rod which had splashed onto it, and preventing its loss to the distance piece. Occasionally, however, a heavier oil than is used in the crankcase, or an oil compounded with tallow is used on the rod, and when this oil is carried into the crankcase it results in contamination, causing frequent crankcase changes. On an electrically-driven machine, crankcase oil changes will be very infrequent unless contamination occurs.

Unless a poisonous gas is being handled, the distance piece covers should be left off on the "operating" side, so that operators can see the packer and its thermometer at a glance, and feel the rod. Recently the author had occasion to work with a gas dangerously high in hydrogen sulfide, and the covers were left on. This method of operation proved so unsatisfactory that the covers were left attached by one bolt, but were rolled out of the way. They could be rolled back into place if necessary in a moment. Any slight leakage was carried away by the vent lines which were tied into a vacuum collecting system. This method of operation proved quite satisfactory, and the thermometers were an excellent index for the operation of the packers.

Lubrication—When ordering lubricators it is possible to have them broken up into several compartments, usually at no extra cost. This may be found of great convenience at a later date if it is advantageous to use oils of a different specification for the various lubrication points.

The lube oil should be filtered. If the compressor is electric driven and not a part of an engine driven machine, then an edge type filter will usually fulfill this requirement. The crankcase oil will usually require a little cooling, and for this the manufacturer will supply a small shell and tube heat exchanger. This may not always be economical to use, particularly if the cooling is to be furnished by dry air coolers. It may well require the jacket water to be cooled cooler than necessary in order to provide the temperature gradient to bring the oil temperature down to the required level, thus calling for more heat exchange surface in the jacket cooler than is necessary.

Therefore the possibility of cooling the oil directly in extended surface coils in the dry air cooler should be explored. This is far more attractive in a direct connected engine driven machine than it is in a compressor only, as there is so much more heat to extract in the former. Crankcase oil temperatures in the order of 150-160 degrees are usually quite acceptable. High oil temperatures are to be avoided principally because of bearing fatigue, the strength of the babbitt decreasing at the fourth power of the temperature. This usually becomes a temperature limiting factor ahead of any destruction to the lubricating oil.

A low oil pressure shutdown switch should always be used to save the bearings and sometimes the shaft in the event of an oil pump failure.

Intercooling—Gas intercooling can be done in shell and tube exchangers near the compressors, or in atmospheric sections placed in a cooling tower, or in dry air coolers with extended surface tubing where the air is circulated by fans. The choice depends upon the individual situation, the availability of cold water, or unused space in a cooling tower.

Dry air coolers are becoming more and more widely used for the ease of installation and the low maintenance involved, also for the simplicity of temperature control. Another factor to be remembered is that they give their full salvage value when it is desired to move them to a new location. Vertical shaft fans directing the flow of air upward will be found more satisfactory, as these are insensitive to wind direction. A number of small fans in parallel will be found preferrable to one large fan (Figure 7), as this will permit automatic startup and shutdown of one or more of the fans by providing automatic thermal control on the outlet temperature of the gas.

Inasmuch as design factors must provide for the highest ambient temperature expected, such control will prove to be quite economical and satisfactory. If the installation is to be near the ocean, copper fins will be preferred over aluminum. The jacket water and lube oil sections can comprise one unit with the oil sections placed beneath or beside the water sections. If the sections have been designed properly using the proper factors, the outlet temperatures of the two will follow each other roughly together as the ambient goes up or down, hence the control point for shutting down the fan motors may be placed in the water outlets, and the oil should never wander too far from the rather wide limits which are acceptable.

OPERATION

Starting—For the initial startup of the plant the procedure is, of course, quite different and more cautious than subsequent ones. After making sure that the compressor crankcase is extremely clean, fill it with lube oil, usually an SAE 30 grade, and pump oil to the pressure system with the hand pump which the manufacturer usually provides on such machines, both to make sure that the bearings are flooded and that the lines are free of air. Through the valve holes and with the piston backed out of the way, watch the oil holes in the cylinder walls while the lubricator is being hand cranked to insure that the oil comes freely and that all the air has been expelled from the lines. The assembly shoulld be once again checked over to make sure that the compressor pistons have been spaced properly (usually with ⅓ of the clearance on the frame end and ⅔ on the head end to provide for expansion of the rod).

Be positive that the compressor valves have been installed properly. Many compressor valves are interchangeable between suction and discharge, hence it is easy for the helper to get the valve in backwards. As there is no safety valve which will protect the cylinder from reversed discharge valves, a nice longitudinal crack usually reveals this error. Check the inlet valve screens and make sure that the valve covers have been pulled down evenly (a torque wrench is preferred) with the jackscrew loose when this is done. Then run the jackscrew down firmly onto the crab or guard to hold the valve tight onto its seat.

Check all the lines through the plant to insure against any closed valves or flange blinds, start the jacket water through the cylinder, check the safety shut-down and overspeed trip for their proper position, make sure that suction and discharge stop valves are closed and that the bleeder or bypass is open, and the unit should be ready for starting.

The unit should be run for not over two minutes, and careful attention should be paid to listening for knocks. During this period the lubricator, set to a heavy feed, should be checked. The unit should then be shut down, the covers removed and all the bearings carefully checked for excessive temperature. This procedure should be repeated for 5- then 10- then 30-minute intervals, after which, if nothing abnormal is apparent, the unit may be idled for several hours. During this time the packers should be watched and frequent gun injections of oil pumped into the packers. In starts after the first one, the unit should be idled for around ten minutes to permit the oil and parts to attain a running temperature before the load is applied. Whether it is initial or subsequent startups, the load is applied to the high pressure cylinders first in most multi-stage plants, dropping down in pressure to the low cylinders which puts all stages on the line before the main stream is brought into the system.

If the cylinder is equipped with a bleeder on the discharge, the load is applied by first cracking the discharge valve until the gas backing through it can be heard whistling through the vent line. It is then safe to close the vent and immediately open the discharge valve fully. If a bypass is used, it is closed before the discharge stop valve is opened to keep the intake safety valve from popping. If the working pressure of the suction piping adjacent to the cylinder is higher than the discharge pressure, then the opening of the discharge may be completed, if desired, before the bypass is closed. The load is applied by gradually opening the intake valve.

In this connection it is often wise to compute the rod load when the cylinder is pumping against discharge pressure with atmospheric pressure on the intake. In rare instances this will be above the allowable, and, although this is usually not something to be too concerned about, due to the generous factors of safety provided by the manufacturer, the condition should not be permitted to exist for any period longer than is necessary.

For the compressor alone, or if it is electric motor driven, we see no advantage to be gained by applying the load a little at a time, such as is recommended for engine driven machines. Full load can be placed on a new cylinder whenever the running gear has shown itself to be in operable shape and has worn itself in for a few hours. When the unit is engine driven, the startup of a new machine should be entirely different beyond the points covered in the foregoing, for the sake of the new engine. Although this is a treatise on compressors, and not engines, we shall note such a startup briefly, inasmuch as the two are so closely interrelated. It has been the custom to start such engine-driven machines, after their running gear was proved, by operating several hours at ¼ load, several hours at ½, several more at ¾ load, then to apply the full load. This is principally for the benefit of seating the engine piston rings. Discussions with ring engineers have indicated that they would prefer to see such rings seated by short applications of full load, rather than continued applications of part load. For example, two 10-minute applications of full load with 15-minute intervals between for cooling, followed by two 20-minute applications of load interspersed with the same quarter hour idle intervals, and so on, seem to be preferred. Actual trial of this type of break in for the engine piston rings has worked out quite well.

Stopping—In shutting down, the cycle is reversed. The suction stop valve is closed first, followed by the closing of the discharge, as simultaneously as possible with the opening of the vent. If the two cannot be worked together, completely close the discharge first, then open the vent. In the event the suction was not closed properly, the discharge safety valve will pop. If a bypass is used instead of a vent the same procedure can be used, substituting the bypass instead of the vent in the above text.

If the suction piping and flanges are designed to stand the discharge pressure, then the bypass may be opened at will after the suction is closed and before the discharge is closed. If, however, the suction piping is rated at a lower working pressure and is protected with a safety valve, then the bypass should not be opened until the discharge stop valve is closed, else the suction safety valve will open. Although there is nothing wrong with this safety valve opening, it is a nuisance.

If the compressor is electric motor driven, the machine may be shut down as soon as the load is removed. If engine driven, the engine should be allowed to run idle for at least 10 minutes after the load is removed so that the parts affected by combustion may readjust their temperatures gradually.

Cylinder Oil—Cylinder and piston rod lubrication is usually no problem if the gas is dry and pressures are nominal. A good 30 grade oil usually will suffice. Oil additives seem to be of no particular advantage to the compressor cylinder or packer. Wet gases or high pressures demand different treatment. In some cases the use of oils of around 60 grade will be all that is necessary to take care of this situation. Steam cylinder oils, compounded with tallow, make an excellent lubricant for such purposes, but their liabilities usually result in ultimately removing them from service. Any slight failure of the double acting wipers permits the compounded oil to be carried into the crankcase, causing the viscosity of the crankcase oil to skyrocket. Further, such compounded oils produce an excessive carbon residue, which is highly undesirable on the discharge valves.

When compressing plant vapors, it is frequently of material advantage to have the condensate uncontaminated with anything which will throw it off color, so that it may be pumped to a white line without having to be rerun. The use of the white viscous oils will permit this to occur with good lubrication, and with considerable saving in steam. The quantity of oil required for cylinder lubrication is surprisingly low. After the rings have seated, cylinders handling dry gases at nominal pressures can usually be cut to around 5 million square feet per gallon feed. We indicate this and subsequent oil figures with great reluctance, for some will find they do well with half this amount, while others could not get along without doubling it. Such figures are only to suggest an order of magnitude. Wet gases and high pressures can demand double this oil rate.

Too much oil will often result in excessive carbon deposit on the discharge valves, particularly if the ratio of compression is on the high side. The adequacy of lubrication can be checked at the outset by frequently removing an intake valve and feeling the cylinder wall for an almost imperceptible oil film. Later it is wise, after a sufficient time has elapsed to make it significant, to insert a micrometer through the valve opening and measure the wear rate of the liner as a check on the lubrication. The new style lubricator tops which are equipped with glycerin or similar visible liquid, through which the lubricant may be pumped under pressure, make the measurement of feed rather difficult. With this type of equipment, the total number of drops per minute for all feeds within one compartment must be counted, and the total oil per day fed from that compartment can be divided in the proportion as the number of drops for any feed bears to the total drops.

Packer lubrication is usually the same as that for the cylinder in quality, though in certain cases, where the oil used in the one location is not the best for the other, we have found it advisable to use different oils for the two, from two different lubricator compartments. This is the unusual case. We have described the "grease gun" fittings on the inboard oil feed previously, and the use of these to lubricate the packer, far in excess of the capacity of the lubricator, will be found highly useful during the break-in period. After break-in, on gases at nominal pressures, lubrication may often be cut as low as 5 million lineal feet per gallon.

Valves—Valves constitute perhaps the major feature of the compressor. The design of the valve must be such that the pressure drop through it will be very small, so called "valve velocities" will be low which means that a sufficient number of valves and the lift area of each must be great enough to keep the velocity of the gas through the valve down within acceptable limits. In reviewing compressor offerings, this is an item which must not be overlooked, and further, it must not be assumed that all manufacturers calculate the valve velocity the same way. This may seem like a surprising statement, but if the reader will divide the cylinder displacement in cubic feet per minute (of both ends of the cylinder) by the lift area of all the suction valves in square feet, he can work out his own valve velocity in lineal feet per minute which good practice likes to limit to 7500.

In order to keep valve velocities (and the resulting losses) down, large cylinders, and many of the pipe line cylinders which compress through low ratios are frequently made with double deck valves. This permits doubling the valve area within the same physical limitations of available cylinder wall area.

Manufacturers normally supply their valves with steel discs, hardened and lapped to a fine finish, with springs to match the lift of the disc and its mass. Independent suppliers pressure the customer to replace the steel discs with those of one of the various varieties of phenolic resin, claiming longer life and lower cost. Sometimes this claim can be met, but all too often the user finds himself installing discs which are materially thicker than those which were originally supplied, resulting in a greatly decreased lift of the valve, and consequent raising of the valve velocity, and a general increase in losses. If thicker discs are used, the valve body must be machined a corresponding amount so that the design lift will be preserved.

In an earlier section we have warned about checking the proper installation of the valves. Earlier compressor valves were made reversible, so that a suction could be made into a discharge readily. This possibility is seldom found in currently made compressors, for the reason that in one of the positions, the center bolt which held the assembly together would fall into the cylinder if it became loose. This disconcerting phenomenon became the source of too many complaints, and designers set to work developing valves on which the center bolts backed away from the cylinder barrel instead of into it, should they come loose. In most cases this was done at the sacrifice of valve interchangeability, which was a doubtful asset in the first place. However, even with this change, it is usally possible to install valves backwards in the cylinder if the mechanic's mind wanders during the operation.

Gaskets—Solid copper gaskets work well on low-pressure cylinders. Where these will not stand the seating pressure, soft steel gaskets seem to be preferred. A few prefer the aluminum gaskets, and these seat well, but have no hoop strength, hence it is necessary to have them made very accurately so that their outside diameters will be as big as the recesses into which they fit, so that the recesses will back them up and make it unnecessary for the gaskets to withstand any pressure in bursting. Gasketing is usually a real problem on high-pressure cylinders, and there are some who are still optimistic enough to believe they can lap the valve into the recess with lapping compound and operate without any gaskets whatever. Slight gasket leaks in high pressure cylinders can quickly slot the recess, causing an appreciable amount of tedious hand work in the repair process, hence it behooves the operators to watch the discharge gas thermometers and feel the valve covers frequently.

Rings—High discharge temperatures can also be caused by rings which are not seating, or which are badly worn. When such trouble is indicated, the valves should be checked first, then, if the trouble is not corrected, the rings should be examined to insure that they are seating, or do not have excessive wear or end gap, or that the grooves are not "veed" in the piston, and that the liner is not scored or out-of-round. At any sign of cylinder trouble it is always wise to make a capacity check of the calculated throughput against the metered throughput. If no external trouble is noticed it is still the wise thing to do to make routine capacity checks at least semi-annually. A "payout" can normally be obtained through proper maintenance if the cylinder is found to deviate from its calculated capacity by more than 10 percent.

High ratios of compression will produce high discharge gas temperatures, such temperatures being readily calculated, or read from charts prepared for this purpose and found in various handbooks. If high ratios of compression are contemplated, the volumetric efficiency of the cylinder should be calculated to insure that the cylinder will not be shut off, that is, that the efficiency will be well above zero. High clearance in cylinders produces a low volumetric efficiency. In every case, the "rod loading" produced by the differential pressure across the piston should be calculated and checked against the maximum allowable loading from the manufacturer's design calculations. Gas discharge temperatures above 300 F. are to be frowned upon, as they accelerate the carbon deposit on the discharge valves, and often build up deposits of carbon in the discharge fittings which eventually results in excessive pressure drop, causing higher horsepower loads and reduced cylinder capacity. Hot cylinder jackets should be regarded with suspicion, as frequently this is an indication of air or gas in the jacket water which is interrupting the water circulation.

Safety Valves—The safety shutdowns should be tested at least monthly to insure that they are always in operable condition. It will be safer to actually make them function as they are supposed to than to develop a synthetic test which leaves any part of their function to conjecture. The safety valves around the cylinder should also be tested at similar intervals. It will be found most convenient to post a schedule for such testing of each item which can be filled out and signed by the responsible party. This will then become a matter of record, which may prove invaluable in the event of accident.

MAINTENANCE

Maintenance on the compressor consists of principally valves, rings and packers. It is wise to know when to anticipate trouble, so that it can be corrected at the operator's convenience, rather than make its own demands upon the operator's inconvenience. It will take some time, keen observation, and good record keeping to work out a maintenance schedule which will permit the maximum safe run on any of the equipment. Compressor operation on dry gases at normal pressures can usually be estimated at a one year run, and may possibly be extended as long as two. High pressure cylinders must usually be checked over in approximately six months. These times are merely to suggest an order of magnitude. Each operator should work out the optimum inspection times for himself for each class of service.

Valve Maintenance—When they need it, valve seats may be ground. If the discs show slight wear, they can in some cases be turned over. Occasionally it may pay to reface valve discs on a magnetic chuck, using a grinder. If this is done, a maximum of around .005-inch is all that should be removed from the disc. Check the valve springs, and replace whenever necessary. Check the completed job with distillate after assembly. Keep a stern watch on the discharge valves for carbon accumulations. This may prove to be the bottleneck which requires first attention.

Ring Maintenance—Check the piston rings for radial wear, which will usually show up by excessive end gap. Three piece segment rings with an internal expander are the type most universally used today. Various mixtures of bronze rings seem to be preferred for quick seating and freedom from scuffing. They may wear more rapidly than iron rings in some instances, but their ease of seating and freedom from scuffing will usually be considered to more than make up for any shorter life experienced. Occasionally phenolic resin rings will work out wonderfully well under difficult circumstances, and at a reduced cost. We have known of such instances. We have also known of instances where, after months of seemingly no wear, they suddenly wore out over night, and the expanders then proceeded to ruin the liners. They are a great help where sour gases are encountered.

The sides of the ring grooves should always be checked carefully for being true, particularly in high-pressure pistons, as this is probably the first point of ring trouble. Grooves have a tendency to "vee" causing blowby and rapid ring wear. Experience has shown it wise to anticipate this trouble and lay in rings which are .025-inch over width. Grooves can be trued up to this oversize, and later on can be turned again in still more increments of .025-inch each. Side clearance in the ring grooves in high pressure cylinders should not exceed .002-inch when a new fit is being made.

Cylinder Liner Maintenance—Cylinder liners should be reconditioned when they are scuffed, out of round, or are worn excessively. If the barrels would wear true, it might be another story, but they invariably wear more at certain points than at others, hence it is wise to recondition when wear reaches the order of around .002-inch per inch of cylinder diameter. After using the practice for many years of reboring to standards of oversize, we have found it better in the overall picture to replace the liner, or spray it back to standard size wherever possible. This permits further use of the piston and does away with the necessity of carrying so many oversizes of piston rings on the shelves. Pistons normally wear much more slowly than liners, and this fortunate circumstance can be made use of. Unfortunately, not all cylinders are lined, and most of them which are not will not stand sufficient cut to install a liner of standard diameter without jeopardizing the maximum working pressure. Many operators are using the metal spray process to good advantage for this work. Pistons may be built up this way, too, although it is not good practice to run a sprayed piston against a sprayed liner. Good success has resulted with spray up to 1000-pound operating pressure, though the higher the pressure the greater the risk. Metal spray work is only as good as the technician who does the work. Our experience has indicated good spray men to be few and far between. If the reader has had sour experiences with spray work, let him regard his workman with a certain amount of skepticism. It is not meant to infer that all spray jobs come out perfect, but in the hands of a good man, better than 90 percent should work out well if the application is a proper one and is well done.

Piston Rod Maintenance—Allowable piston rod wear is an inverse function of pressure, rods at normal pressures seeming to stand a wear of .004 to .006-inch per inch of diameter with good packer operation. Rods hardened to a Rockwell C of around 50 usually show such a low wear rate that the extra cost of the hardening appears well worthwhile. Successful operation of sprayed rods has resulted at pressures up to 1000 pounds. This is found to be a very useful method of reconditioning rods, as it does not call for the rod to be removed from the piston. The removal of the rod is quite a chore in some machines, involving considerable labor, and occasionally resulting in a broken piston, where the interference fit was rather heavy.

The porous nature of the sprayed surface is particularly conducive to good lubrication. Rods can be chromed and ground, but this is a rather expensive procedure. It does produce a good durable surface. When replacing the rod, make sure that the crosshead adjustment is such that the rod runs level through the packing. Make sure, also, that the piston is tight on the rod.

Packer Ring Maintenance—The majority of rings sold for packers are of bronze. Like the piston rings, they seat readily and go to work easily. Various types of phenolic resin rings have been found successful in certain applications, and are quite valuable with sour gases. The manufacturer will specify the packer to be used for the various pressures, the variation occurring in the number of sealing pairs. Most pairs of such rings consist of one radial cut 3-piece ring towards the pressure, and one sealing ring away from the pressure, in the same cage. These sealing rings are usually made either with tangent cuts, in 3 pieces, or they consist of three radial cut rings with 3 small pieces or bridges to seal the cuts, which are usually referred to as "6-piece bridge" rings. From observations of wear characteristics we have a personal preference for the bridge ring, though we appreciate that the trend is the other way. The tangential ring is much easier for the manufacturer to make. Manufacturers differ in the number of breaker rings used, but there should never be less than two in any packer. The garter springs used around the rings are normally made of a stainless material.

The packing should be serviced before the ends of the sealing pairs butt and generous end clearances given. At this time check the wear on the sealing faces of the cages with a straight edge, and grind true if any wear is visible. Replace the sealing pairs of rings when they become thin radially. One side of each ring will be stencilled for mating. Make sure that this stencilled side is not against the seal face of the cage. If packers seal well, their running period may be six months in high pressure service, to perhaps as long as two years in low pressure units. The theory of the breaker rings is quite different from that of the sealing pairs. They are normally made in three radial cut segments, of single thickness sufficient to occupy the whole cage. The designer will provide only enough end clearance so that when the surface against the rod is worn in, the segments will butt. These rings tend to reduce the pressure and duty on the sealing rings, but what is more important, if they are operating properly, they reduce or slow down the back flow of the high pressure gas, trapped in the sealing cages during the discharge stroke, while the suction stroke on the frame end of the cylinder is taking place. When the breaker rings are not functioning properly, the high pressure gas trapped in the sealing cages, in its rush to get back into the cylinder whose pressure has been so suddenly reduced, blow the sealing rings apart momentarily, resulting in rather rapid spring breakage. When packers are found with springs around the sealing rings broken, the packer rings should be checked for adjustment.

Crank Bearing Maintenance—Crank bearings should be checked often enough to insure that the clearance does not become excessive. Even with gas engine pistons working on the same shaft, the much higher loading of the compressor rods will normally make these cranks the most rapidly wearing of all bearings in the machine. In the general case, inspections at six-months intervals is usually worth while. The manufacturers recommendations should be followed in the adjustment of these and the other bearings on the machine. Earlier it was noted that the rod loading limitations should not be exceeded. Many manufacturers will still further curtail the loading when the cylinder is operating single acting, for the sake of the crank bearing which gets no reversal with such operation, and therefore may not be so well lubricated. Single acting limitations should be noted before such operation is contemplated.

Check Cylinder Clearance—When discrepancies between calculated and metered capacities are not corrected by working over compressor valves and rings, it may be found advisable to check the actual cylinder clearance. The manufacturer usually calculates this value, and occasionally he is wrong. Such a measurement is somewhat of a chore, though it can be done with water by sealing all the valves but one intake, and blocking around the piston with grease. The trouble may also lie in the lack of a suction or discharge surge chamber, or the use of one too small. Pulsations can readily cause such discrepancies, and to determine whether or not such effects are troubling the cylinder, the indicator is about the only useful trouble-shooting instrument. The principal difficulty in using the indicator will lie in obtaining an accurate indicator motion. For this it will be best to use an instrument with a reducing motion incorporated within it. Motion from the crosshead should be carried to the indicator with a steel tape, using a sufficiently heavy spring against the tape to insure that there is not the slightest amount of whip in the tape. If necessary, watch the tape with a stroboscope to insure against whip or lost motion. Try to avoid the use of wheels for the same reason. See that the indicator string does not pull at an angle to the tape, and that the tape runs in line with the crosshead. Indicator cards accurately taken will reveal all the pressure disturbances to the cylinder, as well as showing the performance of the rings and valves. In most instances the difficulty in obtaining indicator cards has been amply paid for in accurately locating trouble.

Experience is the priceless ingredient in good compressor operation. It is hoped that the reader may pick up a few points in the foregoing from the experience of others which may prove profitable.

How to Measure Piston Rod Motion in Reciprocating Compressors

Wear of sliding elements (piston rings, rider bands, packing, crosshead shoes) as well as piston rod deflection under load can be measured onstream.

Vern Maddox, Bently Nevada Corp., Houston

SIMPLE, INEXPENSIVE TECHNIQUES can be used to measure piston rod deflection and displacement while a reciprocating compressor is in operation. These measurements can indicate excessive wear in sliding elements and potential piston rod breakage from excessive rod flexure.

To safely optimize the onstream time of reciprocating compressors, two conditions must be known: condition of the sliding elements (piston rings, rider bands, packing, crosshead shoes, etc.) and the amount of piston rod deflection under load. Excessive wear in any of the sliding elements will cause accelerated wear and damage to other sliding and stationary elements. Abnormal piston rod flexure can result in rod breakage that is not only costly in terms of physical machine damage but can also create an extreme safety hazard, especially at high pressures or when toxic or explosive gases are compressed. A measurement system consisting of the basic noncontact proximity system elements plus necessary read-out instruments is all that is required to accurately determine changes in piston rod position (displacement) and cyclic flexural loading (deflection).

Readout instrumentation. The type of readout instrumentation needed to make these measurements basically

Basics of proximity measurement

The full use of the technique described requires an understanding of an eddy current proximity measurement system. You need to know how the measurement works, and what is indicated by the system's output signal.

A basic non-contact eddy current measuring system consists of three elements: (1) a probe that observes any conductive surface; (2) a detector-driver to drive the probe and provide the desired output signal; (3) a well regulated −18-volt DC power supply. Fig. 1 shows these basic elements and how they are connected to form the system. The −18-volt DC from the power supply is applied at the detector-driver input terminals. The detector-driver converts this voltage into a high-frequency signal which is transmitted to the probe through a coaxial cable. The measuring element of the probe is a coil. The coil radiates the high-frequency signal into the surrounding area as a magnetic field. If there is no conductive material to intercept the magnetic field, there is no loss of the high frequency signal. Because the magnetic field is only affected by conductive material, any nonconductive material between the probe tip and the observed surface will have no effect on the measurement. When a conductive surface approaches the probe tip, eddy currents are generated on the surface of the material and power is absorbed. As the surface moves nearer the probe tip, more power is absorbed. This power absorption continues until the gap is approximately 0.015 inches. The total energy radiated by the probe is absorbed by the conductive surface and maximum power is lost. (A gap of 0.015 inches is typical for steel surfaces; this minimum gap varies because of differences in the conductivity of other materials.)

When properly calibrated to the industry standard of 0.2 volts per 0.001 inch peak-to-peak motion, the detector-driver has a linear output voltage proportional to the distance changes between the probe tip and the observed surface (Fig. 2). This voltage change is therefore indicative of the peak-to-peak dynamic motion of the shaft and the average position of the shaft with respect to the probe tip. If the detector-driver output is measured with a DC voltmeter, the signal will indicate the average position of the shaft relative to the probe tip. The same signal measured with an oscilloscope will indicate the peak-to-peak dynamic

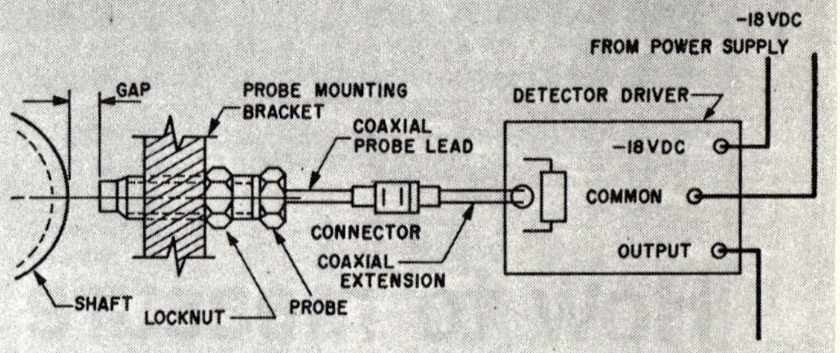

Fig. 1—Basic elements of a proximity measuring system.

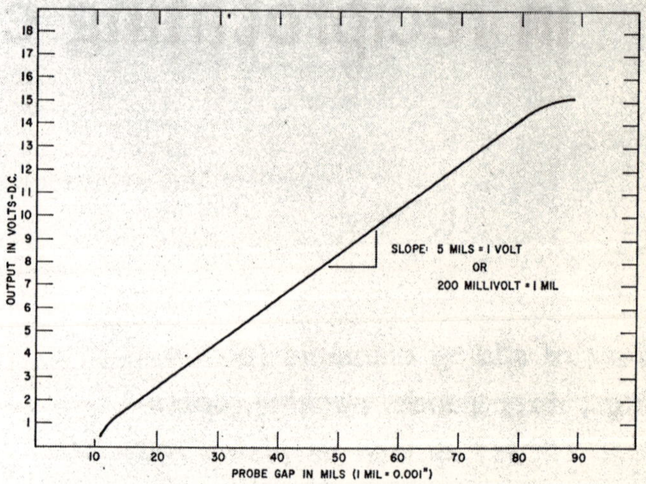

Fig. 2—Typical proximity system calibration curve.

motion, or vibration, of the shaft relative to the probe tip. Fig. 3 shows these combined measurements superimposed on a typical calibration curve. This basic system is used to measure position and dynamic motion on practically all types of rotating machinery and some types of reciprocating machinery.

depends on whether or not probes are permanently mounted, or moved from machine to machine as a portable unit.

The ideal arrangement would be permanently mounted vertical and horizontal probes attached to some fixed reference, such as the stuffing box flange, and continuous monitoring instruments to display piston rod deflection and displacement. Advantages of the permanent monitoring arrangement are continuous observation of the desired measurement signal and adjustable alarm and shutdown set points. On critical machines, an audible warning is essential when unacceptable levels of deflection or displacement are reached.

High-pressure compressors. Reciprocating compressors that operate at extremely high pressures (30,000 to 50,000 psi) are potentially dangerous when piston rod deflection exceeds some known value of peak-to-peak dynamic motion. On this type of machine (commonly referred to as a "hyper" compressor), field experience has shown that an automatic shutdown based on peak-to-peak dynamic motion of the piston rod is justified and necessary. There are many other reciprocating compressors that are hazardous to operate under piston rod malfunction conditions. However, they would be difficult to group into a simple classification, such as the "hyper" compressors. Therefore, the degree of monitoring sophistication required for this general class of potentially dangerous machines must be evaluated on an individual machine basis.

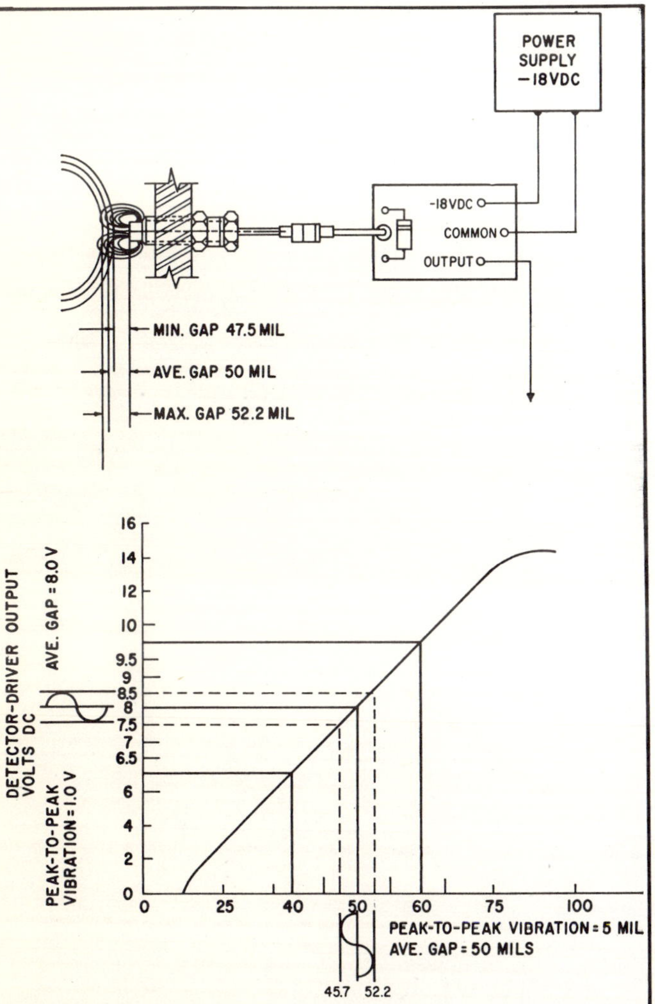

Fig. 3—Combined average position and peak-to-peak dynamic motion measurement signals. Upper figure shows actual arrangement and lower figure shows plot of the motion.

Noncontinuous monitoring. When continuous monitoring cannot be justified on the basis of critical operation or potential safety hazard, the necessary measurements can be obtained by a less sophisticated method. For this general class of compressors, two possible combinations of probes and readout instrumentation exist.

First, probes can be permanently mounted to a fixed reference structure and piston rod motion observed in vertical and horizontal directions. A single probe can be mounted in either direction but valuable rod motion data may be missed since the rod does not necessarily deflect in only one direction. Therefore, it is desirable to observe rod motion with vertical and horizontal probes mounted 90° apart in the same plane and perpendicular to the piston rod surface.

Second, the probes can be made portable and moved from one machine to another as measurements are required. Ordinarily this is done by attaching the probes to some type of magnetic mounting base (such as a dial indicator magnetic base). A method for readjusting the probe tip to rod clearance as the device is moved from one machine to another must be devised. A simple way to accomplish the required adjustability is to attach the probe to the shaft of a nonrotating spindle micrometer head (Fig. 4). The micrometer is not essential. But it is convenient to obtain and use, and mechanically observes changes in probe tip clearance during initial probe gap adjustment.

After deciding whether the probes will be permanently mounted or portable, necessary readout instruments must be considered. With either probe arrangement, the data gathering instruments will be portable for this class of compressors. For complete portability, the readout instrument package should include the −18-volt DC power supply and at least one detector-driver to power the

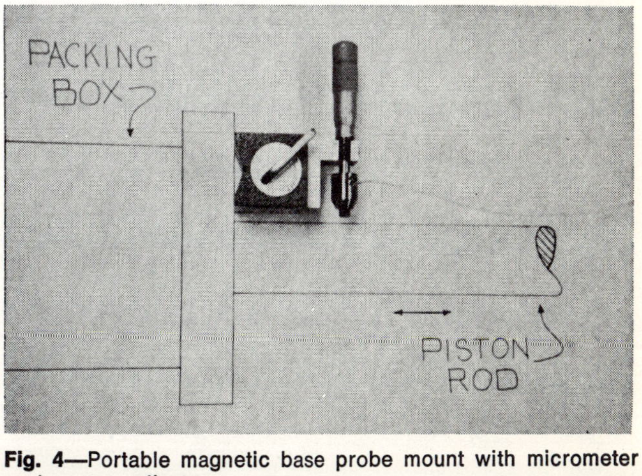

Fig. 4—Portable magnetic base probe mount with micrometer probe gap adjuster.

probes. In addition, this portable package should include a DC voltmeter, a meter to read peak-to-peak dynamic motion in mils (1 mil = 0.001 inches), and be suitable for either battery or 110-volt AC operation. This type of portable unit is simple to use and will provide the required piston rod deflection and displacement data. Also, it could be easily incorporated into a regularly scheduled predictive maintenance program since these data are easily understood and interpreted by engineers or technicians. With a minimum of training on use of the instruments and interpretation of the data, on-the-spot decisions regarding compressor operating conditions should become routine.

The complete self-contained portable package is by far the simplest to transport and use. However, the same data can be obtained with standard instruments normally available to Maintenance departments. The −18-volt DC power supply and detector-driver must still remain an individual part of any non-contact proximity measuring system so further discussion of these components is unnecessary. The more standard types of maintenance instruments used to obtain the necessary measurements are: (1) a good quality DC voltmeter, (2) a laboratory quality DC oscilloscope (typical frequency range of DC to 500 KHz) or (3) a combination of both instruments. An oscilloscope, as described, can be used to measure both piston rod deflection (AC component is the peak-to-peak dynamic motion) and displacement (DC component is average rod position with respect to the probe tip). However, a good quality laboratory oscilloscope is expensive and requires slightly more training to operate. The oscilloscope is a very accurate way to make these measurements and it allows the operator to observe the actual dynamic motion of the piston rod during normal and malfunction operating conditions.

Case history. The dynamic motion wave-forms produced by rod flexure frequently contain valuable malfunction diagnosis information that cannot be observed with other instruments. A good example comes from actual field measurements made on several 35,000 to 50,000 psi "hyper" compressor piston rods. These machines are equipped with permanently mounted vertical and horizontal probes on each rod and a continuous machinery protection monitoring system to observe all probe signals. These particular compressors go through an interesting sequence of amplitude and frequency changes from startup to impending rod failure, that can only be observed with an oscilloscope. When the compressors are started up with a good rod and new packing, oscilloscope traces reveal fairly low amplitude at running speed and a high frequency component identified as packing "squeal" caused by new, tight packing. This high frequency component gradually disappears as the packing "wears in." As the packing begins to wear and clearances increase, the rod has less support and begins to flex, causing an increase in peak-to-peak motion at running speed. This motion continues to increase as a function of packing wear. At some increased rod deflection level, a very interesting and significant event takes place that can only be documented with an oscilloscope. In addition to slowly increasing rod deflection levels, a phase shift occurs between the vertical and horizontal probes. These two probes are mounted in the same measurement plane normal to the rod surface. Therefore, any phase change between the probes implies that the rod is flexing in two directions at the same time, i.e., a vertical deflection peak followed by a horizontal deflection peak. This is obviously a very extreme condition of rod flexure that must result in sudden and catastrophic rod failure. By thoroughly observing and documenting this sequence of events with an oscilloscope, these machines can be operated safely for a maximum length of time. As the example illustrates, the oscilloscope should never be discounted as a valuable diagnostic tool. A thorough understanding of how the machinery behaves and what the basic measurements mean will be the best guidelines for determining when an oscilloscope must be used.

Improve Valve Life in Reciprocating Compressors

Reciprocating compressors are still an important type of machinery in the HPI. And, valve failure continues to be one of the most common problems with these machines. Here's how to identify this problem and what to do about it.

D. M. Whittaker, Shell Canada Ltd.,
Sarnia, Ont., Canada

DURING THE HISTORY of petroleum refining, reciprocating compressors have been used extensively to compress hydrogen for use in many processes. Because of the process importance of the compressor, it is essential that a high degree of mechanical reliability be maintained. Therefore, industries are actively engaged in extensive predictive maintenance programs to ensure maximum reliability. Despite these programs, in-service operating conditions of the compressor often cause critical components such as valve assemblies to fail.

Because reciprocating compressors in refinery service must inadvertently handle a wide variety of flow regimes as illustrated in Fig. 1, the compressor valve assemblies are sometimes very severely loaded. In addition to the inherent design of the compressor valve assembly, flow pattern, purity and moisture content of the hydrogen will dictate the life of the valve assembly. Under extreme conditions, valve life can be limited to a matter of weeks.

Valve failures are typified by broken springs and or broken plates. These items will be discussed separately with suggested improvements to increase valve assembly mechanical reliability.

VALVE SPRING SURGING

A broken spring may be caused by resonance (produced by flow regimes), which may occur in the spring, producing waves which travel along the length of the spring. These waves are successive compressions and extensions moving from coil to coil within each spring. The spring may then be subjected repeatedly to a stress corresponding to stress for solid compression. Because this stress is often above the endurance strength of the spring, repeated contact of the coils will damage the surface. Surging can result in early failures, where calculated stresses for normal loads are apparently safe. The lowest natural frequency (first harmonic) of a spring is given by:

$$\phi = \tfrac{1}{2}\left(\frac{k}{m}\right)^{1/2} \text{ cycles per unit time}$$

where k = spring constant
m = mass of the active coils (lb-sec²/in.)

The volume of the active wire is

$$V \simeq \left(\frac{\pi D_w^2}{4}\right)(\pi D_m)\, N_c \qquad (\text{in.}^3)$$

where D_w = wire size (in.)
D_m = mean coil diameter
N_c = number of coils.

Therefore

$$m = \frac{\rho}{g_o}\,\frac{\pi D_w^2}{4}\,\pi D_m N_c \qquad \text{lb.-sec.}^2/\text{in.}$$

where g_o = 386 in./sec.²
ρ = density (lb./in.³)

Without developing the formula, it may be shown that:

$$\phi = \tfrac{1}{2}\left(\frac{k}{m}\right)^{1/2} = \tfrac{1}{2}\left(\frac{G\,D_w^4}{8D_m^3N_c}\,\frac{(g_o)\,(4)}{(\rho\pi^2D_w^2D_mN_c)}\right)^{1/2}$$

$$= \frac{2.21\,D_w}{N_cD_m^2}\left(\frac{G}{\rho}\right)^{1/2} \text{ cps}$$

where G = modulus of elasticity (psi).

As a general rule, if natural frequency is 12 or more times the rate at which the spring operates, no trouble is anticipated from surging. However, should surge be suspected, modifications should be made, such that natural frequency is 20 or more times the rate at which the spring operates. The additional margin of safety takes into account effects of flow regimes described in Fig. 1. In actual testing, using this approach, valve spring life was increased from 3 months to 12 months. The 12 months represent the period between major shutdowns at which time valve assemblies are re-conditioned. Due to increased spring life, a considerable saving was realized in repair labor and production.

VALVE PLATES

Breakage of metal valve plates may be caused by presence of liquid in the gas stream. For flow patterns described in Fig. 1, the valve plate upon closing, tends to seat unevenly due to the incompressibility of the fluid. Repeated actions of this type can eventually cause failure of the plate.

Valve plates are available in a wide variety of materials, both metallic and non-metallic. However, the non-metallic valve plate has been restricted to applications of low pressure ratios where discharge gas temperatures are less than 300° F.

Recently, valve plate manufacturers have introduced a teflon valve plate impregnated with glass and nylon. The teflon plate has many advantages over the conventional non-metallic plate and metallic plate. These are outlined in Table 1.

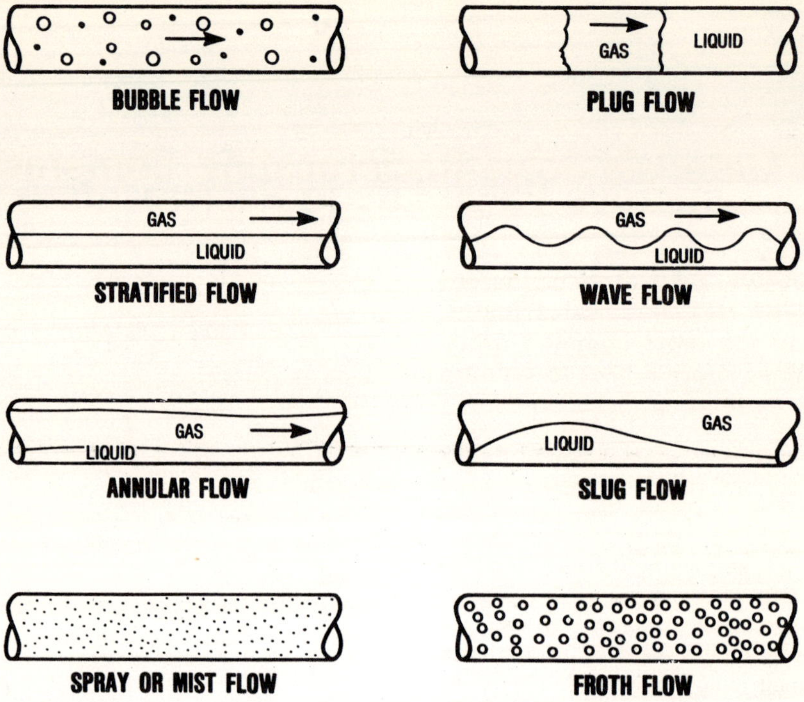

Fig. 1—Types of two phase flow.

TABLE 1

	Teflon valve plate	Conventional non-metallic plate	Metallic plate
Flexural strength..	High	Moderate	Low
Surface coating....	None	Breaks down resulting in failure	None
Valve plate breakage........	Does not damage valve seat	Does not damage valve seat	Damages valve seat
Maximum temp. limit............	400-500°F	200-300°F	Greater
Price..............	Equal to non-metallic	....	Twice non-metallic

These teflon valve plates have been used to replace metal plates where discharge gas temperatures were below 500° F. As a result, valve plate life has been increased from approximately 3 months to 12 months. Isolated cases of valve failure are still reported, but without damage to valve seats. Use of teflon valve plates markedly reduced unscheduled shutdowns with a substantial saving in labor and production.

CONCLUSIONS

Valve assembly failures are typified by both broken springs and broken plates. Spring failure may be caused by surging. By determining natural frequency of the spring and operating frequency of the compressor system, surging can be eliminated through proper choice of springs. Plate failure may be caused by presence of liquid in the gas stream. If the liquid cannot be removed entirely from the system, valve plates can be chosen with high flexural strength such that breakage will be substantially reduced.

Through this judicious choice of valve springs and valve plates, a total of $15,000 in labor and production was saved annually on four compressors, while the need for additional costly liquid separation equipment was eliminated.

ACKNOWLEDGMENT

This paper was originally prepared for the Technical Article Program of the Compressed Air & Gas Institute, Cleveland, Ohio.

LITERATURE CITED

[1] Faires, V. M., *Design of machine elements,* Fourth Edition, MacMillan, New York/Collier-MacMillan, London (1968).
[2] Strauss, W., *Industrial gas cleaning,* Vol. 8, Pergamon Press (1966).

New Materials Solve Reciprocating Compressor Problems

Reciprocating compressor problems with rods, packing, valves and valve parts can be solved with materials now available.

E. F. Barnes, Union Carbide Corp.,
Chemicals and Plastics, Texas City, Texas

CORROSIVE AND EROSIVE GASES in hydrocarbon processing plants cause failures in reciprocating compressor components. New materials are now available that will successfully reduce component failure thereby increasing onstream time.

Rods and packing. In my opinion, one of the most interesting new materials is the development of lubricated Teflon[(1)] compressor rod packing. The most successful use of this material in Union Carbide's Texas City plant has been with metallic backup rings which support and also serve as heat sinks for the Teflon components.

Usually, with Teflon or glass-filled Teflon packing, it is necessary to consider the wear that will occur on the rod. Alternatives include the following: a. a short life rod, b. chrome plated rod, c. surface hardened rod, or d. UCAR[(2)] tungsten carbide coated rod.

I do not like to use metallized rods above 400 psi in compressor rod applications because of the possibility of the coating peeling.

Table 1 lists some of successful applications of rods and packing at our Texas City plant. Initial attempts were made with several rod preparations, usually successful, such as the following:

- Flash chrome plated
- Nitrided
- UCAR detonation gun coating with LW 1 or LW 15.

By mistake, and in similar applications, some Teflon packing was run on metallized and uncoated rods (lubricated with oil) and performance was excellent. We then began selecting applications for use of Teflon on uncoated rods.

The more difficult applications of rods and packing are noted in Table 2 with comments on special considerations. These machines are fully lubricated (not partially lubed or minilubed). Some of these applications represent years of trial of different materials and rod surfaces with accompanying failures. While the Teflon packing can yield savings in maintenance because of longer life than bronze or phenolic packing, justification of a UCAR "D" gun coated and diamond ground rod requires short service life of standard material to consider the increased rod cost.

As Table 2 shows, corrosive service can require stainless rods to prevent coating failure. If an expensive rod is slightly damaged, the UCAR coating can be used to build

TABLE 1—Lubricated Teflon compressor rod applications

	Bore	Stroke	RPM	P_s psig	P_d psig	Rod size	Packing	Rod surface	Comment
1	19¼	14	330	10	40	3¼	R & T Cast Iron Backup	Steel	
2	14	14	330	40	130	3¼	Carbon Filled Teflon	Steel	
3	10¼	14	330	130	720	3¼	As above	Steel	
4	8¼	14	300	100	650	2½	As above	Steel and Metallized No. 2.SS*	Sour Gas
5	18½	9	450	10	40	2¼	As above	Steel**	
6	12½	9	450	40	180	2¼	As above	4140 30Rc	
7	5¾	14	300	250	750	3¼	As above but Bronze Backup	4140 30Rc	Sooty C_2H_4

* 4140 UCAR Flame Plated LW-15 excellent
** Rods 17-4, 15-5, Condition (1150), UCAR "D" Gun Coated LW-15 also

TABLE 2—Difficult compressor rod packing applications

	Bore in.	Stroke in.	RPM	P_s psig	P_d psig	Rod Size, in.	Packing	Rod Surface	Comment
8	4 5/16	25	180	3200	6000	3¼	Carbon Filled Teflon R & T Bronze Backup	LW-15 over 420-S.S. 30Rc	Corrosive Service LW-15 Over 4140 Peels from Corrosion of 4140 Rod (6 mos)
9	5⅛	25	180	600	1900	3¼	Carbon Filled Teflon R & T Bronze Backup	LW-15 over 420 S.S. 30Rc	
10	6⅜	9	300	2200	2400	1⅝	Carbon Filled Teflon R & T Bronze Backup	LW-1 on 420 S.S. 30Rc	Heavy Liquid Wash & Entrainment-Cycle Machine H_2 Service
11	5	14	300	700	2500	2½	Carbon Filled Teflon R & T Bronze Backup	LW-1 over 17-4PH and 410 S.S.	Many years service. No Measurable Rod Wear Hot-H_2 Service
12	5¼	7	450	300	1300	1⅜	Carbon Filled Teflon R & T Bronze Backup	LW-15 420 S.S.	Lots of Mildly Acidic Entrainment.
13	3½	14	300	3200	6000	3½	Carbon Filled Teflon Bronze Backup	Flame Hardened 4140	2+ yr. Life Oil Cooled Packing Case. Stepless Tangent cut Packing.

(1) Trademark of E. I. duPont de Nemours & Co., Inc.
(2) Trademark of Union Carbide Corp.

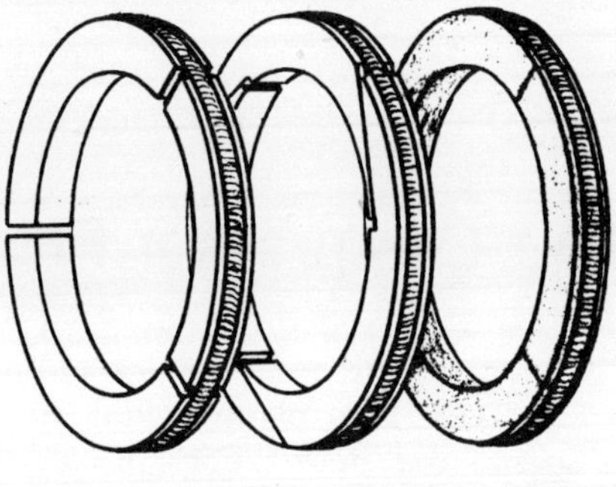

Fig. 1—Conventional rod packing with radial-tangent pair and and backup ring.

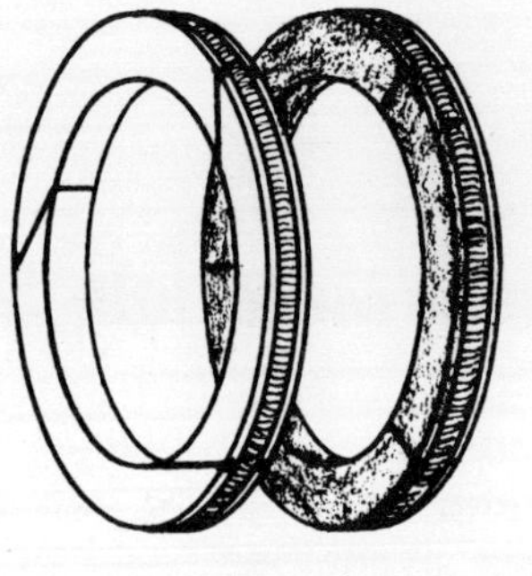

Fig. 2—Modern rod packing design with single active stepless tangent-cut sealing ring and a radial backup ring.

up the rod to its original size or about 0.010 inches per side coating.

For some of the Table 2 rods (item 10), nitrided rods, and flash chrome plated rods have been used for several years. For this location, UCAR coated rods are generally more available.

A comparison of traditional and more modern packing design can be seen in Figs. 1 and 2. Fig. 1 shows rod packing with the conventional radial-tangent pair and a back-up ring which does not seal on the rod.

Fig. 2 shows a single active stepless tangent-cut sealing ring with a radial back-up ring. The advantage of this design is compactness. Item 13 in Table 2 reflects this application.

In services with solid particles and plenty of rod movement, packing failures usually show lots of debris between the plastic components of rod packing. I have often wondered whether or not the Fig. 2 style packing is more resistant to this debris pickup than Fig. 1 packing.

Compressor valves. Besides packing, other compressor components require consideration. Valves are one of the most critical items in a reciprocating compressor. Valves must endure many load cycles, rubbing parts and poorly cushioned impact combined with high gas velocity. This does not include added burdens possible with corrosion, liquids and small foreign abrasive sand blasting particles. Frequently, valve parts show distress from the above while the remainder of the process vessels and pipe are unaffected. Compressor valve springs and discs are sometimes stressed to a higher level than any other part of the materials containing the process.

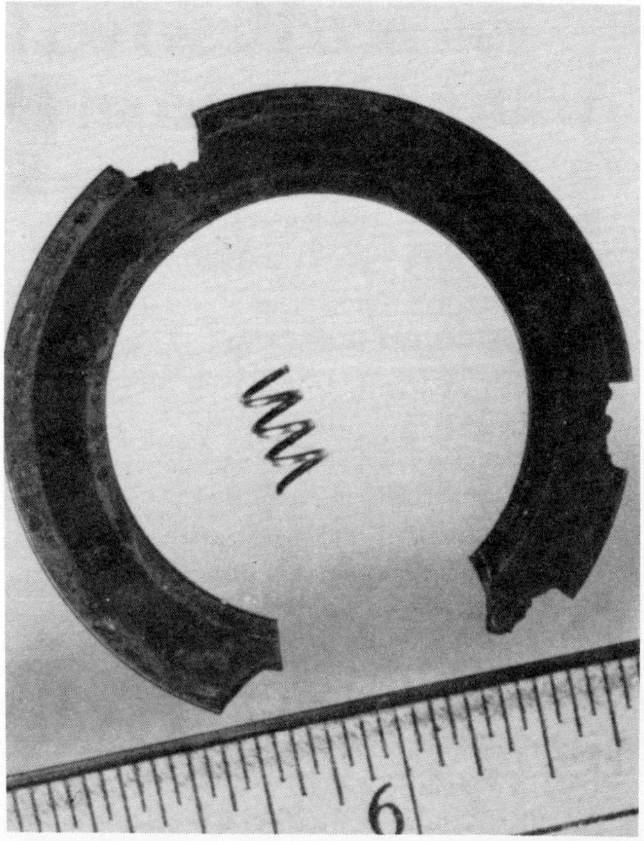

Fig. 3—Failed compressor disc (Note round wire coil spring in center of photo).

A disc manufactured of a material and thickness that is grossly unsuited for the application is usually quickly respecified. However, reasonably persistent valve failures are not as spectacular and are not quickly corrected. Sometimes, because of process continuity considerations, malfunctioning valves are left in service until the amount of damage conceals the beginning cause of failure (Fig. 3 is an example). Failure analysis under these conditions is difficult. Naturally, there are times that piston ring failures or partial failures trigger valve failures and a situation of which comes first—the chicken or the egg, must be answered or remedied. This category of compressor valve disc and spring failures can be intricately intertwined. There can be further denial of logical reasoning to the problem by intermittent exposure to a process component (acid) which greatly aggravates failure and may not be in phase with the occurrence of the damage. A practice in problem solving that I frequently use is to specify more corrosion resistant materials (valve disc and springs) when unexplained and irregular failures occur. With failure analysis of valve components, it is sometimes possible to define a change in materials or geometry to improve life with such a course of action.

A recently available new material has provided improved life for a number of services in this plant's reciprocating compressors. These new compressor valve springs are made of MP-35 N[(3)] and Elgiloy.[(4)] These materials are used in various sour gas and liquids services.[1] Both the physical properties and corrosion resistance of these mate-

(3) MP-35-N—Trademark Standard Pressed Steel Co., Jenkintown, Pa.
(4) Elgiloy—Trademark Elgiloy Co., Elgin, Ill.

rials are impressive. For problem services we have used both types of springs. Some of the services are listed in Tables 1 and 2.

We have used only round wire coil springs. Some valves require a change in the spring load and rate and there is a long delivery time for rectangular cross section springs. While some valves use leaf springs, our immediate problems with these valves have not required any trials of leaf springs of the new materials.

Springs of Titanium-6 Aluminum-5 Vanadium (TI-6 A1-4 V) as well as valve discs have long been long lived in some hydrogen and hydrogen-carbon monoxide compressors. While available literature[9] contains a lot of information on hydrogen fatigue cracking of Ti-6 A1-4 V, this material has been superior in life to most other materials used.

For many process streams containing hydrogen, valve plates of titanium hold up very well. Wear of plates, springs and seats can be lessened by lift and spring change.

When valves of different design are used in the same service, it is not uncommon that one valve design can use less corrosion resistant material than others. In my opinion, the difference can be defined by stress and wear intensities. In my experience, over a longer period, valves of corrosion resistant components outlast other designs. In other words, using corrosion resistant components is a wise investment.

The Reliability Program at Texas City is based on failure analysis. Oil, Metallurgical and Statistical (life) Analyses are usually required for these persistent problems that can sometime yield to changes in material.

LITERATURE CITED

1. Hamby, T. W.; Tuttle, R. N., "Deep, High Pressure Sour Gas is Challenge," *Oil and Gas Journal*, May 12, 1975, page 114.
2. Whittaker, D. M., "Improve Valve Life in Reciprocating Compressors," *Hydrocarbon Processing*, May, 1975, p. 143.
3. Congram, Gary E., "Compressor Maintenance Assures Longer Run Times," *The Oil and Gas Journal*, June 23, 1975, p. 100.
4. Hartwick, W., "Power Requirements and Associated Effects of Reciprocating Compressor Cylinder Ends Deactivated by Internal Bypassing," ASME Paper 75-DGP-9, ASME, United Engineering Center, 345 East 47th Street, New York, N.Y. 10017.
5. Corrosion Resistance of Multiphase Alloy MP-35 N, MECCO, Aug. 14, 1969, UCC, South Charleston, West Va.
6. Treseder, R. S.; Swanson, T. M., "Factors in Sulfide Corrosion Cracking of High Strength Steels," Corrosion, NACE, Feb., 1968, p. 31.
7. Kohut, G. B.; McGuire, W. J., "Sulfide Stress Cracking Causes Failure of Compressor Components in Refinery Service," *Materials Protection*, June, 1968, p. 17.
8. Anon.—"Coping with Chloride while Cat Reforming," NPRA Questions & Answers, No. 7, *Oil and Gas Journal*, June 2, 1975, p. 137.
9. Hydrogen in Metals—No. 2 in American Society for Metals Materials/Metal Working, Technology Series 1974 Library of Congress, Cat. No. 73-86455 ASM.

Foundation Repair Techniques

The most carefully specified reciprocating machinery system can become a problem if the foundation is inadequate. Here's how faulty foundations can be economically and reliably repaired in the field.

E. M. Renfro, Consulting Engineer, Houston

THE DILEMMA of what to do about cracked concrete foundations is no longer the headache it was in the past. Epoxy technology and application techniques have evolved to the state that repairs are quite reliable. Equipment downtime is reduced from weeks or months to days. In some cases, repairs can even be effected without a shutdown.

TYPES OF CRACKING

The tensile strength of concrete is only about 10 per cent of its compressive strength.[1] Because of this weakness in tension, reinforcing steel is embedded in concrete to carry the tensile loads. This addition of steel promotes one form of cracking and at the same time inhibits another form of cracking.

Concrete shrinks slightly on curing. Typical drying shrinkages of reinforced concrete structures are in the range of 0.02-0.03 per cent. The addition of reinforcing steel does not prevent shrinkage, but it does restrict the shrinkage. This shrinkage produces both tension in the concrete and compression in the steel, and given enough steel, cracking of the concrete can result.[2] The resulting cracks are often referred to as curing cracks. They are not considered serious because they are randomly located and there is seldom any movement between the two segments. They cause no problems other than to provide a path for more rapid penetration of lubricating oil spilled later on the concrete. Such damage can easily be prevented by sealing the concrete with the proper epoxy system after foundation construction has been completed and before the equipment has been installed.

Those cracks which appear as a result of externally applied dynamic forces are characterized by continuous progression through the block and by movement between the segments. Once movement begins between the segments, it promotes further propagation through the entire block. These constitute structural failures and are therefore considered serious.

Movement of the block can place the equipment frame in tension. Fortunately, fatigue of the grout bond to the equipment base usually occurs before the casting itself fatigues and cracks. Failure of the grout bond, which allows movement of the machine base, is easily detectable either visually or by dial indicators. If such movement is allowed to continue for long periods of time, crankshaft failure, failure of main bearing supports or both can be the result.

CAUSES OF FAILURE

Failures in concrete machinery foundations can be attributed to improper or inadequate foundation design, poor construction practices and/or poor operating conditions.

The criteria for machinery foundations used by the various design engineers varies from the simplest, where mass is obtained by rules-of-thumb plus allowances for local soil conditions, to a highly sophisticated and detailed analysis. In either case, it is a common practice to place reinforcing steel in the block in accordance with the same fundamental principles that apply to reinforced concrete members in general. This supposition is evidenced by the typical practice of placing reinforcing steel in a symmetrical pattern throughout the block. It is the author's opinion that the placement of reinforcing steel should be with more consideration as to the source and direction of the external forces applied to the foundation. According to this reasoning, the preponderance of reinforcing steel in a reciprocating engine/compressor foundation should be placed in the upper portion of the block in a direction perpendicular to the crankshaft. Weighting the placement of steel in this location would reduce the tendency for cleavage-type failures that sometimes begin at the top of the foundation in the notch below the oil pan and extend through the block to the mat below. These cracks are always parallel to the crankshaft and are caused by forces set up by the power pistons or compressor stages.

Obtaining a good concrete foundation does not necessarily result from good design as a matter of course. Good construction practices must be followed and in detail. After the reinforcing steel has been installed and the forms constructed, the concrete is poured. Concrete must be properly prepared and poured while the mixture is fresh. Ideally, the pour should be monolithic and vibrated in the process to remove voids. Where the mass is quite large or where ambient temperatures are at either extreme, curing temperature control measures must be employed. And finally, moist curing conditions must be maintained to assure proper hydration of the cement. All of this seems to be relatively simple. However, even slight deviation on any one single step can result in a poor quality foundation.

In the United States, and other fully developed countries, we are blessed with the availability of quality construction materials, good construction equipment, highly skilled labor force and an excellent means of transportation. When foundation problems occur, which can be attributed to faulty construction, they can usually be traced to overconfidence or lack of proper supervision.

When construction is carried out in the lesser developed countries or in remote areas of the world, the problems associated with maintaining good construction practices are greatly increased. Good equipment and spare parts may be difficult to find, and then only at great expense. The local labor forces are generally unskilled and there may be language barriers. Local laws may restrict the importation of quality construction materials. Means of transportation may be so poor that delivery time is increased many times. The function of supervision over construction may become an arena for compromises. As a result, makeshift construction sometimes prevails.

There are always good channels of communication between the owners and the design engineers. As a result plants are designed in harmony with the capability of the local operating personnel. Seldom does a contractor have an opportunity to voice local construction capability during the design period. Consequently, plants are often designed with U.S. construction capability in mind.

Many foundations fail because of operating conditions. When equipment is modified to increase its service rating, it may no longer be in balance. Failure to maintain the ignition system can cause severe unbalanced forces in rough running equipment. Failure to regrout when necessary can multiply the forces exerted on the foundation because of vibration impact. Any or all of these conditions which deviate from the original design concept can, in time, damage the underlying foundation.

METHODS OF REPAIR

Because foundation cracks are seldom noticed until they are called to the owner's attention by the regrouting contractor, repairs are usually carried out at the time of regrouting. There are many variations in repair technique. The choice of which is usually dependent upon the access to the foundation, availability of alloy steel stud bolts and the urgency in returning the equipment to service.

Cracks parallel to the crankshaft occur most frequently at a location beginning at the trough below the oil pan and extending through the block to the mat below. Fig. 1 illustrates a method of repairing such cracks by drilling horizontal holes regularly spaced from one end of the foundation to the other end. This series of holes are placed at an elevation of approximately two feet below the oil pan trough. High tensile alloy steel stud bolts are inserted in the holes along with small diameter copper injection and vent tubes. The ends of the hole are then sealed and the bolt tightened to draw the segments of the block back together. Once the bolt is torqued, epoxy is injected into the annular space around the bolt. After most of the air is displaced through the vent tube, the vent tube is plugged and the crack is pressure-injected from the inside out.[3]

This repair method places the concrete in compression which would otherwise be in tension. The compressive condition must be overcome before the crack could possibly reoccur. In this case, the repair relies not only on bond of the adhesive between the two segments but also on the prestressed condition of the concrete. As a result,

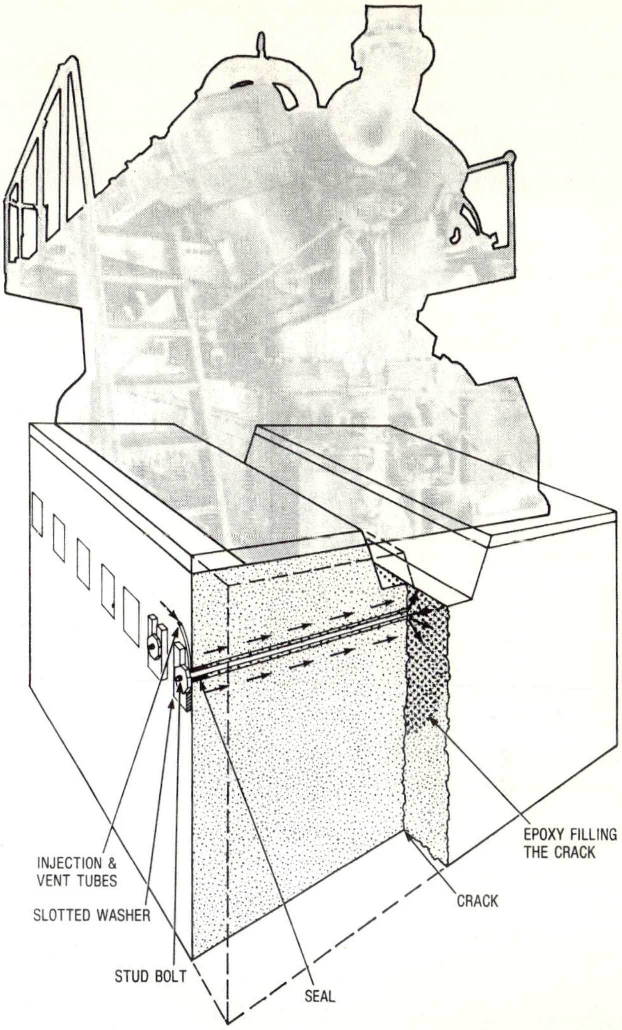

Fig. 1—Method of repairing cracks parallel to the crankshaft. Here, a stud bolt is placed through the foundation to cross the crack.

the repaired foundation is often much stronger than the original foundation. This technique is often used when the concrete is of poor quality.

When access to the foundation is limited or downtime is critical and stud bolts are not immediately available, an alternate method can be used. In this method, holes are drilled after the existing grout has been removed. These holes are drilled into the foundation perpendicular to the crankshaft and at an angle to intersect the crack at an elevation of approximately two feet below the oil pan trough. Heavy reinforcing steel, usually 1-1½ inches in diameter is inserted into the hole along with the small diameter copper injection and vent tubes. The entrance is then plugged and the equipment regrouted. After equipment regrouting, the annular space around the reinforcing steel is filled, the vent tube plugged and the crack pressure-injected from the inside out.

This method does not utilize the prestressing technique. Most of the time, any shortcomings, as a result of not using the prestressing technique, can be overcome by simply adding more steel and reducing the spacing between bars.

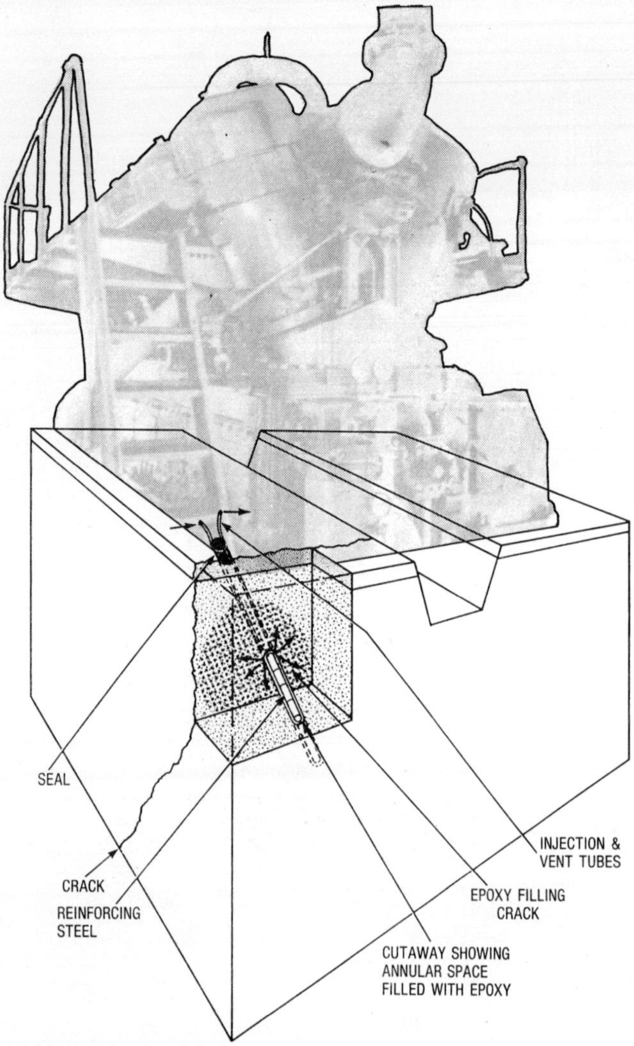

Fig. 2—Method of repairing cracks perpendicular to the crankshaft. Here, reinforcing steel is placed at a position skew with relation to the crankshaft to cross the crack.

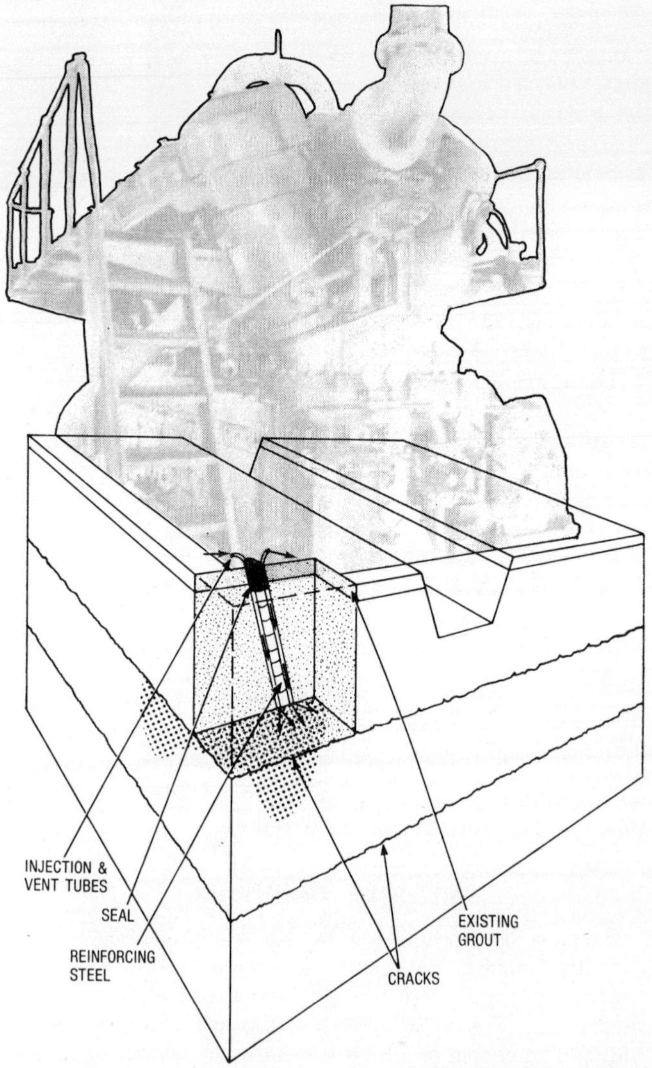

Fig. 3—Method of repairing cracks which develop between concrete pours. Note the reinforcing steel crossing the crack at a location directly below the load bearing area.

A similar procedure for repairing cracks perpendicular to the crankshaft can be carried out by rotating the drilling angle 90 degrees (Fig. 2). The reinforcing steel in this case would be in a position skew to the crankshaft.[4]

Fig. 3 illustrates a method of repairing cracks between concrete pours. These cracks often appear between the block and the mat below or between the pours when blocks are so large that a monolithic pour can not be easily made. The technique used in this case is essentially the same as the alternate method for Fig. 1.

As one word of caution to equipment owners, these techniques require special epoxy formulations and careful technique. It is the author's opinion that repairs should be carried out by experienced specialists using products manufactured specifically for such applications. As an explanation, there are approximately 100 different types of curing agents available to the epoxy formulator. Only a few impart properties to the cured epoxy resin which are satisfactory for this service. The choice of curing agent is based on such criteria as pot life time, non-shrink capability, ability to bond through oil films, creep resistance at the operating temperature, etc. Because all jobs are slightly different, the applicator should be familiar with products available and have experience in their use.

LITERATURE CITED

[1] Waddell, Joseph J., "Concrete Construction Handbook," McGraw-Hill, New York, 1968, p. 6-12.
[2] Waddell, Joseph J., "Concrete Construction Handbook," McGraw-Hill, New York, 1968, p. 9-4.
[3] Renfro, E. M., "Five Years with Epoxy Grouts," 19th annual meeting of the Gas Compressor Institute, Liberal, Kansas, Preprint, April 4-5, 1972.
[4] Exxon Chemical Co., U. S. A., product publication number OFC-A74-1500, 1974, p. 12.

Part VI.

Axial Compressors

Axial Compressors: Sizing, Performance

The following procedure for sizing an axial compressor is intended to serve for approximating purposes only on applications requiring air with an inlet pressure of 14.7 psia and a temperature of 100°F. As a guide in following this procedure, consider as an example a compressor needed to handle 70,000 cfm of air at the above conditions and required to boost the air pressure to 57 psia, such as would be required for a large cat cracker.

Step 1. With the volume and required discharge pressure given, select the compressor size and number of stages from Figure 1. With 70,000 cfm and a discharge pressure of 57 psia, the compressor size would be a 900 casing size with eleven stages.

Step 2. With a known volume, obtain the approximate speed from Figure 2. With 70,000 cfm, the speed would be between 5,000 and 5,100 rpm.

Step 3. With a known required discharge pressure in psia, obtain the hp required per 100 cfm from Figure 3. With a discharge pressure of 57 psia required, the hp per 100 cfm would be 12.65 and 12.65 (700) equaling a total hp requirement of 8,855.

Step 4. With a known compressor selection and the number of stages, obtain dimensions from Figure 4. The 900 casing size compressor with eleven stages would have overall box dimensions of 151 inches long, 118 inches wide (base) and 96 inches high. The compressor weight is approximately 40,500 pounds.

Step 5. The approximate price of an axial compressor with a motor-gear (Figure 5) or with steam-turbine drive (Figure 6) may be obtained. Figure 5 prices (based on 1963 U.S. dollars) include axial compressor, synchronous motor with 0.8 power factor, suitable speed-increasing gear, lubrication system and baseplate for the compressor and drive. Figure 6 prices include an axial compressor, multivalve steam turbine, baseplate and lubrication system.

Example: The 70,000 cfm axial compressor previously selected with motor-gear drive would

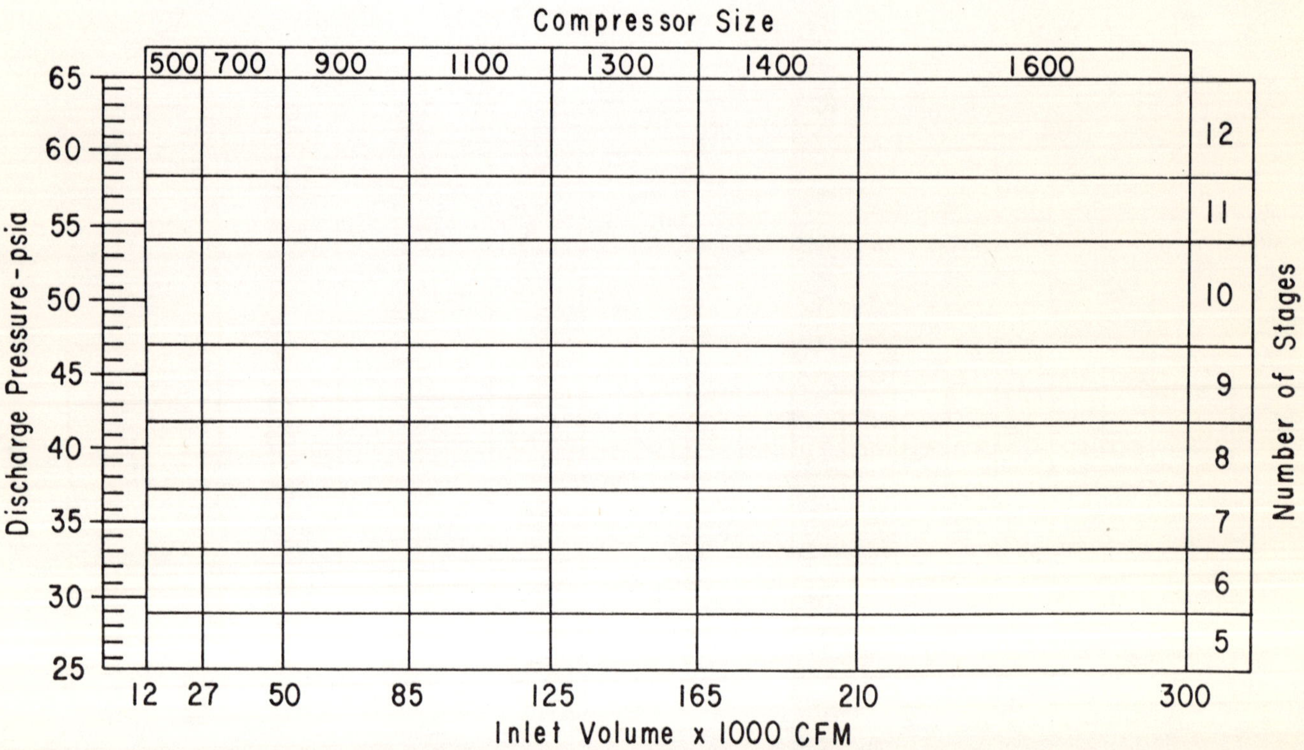

Fig. 1—Use this chart to select compressor size.

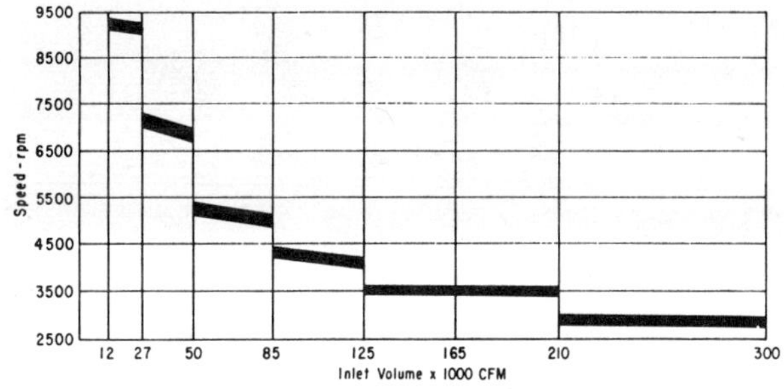

Fig. 2—Use this chart to find compressor speed.

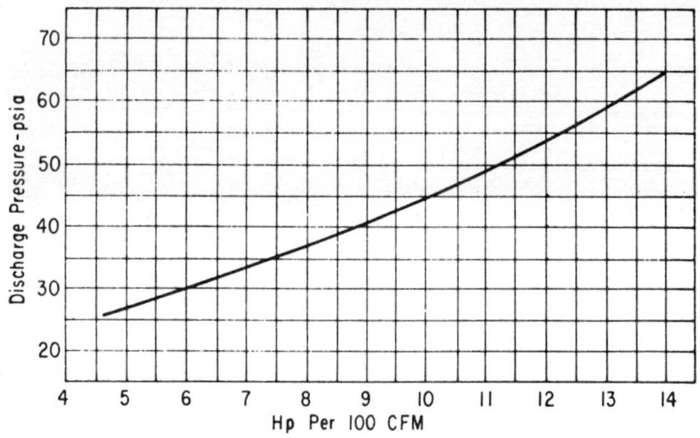

Fig. 3—Use this chart to find horsepower required.

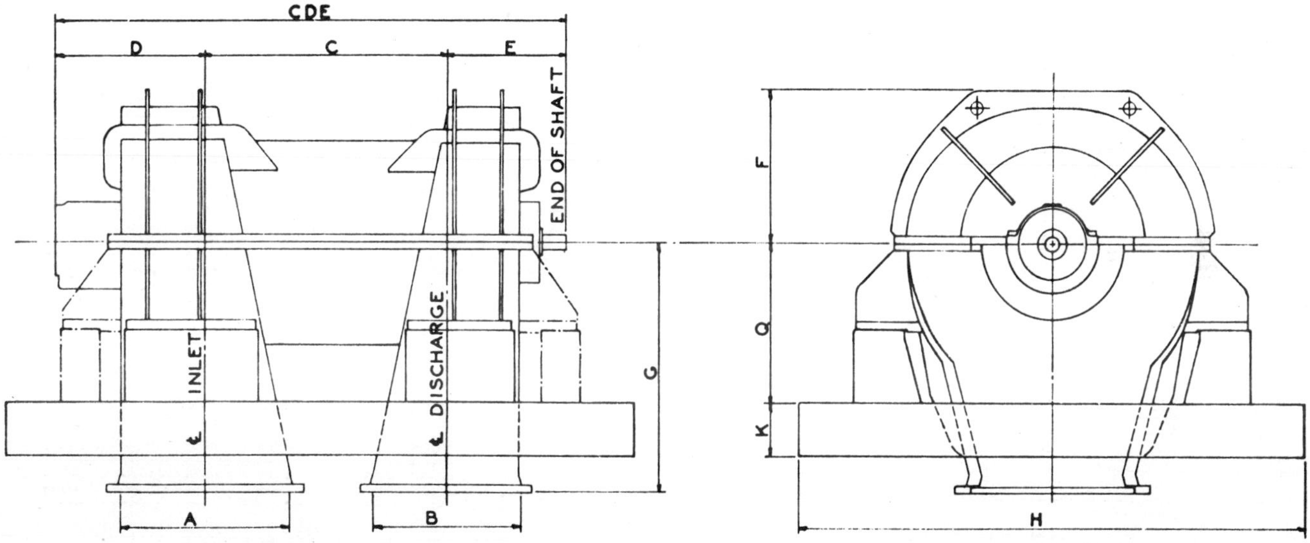

TABULATION FOR FIVE STAGE AXIAL COMPRESSOR

CASING SIZE	A	B	C	D	E	F	G	H		K	Q	CDE	COMPR. WT. ONLY – LBS.	FOR EACH ADDITIONAL STAGE: ADD TO "C" & "CDE"	FOR EACH ADDITIONAL STAGE: ADD TO "COMPR. WT."
500+	24	18	20½	32	21	18	33	52		12	36	73½	5000	3	500
700+	36	30	51	32	25	31	60	72		12	27	108	20000	4½	1000
900	48	36	63	32	20	36	60	118		12	36	115	31500	6	1500
900*			72½									124½	40000	7½	2000
1100	60	42	69	32	28½	40	72	124		12	36	129½	34300	6½	1600
1100*			76									136½	46800	7½	3400
1300	66	54	76	32	26	45	78	124		12	36	134	62000	6	2000
1300*						60							69000	7	3000
1400	78	54	76	32	26	45	84	138		12	36	134	62000	8	3000
1700	80	72	104½	32	32	54½	84	192		18	42	168½	125000	8	5000

* MOVEABLE BLADES
\+ END MOUNTED (— — —)

Fig 4—Use this table to find compressor dimensions and weight.

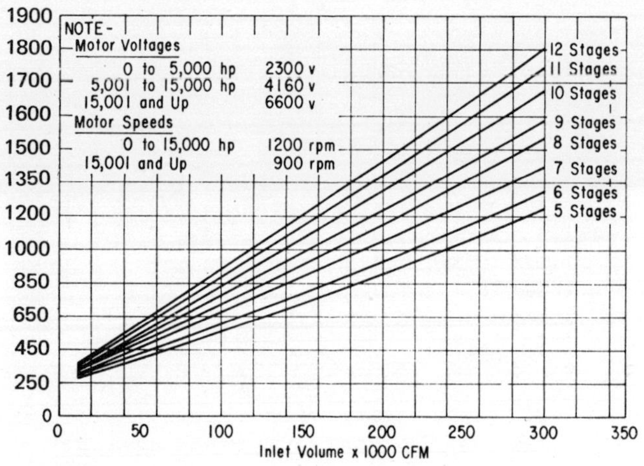

Fig. 5—Axial compressor price with motor-gear drive.

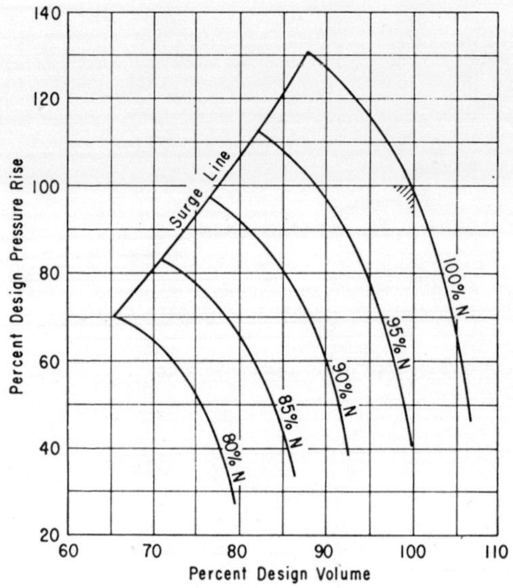

Fig. 7—Typical axial compressor performance curve.

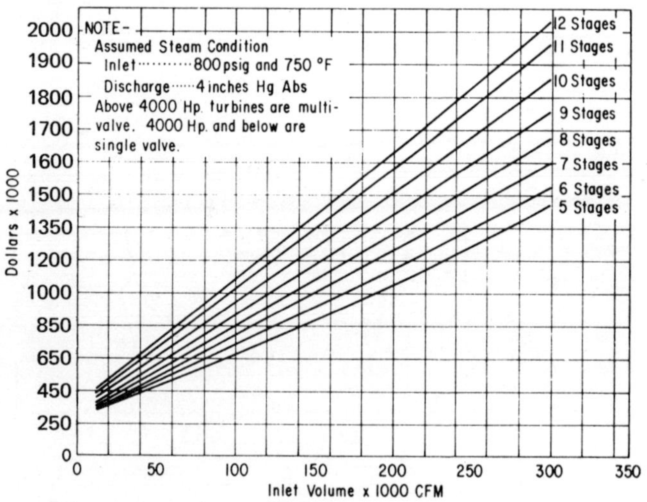

Fig. 6—Axial compressor price with steam-turbine drive.

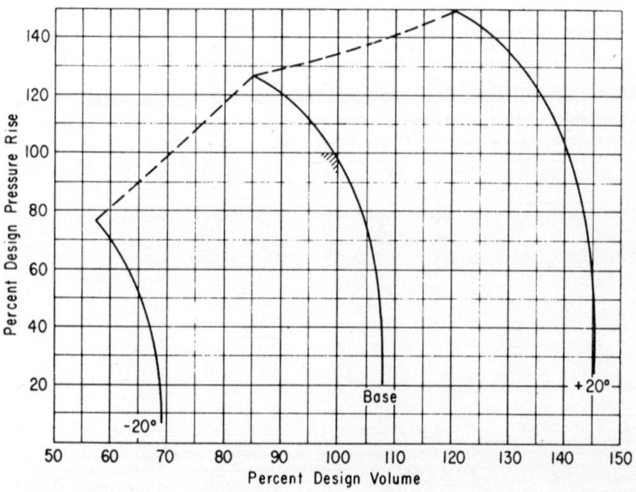

Fig. 8—Expected performance of axial compressor with full stator blade control at constant speed.

have a price of $700,000 as indicated in Figure 5. The price with turbine is $850,000 (Figure 6).

Axial Compressor Performance

The typical axial performance curve is shown in Figure 7. Note the wide variation in pressure with a relatively small capacity change for a constant-speed line. This makes the axial compressor ideally suited for a base-load machine.

Also, the steep curve and high pressure rise at part-load points allow parallel operation with other machines. The machines do not have to be matched too closely in design pressure ratio since the axial's discharge pressure will adjust to that of the parallel machine without danger of surge.

Some applications require a wider volume range than that offered by the basic axial compressor. In these applications, adjustable inlet guide vanes or full or partial stator blade control on axial compressors are available to provide the wider range of operating conditions. Notice in Figure 8 that at 100% speed, the capacity of the unit can be varied appreciably by adjusting all stator blades.

If less variation in operating range is required, partial adjustment of the stator blades, or merely inlet guide vane control may suffice. Any of these methods of range extension can be achieved with the machine in operation. Adjustment can be done manually, or the unit can be provided with an automatic control system to change adjustment as process requirements vary.

Part VII.

Rotary Compressors

Rotary Compressors—An Introduction

The rotary compressor is the least known and least popular type of gas compressors. Rotary compressors have been proposed as the answer to the compressor dilemma—a machine suitable for variable compression ratios and a wide capacity range, one that operates at electric motor speeds and has a potential maintenance history equal to that of a centrifugal compressor.

This chapter considers the rotary compressor to be an adiabatic or isentropic machine. It has no valves. It operates by confining a volume of gas in a rotating pocket (Figure 1). The confined gas pockets converge into the discharge chamber. The action is that simple. The speed of the gas is nominal and not great enough to cause polytropic effects such as are experienced in centrifugal compressors when near sonic gas velocities impinge against the vane and volute obstructions. Why then does the discharge temperature exceed the adiabatic rise?

The premise of this section is that this type of machine has three aerodynamic losses. The first loss is that incurred in charging and exhausting the pockets which increases the compression ratio within the pocket. The second loss concerns the leakage between the rotors and between the rotors and the case. The third loss is the thermal burden caused by the intake warm-up.

The pocket charging losses are considered to be a function of the rotor tip speed. The gas must follow in the wake of the pocket tip which is continuously opening in the inlet chamber. It is necessary to expend a certain head to match the rotor speed. This is equivalent to the Net Positive Suction Head (NPSH) required to charge a centrifugal pump.

Table 1 shows the pressure differential that must be sacrificed to attain a velocity equivalent to the rotor tip speed. The sonic velocity of the air considered herein is 1,130 fps at a suction temperature of 60°F. These velad losses are applied to Equation 1 to produce the "B" correction factor. This modifies the visual, line pressure R_C to the more realistic intrinsic pressure differential as it exists in the converging screws. This example considers two cases: one where $R_C = 2$ or $P_2 = 29.4$ and the other where $R_c = 3$ or $P_2 = 44.1$ with $P_1 = 14.7$.

$$B = (1.0 + \Delta P_2/P_2)/(1 - \Delta P_1/P_1) \tag{1}$$

$$\eta ad = (R_C\sigma - 1)/(BR^{\sigma} - 1) \tag{2}$$

These equations apply the "B" correction to R_C and give the compression efficiency in reference to the minimal adiabatic power requirement. The compression efficiencies are given in Table 2 for the two cases and for a single K and for two K of resistance. The full gamut of tip speeds is shown in reference to the sonic velocity of the gas handled. Most rotary compressors operate well below 0.1 Mach as shown in Table 3. The helical screw compressors are rated at 0.3 Mach; some may go to 0.4 Mach. Certain tests have been made with tip speeds as great as 0.7 Mach.

The compression efficiencies produced from one K of resistance are too optimistic. Those produced from two K are too conservative. A compromise

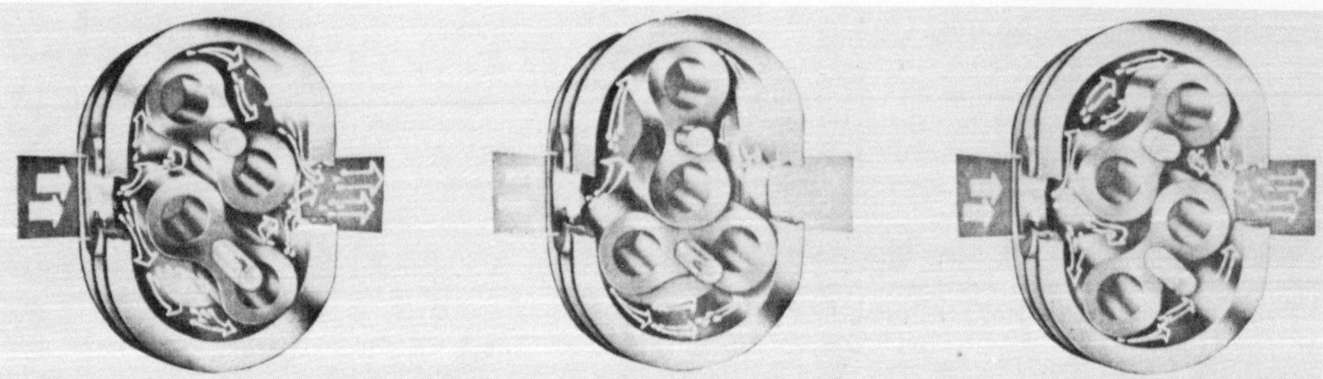

Fig. 1—The above three positions of a two-lobe rotary compressor give an animated view of the gas flow through the machine.

that uses 2 velads of suction resistance and one velad of discharge resistance appears to be realistic. This resolution of compressor performance does not have the supporting test data that should be allocated to such a project. It does present the rudiments and suggest the outline for such a test program. In the interim it offers the most comprehensive procedure for projecting the performance of high speed rotary compressors.

The velad hypothesis used in developing the compression efficiencies in Table 2 is best understood by an example application. Using the foregoing data, with rotor tip speed of 0.45 Mach which is 510 fps, the velad is $V^2/2g$ = 510(510)/64.4 = 4,025 ft/K. The specific volume of air at the given standard conditions is 13.1 cf/lb. The gas head per psi is 144 (13.1) = 1,870 ft/psi. Each suction velad represents a pressure drop of 4,025/1,870 = 2.15 psi/K.

Presume that one velad is consumed in overcoming the resistance of the compressor chamber, entering and dead heading into the rotor pocket. The absolute velocity of the air must also match that of the rotor tip. This totals 2 velads for the charging operation or 2 (2.15) = 4.30 psi. This loss is expressed as a decimal fraction, 4.3/14.7 = 0.293 or 29.3 percent of the suction system pressure.

Table 1
Charging and Exhausting Pressure Losses for Rotary Type Compressors

Tip Speed		Velocity Head	Differential Pressure		
			Suction	Discharge	
Mach	fps	$U^2/2g$	PSI	$R_c = 2$	$R_c = 3$
0.71	807	10,000	5.35	8.69	11.70
0.45	510	4,025	2.15	3.50	4.70
0.38	430	2,875	1.50	2.46	3.32
0.26	296	1,360	0.71	1.16	1.57
0.14	162	400	0.21	0.34	0.46
0.10	108	180	0.10	0.15	0.21

Table 2
Compression Efficiencies for Helical Screw Type Compressors

Tip Speed Mach No.	$R_c = 2.0$		$R_c = 3.0$	
	$K = 1.0$	$K = 2.0$	$K = 1.0$	$K = 2.0$
0.71	44%	22%	58%	30%
0.45	68	50	78	62
0.38	76	61	84	71
0.26	88	72	92	80
0.14	95	93	97	95
0.10	98	96	99	98

Table 3
Summary of Rotary Displacement Compressor Performance Data

	Helical Screw	Spiral Axial	Straight Lobe	Slide-Vane	Liquid-Liner
Configuration, Features (Male x Female)	4 x 6	2 x 4	2 x 2	8 Blades	16 Sprockets
Max. Displacement, icfm	20,000	13,000	30,000	6,000	13,000
Max. Dia., in.	25	16	28	33	48
Min. Dia., in.	4	6	10	5	12
Limiting Tip Speed, y Mach	0.30	0.12	0.05	0.05	0.06
Normal Tip Speed, y Mach	0.24	0.09	0.04	0.04	0.05
Max. L/d, Low Pressure	1.62	2.50	2.50	3.00	1.1
Normal L/d, High Pressure	1.00	1.50	1.50	2.00	1.0
V Factor for Volumetric Eff.	7	3	5	3	3
X Factor for Displacement	0.0612	0.133	0.27	0.046	0.071
Normal Overall Eff., %	75	70	68	72	50
Normal Mech. Eff. at +/− 100 hp, %	90	93	95	94	90
Normal Ratio of Compression, Rc	2/3/4	3	1.7	2/3/4	5
Normal Blank-off, Rc	6	5	5	7	9
Displacement Form Factor, A	0.462	1.00	2.00	0.345	0.535

Note: Mach = 1,250 fps air at 185°F.

It is further presumed that the *built-in* clearance releases the trapped air at or very close to the discharge chamber pressure, so that no overcharging or undercharging exists in the exhaust pocket. The absolute velocity of the gas leaving the rotor and entering the discharge chamber represents another velad expenditure. The analytical difference between the suction and discharge velad is the change in the gas specific volume. The discharge v_{sd} is $v_{os}/R_c{}^{1/k}$ = 13.1/2.19 = 5.95 cf/lb, and gas head per psi is 144 (5.95) = 857 ft/psi. The gas is dispelled into the discharge chamber without the benefit of any convergence devices to recover the velocity energy of 4,025/857 = 4.7 psi per velad. This energy is wasted and represents 4.7/44.1 = 0.106 decimal fraction of the discharge system. This data is applied to Equation 1 to evaluate B = (1 + 0.106)/(1 − 0.293) = 1.106/0.707 = 1.565. This makes the intrinsic BR_c = 4.7. This value applied to Equation 2 makes the adiabatic compression efficiency η_{ad} = $(3.0^{\sigma} - 1)/(4.7^{\sigma} - 1)$ = 0.37/0.555= 66.5 percent.

This procedure can be resolved into the following simplified equations:

$$\theta_i = 2.5\underline{m}(U^2)/T(10)^5 \quad (3)$$

where m represents the molecular weight of the gas, U is the male rotor tip speed in fps and T is the gas temperature °R. This equation includes two velads of resistance. The exhaust loss equation below includes one velad of resistance.

$$\theta_e = \theta_i/2\,(R_c)^{\sigma} \quad (4)$$

The two resistances, θ_i and θ_e are used to evaluate the intrinsic pressure correction factor "B",

$$B = (1.0 + \theta_e)/(1.0 - \theta_i). \quad (5)$$

Leakage

Leakage is the next step to evaluate. The rotary compressor configuration offers three longitudinal lines of escape. Two lines follow each rotor and the casing, and the third line is the addendum of the converging rotors. The internal leakage for rotary machinery follows the equation:

$$W_L = 23L\,P_2G/(T_2)^{0.5}\ \text{lb/min} \quad (6)$$

The displacement of a rotary machine is:

$$QD = d\,LUX,\ \text{cfm} \quad (7)$$

The capacity of the machine is dependent upon the charging pressure. Equation 3 has established the percentage of the suction pressure that is lost in the charging operation. The volumetric efficiency simply follows as:

$$E_{vr} = 100 - (\theta_i + W_s/R_c^{\sigma}). \quad (8)$$

Where the suction charging loss is not available, use the approximate expression, E_{vr} = 98 − $V\,R_c$. Table 3 gives an average V factor for each type of rotary compressor.

The displacement multiplied by the volumetric efficiency gives the capacity of the compressor. The capacity divided by the specific volume gives the capacity in terms of pounds per minute:

$$W = 0.093\,dm\,E_{vr}\,LUX\,P_1/T_1. \quad (9)$$

The *slip leakage* is the ratio of (Equation 6/Equation 9) which reads:

$$W_s = 30.7\,(0.577G + 0.0038)\,R_c{}^{1.9}\,d/m\,E_{vr}\,LXU^{0.5} \quad (10)$$

The manufacturer usually determines the volumetric efficiency in a unique manner. The speed required to sustain the design pressure at zero flow is known as the *slippage*. The decimal fraction of the slippage speed divided by the rated speed is the volumetric efficiency. For example, the static slippage for an 8-inch, straight-lobe compressor is 120 rpm at 3 psig and 144 rpm at 10 psig. The percent slip in reference to the rated speed of 1,200 rpm is (120/1,200) or 10 percent. The manufacturers' volumetric efficiency is 90 percent for the 3-psig static system. The slippage, N_s*, for the 10-psig system is 12 percent and the manufacturers' volumetric efficiency is 88 percent.

This method does not consider the dynamic loss which is evaluated by Equations 3, 8 and 10. The tip speed loss in most other types of rotary compressors is so low that it is negligible. Table 4 shows that the tip speed must exceed 0.12 Mach before it is significant. These *zero-flow* speeds provide valuable checks on the clearance gap and a means of estimating the *zero-flow* pressure. Having the percent of slip speed, N_s*, the likely discharge pressure is

$$R_c{}^{1.9} = N_s{}^*m\,L\,E_{vr}\,X\,(U_s{}^{0.5})\,/30.7d\,(0.577G + 0.0038).$$

Using G as the actual gap = 0.0133 and G^* as the effective gap = 0.0095, U_s = rpm(d)/229 = 144(8)/229 = 5, N_s* = 0.12, E_{vr} = 0.98 assume L = d; X = 0.265, $U_s{}^{0.5}$ = 2.23 and m = 29:

$$R_c{}^{1.9} = 0.12\,(29)\,0.98\,(0.265)\,2.23/30.7\,(0.0095) = 7.00$$

$$R_c = 2.80,\ P_2 = 2.80(14.7) = 41.2\ \text{psia or 26.5 psig.}$$

Having the speed and the discharge pressure, it is possible to solve for the clearance gap G in the same equation.

The average clearance gap G for small (3 to 4-inch) rotors is 0.010, 0.020 for 13-inch rotors and 0.030 for 20-inch and larger rotors. A_e is a geometric coefficient based upon the open-end area divided by the square of the male rotor diameter, d is the diameter and L is the length of the rotor, both in inches. The A_e and X coefficients for the various rotary configurations are given in Table 3 (X = $0.1235A_e$). The exponent, 1.9, is an average value suitable for diatomic and heavy polyatomic gases. The complete exponent is $1 + (k + 1)/2k$, where k is the ratio of specific heat at the mean compression temperature.

Table 4 gives the percent leakage for tip speed variation of 0.1 to 0.35 Mach in reference to the total weight flow for a typical 6-inch helical type rotary compressor having a nominal clearance gap of 0.0133 inch. The leakages were calculated from Equation 10. A variation in the diameter or the gap increases the leakage in direct proportion. The R_c is the most significant factor that affects the leakage.

Table 5 also shows the amount of *warm-up* resulting from bypass leakage to the incoming flow. The percent of power loss is directly proportional to the increase in the Rankine temperature. This leakage percentage is also shown in Table 5. A simple expression for evaluating the leakage *warm-up* effect for each percentage of leakage is

$$X_t = 0.12(R_c) + 0.02. \tag{11}$$

Equations 10 and 11 offer a means of evaluating the leakage and *warm-up* losses. Neither are, in reality, dynamic losses and should not be corrected by a gas head alteration. The leakage and the thermal burden directly affect the horsepower and can be applied as an efficiency factor in the denominator of the power equation. The leakage effect is given in Equation 12. The thermal effect of the suction *warm-up* is given in Equation 13.

$$\eta_s = 1 - W_s. \tag{12}$$

$$\eta_t = 1 - W_s[(0.12\,R_c) + 0.02] \tag{13}$$

$$\text{Adhp} = 0.0468(R_c^{\sigma} - 1)\,W\,Z\,T\,/m\sigma. \tag{14}$$

The basic adiabatic horsepower is the same for all types of compressors, namely:

Where W is the gas flow in pounds per minute, Z is the compressibility factor and σ is $(k - 1)/k$.

$$\text{Gas hp} = \text{Adhp}/\eta_{ad}\,(\eta_s)\eta_t. \tag{15}$$

Table 4
Adiabatic Slip and Thermal Efficiencies

Mach No. (Tip Speed)	Efficiencies	R_c 2.0	2.5	3.0	4.0
0.35 M* (392 fps)	η_{ad}	60.5	66.3	69.5	73.4
	η_s	96.0	93.8	91.3	84.8
	η_t	99.0	98.3	96.7	92.4
	η_o	57.5	61.0	61.4	57.5
0.26 M* (300 fps)	η_{ad}	73.3	78.0	80.5	83.5
	η_s	95.9	93.7	91.2	84.5
	η_t	98.9	98.0	96.7	92.2
	η_o	69.5	71.7	71.0	65.2
0.12 M* (136 fps)	η_{ad}	93.5	94.5	95.6	96.2
	η_s	94.5	91.4	88.0	79.1
	η_t	98.5	97.2	95.4	89.5
	η_o	87.1	84.0	80.3	68.1
0.10 M* (113 fps)	η_{ad}	96.0	96.8	97.3	97.6
	η_s	94.1	90.9	87.2	77.6
	η_t	98.7	97.1	95.6	88.8
	η_o	89.2	85.5	81.2	67.3

Note: η_o is the over-all efficiency.

Table 5
Anticipated Leakage Percentage of Total Weight Flow

Mach No. (Tip Speed)	R_c 2.0	2.5	3.0	4.0
0.35 (392 fps)	4.0	6.2	8.7	15.2
0.26 (300 fps)	4.1	6.3	8.8	15.5
0.12 (136 fps)	5.5	8.6	12.0	20.9
0.10 (113 fps)	5.9	9.1	12.8	22.4
Average	5.0	7.5	10.6	18.5
(a) Temp Rise	7°F	12°F	22°F	48°F
Power Loss	1.3%	2.3%	4.1%	8.9%

Note: (a) The above leakage raises the suction temperature by the amount shown.

$$Bhp = \text{Gas hp} + \text{friction (gas hp)}^{0.4} \tag{16}$$

Table 4 gives the magnitude of the three principal losses for a 6-inch helical screw rotary air compressor having an average clearance gap of 0.0133 inch. The composite efficiencies (η_o) compare favorably with catalog data. It is essential that the clearance gap be reduced in order to maintain a reasonable efficiency at R_c values in excess of 3. The penalty effect of the greater tip speed is amplified by the charging and expelling loss which are included in the adiabatic efficiency.

It is possible to extrapolate these efficiencies for other conditions. For example, determine the efficiency of a 10-inch straight-lobe compressor having a clearance gap of 0.030 inch operating at a tip speed of 0.05 Mach and at 2.0 R_c. The adiabatic efficiency tends toward an asymptotic minimum value of 98.5 percent. The slip loss varies as the effective gap, G^*, and the square root of the change in tip speed: $([0.030\ (0.577) + 0.0038]/[0.0133\ (0.577) + 0.0038])(0.05/0.10)^{0.5}$ times 5.9 percent leakage, which is the closest approach given in Table 5. The leakage for the new condition is $(0.0211/0.0115)0.707(5.9) = 7.65$ percent and the new η_s is 92.35 percent. The new *warm-up* effect is $7.65(0.24 + 0.02) = 2.00$ percent thermal leakage loss, and $\eta_t = 98.00$ percent. The new overall efficiency is $0.985(0.9235)0.980 = 89.15$ percent.

The overall efficiency as given in Table 4 is the product of the adiabatic efficiency n_{ad} (see Equation 1), the slippage efficiency n_s (see Equation 12) and the thermal *warm-up* efficiency n_t (see Equation 13). The latter two efficiencies are developed in the preceding sentences. The adiabatic efficiency is assumed to qualify for the minimum asymptotic value of 98.5 percent.

Helical Screw Compressors

The helical screw compressor is the only rotary unit that operates at tip speeds in excess of 0.12 Mach. Table 3 shows five distinctly different categories of rotary compressors. The spiral axial operates at 0.12 Mach and the other three operate at or about 0.05 Mach. The latter group can be categorically assigned a composite efficiency of 90 percent, except for the sliding-vane and liquid-liner unit, operating above 4 R_c.

The helical screw compressor is the most versatile form of rotary compressors. An exploded view of the unit is shown in Figure 2. A profile of the four-lobe male and the matching six-lobe female configuration is shown in Figure 3. The optimized performance of the helical compressor can be projected on a Baljé chart (see Figure 4). The optimum performance of a centrifugal compressor lies along a D_sN_s slope of 150 and a pressure coefficient of 0.55, with the island peak at $D_s = 1.5$ and $N_s = 100$ as shown in Figure 4. The similar optimum performance of a helical compressor has a D_sN_s slope of 26 with the peak performance island at $D_s = 2.0$ and $N_s = 13$. The D_sN_s slope of 26 is identified as having a pressure coefficient of 4.

Fig. 2—An exploded view of a helical screw rotary compressor. (Courtesy of Beloit Power Systems, Inc.)

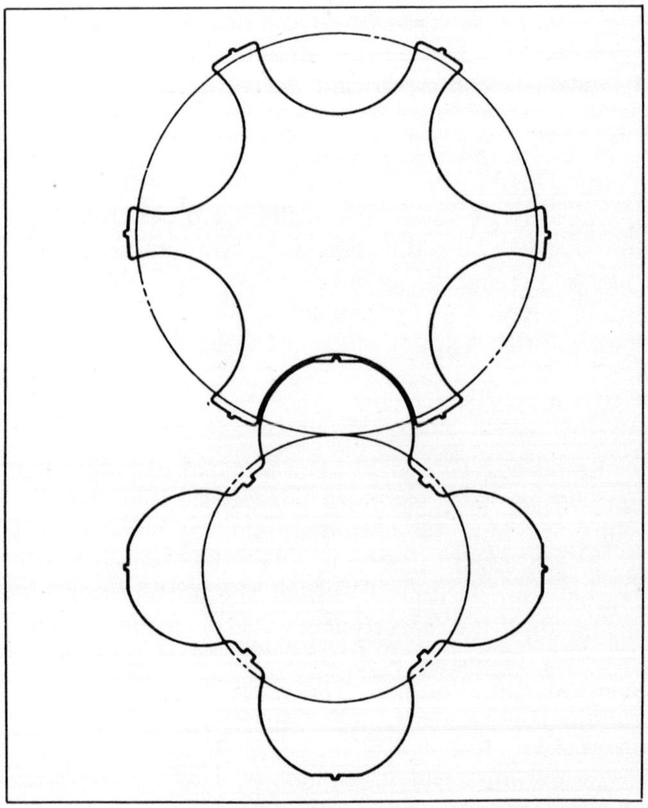

Fig. 3—Rotor profiles of the four-lobe male and the six-lobe female. Built-in clearance exists between the two rotors. (Courtesy of Beloit Power Systems, Inc.)

This means that the helical machine is capable of attaining the same head as a centrifugal compressor

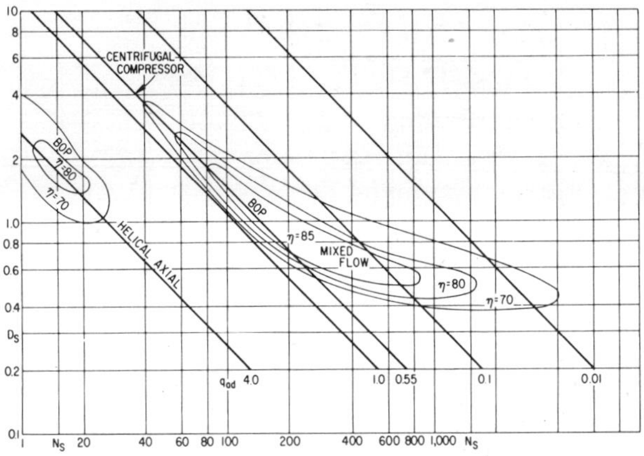

Fig. 4—The optimized performance of the helical compressor can be projected as seen on the Baljé chart. The optimum performance of a centrifugal compressor lies along a D_sN_s slope of 150 and a pressure coefficient of 0.55, with the island peak at D_s = 1.5 and N_s = 100.

at (0.55/4) = 14 percent of the centrifugal compressor's tip speed. In other words, the helical compressor can develop 7.3 times the head of a centrifugal compressor operating at the same tip speed. This *low specific speed* characteristic is highly desirable for compressing small volumes of gas to high pressures. The helical screw compressor has performance characteristics and flexibilities that approach the piston compressor. It operates at moderate speed more comparable with electric motors. It has compression efficiencies comparable to centrifugal compressors. It should have maintenance costs that compare favorably with the centrifugal machines which are less than one-third the cost of maintaining piston machines. With these basic advantages favoring the rotary machine, it might be expected to take over a stronger position in the market. There are deterrents that impede such a take-over, like noise, intrinsic leakage, rotor deflection, casing strength and the technological deficiencies described earlier.

The most successful helical screw compressor applications are delivering air for general construction and road building purposes, mainly for operating pneumatic tools. The size of these portable units run from 600 to 1,200 icfm at 125 psig and require 250 to 500 hp. The key to their success is the moderate tip speed of 0.11 Mach and lower. The second factor is the technique of internal oil flooding. The intake air is sprayed with oil at a ratio of one part per thousand, or 7.5 gallons per thousand cubic feet. This flooding absorbs about two-thirds of the heat of compression, reducing a normal 350° F discharge temperature to about 100° F above ambient or 170° F. This eliminates the need for an after-cooler. The oil is recovered from the separator, cooled and returned to the compressor suction. The sheep-skin filters are quite efficient, less than 0.1 percent of the oil is lost. This is equivalent to one gallon per 60,000 hp-hr of operation. The oil also functions as a lubricant and a sealant to minimize the bypass leakage. The latter is perhaps the most important function of all in that it silences the shrill bypassing noise.

The helical unit was originally considered to be suitable only for dry gas. Condensate and grit particles could not be tolerated. These restrictions have been relaxed in recent years. The helical machines are being used to handle the waste gas from a hydrogenation process. This is a most difficult service where occasional lumps of a very viscous coagulate must pass through the machine. The helical compressor is not without its problems, but the helical units perhaps have less difficulties than those which would be experienced from a piston machine in the same service.

The helical machine can be operated without a driving gear. The lobe contact does not experience a locking interference which can occur in other lobe models. Timing gears are used to drive and also avoid this interference. The male lobe is the driver. The female rotates in the opposite direction (see Figure 3). The containment wrap requires 300° of rotation. The pitch of the helical screw is about 45°.

Built-In Clearances

The transition of the Lysholm design helical screw unit to the present Svenska Rotor Maskiner (SRM) form included the *built-in clearance* feature. The space between the male lobe and the female recess, as illustrated in the cross-sectional view of Figure 3, is the built-in clearance. When the gas retained in this clearance volume is returned to the suction chamber, the respective pressures should balance. When the clearance volume is oversized, there is an excessive pressure and volume returned to the suction chamber. The results are equivalent to another form of bypass. Conversely, when the clearance volume is undersized, the trapped gas pressure is less than the suction chamber pressure. This reduced pressure is undoubtedly an aid to the charging operation.

Perhaps the most suitable and neglected application is its likely potential as a gas motor or still better as a cryogenic expander. Referring to the Baljé chart where the helical compressor had a D_sN_s characteristic slope of 26 and q_{ad} factor of 4, the equivalent rotary expander key performance factor (tip speed/spouting velocity) should be 0.08. This is equivalent to an 8.5-inch rotor running at 3,600 rpm matching the peak efficiency of a 4.5-inch radial in-flow impeller turning at 51,000 rpm.

Spiral Axial Compressors

The spiral axial compressor has a top speed equal to the slowest helical screw machine. A sectional view of the spiral machine is similar to Figure 2. The envelope volume is more than twice the capacity of the helical machine having the same *d*, *L* and *N* dimensions. The machine has a timing gear drive and runs dry without the rotors making contact. It provides 20 to 30-psig air for manufacturing operations.

Straight-Lobe Compressors

The straight-lobe rotary compressor operates at the relatively low tip speed of 50 fps. This is equivalent to a 10-inch rotor driven by a six-pole motor or a 13-inch rotor running at eight-pole speed. An animated view of gas flow through the machine is shown in Figure 1. This machine can accommodate a considerable quantity of liquid, perhaps as much as 50 parts per thousand. Figure 5 is a sectional elevation view.

The public utilities use the straight-lobe compressors as displacement meters, low-pressure pipe line boosters, vacuum gas-well gathering systems and as a supercharger for a gas or diesel engines. This type of compressor offers the most efficient form of vacuum service.

Optimum single-stage performance is realized at 23 inches of mercury and 28 inches for two-stage operation. It is used for desalinization and other evaporative processes where large volumes of vapor are removed at high vacuums.

Slide-Vane Compressors

The slide-vane compressor is illustrated in Figure 6. The rotor runs eccentrically within a cylinder. Radial slots in the rotor carry sliding vanes which form a series of longitudinal cells. The centrifugal spin of the rotor holds the vanes against the cylinder wall. The cell volume diminishes as the rotor approaches the discharge chamber, which is in the lower right-hand corner of Figure 6. A single-stage slide-vane compressor can pull a 28-inch Hg vacuum or can pump air to 50 psig. A two-stage unit can compress air at 250 psig. The unit is not suited for handling saturated and super-saturated vapor. It is particularly inapt with cold jackets. This increases the cylinder wall condensation. The best operation requires a warm jacket and the use of a lubricant heavily loaded with a solvent resistant inhibitor such as rapeseed oil, lanolin or tallow. The manufacturers advocate rather generous quantities of cold jacket-water and lubricant. They prescribe 10 drops per minute for a 10-inch and smaller bore cylinder, plus one drop per minute per shaft seal. This is equivalent to 2 to 3 pints per day of SAE 30 lube oil or about 3,000 bhp-hr per gallon.

The oil rate is doubled for the 20-inch cylinder and tripled for cylinders over 30 inches in diameter. The vanes are made of laminated phenolic resins. Their life and cylinder wear are sensitive to the quantity and quality of the lube oil applied. The most successful application of the slide-vane compressor has been with the 100 hp, 360 cfm, portable construction-type air compressor. The rotors are flooded with lubricant using the same quantity and technique described for the larger helical compressors.

These oil rates are on the order of 10 times that required for equivalent piston machinery service.

Fig. 5—Sectional view of a two-lobe rotary compressor. Note timing gear at left and outside bearings. (Courtesy of Fuller Company, a GATX Company.)

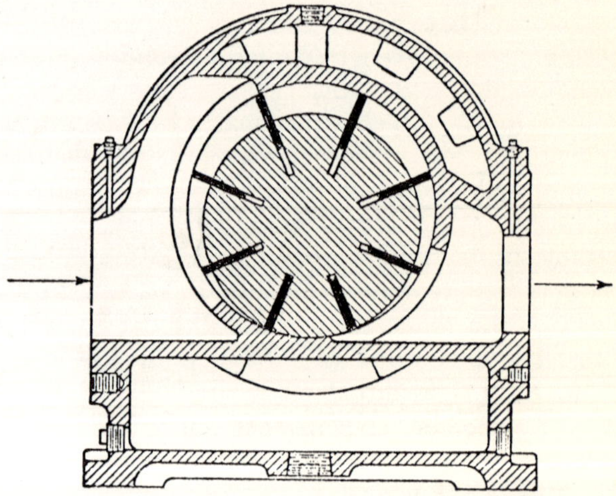

Fig. 6—End view of a sliding-vane compressor. The rotor turns clockwise, closing the annular cells into the discharge chamber. (Courtesy of Allis-Chalmers Corporation.)

Table 6 was compiled from 20 years data accumulated on reciprocating compressors by a major California oil company's gas plants. The gas engine oil consumption varies from 4,000 (two-cycle) to 10,000 bhp-hr per gallon, (four-cycle).

Jacket-Water Requirements

The slide-vane and piston compressors have common difficulties in their lubrication and jacket temperature. There is a scarcity of credible reference data on both subjects and for that reason these data are included. The quantity of heat absorbed by the jacket-water is best determined from these equations:

Table 6
Quantity of Lube Oil Required for Compressor Cylinder Lubrication

Range of cylinder diameters, in.	Nominal discharge pressure, psig	Film thickness, micro-inches	bhp-hr per gal. (thousands)	Compressor lube oil, pints/(cyl.) (100 bhp of load) Pints/day	Drops/min.*
24-36	80	16	13	1.5	13
15-23	200	17	19	1.0	9
10-14	800	23	24	0.8	7
7-9	2,000	24	32	0.6	5
4-7	8,000	47	24	0.8	7
3-5	20,000	57	27	0.7	6

*Based on 12,500 drops per pint of 5,000 SSU SAE 40 lube oil at 75°F with vacuum-type lubricator. Feed includes rod packers, e.g., 300-hp, 16-inch cylinder requires 3 pints/day and 27 drops/minute. Reduce drop count to one-third of above for pressure-type lubricator.

$$H_j = 4\,(t_{ag} - t_{aw}) + 100\text{, Btu/bhp-hr.} \quad (17)$$

$$Q_{gpm} = H_j\,(\text{bhp})/500\,(\Delta t). \quad (18)$$

The letters t_{ag} represent the average of the gas temperature, $(t_{out} + t_{in})/2$, and t_{aw} is the average jacket-water temperature in the cylinder. The heat rejection is 3 to 12 percent of the mechanical heat equivalent, 2,545 Btu per bhp-hr. Statements advocating 20 and even 30 percent of the total power-heat equivalent are absurd. One oil company has operated its entire system of gas compressors with dry jackets for over 15 years (2). They have had no detrimental experience. There has never been a creditable reduction in the power as a result of jacket-water heat rejection (2).

There has been considerable damage experienced as a result of a cold cylinder jacket water system in handling wet gases at or gases close to the saturation line. The cold liner causes the liquids in the gas to condense and wash off the lubricant, reducing the life of the piston rings or the sliding-vanes to a matter of weeks. In instances of this order, it is advisable to reduce the water circulation so that the temperature leaving the cylinder is within 10° F of t_{ag}. The Δt or jacket-water temperature rise in Equation 18 should be 2° and 5° F for a warm (140° F) system.

Liquid-Liner Compressors

The operation of the liquid liner rotary compressor is illustrated in Figure 7. The sprocket blades throw the sealant (usually water) in a cylindrical form against the internal housing. This cylindrical liner contains the gas flow from the large

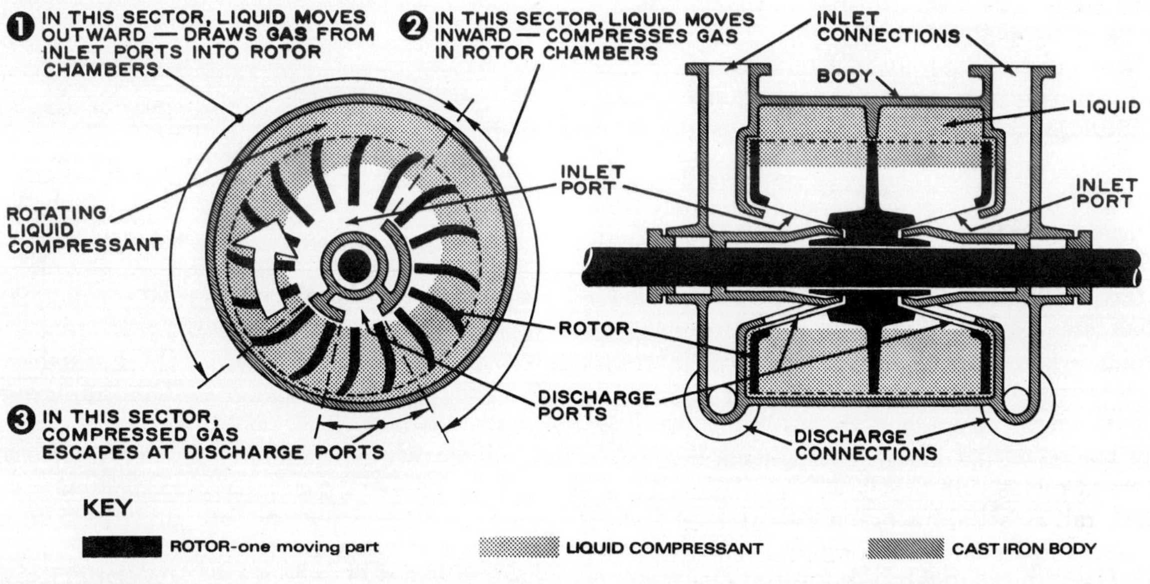

Fig. 7—A sectional and end view of a liquid-liner type compressor. (Courtesy of Nash Engineering Company.)

eccentric entrance to the confinement of the discharge port. This compressor is used to handle the more corrosive and volatile gases in chemical plants. It is indispensible for handling exothermic gases like chlorine, oxygen and acetylene. The liquid-liner compressor is used to handle priming liquids and fibrous material matrix. It is used for vacuum fibrous-plastic molding where material carry-over would interfere with the rotation of the lobe elements.

The liquid-liner compressor offers an absolutely oil-free, 125-psig, single-stage air compression service. Despite a poor nominal efficiency of 50 percent, it is still 10 times more efficient than a steam ejector and is being used as a replacement. It is used as a vacuum pump to remove inert gases from condenser systems and condensate from steam heating plants.

The compressor requires considerable quantities of cold coolant to maintain the discharge within 20° F of ambient or the feed temperature. The manufacturers advocate complete rejection of all mechanical energy expended for the compression. This is an exaggerated requirement and a rather costly auxiliary, especially when the coolant is not permitted to rise above 85° F. A 200-hp liquid-liner compressor requires 100 gpm of coolant at a 10° F differential. Frequently it is necessary to provide a small cooling tower for the coolant service. When an exotic coolant like sulfuric acid is used, it is necessary to provide a special pumping and cooling system.

The liquid-liner rotary compressor has an average composite adiabatic and leakage efficiency of 45 percent at the normal rated tip speed of 66 fps. This efficiency is improved to 63 percent by reducing the tip speed to 30 fps and drops to 40 percent at 80 fps. Judging from the data in Table 2, the charge and exhaust loss should not exceed 2 percent. The bypass is estimated to be 3 percent. The mechanical losses should not exceed 5 percent, making a total loss of 10 percent or an ultimate efficiency of 90 percent. The churning power of the sprockets to support the liquid-liner is estimated to be 80 percent of the adiabatic compression power. A rather crude generalized overall compression and mechanical efficiency is

$$\eta_{oL} = 90/[100 + (fps)^{1.10}(\text{Sp. Gr.})^2/(CE)]. \qquad (19)$$

It has been suggested that the churning loss varies as the square of the specific gravity (*Sp. Gr.*) and inversely as the impeller viscous drag (*CE*) as projected by the Hydraulic Institute. Typical viscosity effects are shown in Table 7. The Mach number criterion is not valid for the liquid-liner

Table 7
Effect of Coolant Viscosity on Sprocket Drag for Liquid-Liner Compressors

Viscosity SSU	Correction Factor "CE"
< 40	1.00
85	0.9
200	0.8
600	0.7
1,000	0.6
1,800	0.5
2,500	0.4

compressor. The hydraulic resistances control the critical velocities.

Concerning the quantity of coolant required for a liquid-liner compressor, the 90 percent mechanical efficiency factor in Equation 19 should be corrected to 100 percent because the mechanical losses are not obvious in the gas temperature. The closure problem after this paragraph deals with a 550 adhp example, which has an overall efficiency of

$$\eta_{oL} = 90/[100 + (71)^{1.1}] = 0.432.$$

The motor load is 550/0.432 = 1,270 bhp, and the motor nameplate rating should be 1,400 to include a 1.10 service factor. The cooling burden is $550/[100/(100 + 71^{1.1})] = 1{,}140$ hp. The coolant required for a 20° F temperature rise is

$$Q_{gpm} = H_j(\text{bhp})/500\Delta_t = 1{,}145\,(2{,}545)/500(20) = 290 \text{ gpm}.$$

The data given in Table 8 require several explanations. All machines are operated at the normal tip speeds given in Table 3. The adiabatic losses for charging and exhausting the helical screw compressor are appreciably higher than the other machines. The greater tip speed accounts for this loss. The slippage is controlled by the displacement form factor and the tip speed, which favors the greater displacement, lower speed machines. The overall effect is contrary to the normal overall efficiencies given in Table 4. This implies that the clearances for the axial and the straight-lobe machines are considerably greater than 0.025 inch which is nominal for the helical machine.

It is interesting to note that the rotors for the first three types of compressors are almost equal despite a 6 to 1 difference in speed. The power re-

Table 8
Various Rotary Compressors Applied to a Common Service

	Type				
	Helical Screw	Spiral Axial	Straight-Lobe	Slide-Vane	Liquid-Liner
(g) Suction Loss, θi	9.35	1.32	0.89	0.90	1.40
(h) Discharge Loss, θe	7.35	1.04	0.70	0.70	1.10
(i) Intrinsic Corr., B	1.185	1.023	1.016	1.016	1.025
(j) Adiabatic Eff., ηad	85.6	97.7	98.5	98.5	97.9
(k) Slippage, Ws, %	28.5	16.6	11.8	11.8	3.0*
(l) Slip Eff., ηs %	71.5	83.4	88.2	88.2	97.0
(m) Thermal Eff., ηt %	89.2	93.7	95.8	95.5	42.5†
(n) Volumetric Eff., Evr	68.0	85.7	89.1	89.9	96.6
(o) Displacement, cfm	14,700	11,650	11,220	11,120	10,370
(p) Rotor Dia., inches	26.6	26.2	27.0	65.0	45.5
(q) Com'l. Size, d x L	25 x 25	22 x 33	22 x 33	46 x 92**	43 x 48††
(r) Speed, rpm	3,500	1,250	593	284	378
(s) Motor, hp	1,100	800	750	750	1,400
(t) Service Factor	1.09	1.11	1.10	1.12	1.10
(u) Disch. Temp., °F	309	270	262	263	120

* Estimate
† By Equation 5.15
**Twin 32.5 x 65 or Triplet 26.5 x 33 (667 rpm) are more realistic
††Twin 32 x 32 (613 rpm) alternate (p) where $L = d$.

quirement is the most significant criteria and represents the payout for the service rendered. A premium of one-third of the power cost plus perhaps an equal maintenance saving should be worth some consideration.

Figure 8 shows the compression efficiencies for four types of rotary compressors as given by catalog data. It offers excellent basis for appraisal.

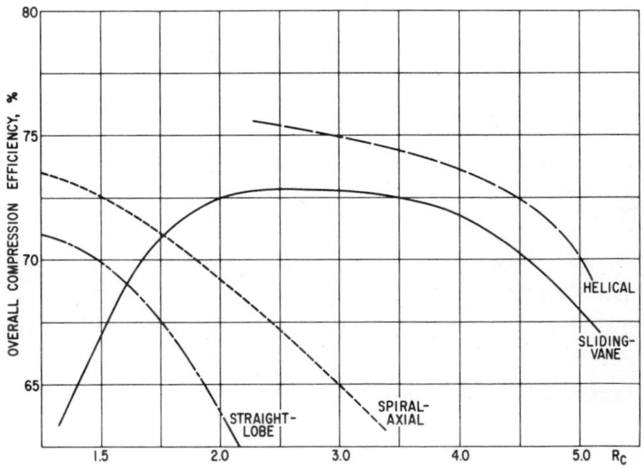

Fig. 8—Compression efficiency of rotary compressor as compiled from catalog data.

Rating of Rotary Machinery

The example application presumes a volume of 10,000 icfm of 18 m (k = 1.28) natural gas (Z = 1) to be compressed from 10 psia and 100° F to 30 psia. What size rotary compressor and motor driver is required using the normal rating criteria given in Table 3 and waiving the 1.7 R_c limitation for the straight-lobe unit? The rotor clearance G is assumed to be 0.024 inch and d is assumed to equal L.

Calculations

(a) Sonic $\tau = 224\ (kT/m)^{0.5} = 224(1.28 \times 560/18)^{0.5}$ = 1,420 fps.

(b) (yM^*) = 0.24 (Table 3); $y\tau$ = 0.24 (1,420 = 341 fps = U.

(c) k = 1.28; σ = 0.219; R_c = 30/10 = 3.0, R_c^{σ} = 1.273.

(d) $v_s = 10.73\ ZT/mP = 10.73(560)/18(10) = 33.3$ cf/lb.

(e) $W = icfm/v_s = 10{,}000/33.3 = 300$ lb/min.

(f) Adhp= $0.0468\ (R_c^{\sigma} - 1)\ WZT/(m\sigma) = 0.0468(0.273)$ $300(560)/18(0.219) = 545$ hp (by Equation 14).

Note: These calculations are common to all five model applications. The helical case is developed in detail. Only significant variables are shown for the other three dry machines. Highlights are given in Table 8 for the liquid-liner model.

(g) $\theta_i = 2.5(18)341^2/560(10)^5 = 0.0935$.

(h) $\theta_e = 0.0935/2(1.273) = 0.0327$.

(i) $B = (1.0 + \theta_e)/(1.0 - \theta_i) = 1.0327/0.9065 = 1.14$.

(j) $\eta_{ad} = (R_c^\sigma - 1)/[(BR)^\sigma - 1] = 0.273/0.310 = 0.876$.

(k) $W_s = 30.7(0.577G + 0.0038)R_c{}^{1.9}d/m\ L\ X\ U^{0.5}$, where $D = L$, $R_c{}^{1.9} = 8.1$ and $U^{0.5} = 18.5$. $W_s = 30.7(0.0176)8.1/18(18.5)\ 0.0612$; $W_s = 0.215$.

(l) $\eta_s = 1.0 - 0.215 = 78.5$ percent.

(m) $\eta_t = 1.0 - W_s(0.12\ R_c + 0.02) = 1.0 - 0.082 = 91.8$ percent.

(n) $E_{vr} = 1.0 - (\theta_i + W_s)/R_c^\sigma = 1.0 - (0.0935 + 0.215)/1.273 = 1.0 - 0.2625 = 73.75$ percent.

(o) $QD = Q/E_{vr} = 10{,}000\ icfm/0.7375 = 13{,}550$ cfm. $QD = d\ L\ U\ X$, then

(p) $dL = QD/UX = 13{,}550/341(0.0612) = 650$. If $d = L$, then $d = 25.5$ inches, or if $d = 24$ inches, then $L = 27$ inches. Table 3 indicates that a machine of these dimensions should have 13,550 cfm displacement. The maximum displacement of 20,000 cfm is possible for a 25 x 25 helical compressor at 0.37 Mach (525 fps).

(q) The motor load is $545/\eta_{ad}\eta_s\eta_t = 545/0.876(0.785)\ 0.918 = 865$ hp. This would impose an 86.5 load factor on a 1,000-hp motor.

(r) The rotor speed is $341(229)/24 = 3{,}250$ rpm for a 24-inch rotor. A two-pole speed motor could be applied to a rotor, $341(229)/3{,}550 = 22$-inch rotor, 29.5 inches long.

(s) The estimated discharge temperature is determined in the following manner: $\sigma' = 0.129/\eta_{ad}\ \eta_t, = 0.219/0.876(0.918) = 0.272$. $T_e = T_i\ R_c^{\sigma'} = 560(3.0^{0.272}) = 560(1.35) = 755\ R$ or 295° F.

(t) A check for the discharge temperature from the adiabatic approach follows. $T_{ad} = 560(1.273 - 1.0) = 153\Delta_t$. $T_e = 153/0.876(0.918) = 190°$ F above 100° F at the inlet, makes the discharge temperature 290° F.